Ergebnisse der Mathematik und ihrer Grenzgebiete

3. Folge · Band 21

A Series of Modern Surveys in Mathematics

Siegfried Bosch
Werner Lütkebohmert
Michel Raynaud

Néron Models

Springer-Verlag Berlin Heidelberg New York
London Paris Tokyo Hong Kong

Siegfried Bosch
Werner Lütkebohmert
Westfälische Wilhelms-Universität
Mathematisches Institut
Einsteinstraße 62
D-4400 Münster

Michel Raynaud
Université de Paris-Sud
Bât. 425, Campus Universitaire
F-91405 Orsay

Mathematics Subject Classification (1980):
14Kxx, 14L15, 14H40, 14B12

ISBN 978-3-540-50587-7 Springer-Verlag Berlin Heidelberg New York
ISBN 978-0-387-50587-9 Springer-Verlag New York Berlin Heidelberg

Library of Congress Cataloging-in-Publication Data
Bosch, S. (Siegfried), 1944-
Néron models / Siegfried Bosch, Werner Lütkebohmert, Michel Raynaud. p. cm.
(Ergebnisse der Mathematik und ihrer Grenzgebiete; 3. Folge, Bd. 21) Includes bibliographical references.
ISBN 0-387-50587-3 (U.S.: alk. paper)
1. Néron models. 2. Abelian varieties. I. Lütkebohmert, Werner. II. Raynaud, Michel.
III. Title. IV. Series. QA564.B587 1990 516.3'5-dc20 89-21963 CIP

Typesetting: Asco Trade Typesetting Ltd., Hong Kong
2141/3140-543210 Printed on acid-free paper

à la mémoire d'André Néron

Preface

Néron models were invented by A. Néron in the early 1960's with the intention to study the integral structure of abelian varieties over number fields. Since then, arithmeticians and algebraic geometers have applied the theory of Néron models with great success, usually without going into the details of Néron's construction process. In fact, even for experts the existence proof given by Néron was not easy to follow. Quite recently, in connection with new developments in arithmetic algebraic geometry, the desire to understand more about Néron models, and even to go back to the basics of their construction, was reactivated. We have taken this as an incentive to present a treatment of Néron models in the form of a book.

The three of us have approached Néron models from different angles. The senior author has been involved in the developments from the beginning on. Immediately after the discovery of Néron models, it was one of his first assignments from A. Grothendieck to translate Néron's construction to the language of schemes. The other two authors worked in the early 1980's on the uniformization of abelian varieties, thereby finding a rigid analytic approach to Néron models. It was at this time that we realized that we had a common interest in the field and decided to write a book on Néron models and related topics.

At first we had the idea of covering a much wider variety of subjects than we actually do here. We wanted to start with a presentation of the construction of Néron models, on an elementary level and understandable by beginners, and then to continue with a general structure theory for rigid analytic groups, with the intention of applying it to the discussion of uniformizations and polarizations of abelian varieties. However, it did not take long to realize that an appropriate treatment of Néron models would require a book of its own. So we changed our plans; colleagues watching the project encouraged us in doing so. Now, having finished the manuscript, we hope that the "elementary" part of the book, which consists of Chapters 1 to 7, is, indeed, understandable by beginners.

We are, of course, indebted to Néron for the original ideas leading to the construction of Néron models, and to the work of Grothendieck which provides language and methods of expressing these ideas in an adequate context. There are other sources from which we have borrowed, most noteworthy the work of A. Weil as well as various contributions of M. Artin.

In preparing this book we received help from many sides. We thank the Deutsche Forschungsgemeinschaft for its constant support during the entire project. Similarly we wish to thank the Centre National de la Recherche Scientifique, as well as the Institute des Hautes Etudes Scientifiques for its hospitality. Finally, we are indebted to our home universities and Mathematics departments in Münster and

Orsay for their interest in the project, for their help whenever possible, and for granting sabbaticals during which substantial work on the subject was done. Also we thank the Heinrich-Hertz-Stiftung.

During the project Dr. W. Heinen from Münster was of invaluable help to us; he proofread the manuscripts and set up the index. We thank him heartily for his work. Last but not least, our thanks go to the publishers for their cooperation.

Münster and Orsay
June 1989

Siegfried Bosch
Werner Lütkebohmert
Michel Raynaud

Table of Contents

Introduction

Let K be a number field, S the spectrum of its ring of integers, and A_K an abelian variety over K. Standard arguments show that A_K extends to an abelian scheme A' over a non-empty open part S' of S. Thus A_K has good reduction at all points s of S' in the sense that A_K extends to an abelian scheme or, what amounts to the same, to a smooth and proper scheme over the local ring at s. In general, one cannot expect that A_K also has good reduction at the finitely many points in $S - S'$. However, one can ask if, even at these points, there is a notion of "good" models which generalizes the notion of good reduction. It came as a surprise for arithmeticians and algebraic geometers when A. Néron, relaxing the condition of properness and concentrating on the group structure and the smoothness, discovered in the years 1961–1963 that such models exist in a canonical way; see Néron [2], see also his lecture at the Séminaire Bourbaki [1]. Gluing these models with the abelian scheme A', one obtains a smooth S-group scheme A of finite type which may be viewed as a best possible integral group structure over S on A_K. It is called a Néron model of A_K and is characterized by the universal property that, for any smooth S-scheme Z and any K-morphism $u_K : Z_K \longrightarrow A_K$, there is a unique S-morphism $u : Z \longrightarrow A$ extending u_K. In particular, rational points of A_K can be interpreted as integral points of A.

Néron himself used his models to study rational points of abelian varieties over global fields, especially their heights. In his paper [3], he shows that the local height contribution at a non-archimedean place can be calculated on the local Néron model in terms of intersection multiplicities between divisors and integral points.

Before Néron's discovery, in 1955, Shimura systematically studied the reduction of algebraic varieties over a discrete valuation ring R, in the affine, projective, as well as in the "abstract" case; see Shimura [1]. In particular, he defined the specialization of subvarieties as well as the reduction of algebraic cycles. In the years 1955 to 1960, several other authors became interested in the reduction of abelian varieties, either in the abstract form or in the form of Albanese and Picard varieties. Koizumi [1] proved that if an abelian variety A_K over K extends to a proper and smooth R-scheme A, then the group structure of A_K also extends. Furthermore, it follows from Koizumi and Shimura [1] that A is essentially uniquely determined by A_K. The latter corresponds to the fact that A is a Néron model of A_K and therefore satisfies the universal mapping property characterizing Néron models. Igusa [1] showed that the Jacobian of a curve with good reduction has good reduction. He also considered the case where the reduction of the curve has an ordinary double point as singularity.

Concerning the reduction of elliptic curves, a systematic investigation of degenerate fibres was carried out by Kodaira [1] for the special case of holomorphic fibrations of smooth surfaces by elliptic curves. Among other things, he classified the possible diagrams of the fibres for minimal fibrations by using the intersection form.

On the other hand, starting with an elliptic curve over the field of fractions of an arbitrary Dedekind ring R, equations of Weierstraß type can provide natural R-models, even at bad places. It seems certain that, at least in characteristic different from 2 and 3, the minimal Weierstraß model was known to arithmeticians at the time Néron worked on his article [2]. However, it was Néron's idea to consider minimal models which are regular and proper, but not necessarily planar. In [2], after constructing Néron models for general abelian varieties, he turns to elliptic curves, shows the existence of regular and proper minimal models, and works out their different types. The classification of special fibres which he obtains is the same as Kodaira's. In order to pass to the "Néron model" as considered in the case of general abelian varieties, one has to restrict to the smooth locus of the corresponding regular and proper minimal model. Furthermore, the identity component coincides with the smooth part of the minimal Weierstraß model.

In his paper [2], Néron uses a terminology which is derived from that in Weil's Foundations of Algebraic Geometry [1]. The terminology has earned its merits when working with varieties over fields. However, applying it to a relative situation, even if the base is as simple as a discrete valuation ring, one cannot avoid a number of unpleasant technical problems. For example, since there are two fibres, namely the generic and the special fibre, it is necessary to work with two universal domains, one for each fibre. Both domains have to behave well with respect to specialization, and so on. Clearly, Weil's terminology was not adapted to handle problems of this kind.

Néron's paper appeared at a time when Grothendieck had just started a revolution in algebraic geometry. With his theory of schemes, he had developed a new machinery, specially designed for treating problems in relative algebraic geometry. Néron knew of this fact, but he did not want to abandon the framework in which he was used working. In the introduction to his article [2], he says that the notion of a scheme over a commutative ring will frequently intervene in his text, in a more or less explicit way. However—and now we quote—"faute d'être suffisamment accoutumé à ce langage, nous avons estimé plus prudent de renoncer à son emploi systématique, et d'utiliser le plus souvent un langage dérivé de celui des *Foundations* de Weil ... ou de celui de Shimura ..., laissant les spécialistes se charger de la traduction."

Certainly, a few specialists did the translation, but mainly for themselves and without publishing proofs. It was only about 20 years later, in 1984, at the occasion of a conference on Arithmetic Algebraic Geometry, that M. Artin wrote a Proceedings article [9] explaining the construction of Néron models from a scheme viewpoint. So, at Néron's time, the situation remained somewhat mysterious. On the one hand, it was very hard to follow Néron's arguments concerning the construction of his models. On the other, arithmeticians were able to use the notion of Néron models with great success, for example, in the investigation of Galois cohomology of abelian varieties. Since Néron models are characterized by a simple universal property, it is possible to work with them without knowing about the actual construction process.

After Néron's work, substantial progress on the structure of Néron models was achieved with the so-called semi-abelian reduction theorem. It states that, up to finite extension of the ground field, Néron models of abelian varieties are semi-abelian. A first proof of this result was carried out by Grothendieck during the fall of 1964; he explained it in a series of letters to Serre, using regular models for curves and l-adic monodromy. The proof was published later in [SGA 7_{I}]. Independently, Mumford was able to obtain the semi-abelian reduction theorem via his theory of algebraic theta-functions, at least for the case where the residue characteristic is different from 2; for this proof see the Appendix II to Chai [1]. The behavior of a Néron model with respect to base change can be difficult to follow; however, in the semi-abelian case it is particularly simple because the identity component is preserved.

In the late sixties, Raynaud [6] further developed the relative Picard functor over discrete valuation rings R in such a way that, in quite general situations, the Néron model of the Jacobian of a curve could be described in terms of the relative Picard functor of a regular R-model of this curve. Using Abhyankar's desingularization of surfaces, one thereby obtains, at least in the case of Jacobians, a second method of constructing Néron models which is largely independent of the original construction given by Néron.

Today, using the relative Picard functor, the semi-abelian reduction theorem is viewed as a consequence of the corresponding semi-stable reduction theorem on curves; see, for example, Artin and Winters [1], or see Bosch and Lütkebohmert [3] for an approach through rigid analytic uniformization theory. To a certain extent, the semi-abelian reduction theorem has changed the view on the reduction of abelian varieties. Namely, it is sometimes enough to work with semi-abelian models and to consider the corresponding monodromy at torsion points. As an example, we refer to Faltings' proof [1] of the Mordell conjecture.

On the other hand, there are questions where, in contrast to the above, Néron models are involved with all their beautiful structure, with their Lie algebra, and with their group of connected components. An example is given by the precise form of the Taniyama-Weil conjecture on modular elliptic curves over $\mathbb{Q}$; cf. Mazur and Swinnerton-Dyer [1].

For further applications of Néron models, we refer to the work of Ogg [1] and Shafarevich [1] concerning moderately ramified torsors over function fields. This was extended by Grothendieck to arbitrary torsors; cf. Raynaud [1].

It should also be noted that the Néron model is of interest when studying the Shafarevich-Tate group Ш. Namely, let A be the Néron model over a Dedekind scheme S of an abelian variety A_K where K is the field of fractions of S. Then Ш is the group of "locally trivial" torsors under A_K, a group which is closely related to the group $H^1(S, A)$. In this way the Néron model is involved in questions concerning the group Ш. For example, concerning its conjectural finiteness in the global arithmetic case.

Finally, to give another application involving torsors under abelian varieties, we mention that Tate studied in [1] the group $H^1(K, A_K)$, where A_K is an abelian variety over a local field K of characteristic 0 having a finite residue field. He used the compact group $\hat{A}_K(K)$ (where $\hat{A}_K$ is the dual abelian variety of A_K) as well as its Pontryagin dual. Later, when the theory of Néron models was available, there appeared some variants of this work for algebraically closed residue field; cf. Bégueri [1] and Milne [2]. Here the Néron model of A_K, in particular, its proalgebraic structure plays an important role.

The aim of the present book is to provide an exposition of the theory of Néron models and of related methods in algebraic geometry. Using the language and techniques of Grothendieck, we describe Néron's construction, discuss the basic properties of Néron models, and explain the relationship between these models and the relative Picard functor in the case of Jacobians. Finally, using generalized Néron models which are just locally of finite type, we study Néron models of not necessarily proper algebraic groups.

We now describe the contents in more detail. Chapter 1 is meant as a first orientation on Néron models. The actual construction of Néron models in the local case takes place in Chapters 3 to 6. Instead of just using Grothendieck's [EGA] as a general reference, we have chosen to explain in Chapter 2 some of the basic notions we need. So, for the convenience of the reader, we give a self-contained exposition of the notion of smoothness relating it closely to the Jacobi criterion. A discussion of henselian rings, an overview on flatness, as well as a presentation of the basics on relative rational maps follows. Also, at the beginning of Chapter 6, we have included an introduction to descent theory.

In Chapter 3, we start the construction of Néron models with the smoothening process. Working over a discrete valuation ring R with field of fractions K, this process modifies any R-model X (of finite type and with a smooth generic fibre X_K) by means of a sequence of blowing-ups with centers in special fibres to an R-model X' such that each integral point of X lifts to an integral point of the smooth locus of X'. This leads to the construction of so-called weak Néron models. Since there is a strong analogy between the smoothening process and the technique of Artin approximation, we have included the latter, although it is not actually needed for the construction of Néron models.

Next, in Chapter 4, we look at group schemes. We consider a smooth K-group scheme of finite type X_K admitting a weak Néron model X and show that the group law on X_K extends to an R-birational group law on X if we remove all non-minimal components from the special fibre of X; the minimality is measured with respect to a non-trivial left-invariant differential form of maximal degree on X_K. In Chapter 5, working over a strictly henselian base and following ideas of M. Artin, we associate to the R-birational group law on X an R-group scheme. The latter is, by a generalization of a theorem of Weil for rational maps from smooth schemes into group schemes, already the Néron model of X_K. The generalization to an arbitrary discrete valuation ring is done in Chapter 6 by means of descent. After we have finished the construction of Néron models in Chapter 6, we discuss their properties in Chapter 7.

The next topic to be dealt with is the relative Picard functor and, in particular, its relationship to Néron models in the case of Jacobians of curves. Since there seems to be no systematic exposition of the relative Picard functor $\mathrm{Pic}_{X/S}$ available which takes into account developments after Grothendieck's lectures [FGA], we thought it necessary to include a chapter on this topic. In Chapter 8 we explain the various representability results for $\mathrm{Pic}_{X/S}$ in terms of schemes or algebraic spaces, mainly due to Grothendieck [FGA] and Artin [5]. From this point on, due to lack of space, it was impossible to give detailed proofs for all the results we mention. It is our strategy to list the important results, to prove them whenever possible without too much effort, or to sketch proofs otherwise. In any case, we attempt to give precise references and to point out improvements which have appeared in the subsequent literature.

The same can be said for the first half of Chapter 9 where we deal with relative Jacobians of curves. Among other things, modulo some considerations contained in Chapter 7, we show here how to derive the semi-abelian reduction theorem for Néron models from the semi-stable reduction theorem for curves. A proof of the latter theorem has not been included in the book since a detailed discussion of models for curves and of related methods would be a topic of its own, too large to be dealt with in the present book. Instead, for a proof using Abhyankar's desingularization, we refer to Artin and Winters [1] or, for a proof using rigid geometry, to Bosch and Lütkebohmert [1]. Finally, in Sections 5 to 7 of Chapter 9, we compare the Néron model with the relative Picard functor in the case of Jacobians. As an application, we show how to compute the group of connected components of a Néron model.

The book ends with a chapter on Néron models of commutative, but not necessarily proper algebraic groups. In the local case, we prove a criterion for a smooth commutative K-group scheme X_K of finite type to admit a Néron model which, over an excellent strictly henselian base, amounts to the condition that X_K does not contain subgroups of type $\mathbb{G}_a$ or $\mathbb{G}_m$. We also indicate how to globalize this result. In doing so, it is natural to admit Néron models which are locally of finite type (lft), but not necessarily of finite type. This way we can construct Néron models for tori as well as study the same problem for K-wound unipotent groups. Since our investigations seem to have few applications at the moment and, since some of the statements are still at a conjectural stage, we have chosen only to give short indications of proofs.

Bibliographical references are given by mentioning the author, with a number in square brackets to indicate the particular work we are referring to. An exception is made for Grothendieck, where we also use the familiar abbreviations [FGA], [EGA], and [SGA], as listed at the beginning of the bibliography. Cross references to theorems, propositions, etc., like Theorem 1.3/1, usually contain the number of the chapter, the section number, and the number of the particular result. For references within the same section, the chapter and the section numbers will not be repeated.

Chapter 1. What Is a Néron Model?

This chapter is meant to provide a first orientation to the basics of Néron models. Among other things, it contains an explanation of the context in which Néron models are considered, as well as a discussion of the main results on the construction and existence, including some examples.

We start by looking at models over Dedekind schemes. In particular, the notion of étale integral points is introduced, and models of finite type satisfying the extension property for étale integral points are considered. For a local base, the existence of such models is characterized in terms of a boundedness condition. Then, in Section 1.2, we define Néron models and prove some elementary properties which follow immediately from the definition. We also discuss the relationship between global and local Néron models as well as a criterion for a smooth group scheme of finite type to be a Néron model. Next, in Section 1.3, we state the main existence theorem for Néron models in the local case and explain the skeleton of its proof, anticipating some key results which are obtained in later chapters.

In Section 1.4, we discuss the case of abelian varieties. More precisely, we study the notion of good reduction and show how the existence of local Néron models leads to the existence of global Néron models. In Section 1.5, in order to provide some explicit examples, we consider elliptic curves. In particular, we compare the Néron model with the minimal proper and regular model and with the minimal Weierstraß model. The chapter ends with a look at Néron's article [2] which serves as a basis for the construction of Néron models. For this section, a certain familiarity with the contents of later Chapters 3 to 6 is advisable.

1.1 Integral Points

When dealing with Néron models, one usually works over a base scheme S which is a *Dedekind scheme*, i.e., a noetherian normal scheme of dimension ≤ 1. The local rings of S are either fields or discrete valuation rings. For example, S can be the spectrum of a Dedekind domain. We will talk about the *local case* if S consists of a local scheme and, thus, is the spectrum of a discrete valuation ring or even of a field; the general case will be referred to as the *global case*. Any Dedekind scheme S decomposes into a disjoint sum of finitely many irreducible components S_i with a generic point η_i each. We set $K := \bigoplus k(\eta_i)$, so K is the *ring of rational functions on* S. Furthermore, the affine scheme Spec K is referred to as the *scheme of generic points of* S. If S is connected—and this is the case to keep in mind—there is a unique generic point $\eta \in S$. Its residue field is K and we can identify η with the associated geometric point $\operatorname{Spec} K \longrightarrow S$. It is only for technical reasons that we do not require Dedekind schemes to be connected.

There are three examples of Dedekind schemes, which are of special interest. To describe the first one, let K be a number field, i.e., a finite extension of $\mathbb{Q}$, and let R be the ring of integers of K. Then set $S = \operatorname{Spec} R$. Similarly, we can consider an algebraic function field K of dimension 1 over a constant field k and define S to be the normal proper k-curve associated to K. In both cases, S is a Dedekind scheme. On the other hand, we can start with a normal noetherian local scheme of dimension 2 and remove the closed point from it. Also this way we obtain a Dedekind scheme.

Now let S be an arbitrary Dedekind scheme with ring of rational functions K and consider an S-scheme X. We define its *generic fibre* (or, more precisely, its *scheme of generic fibres*) by $X_K := X \otimes_S K$, viewed as a scheme over K. Conversely, if we start with a K-scheme X_K, any S-scheme Y extending X_K, i.e., with generic fibre $Y_K = X_K$, will be called an *S-model of X_K*. There is an abundance of such models. For example, any change of Y (such as blowing up or removing a closed subscheme) which takes place in fibres disjoint from X_K, will produce a new S-model of the same K-scheme X_K. On the other hand, X_K can be viewed as an S-model of itself. In the local case, the latter is even of finite type over S if X_K is of finite type over K.

The main problem we will be concerned with when studying the existence of Néron models is to construct S-models X of X_K which satisfy certain natural properties. One of them is the extension property concerning étale integral points, or just étale points, as we will say; for the notion of étale see Section 2.2.

Definition 1. *Let X be a scheme over a Dedekind scheme S. Then we say that X satisfies the extension property for étale points at a closed point $s \in S$ if, for each étale local $\mathcal{O}_{S,s}$-algebra R' with field of fractions K', the canonical map $X(R') \longrightarrow X_K(K')$ is surjective.*

Each étale local $\mathcal{O}_{S,s}$-algebra is a discrete valuation ring again. In fact, it can be seen from Chapter 2, in particular, from 2.4/8 and 2.3/9, that the étale local $\mathcal{O}_{S,s}$-algebras R' correspond bijectively to the (faithfully flat) extensions of discrete valuation rings $\mathcal{O}_{S,s} \subset R'$ with the properties that a uniformizing element of $\mathcal{O}_{S,s}$ is also uniformizing for R', that the extension of fraction fields of $\mathcal{O}_{S,s} \subset R'$ is finite and separable, and that the residue extension of $\mathcal{O}_{S,s} \subset R'$ is finite and separable. So we conclude from the valuative criterion of separatedness [EGA II], 7.2.3, that the map $X(R') \longrightarrow X_K(K')$ is injective if X is separated over S. Furthermore, the extension property for étale points as formulated in Definition 1 is similar to the one occurring in the valuative criterion of properness [EGA II], 7.3.8; the only difference is that we restrict ourselves to valuation rings R' which are étale over $\mathcal{O}_{S,s}$.

Instead of considering all étale local $\mathcal{O}_{S,s}$-algebras R' one can just as well apply limit arguments and work with a strict henselization R^{sh} of $\mathcal{O}_{S,s}$. The latter is the inductive limit over all pairs (R', α) where R' is an étale local $\mathcal{O}_{S,s}$-algebra and where α is an R-homomorphism from R' into a fixed separable algebraic closure of the residue field $k(s)$; see Section 2.3. Then, if K^{sh} is the field of fractions of R^{sh}, it follows that X satisfies the extension property for étale points at $s \in S$ if and only if the map

$X(R^{sh}) \longrightarrow X_K(K^{sh})$ is surjective. Furthermore, let us mention that X satisfies the extension property for étale points at $s \in S$ if and only if $X \otimes_S \mathcal{O}_{S,s}$, viewed as a scheme over $\mathcal{O}_{S,s}$, does.

A simple method for constructing S-models of finite type is the method of *chasing denominators*. It applies to the case where S is affine, say $S = \operatorname{Spec} R$, and where X_K is affine of finite type over K (resp. projective over K). The resulting models are affine of finite type over R (resp. projective over R). To explain the affine case, let X_K be the spectrum of a ring

$$A_K = K[t_1, \ldots, t_n]/I_K ;$$

i.e., of a quotient of a free polynomial ring by an ideal I_K. Then I_K is generated by finitely many polynomials $f_1, \ldots, f_r$, which we may assume to have coefficients in R. So set

$$A := R[t_1, \ldots, t_n]/I ,$$

where I is the ideal generated by $f_1, \ldots, f_r$. Then $X := \operatorname{Spec} R$ is an R-model of finite type of X_K. Furthermore, since a module over a valuation ring is flat as soon as there is no torsion, we see that X will be flat over R if we saturate I; i.e., if we set

$$I := I_K \cap R[t_1, \ldots, t_n].$$

Then, by its definition, X is just the schematic closure of X_K in the affine n-space over R; for the notion of schematic closure see Section 2.5. Finally, the projective case is completely analogous; here one works with the Proj of homogeneous coordinate rings.

If X_K is projective, any R-model X obtained by chasing denominators is projective and, thus, satisfies the extension property for étale points by the valuative criterion of properness. If X_K is just of finite type, but not projective, the construction of an S-model of finite type satisfying the extension property for étale points can be quite complicated or even impossible as the example of the affine n-space $\mathbb{A}^n_K$ shows. As a necessary condition in the local case, we will introduce the notion of *boundedness*.

So assume that S consists of a discrete valuation ring R with field of fractions K. Furthermore, consider a faithfully flat extension of discrete valuation rings $R \subset R'$ and let K' be the field of fractions of R'. Then R and R' give rise to absolute values on K and on K'; we denote them by $|\ |$ assuming that both coincide on K. For us the case where R' is a strict henselization R^{sh} of R will be of interest. Now, for any K-scheme X_K, for any point $x \in X_K(K')$, and for any section g of $\mathcal{O}_{X_K}$ being defined at x, we may view $g(x)$ as an element of K' so that its absolute value $|g(x)|$ is well-defined. In particular, it makes sense to say that g is bounded on a subset of $X_K(K')$. Applying this procedure to the coordinate functions of the affine n-space $\mathbb{A}^n_K$, we arrive at the notion of a *bounded subset of* $\mathbb{A}^n_K(K')$.

Definition 2. *As before, let $R \subset R'$ be a faithfully flat extension of discrete valuation rings with fields of fractions K and K'. Furthermore, let X_K be a K-scheme of finite type and consider a subset $E \subset X_K(K')$.*

(a) *If X_K is affine, E is called bounded in X_K if there exists a closed immersion $X_K \hookrightarrow \mathbb{A}^n_K$ mapping E onto a bounded subset of $\mathbb{A}^n_K(K')$.*

(b) *In the general case, E is called bounded in X_K if there exists a covering of X_K by finitely many affine open subschemes $U_1, \dots, U_s \subset X_K$ as well as a decomposition $E = \bigcup E_i$ into subsets $E_i \subset U_i(K')$ such that, for each i, the set E_i is bounded in U_i in the sense of* (a).

It should be kept in mind that the definition of boundedness takes into account the choice of valuation rings $R \subset R'$ and, thereby, the choice of particular valuations on K and K', although the latter is not expressed explicitly when we say that a subset $E \subset X_K(K')$ is bounded in X_K.

If X_K is affine, say if $X_K = \operatorname{Spec} A_K$, condition (a) of the definition means that there are elements $g_1, \dots, g_n \in A_K$ generating A_K as a K-algebra which, as maps $X_K(K') \longrightarrow K'$, are bounded on E. The latter is equivalent to the fact that each $g \in A_K$ is bounded on E and it is easily seen that, in the affine case, conditions (a) and (b) of the definition are equivalent. Moreover, if there is one closed immersion $X_K \hookrightarrow \mathbb{A}^n_K$ mapping E onto a bounded subset of $\mathbb{A}^n_K(K')$, it follows that the latter property is enjoyed by all closed immersions of type $X_K \hookrightarrow \mathbb{A}^m_K$.

We want to show that condition (b) of Definition 2 is independent of the particular affine open covering $\{U_i\}$ of X.

Lemma 3. *Let $R \subset R'$ be a faithfully flat extension of discrete valuation rings with fields of fractions K and K'. Furthermore, let X_K be a K-scheme of finite type and consider a subset $E \subset X_K(K')$. If there exists a finite affine open covering $\mathfrak{U} = \{U_i\}$ of X_K such that condition* (b) *of Definition* 2 *is satisfied, then the latter condition is satisfied independently of the particular covering $\mathfrak{U}$. More precisely, given any finite affine open covering $\mathfrak{V} = \{V_j\}$ of X_K, there is a partition $E = \bigcup F_j$ into subsets $F_j \subset V_j(K')$ such that F_j is bounded in V_j for each j.*

Proof. Since conditions (a) and (b) of Definition 2 are equivalent in the affine case, we may assume that $\mathfrak{V}$ is a refinement of $\mathfrak{U}$. Now pick an element $U_i \in \mathfrak{U}$, say $U_i = \operatorname{Spec} A$, and let it be covered by the elements $V_1, \dots, V_r \in \mathfrak{V}$. Then we may assume that V_ρ is of type $\operatorname{Spec} A_{f_\rho}$, $\rho = 1, \dots r$, where $f_1, \dots, f_r$ generate the unit ideal in A. So there is an equation $\Sigma a_\rho f_\rho = 1$ with coefficients $a_\rho \in A$. Let E_i be a bounded subset of $U_i(K')$. Then it follows from the equation representing the unit 1 that

$$\varepsilon := \inf\{\max\{|f_\rho(x)| \,;\, \rho = 1, \dots, r\} \,;\, x \in E_i\}$$

is positive. Therefore, setting

$$F_\rho = \{x \in E_i \,;\, |f_\rho(x)| \geq \varepsilon\}\ ,$$

we have $E_i = F_1 \cup \dots \cup F_r$, and each F_ρ is bounded in $V_\rho = \operatorname{Spec} A_{f_\rho}$. Proceeding in the same way with all $U_i \in \mathfrak{U}$, we see that $\mathfrak{V}$ satisfies condition (b) of Definition 2 if $\mathfrak{U}$ does. □

We want to give two immediate applications of the above lemma, the first one saying that the image of a bounded set is bounded again and the second one that the notion of boundedness, in some sense, is compatible with extensions of the field K.

Proposition 4. *Let $R \subset R'$ be a faithfully flat extension of discrete valuation rings with fields of fractions K and K' and consider a K-morphism $f: X_K \longrightarrow Y_K$ between K-schemes of finite type. Then, for any bounded subset $E \subset X_K(K')$, its image under $X_K(K') \longrightarrow Y_K(K')$ is bounded in Y_K.*

Proposition 5. *Let $R \subset R'$ be a faithfully flat extension of discrete valuation rings with fields of fractions K and K'. Furthermore, let X_K be a K-scheme of finite type. Then a subset $E \subset X_K(K')$ is bounded in X_K if and only if the corresponding subset $E' \subset X_{K'}(K')$ is bounded in $X_{K'}$.*

Both assertions are obvious in the affine case; the reduction to this case is done using Lemma 3. Next we want to show that properness always implies boundedness.

Proposition 6. *Let $R \subset R'$ be a faithfully flat extension of discrete valuation rings with fields of fractions K and K', and consider a proper K-scheme X_K. Then any subset $E \subset X_K(K')$ is bounded in X_K.*

Proof. Let us begin with the remark that the notion of boundedness as introduced in Definition 2 works just as well without the discreteness assumption if we restrict to faithfully flat extensions of valuation rings $R \subset R'$ corresponding to valuations of height 1 on K and K'. The above mentioned properties of boundedness remain true. So, for the purposes of the present proposition, we may extend the valuation of K' to an algebraic closure of K' and thereby assume that K' is algebraically closed.

Due to Chow's lemma [EGA II], 5.6.1, there is a surjective K-morphism $Y_K \longrightarrow X_K$, where Y_K is projective. Then, using Proposition 4, we see that it is enough to look at the case where X_K is projective or, more specifically, where $X_K = \mathbb{P}^n_K$ and where $E = \mathbb{P}^n_K(K')$. To do this, fix a set of homogeneous coordinates on $\mathbb{P}^n_K$ and consider the associated standard covering of $\mathbb{P}^n_K$. For $i = 0, \dots n$, let $U_i \simeq \mathbb{A}^n_K$ be the affine open part of $\mathbb{P}^n_K$ where the i-th coordinate does not vanish. Writing points $x \in \mathbb{P}^n_K(K')$ in homogeneous coordinates in the form $x = (x_0, \dots, x_n)$ with $x_0, \dots, x_n \in K'$, we can set

$$E_i := \{x = (x_0, \dots, x_n) \in \mathbb{P}^n_K(K') \, ; \, |x_i| = \max(|x_0|, \dots, |x_n|)\} \, .$$

Then $\mathbb{P}^n_K(K') = \bigcup E_i$ with $E_i \subset U_i(K')$ being bounded in U_i. So it follows that $\mathbb{P}^n_K(K')$ is bounded in $\mathbb{P}^n_K$. □

If X_K is a closed subscheme of $\mathbb{A}^n_K$, and if X is its schematic closure in $\mathbb{A}^n_R$, the image of the canonical map

$$X(R') \longrightarrow X_K(K') \subset \mathbb{A}^n_K(K')$$

consists of those points in $X_K(K')$ whose coordinates are bounded by 1. In particular, multiplying coordinate functions on $\mathbb{A}^n_K$ by suitable constants, we can always assume that the image of $X(R') \longrightarrow X_K(K')$ contains a given subset $E \subset X_K(K')$

provided E is bounded in X_K. So, for affine schemes, we see that the following characterization of boundedness is valid:

Proposition 7. *Let $R \subset R'$ be a faithfully flat extension of discrete valuation rings with fields of fractions K and K'. Furthermore, let X_K be a K-scheme (resp. an affine K-scheme) of finite type. Then a subset $E \subset X_K(K')$ is bounded in X_K if and only if there is an R-model (resp. an affine R-model) X of X_K of finite type such that the image of the canonical map $X(R') \longrightarrow X_K(K')$ contains E.*

In particular, taking for R' a strict henselization R^{sh} of R and for K' the field K^{sh} of fractions of R^{sh}, there is an R-model (resp. an affine R-model) X of X_K of finite type satisfying the extension property for étale points if and only if $X_K(K^{sh})$ is bounded in X_K.

Proof. If, in the general case, $E \subset X_K(K')$ is bounded in X_K, one considers an affine open covering $\{U_{i,K}\}$ of X_K and a decomposition $E = \bigcup E_i$ into subsets $E_i \subset U_{i,K}(K')$ which are bounded in $U_{i,K}$. Then one can find an affine R-model U_i of each $U_{i,K}$ such that E_i belongs to the image of $U_i(R') \longrightarrow U_{i,K}(K')$. Gluing the U_i along the generic fibre, one ends up with an R-model X of X_K such that the image of $X(R') \longrightarrow X_K(K')$ contains E.

Remark 8. If X_K is a separated K-scheme, the R-model X we obtain in Proposition 7 will not, in general, be separated. It requires substantial extra work to modify X in such a way that it becomes separated; see 3.5/6.

Using the approximation theorem of Greenberg [2], we want to add here a non-trivial criterion for boundedness.

Proposition 9. *Let R be an excellent henselian discrete valuation ring with field of fractions K and let X_K be an open subscheme of a K-scheme $\bar{X}_K$ of finite type. Furthermore, consider a subset $E \subset X_K(K)$ which is bounded in $\bar{X}_K$. Then, if $(\bar{X}_K - X_K)(K) = \emptyset$, the set E is bounded in X_K, too.*

Proof. We may assume that $\bar{X}_K$ is affine. Let $\bar{X} = \operatorname{Spec} \bar{A}$ be an affine R-model of $\bar{X}_K$ such that each point of E extends to an R-valued point of $\bar{X}$. Furthermore, let Z be the schematic closure of $\bar{X}_K - X_K$ in $\bar{X}$ so that $X_K = \bar{X}_K - Z_K$. Therefore $Z(K)$ and, thus, also $Z(R)$ are empty. Now fix a uniformizing element π of R and set $R_n = R/(\pi^n)$. It follows then from Greenberg [2], Cor. 2, that $Z(R_n)$ is empty if n is large enough. Therefore, if Z is defined in $\bar{X}$ by the elements $f_1, \ldots, f_r \in \bar{A}$, we must have

$$\max\{|f_1(x)|, \ldots, |f_r(x)|\} > |\pi^n|$$

for all $x \in X_K(K)$.

Using the latter fact, it is easy to show that $E \subset X_K(K)$ is bounded in X_K. Namely set

$$E_i = \{x \in E; |f_i(x)| > |\pi^n|\}.$$

Then E is the union of the E_i and X_K is the union of the affine open subschemes $\operatorname{Spec} \bar{A}_K[f_i^{-1}]$. Furthermore, since E_i is bounded in $\bar{X}_K$, it is obvious that E_i is bounded in $\operatorname{Spec} \bar{A}_K[f_i^{-1}]$. Thus E is bounded in X_K. □

Each separated K-scheme of finite type X_K admits a compactification; i.e., there is a proper K-scheme $\bar{X}_K$ containing X_K as a dense open subscheme; cf. Nagata [1], [2]. If there exists a compactification with $(\bar{X}_K - X_K)(K) = \emptyset$, we say that X_K has no rational point at infinity. Using this terminology, we can conclude from Propositions 6 and 9:

Corollary 10. *Let R be an excellent discrete valuation ring with field of fractions K and let X_K be a separated K-scheme of finite type with no rational point at infinity. Then $X_K(K)$ is bounded in X_K.*

1.2 Néron Models

In the following, let S be a Dedekind scheme with ring of rational functions K. Considering a smooth and separated K-scheme X_K of finite type, we are interested in constructing S-models X of X_K which are smooth, separated, and of finite type over S. Furthermore, we may ask if among all such models X one can select a minimal one; i.e., an S-model X such that for any other S-model Y of this type there is a unique morphism $Y \longrightarrow X$ restricting to the identity on the generic fibre. Requiring this mapping property for arbitrary smooth S-schemes Y, we arrive at the notion of Néron models.

Definition 1. *Let X_K be a smooth and separated K-scheme of finite type. A Néron model of X_K is an S-model X which is smooth, separated, and of finite type, and which satisfies the following universal property, called Néron mapping property:*

For each smooth S-scheme Y and each K-morphism $u_K : Y_K \longrightarrow X_K$ there is a unique S-morphism $u : Y \longrightarrow X$ extending u_K.

The restriction to schemes of finite type is not really necessary. In Chapter 10 we will consider Néron models, so-called Néron lft-models, which are locally of finite type (by the smoothness condition), but not necessarily of finite type. However, adding the finiteness condition simplifies things to a certain extent. In many important cases, Néron models are automatically of finite type; see, for example, the case of abelian varieties.

As a first step towards Néron models, we will have to consider a weaker form, so-called *weak Néron models* of X_K. Thereby we understand smooth S-models X of finite type which satisfy the extension property for étale points 1.1/1; see also 3.5/1 for the definition we will work with in later chapters.

We want to list some elementary properties of Néron models which follow immediately from the definition.

Proposition 2. *Let X be a smooth and separated S-scheme which is a Néron model of its generic fibre X_K.*

(a) *X is uniquely determined by X_K, up to canonical isomorphism.*

(b) *X is a weak Néron model of its generic fibre; in particular, it satisfies the extension property for étale points.*

(c) *The formation of Néron models commutes with étale base change; i.e., if $S' \longrightarrow S$ is an étale morphism and if K' is the ring of rational functions on S', then $X_{S'} = X \times_S S'$ is a Néron model over S' of the K-scheme $X_{K'} = X_K \times_K K'$.*

Proof. Assertion (a) follows immediately from the Néron mapping property. The same is true for assertion (b) (modulo a limit argument as provided by Lemma 5 below); one has to apply the Néron mapping property to schemes Y which are étale over S. To verify assertion (c), we only have to show the Néron mapping property for $X_{S'}$. So consider a smooth S'-scheme Y' and a K'-morphism $Y'_{K'} \longrightarrow X_{K'}$. Composing the latter morphism with the projection $X_{K'} \longrightarrow X_K$, we obtain a K-morphism $Y'_{K'} \longrightarrow X_K$ which uniquely extends to an S-morphism $Y' \longrightarrow X$ since X is a Néron model of X_K; namely, Y' is smooth over S since the composition of the structural morphism $Y' \longrightarrow S'$, which is smooth, with the étale morphism $S' \longrightarrow S$ is smooth again. Now $Y' \longrightarrow X$ yields an S'-morphism $Y' \longrightarrow X_{S'}$ and the latter is a unique extension of the K'-morphism $Y'_{K'} \longrightarrow X_{K'}$. □

Next, we mention that the notion of Néron models is local on the base:

Proposition 3. *Let S be a Dedekind scheme and let (S_i) be an open covering of S. Furthermore, let X be an S-scheme. Then X is a Néron model of its generic fibre if and only if, for each i, the same is true for the S_i-scheme $X \times_S S_i$.*

In the above assertion, one can replace the open subschemes $S_i \subset S$ by the localizations of S at closed points. However, then it is necessary to require the scheme we start with to be of finite type.

Proposition 4. *Let S be a Dedekind scheme and let X be an S-scheme of finite type. Then the following assertions are equivalent:*

(a) *X is a Néron model of its generic fibre.*

(b) *For each closed point $s \in S$, the $\mathcal{O}_{S,s}$-scheme $X \times_S \operatorname{Spec} \mathcal{O}_{S,s}$ is a Néron model of its generic fibre.*

If we want to verify the implication (a) $\Longrightarrow$ (*b*), we cannot just apply an argument of base change as provided by Proposition 2 (c). The reason is that $\operatorname{Spec} \mathcal{O}_{S,s}$ is a limit of open subschemes of S but not, in general, an étale extension of S. So we will have to combine limit arguments with arguments of base change. Let us mention the necessary facts on limits.

Lemma 5 ([EGA IV_3], 8.8.2). *Let S be a base scheme and let s be a point of S.*

(a) *Let X and Y be S-schemes which are of finite presentation. Then the canonical map*

$$\varinjlim \operatorname{Hom}_{S'}(X \times_S S', Y \times_S S') \longrightarrow \operatorname{Hom}_{\mathcal{O}_{S,s}}(X \otimes_S \mathcal{O}_{S,s}, Y \otimes_S \mathcal{O}_{S,s})$$

is bijective, the direct limit being taken over all open neighborhoods S' of s in S.

(b) *Let $X_{(s)}$ be an $\mathcal{O}_{S,s}$-scheme of finite presentation. Then there are an open neighborhood S' of s in S and an S'-scheme X' of finite presentation such that $X' \otimes_{S'} \mathcal{O}_{S,s}$ is isomorphic to $X_{(s)}$.*

Proof of Proposition 4. To verify the implication (a) $\Longrightarrow$ (*b*), pick a point $s \in S$ and write $X_{(s)} = X \otimes_S \mathcal{O}_{S,s}$. Let K be the field of fractions of $\mathcal{O}_{S,s}$. It is only to show that $X_{(s)}$ satisfies the Néron mapping property. So consider a K-morphism $u_K : Y_{(s),K} \longrightarrow X_{(s),K}$ where $Y_{(s)}$ is a smooth $\mathcal{O}_{S,s}$-scheme; we may assume that $Y_{(s)}$ is of finite type and, thus, of finite presentation over $\mathcal{O}_{S,s}$. Then we can extend $Y_{(s)}$ to a scheme Y' over a connected open neighborhood $S' \subset S$ of s and, taking S' small enough, we may even suppose that Y' is smooth just as $Y_{(s)}$ is; cf. the definition of smoothness in 2.2/3. Using the fact that $X' := X \times_S S'$ is a Néron model of its generic fibre, it follows that u_K extends uniquely to an S'-morphism $u' : Y' \longrightarrow X'$. Then $u' \otimes_{S'} \mathcal{O}_{S,s} : Y_{(s)} \longrightarrow X_{(s)}$ is a unique $\mathcal{O}_{S,s}$-morphism extending u_K. So $X_{(s)}$ is a Néron model of its generic fibre.

The opposite implication (b) $\Longrightarrow$ (a) is obtained similarly. Let K be the ring of rational functions on S and consider a K-morphism $u_K : Y_K \longrightarrow X_K$ where Y is a smooth S-scheme. Again we may assume that Y is of finite type and, thus, of finite presentation over S. Then condition (b) implies that, over a neighborhood $S(s)$ of each closed point $s \in S$, the morphism u_K extends uniquely to an $S(s)$-morphism $u(s) : Y \times_S S(s) \longrightarrow X \times_S S(s)$. Gluing all $u(s)$ yields a unique S-morphism $u : Y \longrightarrow X$ extending u_K. Since the smoothness and the separatedness of the $\mathcal{O}_{S,s}$-scheme $X \otimes_S \mathcal{O}_{S,s}$ imply the smoothness and separatedness of X over a neighborhood of s, we see that X is a Néron model of X_K. □

In the situation of condition (a) of Proposition 4 we will say that X is a *global* Néron model of the generic fibre X_K whereas in the situation of condition (b) the schemes $X \times_S \operatorname{Spec} \mathcal{O}_{S,s}$ will be called the *local* Néron models of X_K. Thus we see that if X_K admits a global Néron model, all its local Néron models exist. The converse of this assertion is not true as we will see in 10.1/11.

A further consequence of the Néron mapping property is the fact that Néron models respect group schemes.

Proposition 6. *Let X be a smooth and separated S-scheme which is a Néron model of its generic fibre X_K. Assume that X_K is a K-group scheme. Then the group scheme structure of X_K extends uniquely to an S-group scheme structure on X.*

Remark 7. When dealing with group schemes, the separatedness occurring as a condition in Definition 1 is superfluous. Indeed, a group scheme is separated over its base as soon as the unit section is a closed immersion; cf. 7.1/2. So group schemes over fields are automatically separated. Furthermore, let X be a smooth S-group scheme of finite type which satisfies the Néron mapping property. In order to show that X is separated over S, we may apply Proposition 4 and thereby assume that S is local. Then, due to the Néron mapping property, the unit section $\operatorname{Spec} K \longrightarrow X_K$ of the generic fibre X_K extends uniquely to a section $S \longrightarrow X$, namely to the unit section of X. It follows that the latter is a closed immersion, as can be seen from 7.1/1 and its proof. Thus X is separated as claimed.

Although Néron models have been defined within the setting of schemes, their importance seems to be restricted to group schemes or, more generally, to torsors under group schemes as we will see in Chapter 6. For example, $\mathbb{P}^1_K$ admits $\mathbb{P}^1_S$ as a smooth and separated S-model which, due to the properness, satisfies the extension property for étale points. But $\mathbb{P}^1_S$ is not a Néron model of its generic fibre since not all K-automorphisms of $\mathbb{P}^1_K$ extend to S-automorphisms of $\mathbb{P}^1_S$; cf. 3.5/5. The situation is much better in the group scheme case as can be seen from an extension theorem of Weil for rational maps into group schemes; cf. 4.4/1:

Let $u: Y \dashrightarrow X$ be a rational map between S-schemes where Y is smooth and where X is a smooth and separated S-group scheme. Then, if u is defined in codimension ≤ 1, it is defined everywhere.

Using this result, one can show without difficulties that abelian schemes over S, i.e., proper and smooth S-group schemes with connected fibres, provide examples of Néron models.

Proposition 8. *Let X be an abelian scheme over S. Then X is a Néron model of its generic fibre X_K.*

Proof. Let Y be a smooth S-scheme and let $u_K: Y_K \longrightarrow X_K$ be a K-morphism. We claim that u_K extends to a rational map $u: Y \dashrightarrow X$ with a domain of definition $V \subset Y$ which is S-dense; i.e., which is dense in each fibre of Y over S. Namely, consider a closed point $s \in S$ and a generic point ζ of the fibre over s in Y. Then the local ring $\mathcal{O}_{Y,\zeta}$ is a discrete valuation ring; cf. 2.3/9. So the valuative criterion of properness implies that u_K extends to a morphism $\operatorname{Spec} \mathcal{O}_{Y,\zeta} \longrightarrow X$ or, using Lemma 5, to a rational map $Y \dashrightarrow X$ which is defined in a neighborhood of ζ. Therefore u is defined in codimension ≤ 1 and, thus, by Weil's extension theorem, it is defined everywhere. The uniqueness of the extension follows from the separatedness of X. □

We have seen that Néron models satisfy the extension property for étale points. On the other hand, using a similar argument as the one given in the above proof, one can show that a smooth and separated group scheme satisfying the extension property for étale points is already a Néron model; see also 7.1/1.

Criterion 9. *Let X be a smooth and separated S-group scheme of finite type. Then X is a Néron model of its generic fibre if and only if X satisfies the extension property for étale points.*

Describing the necessary steps of the *proof*, we mention first of all that, due to Proposition 4, the criterion has only to be verified in the local case. So assume that

S is a local scheme. Then one has to use the fact that X, as a weak Néron model of its generic fibre, satisfies the so-called weak Néron mapping property; cf. 3.5/3. The latter means that each K-morphism $u_K: Y_K \longrightarrow X_K$ extends to an S-rational map $u: Y \dashrightarrow X$; i.e., to a rational map which is defined on an S-dense open subscheme of Y. So, just as in the case of abelian schemes, the if-part of the assertion is reduced to Weil's extension theorem for morphisms into group schemes. □

1.3 The Local Case: Main Existence Theorem

As we have seen in 1.2/4, the existence of a Néron model over a global Dedekind scheme S implies the existence of the local Néron models at closed points of S. In fact, if global Néron models are to be constructed, the first step is to obtain all local ones. Then one can try to glue them in order to build a global model; see Section 1.4 for the case of abelian varieties. The purpose of the present section is to present the existence theorem for Néron models in the local case.

Theorem 1. *Let R be a discrete valuation ring with field of fractions K, with a strict henselization R^{sh}, and with field of fractions K^{sh} of R^{sh}. Let X_K be a smooth K-group scheme of finite type. Then X_K admits a Néron model X over R if and only if $X_K(K^{sh})$ is bounded in X_K.*

In particular, since properness implies boundedness, abelian varieties admit Néron models in the local case:

Corollary 2. *Let A_K be an abelian variety over the field of fractions K of a discrete valuation ring R. Then A_K admits a Néron model over R.*

The only-if-part of Theorem 1 is a trivial consequence of 1.1/7 since Néron models are of finite type. The proof of the if-part, however, is more complicated and will be carried out in Chapters 3 to 6, each one of them dealing with a certain aspect of the construction of local Néron models. At this place we have to content ourselves with a simplified description of the necessary steps.

We start the construction by choosing a separated R-model X of X_K of finite type which satisfies the extension property for étale points. If X_K is projective, we can take for X the schematic closure of X_K in a projective n-space over R. Similarly, if X_K is affine, we may use the boundedness condition and take for X the schematic closure of X_K in a suitable affine n-space over R. In the general case we use 1.1/7. Since the model X obtained from 1.1/7 might not be separated and since we want to avoid the result 3.5/7 saying that a separated R-model can be found, we will generalize the situation slightly in Chapters 3 and 4 by working with a finite family (X_i) of *separated* R-models of X_K such that the canonical map

$$\coprod X_i(R^{sh}) \longrightarrow X_K(K^{sh})$$

is surjective.

For simplicity, let us consider a separated R-model $X^{(1)}$ of finite type of X_K satisfying the extension property for étale points. Then we apply the so-called smoothening process to $X^{(1)}$, which will be explained in Chapter 3. Thereby we obtain a proper R-morphism $X^{(2)} \longrightarrow X^{(1)}$ consisting of a sequence of blowing-ups with centers in special fibres. It has the property that each R^{sh}-valued point of $X^{(1)}$ lifts to an R^{sh}-valued point of $X^{(2)}$ which factors through the smooth locus $X^{(2)}_{\mathrm{smooth}}$ of $X^{(2)}$; cf. 3.1/3. Thus $X^{(3)} := X^{(2)}_{\mathrm{smooth}}$ is a *smooth* R-model of finite type of X_K which satisfies the extension property for étale points. In other words, $X^{(3)}$ is a weak Néron model of X_K. It satisfies the so-called weak Néron mapping property which means that, for each smooth R-scheme Y and each K-morphism $u_K : Y_K \longrightarrow X^{(3)}_K$, there is an R-rational extension $u : Y \dashrightarrow X^{(3)}$; i.e., a rational extension which is defined on an R-dense open part of Y; cf. 3.5/3. Hence $X^{(3)}$ satisfies certain aspects of a Néron model. However, weak Néron models are not unique and it might be that the group structure of X_K does not extend to a group scheme structure on $X^{(3)}$. Thus, one cannot expect that $X^{(3)}$ is already a Néron model of X_K.

In general, it is necessary to modify $X^{(3)}$. This can be done by using the group structure on X_K; cf. Section 4.3. To simplify the notation, write X instead of $X^{(3)}$. Furthermore, let π be a uniformizing element of R, and let $k = R/\pi R$ be the residue field of R. Fixing a non-trivial left-invariant differential form ω on X_K of degree $d = \dim X_K$, we define its π-order over each component Y_k of the special fibre X_k of X. Namely, let η be the generic point of Y_k. Then $\mathcal{O}_{X,\eta}$ is a discrete valuation ring with uniformizing element π. Since the sheaf of relative differential forms $\Omega^d_{X/R}$ is a line bundle, there is an integer n such that $\pi^{-n}\omega$ extends to a generator of $\Omega^d_{X/R}$ at η, and we can set $\mathrm{ord}_{Y_k} \omega := n$. Then the ω-minimal components of X_k, i.e., those components for which the π-order of ω is minimal, are uniquely determined by X_K up to R-birational isomorphism. They occur in each weak Néron model of X_K and have to be interpreted as the components which have largest volume. More precisely, any isomorphism $u_K : X_K \longrightarrow X_K$, which leaves ω invariant, extends to an R-rational map $X \dashrightarrow X$ which maps the ω-minimal components of X_k birationally onto each other; cf. 4.3/2. So if X' is the open subscheme obtained from X by removing all non-minimal components of the special fibre X_k, the isomorphism u_K gives rise to an R-birational map $X' \dashrightarrow X'$ which even is an open immersion on its domain of definition; see 4.3/1 (ii). Applying this argument to general translations on X_K, one can realize that the group multiplication $m_K : X_K \times X_K \longrightarrow X_K$ extends to an R-birational map $m : X' \times X' \dashrightarrow X'$. In fact, m defines a so-called R-birational group law on X'; cf. 4.3/5. The R-scheme X' is, as we will see in the end (cf. 4.4/4), already an R-dense open subscheme of the Néron model we are going to construct, although X' will not, in general, satisfy the extension property for étale points any more.

Now a Néron model of X_K can be derived from X' by considering its "saturation" under the birational group law. There is a standard procedure, first invented by Weil for the case where the base consists of a field and then generalized by A. Néron and M. Artin, which associates group schemes to R-birational group laws. We will explain it in Chapter 5 for the case where the base ring R is strictly henselian; the generalization to an arbitrary discrete valuation ring is done in Chapter 6 by means of descent. Thereby we will see, cf. 5.1/5, that X' can be enlarged to an R-group scheme X'' which is an R-model of X_K of finite type and which has the property that the group multiplication on X'' restricts to the R-birational group law m on X'. Then one uses a translation argument to show that X'' satisfies the extension property for étale points so that X'' is a Néron model of X_K by Criterion 1.2/9.

1.4 The Global Case: Abelian Varieties

In the preceding section we have discussed the existence of Néron models in the local case. If a global Néron model is to be constructed, one has to find a way to glue the local Néron models. The problem is that the resulting global model might not be of finite type again, a property which is necessary for Néron models. However, as we want to show in the present section, when dealing with abelian varieties the gluing works well and we do obtain global Néron models this way. To start with, let us state Proposition 1.2/4, which describes the relationship between local and global Néron models, in a form which is more useful for applications.

Proposition 1. *Let S be a Dedekind scheme with ring of rational functions K and let X_K be a smooth and separated K-scheme of finite type. Then the following assertions are equivalent*:

(a) *There exists a global Néron model X of X_K over S.*

(b) *There exists a dense open subscheme $S' \subset S$ such that X_K admits a Néron model over S' as well as local Néron models at the finitely many closed points of $S - S'$.*

Proof. The implication (a) $\Longrightarrow$ (b) is trivial, due to 1.2/3 and 1.2/4. To obtain the opposite, we may assume that S is connected. Let $s_1, \ldots, s_r$ be the closed points which form the complement of S' in S and let X' be a Néron model of X_K over S'. Furthermore, let $X_{(s_i)}$ be a local Néron model of X_K over the ring $\mathcal{O}_{S,s_i}$. Then, using 1.2/5, $X_{(s_i)}$ extends to a smooth and separated scheme of finite type X_i over a suitable open neighborhood S_i of s_i. Since X_i and X' coincide at the generic point of S, both must coincide over a non-empty open part of S'. Removing finitely many closed points from S_i, we may assume that $S_i \cap (S - S') = \{s_i\}$ and that X_i coincides with X' over $S' \cap S_i$. But then we can glue each X_i with X' over $S' \cap S_i$ to obtain a smooth and separated S-model X of finite type satisfying $X \times_S S' = X'$ and $X \otimes_S \mathcal{O}_{S,s_i} = X_{(s_i)}$. Thus X is a global Néron model of X_K by 1.2/4. □

Now consider a connected Dedekind scheme S with field of rational functions K and an abelian variety A_K over K. One says that A_K has *good reduction at a closed point* $s \in S$ if A_K extends to a smooth and proper scheme $A_{(s)}$ over $\mathcal{O}_{S,s}$. We want to show that $A_{(s)}$ is automatically an abelian scheme in this case and, thus, a Néron model of A_K.

Proposition 2. *Let S be a connected Dedekind scheme with field of fractions K and let A_K be an abelian variety over K. Assume that A_K extends to an S-scheme A which is smooth and proper. Then A is an abelian scheme under a group structure which extends the given group structure on A_K. In particular, A is a Néron model of A_K.*

Proof. Using 1.2/4 we may assume that we are in the local case where S consists of a discrete valuation ring. Since A is proper, the valuative criterion of properness shows that A is already a weak Néron model of A_K. Furthermore, the special fibre A_k of A is connected by [EGA III_1], 5.5.1. Therefore A_k has to be viewed as an ω-minimal component, with ω being a generating differential form of degree $\dim A_K$ on A_K; use the weak Néron mapping property 3.5/3 and the result 4.3/1. On the other hand, we know from 1.3/2 that A_K admits a Néron model X. Thus, by the Néron mapping property, there is a canonical S-morphism $A \longrightarrow X$ which is an open immersion by 4.3/1 (ii) or 4.4/1. Because A is proper, its image is closed in X. However, X is connected due to the fact that X is flat over S, with the generic fibre $X_K = A_K$ being connected. So $A \longrightarrow X$ is an isomorphism and A is a Néron model of A_K. Thus, applying the Néron mapping property, the group structure of A_K extends to a group scheme structure on A and A is seen to be an abelian scheme. □

In order to apply Proposition 1 in the case of abelian varieties A_K, we have to show that A_K has good reduction at almost all closed points of S and even more: that A_K extends to an abelian scheme A' over a dense open subscheme S' of S. Looking at a simple example, assume that the characteristic of K is different from 2 and consider the case where A_K is an elliptic curve in $\mathbb{P}^2_K$ given by an equation in Weierstraß form

$$y^2z = x^3 + \beta xz^2 + \gamma z^3$$

with a non-zero discriminant $\Delta = 4\beta^3 + 27\gamma^2$. Then the elements β, γ, Δ, and Δ^{-1} belong to almost all local rings $\mathcal{O}_{S,s}$ at closed points $s \in S$. So there exists a non-empty open subscheme $S' \subset S$ such that β, γ, and Δ extend to sections in $\mathcal{O}_S(S')$ and such that Δ and 2 are invertible in $\mathcal{O}(S')$. Consequently, A_K extends to a smooth projective family A' of elliptic curves in $\mathbb{P}^2_{S'}$. Then A' is an abelian scheme extending A_K as we have shown in Proposition 2. Alternatively, we can apply limit arguments of type 1.2/5 and see directly that, after a possible shrinking of S', the scheme A' gives rise to an abelian scheme over S'. In principle, the same reasoning applies to any abelian variety A_K over K.

Theorem 3. *Let S be a connected Dedekind scheme with field of fractions K and let A_K be an abelian variety over K. Then A_K admits a global Néron model A over S. Furthermore, let S' be the subset of S consisting of the generic point and of all closed points in S where A_K has good reduction. Then S' is a dense open subscheme of S and $A \times_S S'$ is an abelian scheme over S'.*

Proof. We have to show that A_K extends to a smooth and proper scheme over a neighborbood of the generic point of S as well as over a neighborhood of each closed point of S where A_K has good reduction. Then all such schemes are abelian schemes by Proposition 2 and, using the Néron mapping property, they can be glued to give an abelian scheme over S'. Furthermore, due to the existence of local Néron models 1.3/2, we conclude from Proposition 1 that A_K admits a global Néron model A.

In order to show that A_K extends to a smooth proper scheme over a non-empty open part of S, choose a closed embedding $A_K \longrightarrow \mathbb{P}_K^n$ into some projective n-space and consider the schematic closure A of A_K in $\mathbb{P}_S^n$. Then A is smooth over the generic point of S and, thus, smooth over an open neighborhood S'' of this point. So $A'' = A \times_S S''$ is a smooth projective S''-model of A_K. Alternatively, we can use 1.2/5 to extend A_K to a scheme A'' of finite type over an open neighborhood S'' of the generic point in S. If S'' is small enough, A'' will be smooth and, by [EGA IV_3], 8.10.5, also proper. The same argument applies if we consider a closed point $s \in S$ where A_K has good reduction. Namely, then A_K extends to a smooth and proper scheme $A_{(s)}$ over $\mathcal{O}_{S,s}$ and we can extend the latter over an open neighborhood of s. □

It follows from the valuative criterion of properness that any K-rational map $u_K : Y_K \dashrightarrow A_K$ from a smooth K-scheme Y_K into an abelian variety A_K is defined in codimension 1 and, thus, is defined everywhere by Weil's extension theorem 4.4/1. Thereby it is seen that, in the case of abelian varieties, the Néron mapping property can be strengthened.

Proposition 4. *Let S be a connected Dedekind scheme with field of fractions K and let A_K be an abelian variety over K with Néron model A over S. Then, for each smooth S-scheme Y, and for each K-rational map $u_K : Y_K \dashrightarrow A_K$, there is a unique S-morphism $u : Y \longrightarrow A$ extending u_K.*

For further generalizations of this result see 8.4/6 and 10.3/1.

1.5 Elliptic Curves

In order to illustrate the construction of Néron models, we want to look at Néron models of elliptic curves. In this particular case, the procedure of construction can be made quite explicit. The reader who is interested in a more profound discussion of models of elliptic curves is referred to Kodaira [1], Néron [2], and Tate [2]. In our terminology, an elliptic curve will always be understood to have a rational point.

We will work over a base scheme S consisting of a strictly henselian discrete valuation ring R with field of fractions K and with an *algebraically closed* residue field k. First we want to clarify the interdependence between Néron models and regular and proper minimal models of elliptic curves over K. So consider an elliptic curve E_K over K. Then E_K admits a Néron model, as we have stated in 1.3/2. It also admits a proper minimal model. By the latter we mean a proper flat R-model E which is a regular scheme and which is minimal among all models E' of this type in the sense that each R-morphism $E \longrightarrow E'$ which is an isomorphism on generic fibres is an isomorphism itself. So there are no irreducible components of the special fibre of E which can be contracted without loosing the regularity of E. Regular and proper minimal models of curves are unique; see Abhyankar [1] and Lipman [1] for the existence of regular and proper models and Lichtenbaum [1], Shafarevich [1], or Néron [2] for the existence of regular and proper minimal models.

Proposition 1. *Assume that R is a strictly henselian discrete valuation ring. Let E be a regular and proper minimal model over R of the elliptic curve E_K. Then the smooth locus of E is a Néron model of E_K.*

Proof. Write E' for the smooth locus of E. It follows from 3.1/2 that each R-valued point of E factors through E'. So, by the valuative criterion of properness, we see that E' satisfies the extension property for étale points and, thus, is a weak Néron model of E_K. Furthermore, it follows from 2.3/5 that all k-valued points of the special fibre E'_k lift to R-valued points of E'.

Fix an invariant differential form ω of degree 1 on E_K. We claim that all components of the special fibre E'_k are ω-minimal. To see this, consider two components X_1 and X_2 of E'_k and two k-valued points $y_k \in X_1$ and $z_k \in X_2$. Lift them to R-valued points y, z of E' and restrict them to K-valued points $y_K, z_K \in E_K$. Then the translation by $z_K y_K^{-1}$ is a K-isomorphism of E_K mapping y_K to z_K. Due to the uniqueness of regular and proper minimal models, this isomorphism extends to an R-isomorphism of E and, thus, of E', mapping y onto z. So there are R-isomorphisms of E' which operate transitively on the components of the special fibre E'_k and which leave ω invariant. Consequently, all components of E'_k must be ω-minimal; cf. 4.3/1.

Now, as explained in Section 1.2 or, in more detail, in Section 4.3 and Chapter 5, the group structure on E_K extends to an R-birational group law on E' and, then, to a group scheme structure on a bigger R-scheme E'' containing E' as an R-dense open subscheme; cf. 5.1/5. However, using the fact that all translations by K-valued points on E_K extend to isomorphisms on E', and to the translations by the corresponding R-valued points on E'', it follows that E' and E'' coincide. So E' is a Néron model of E_K. □

If E is a proper and flat R-model of an elliptic curve E_K over K, then E is smooth over R at all points of the generic fibre. Furthermore, E is smooth at a point x of the special fibre E_k if and only if this fibre is smooth over k at x, or equivalently since k is algebraically closed, if and only if E_k is regular at x. So, in order to pass to the smooth locus of E, one removes all irreducible components with multiplicities > 1 from E_k as well as from the remaining part of E_k all singular points; the latter form a finite set. For algebraically closed residue field k, special fibres of regular and proper minimal models of elliptic curves have been classified by Néron [2], see also Kodaira [1]; there is only a finite list of possible types. An algorithm to compute the type of the special fibre from a given equation for E_K has been given in Tate [2].

If one is interested in a Néron model E of an elliptic curve E_K and not so much in its regular and proper minimal model, one can construct E directly without too much effort starting out from an equation describing E_K in $\mathbb{P}_K^2$, at least when the residue characteristic of K is different from 2 and 3. To do this, one classifies Weierstraß equations into a finite list of types, according to certain conditions involving the values of their coefficients, discriminants, and j-invariants. Then one can construct the Néron model E by direct computation in each of these cases. To demonstrate this, assume that R is a strictly henselian discrete valuation ring with residue characteristic char k different from 2 and 3 and consider an elliptic curve E_K over K, defined in $\mathbb{P}_K^2$ by an equation in Weierstraß form

$$(*) \qquad y^2 z = x^3 + \beta x z^2 + \gamma z^3 .$$

Then discriminant Δ and j-invariant j are given by

$$\Delta = 4\beta^3 + 27\gamma^2 , \qquad j = 2^6 \cdot 3^3 \cdot 4\beta^3/\Delta .$$

Viewing E_K as a group scheme, we assume that the point $(0,1,0)$ defines the unit section of E_K. Let π be a uniformizing element of R, and let $v: K \longrightarrow \mathbb{Z}$ be the additive valuation given by R which satisfies $v(\pi) = 1$. We need some elementary properties of the equation $(*)$.

Lemma 2. *For $n \in \mathbb{Z}$, the change of homogeneous coordinates in $\mathbb{P}_K^2$*

$$(x, y, z) \longmapsto (\pi^{-2n}x, \pi^{-3n}y, z)$$

induces on the equation of E_K the change

$$\beta \longmapsto \pi^{4n}\beta , \qquad \gamma \longmapsto \pi^{6n}\gamma , \qquad \Delta \longmapsto \pi^{12n}\Delta .$$

Lemma 3. (a) *If $v(j) \geq 0$, then $v(\Delta) = \min(v(\beta^3), v(\gamma^2))$. In particular, $v(\Delta) \equiv 0\ (2)$ or $v(\Delta) \equiv 0\ (3)$.*

(b) *If $v(j) < 0$, then $v(\Delta) > v(\beta^3) = v(\gamma^2)$. In particular, $v(\beta) \equiv 0\ (2)$ and $v(\gamma) \equiv 0\ (3)$.*

Making a change of coordinates as described in Lemma 2, we can assume that the coefficients β and γ of $(*)$ belong to R and, furthermore, that $\min(v(\beta^3), v(\gamma^2))$ is minimal. Thereby we arrive at a so-called minimal Weierstraß equation of E_K; i.e., at a Weierstraß equation with coefficients in R such that $v(\Delta)$ is minimal. We list the possible cases which remain.

Lemma 4. *Let the equation $(*)$ be a minimal Weierstraß equation for E_K. Then, if $v(j) \geq 0$, we have $v(\Delta) \in \{0, 2, 3, 4, 6, 8, 9, 10\}$. Furthermore, if $v(j) < 0$, either $v(\beta) = v(\gamma) = 0$, or $v(\beta) = 2$ and $v(\gamma) = 3$.*

Using Néron's symbols as introduced in his table [2], p. 124/125, the possibilities for a minimal Weierstraß equation for E_K as mentioned in the above lemma split into the following subcases; see also the table in Tate [2], p. 46.

$$\text{(a)} \qquad v(j) \geq 0 , \quad v(\Delta) = 0$$

$$(\text{b}_m) \qquad v(j) = -m < 0 , \quad v(\beta) = v(\gamma) = 0$$

(c1) $\quad v(j) \geq 0\,, \quad v(\Delta) = 2$

(c2) $\quad v(j) \geq 0\,, \quad v(\Delta) = 3$

(c3) $\quad v(j) \geq 0\,, \quad v(\Delta) = 4$

(c4) $\quad v(j) \geq 0\,, \quad v(\Delta) = 6$

(c5$_m$) $\quad v(j) = -m < 0\,, \quad v(\beta) = 2\,, \quad v(\gamma) = 3$

(c6) $\quad v(j) \geq 0\,, \quad v(\Delta) = 8$

(c7) $\quad v(j) \geq 0\,, \quad v(\Delta) = 9$

(c8) $\quad v(j) \geq 0\,, \quad v(\Delta) = 10$

Now, to construct a Néron model of E_K, one proceeds as follows. One chooses a minimal Weierstraß equation for E_K and uses it for the definition of an R-model $\bar{E}$ of E_K in $\mathbb{P}^2_R$. Let E^0 be the smooth part of $\bar{E}$. Then one verifies by direct computation, or by using general properties of planar cubics, that E^0 is a smooth R-group scheme extending E_K. In fact, we will see that it is the so-called identity component of the Néron model of E_K. There are three possibilities which we characterize by the first letters of Néron's symbols:

(a) $v(\Delta) = 0$. Then $\bar{E}$ is smooth, so $E^0 = \bar{E}$ is an abelian scheme extending E_K. It follows that E_K is an elliptic curve with good reduction and that $\bar{E}$ is its Néron model.

(b) $v(\Delta) > 0$ and $\min(v(\beta), v(\gamma)) = 0$. Then $\bar{E}$ is not smooth; the special fibre of E^0 is the multiplicative group $\mathbb{G}_{m,k}$.

(c) $v(\Delta) > 0$ and $\min(v(\beta), v(\gamma)) > 0$. Also in this case, $\bar{E}$ is not smooth; the special fibre of E^0 is the additive group $\mathbb{G}_{a,k}$.

Consider the invariant differential $\omega = \dfrac{dx}{y}$ on E_K. Then ω has π-order 0 over E^0. We claim that, for the construction of the Néron model of E_K, it is enough to extend E^0 into a weak Néron model E of E_K with the property that the special fibre of E consists of ω-minimal components, all of them being isomorphic to E^0_k.

Lemma 5. *Let $E^1, \ldots, E^r$ be smooth and separated R-models of E_K. Assume that, for all ρ, the special fibre E^ρ_k, as a k-scheme, is isomorphic to E^0_k, that ω has π-order* 0 *over E^ρ_k, and that the canonical map $\coprod_{\rho=0}^r E^\rho(R) \longrightarrow E_K(K)$ is bijective. Then, gluing the E^ρ along the generic fibre E_K, we obtain a Néron model E of E_K. Furthermore, E^0 is the identity component of E.*

Proof. It is clear that E is a smooth R-model of finite type of E_K which satisfies the extension property for étale points 1.1/1. So E is a weak Néron model of E_K. Furthermore, E is separated since, for $\rho \neq \tau$, the intersection of $E^\rho \times_R E^\tau$ with the diagonal in $E \times_R E$ is just E_K. By the assumption on the π-order of ω, all components of the special fibre E_k are ω-minimal. So, denoting by N the Néron model of E_K, we have an open immersion $E \hookrightarrow N$ by 4.3/1 or 4.4/4. Then E^0 must coincide with the identity component N^0 of the Néron model N. Thereby we see that the special fibre N_k consists of $r + 1$ copies of E^0_k which, in case (c) is the affine

1-space $\mathbb{A}_k^1$, and in case (b) is $\mathbb{A}_k^1$ minus the origin. Since the same is true for E, we conclude from the special type of E_k^0 that $E \hookrightarrow N$ is bijective. So E is a Néron model of E_K. □

In each of Néron's cases, a Néron model E of E_K can be constructed via the above lemma. To show how to proceed, we will look at the cases (c1) and (c2) which are quite simple, as well as at case (b_m) which is more complicated. First note that $e_k := (0, 1, 0) \in \bar{E}(k)$ is a non-singular point of the special fibre of $\bar{E}$; in fact, it is the unit section of E_k^0. So the singularities of $\bar{E}_k$ belong to the affine part $\bar{E}_z$ of $\bar{E}$ which is described in $\mathbb{A}_R^2$ by the equation

$$(**) \qquad y^2 = x^3 + \beta x + \gamma \, .$$

There is precisely one singularity p_k in $\bar{E}_{z,k}$ in the cases (b) or (c); it corresponds to a multiple zero of the right-hand side of (**). So, in order to apply Lemma 5, we have to concentrate on R-models $E^\dagger$ of E_K such that the image of $E^\dagger(R) \longrightarrow E_K(K)$ consists of K-valued points which, in $\bar{E}$, specialize into the singular point p_k.

Case (c1). Then $v(\beta) \geq 1$ and $v(\gamma) = 1$ by Lemma 3; hence $p_k = (0, 0)$, using affine coordinates of $\bar{E}_{z,k}$. Since

$$\{(x, y) \in \bar{E}_z(K); v(x) > 0, v(y) > 0\} = \varnothing \, ,$$

it follows from Lemma 5 that $E^0 = \bar{E} - \{p_k\}$ is the Néron model of E_k. Also it is easily checked that the minimal Weierstraß model is regular and, thus, coincides with the regular and proper minimal model. □

Case (c2). We have $v(\beta) = 1$ and $v(\gamma) \geq 2$ by Lemma 3. Again, $p_k = (0, 0)$ is the singular point of $\bar{E}_{z,k}$. Thus all points $(x, y) \in \bar{E}_z(K)$ which do not extend to R-valued points of E^0 must satisfy $v(x) \geq 1$ and $v(y) \geq 1$. Use $\hat{x} := \pi^{-1}x$ and $\hat{y} := \pi^{-1}y$ as new coordinates and let E^1 be the R-model of E_K obtained by gluing

$$\operatorname{Spec} R[\hat{x}, \hat{y}]/(\hat{y}^2 - \pi\hat{x}^3 - \pi^{-1}\beta\hat{x} - \pi^{-2}\gamma)$$

along its generic fibre to E_K. Then all points $(x, y) \in \bar{E}_z(K)$, which satisfy $v(x) \geq 1$ and $v(y) \geq 1$, extend to R-valued points of E^1. In addition, E^1 is smooth and separated and has special fibre $E_k^1 \simeq \mathbb{A}_k^1 \simeq E_k^0$ as required. Furthermore, since $\hat{x}$ and $\hat{y}$ do not vanish at the generic point of E_k^1, we see that $\omega = dx/y = d\hat{x}/\hat{y}$ is of π-order 0 over E_k^1. Thus Lemma 5 can be applied. The Néron model of E_K is obtained by gluing E^0 and E^1 along the generic fibre E_K; its special fibre consists of two components. □

We mention here that the process of replacing a variable x by $\hat{x} = \pi^{-1}x$ is a special case of a dilatation, a technique to be applied systematically when we work out the smoothening process in Chapter 3. In fact, the method we have used above for the construction of E is a very explicit form of the smoothening process. It has to be applied in a similar way for treating the remaining cases.

Case (b_m). This case is of special interest if R is complete; then E_K is a so-called Tate elliptic curve. We have $v(j) = -m < 0$, $v(\beta) = v(\gamma) = 0$, and, hence, $v(\Delta) = m > 0$. Furthermore, $E_k^0 \simeq \mathbb{G}_{m,k}$. Let us write

$$P(x) = x^3 + \beta x + \gamma$$

for the right-hand side of (**) and $\bar{P}(x)$ for the polynomial obtained from $P(x)$ by reducing coefficients from R to k. Then $\bar{P}(x)$ has a single root $\bar{a} \in k$ and a double root $\bar{b} \in k$. So $p_k = (\bar{b}, 0)$ is the singular point of $\bar{E}_{z,k}$ and all points $(x, y) \in \bar{E}_z(K)$ which do not extend to R-valued points of E^0 must reduce to p_k.

The root $\bar{a}$ lifts to a root $a \in R$ of $P(x)$ since R is strictly henselian. Set $Q(x) := P(x)/(x - a)$. Then $Q(x)$ has coefficients in R and $\bar{Q}(x) = (x - \bar{b})^2$ is the polynomial obtained from it by reducing coefficients from R to k. Extending the valuation v from K to the algebraic closure K^{alg}, the root $\bar{b}$ lifts to two roots b_1, $b_2 \in K^{\text{alg}}$ of $Q(x)$, where $v(a - b_i) = 0$ for $i = 1, 2$. Thus, the discriminant of $P(x)$, which is Δ, coincides with the discriminant of $Q(x)$, up to a unit in R. Since $v(\Delta) = m$, we have

$$v(b_1 - b_2) = m/2 .$$

Furthermore, since R is strictly henselian, the extension of v from K to $K(b_1, b_2)$ is unique, just as for complete fields. So $v(b_1) = v(b_2)$. Using an inductive argument on m, interpreted as the value of the discriminant of $Q(x)$, we want to construct R-models $E^1, \ldots, E^{m-1}$ which, together with E^0, will satisfy the conditions of Lemma 5.

To do this, choose an arbitrary lifting $b \in R$ of $\bar{b}$ and use $x - b$ as a new variable instead of x; denote it by x again. The effect is that the singular point $p_k = (\bar{b}, 0)$ is transformed into the origin $(0, 0)$ this way. We will denote transformed polynomials and roots by $P(x)$, $Q(x)$, a, b_i, etc., again, so that

$$P(x) = (x - a)Q(x) , \qquad Q(x) = (x - b_1)(x - b_2)$$

where now

$$v(a) = 0 , \qquad v(b_1) = v(b_2) \geq 1/2 .$$

For $m = 1$ we obtain $v(b_1 - b_2) = 1/2$ and, hence, $v(b_i) = 1/2$. Then each $x \in R\pi$ satisfies $v(P(x)) = 1$ and we see that $P(x)$ cannot have a square root in R. So there are no points $(x, y) \in \bar{E}_z(K)$ satisfying $v(x) \geq 1$ and $v(y) \geq 1$, and we can conclude from Lemma 5 that, in this case, E^0 is already the Néron model of E_K. Furthermore, the minimal Weierstraß model is regular in this case.

If $m > 1$, we use $\pi^{-1}x$ and $\pi^{-1}y$ as new variables, writing x and y for them again. Then, looking for points $(x, y) \in \bar{E}_z(K)$ satisfying $v(x) \geq 1$ and $v(y) \geq 1$, we have to look for integral solutions of the equation

$$y^2 = (a - \pi x) \cdot Q(x) ,$$

where we have written $Q(x)$ instead of $\pi^{-2}Q(\pi x)$ again. This way the discriminant of $Q(x)$ has been divided by π^2 so that its value is now $m - 2$. Assume $m = 2$. Then

$$\operatorname{Spec} R[x, y]/(y^2 - (a - \pi x) \cdot Q(x))$$

is smooth over R. Gluing it along its generic fibre to E_K, we obtain an R-model E^1 as required in Lemma 5. Namely, the special fibre of E^1 is

$$\operatorname{Spec} k[x, y]/(y^2 - \bar{a}\bar{Q}(x))$$

with $\bar{Q}(x)$ having two distinct roots in k. So it is $\mathbb{P}_k^1$ minus two closed points and, thus, isomorphic to E_k^0. That the differential ω has π-order 0 over E_k^1, is easily checked. So, for $m = 2$, the Néron model is obtained by gluing E^0 and E^1 along the generic fibre E_K; its special fibre consists of two components.

If $m > 2$, the polynomial $\bar{Q}(x)$ has a root of multiplicity 2 and the scheme

$$\operatorname{Spec} R[x, y]/(y^2 - (a - \pi x)\cdot Q(x))$$

is not smooth over R; its special fibre consists of two affine lines intersecting each other. Removing the intersection point, we can construct two R-models E^1 and E^2 of E_K with special fibre isomorphic to E_k^0 each. If $m = 3$, one is reduced to the case considered above where the discriminant of $Q(x)$ has value 1. Thereby it is seen that E^0, E^1, E^2 satisfy the conditions of Lemma 5. If $m > 3$, the value of the discriminant of $Q(x)$ is > 1 and can be reduced by 2 again as shown above. One continues this way until the value of the discriminant of $Q(x)$ is 1 or 0. Thereby one constructs R-models $E^1, \ldots, E^{m-1}$ of E_K which, together with E^0 satisfy the conditions of Lemma 5. So the special fibre of the Néron model E of E_K consists of m components. With a little bit of extra work one can show that the group E_k/E_k^0 is cyclic of order m. Also, by means of the arguments we have given, one can determine the regular and proper minimal model of E_K. Its special fibre consists of a chain of m projective lines forming a loop (if $m > 1$) or of a rational curve with a double point (if $m = 1$). In particular, we can thereby see that the regular and proper minimal model of E_K will not be planar if $m > 3$, because a planar cubic cannot have more than 3 components. □

It is useful to look at Tate elliptic curves also from the rigid analytic viewpoint. So let R be a complete discrete valuation ring. We do not need that R is strictly henselian or that the residue field k is perfect. An elliptic curve E_K over K is called a Tate curve if, in the sense of rigid analytic geometry, it can be represented as a quotient $\mathbb{G}_{m,\mathrm{rig}}/q^{\mathbb{Z}}$ where $\mathbb{G}_{m,\mathrm{rig}}$ is the analytification of the multiplicative group $\mathbb{G}_{m,K}$ and where $q \in K^*$ satisfies $m := v(q) > 0$. The quotient $\mathbb{G}_{m,\mathrm{rig}}/q^{\mathbb{Z}}$ can be thought of as being constructed by gluing m annuli of type $\{x \in \mathbb{G}_{m,\mathrm{rig}}; |\pi| \le |x| \le 1\}$ in a cyclical way. Using this covering, we can extend $\mathbb{G}_{m,\mathrm{rig}}/q^{\mathbb{Z}}$ into a formal scheme X whose special fibre X_k is a projective line with a double point if $m = 1$ and a chain of m projective lines forming a loop if $m > 1$.

Choosing an effective Cartier divisor D on X whose support is contained in the smooth locus of X and which is very ample on all components of X_k and on the generic fibre X_{rig}, one constructs a projective embedding of X and, thus, an R-model E' of E_K whose formal completion is X. Then it turns out that the smooth locus E of E' is a Néron model of E_K. The special fibre E_k coincides with the smooth locus of X_k and, thus, is an extension of $\mathbb{G}_{m,k}$ by $\mathbb{Z}/m\mathbb{Z}$. See Bosch and Lütkebohmert [3] for a generalization of the construction to abelian varieties.

1.6 Néron's Original Article

We want to give here some analysis of Néron's article "Modèles minimaux des variétés abéliennes sur les corps locaux et globaux"[2] which appeared in 1964 and which serves as a basis for the construction of Néron models as done in this book; see also the lecture [1] given by Néron in 1961 at the Séminaire Bourbaki. Consider an abelian variety A_K over a local field K and think of it as being embedded into a projective space $\mathbb{P}_K^N$. Let X be the schematic closure of A_K in $\mathbb{P}_R^N$ where R is the discrete valuation ring of integers of K. Then X is an R-model of A_K on which integral points might not be read as nicely as possible. Moreover, it will be likely that the group structure of A_K does not extend to the smooth part of X. To obtain R-models of A_K which do not have these disadvantages, Néron had to apply a series of substantial modifications to X and, in doing so, he had to overcome a lot of technical difficulties.

His article is divided into three chapters. The first one develops a language of varieties over discrete valuation rings, taking Weil's "Foundations" [1] as point of departure. The main results are "Théorème 3" on p. 57, which corresponds to our smoothening process (see 3.1/3), and, as a corollary, "Théorème 6" on p. 61, which yields the existence of weak Néron models (see 3.5/2). In the second chapter, one finds the construction of Néron models for abelian varieties or, more generally, for torsors under abelian varieties; Néron uses the terminology "modèle faiblement minimal". The existence of Néron models is asserted in "Théorème 2" on p. 79 for the local case and in "Théorème 4" on p. 87 for the global case. Finally, the third chapter, which is fairly independent of the others, contains the construction of regular proper minimal models for elliptic curves.

Néron's article has to be viewed as a contribution to relative algebraic geometry over a discrete valuation ring; the applications he gives in the global case are easily deduced from the local case. Concerning the construction of Néron models, Chapters 1 and 2 of his article are quite difficult to read. To a substantial extent, this is due to the fact that they are very technical and also to the fact that the terminology Néron applies is not commonly used; it has been abandoned since.

To give some impression of his terminology, let us explain the basic setting considered by Néron. We start with a discrete valuation ring R with maximal ideal $\mathfrak{p}$. Denote by K the field of fractions as well as by k the residue field of R. The latter is assumed to be perfect. Néron, familiar with the notion of generic points in the sense of Weil's "Foundations" [1], works with universal domains on two levels. First he chooses a universal domain $\mathfrak{k}$ for the residue field k and then a universal domain $\mathfrak{K}$ for the field of fractions K. The latter is done in such a way that $\mathfrak{K}$ is a universal domain of the field of fractions of a ring $\mathfrak{R}$ which serves as an "integral" universal domain. To define $\mathfrak{R}$ in the equal characteristic case, he considers a lifting of k to the completion of R as well as a uniformizing element T of R and takes for $\mathfrak{R}$ the formal power series ring $\mathfrak{k}[[T]]$. In the unequal characteristic case, he sets $\mathfrak{R} = \hat{R} \otimes_{W(k)} W(\mathfrak{k})$ where $\hat{R}$ is the completion of R and where W indicates rings of Witt vectors. The interference of Witt vectors is the main reason why the residue field k is assumed to be perfect. Then he works with relative schemes over R, so-called $\mathfrak{p}$-varieties. To be precise, a $\mathfrak{p}$-variety corresponds to a flat R-scheme of finite type; its points have values in the universal domains $\mathfrak{K}$ or $\mathfrak{k}$ or, when considering integral points, in the subring $\mathfrak{R}$ of $\mathfrak{K}$. Such a $\mathfrak{p}$-variety is called $\mathfrak{p}$-simple if it is regular; it is called simple modulo $\mathfrak{p}$ at a point of the special fibre if it is smooth over R at this point. For both notions, Néron discusses the Jacobi criterion.

In the following, we want to examine Néron's approach to the smoothening process as presented in his Chapter 1, without pursuing his terminology any further; we will use the language of schemes, as generally applied in this book. Let X be a flat R-scheme of finite type with a smooth generic fibre X_K and consider R'-valued points of X where R' is a discrete valuation ring over R having same uniformizing element as R. (So R' is of ramification index 1 over R, since the residue field k of R is perfect.) For such points $x \in X(R')$, Néron defines the integer $l(x, X)$ which measures the defect of smoothness of X along x; see his section n°17 starting on p. 35 or our section 3.3. He shows that $l(x, X)$ is bounded as a function of x. Then he works out the smoothening process by relying on two techniques: the first one is a generic smoothening and the second is the theory of pro-varieties.

The generic smoothening can be formulated as follows:

Let $u : \operatorname{Spec} R' \longrightarrow X$ be an R'-valued point of X where R' is as above. Reducing modulo the maximal ideal $\mathfrak{p}$ of R, one obtains a morphism $\bar{u} : \operatorname{Spec} k' \longrightarrow X_k$. Let $\bar{Y}$ be the closure of its image and let $f : \tilde{X} \longrightarrow X$ be the blowing-up of $\bar{Y}$ on X. Then, if $\tilde{u} : \operatorname{Spec} R' \longrightarrow \tilde{X}$ is the lifting of u to $\tilde{X}$, one has

$$l(\tilde{u}, \tilde{X}) < \max(l(u, X), 1) .$$

In particular, after a finite repetition, one ends up with a factorization of u through the smooth locus of a blowing-up of X.

The statement may be viewed as an individual smoothening for R'-valued points x of X. In order to obtain some form of smoothening which works simultaneously for several x and R', Néron relies on the technique of pro-varieties; this is one of the most delicate points in his construction. To give a sketch of his approach, consider an affine open part of X and thereby suppose that X is embedded into an affine space $\mathbb{A}_R^N$. Using the coefficients of formal series in $\mathfrak{k}[[T]]$ in the equal characteristic case and Witt coordinates in the unequal characteristic case, Néron introduces on the set of $R/\mathfrak{p}^n$-valued points of $\mathbb{A}_R^N$ a structure of k-variety ${}^n\mathbb{A}_k^N$. Since X has a smooth generic fibre, the image of $X(R)$ in ${}^n\mathbb{A}_k^N$ gives rise to a constructible subset ${}^nX$ and one obtains a projective system of morphisms

$$\cdots \longrightarrow {}^{n+1}X \longrightarrow {}^nX \longrightarrow \cdots$$

defining a k-pro-variety.

The possibility of parametrizing solutions of X modulo $\mathfrak{p}^n$ by a k-variety or, more specifically, of points of X with values in the completion $\hat{R}$ of R by a k-pro-variety, had been systematically studied by M. Greenberg [1] within the context of schemes and representable functors; see also Serre [3]. The technique is referred to as the Greenberg functor. However, since Néron did not use the language of functors, he gave proofs of his own for the facts he needed.

Let us return to the situation of a generic smoothening as above where we consider a blowing-up $f : \tilde{X} \longrightarrow X$ with center $\bar{Y}$. Then there is an induced morphism ${}^{n}f : {}^{n}\tilde{X} \longrightarrow {}^{n}X$ for each n and, taking limits over n, a bijection $\tilde{X}(\hat{R}) \simeq X(\hat{R})$. To obtain a simultaneous smoothening, Néron has to consider partial inverses of the maps ${}^{n}f$. More precisely, for each n, there is a constructible subset ${}^{n}Y$ of ${}^{n}X$ given by the points in $X(\hat{R})$ which reduce to points of $\bar{Y}$ and he shows that there is a constructible map ${}^{n+1}Y \longrightarrow {}^{n}\tilde{X}$ such that the diagram

$$\begin{array}{ccc} {}^{n+1}Y & \longrightarrow & {}^{n}\tilde{X} \\ \downarrow & & \downarrow {}^{n}f \\ {}^{n}Y & \hookrightarrow & {}^{n}X \end{array}$$

commutes. (In the case of Witt coordinates, a map of type ${}^{n+1}Y \longrightarrow {}^{n}\tilde{X}$ involves radicial morphisms of extracting p-th roots. Later, to overcome this kind of difficulties, Serre [2] worked with quasi-algebraic varieties.)

Now set $l = \max l(x, X)$ where the maximum is taken over all R'-valued points of X and let Z be an irreducible component of ${}^{l}X$. Combining blowing-ups and shiftings as above, Néron shows the following assertion: there exists a non-empty open part U of Z such that there is a simultaneous smoothening of X with respect to all points of $X(R')$ whose image in ${}^{l}X$ is already contained in U. Using this assertion, he can finish the smoothening process by a constructibility argument; cf. his "Théorème 3" on p. 57.

The proof we will give for the existence of the smoothening process is basically the same as Néron's, except for the fact that we can avoid using pro-varieties and the Greenberg functor. We do this by establishing a more precise form of the generic smoothening; cf. 3.3/5. Namely, as we will see, considering the blowing-up $f : \tilde{X} \longrightarrow X$, there exists a non-empty open subscheme $V \subset \bar{Y}$, described in terms of differential calculus, such that, for each R'-valued point v of X whose special fibre factors through V, and for the lifting $\tilde{v}$ of v to $\tilde{X}$, we have

$$l(\tilde{v}, \tilde{X}) < \max(l(v, X), 1) .$$

Then it is possible to end the smoothening process directly by a constructibility argument without looking at solutions of X modulo higher powers of $\mathfrak{p}$.

At the end of Néron's Chapter 1, there is the discussion of what we call weak Néron models and the measuring of the size of their components. The latter is done with respect to a non-zero differential form ω of maximal degree of X_K. The smoothening process implies that, up to birational equivalence, there are only finitely many components of "maximal volume" with respect to ω. The arguments are the same as we will present them later at the corresponding places in our Chapters 3 and 4.

Let us dicuss now Néron's Chapter 2. It starts with the definition of torsors, or principally homogeneous spaces in his terminology. The definition is given in terms of ternary laws of composition in such a way that the underlying group of the torsor is hidden. Presumably this is done in order not to separate the construction of Néron models into the group case and the case of a torsor under a group scheme. So consider a torsor X_K under an abelian variety A_K over K and a projective R-model X' of X_K. Néron applies the smoothening process to X', restricts to the smooth locus, and removes from the special fibre all irreducible components which do not have maximal volume. The volume is measured with respect to a non-zero invariant differential form of maximal degree on X_K; write X'' for the resulting R-model of X_K. Then he shows that the structure of torsor on X_K extends to a birational law of torsor on X''.

The next step is to show that finitely many "translates" of X'' (defined over certain unramified extensions of R) cover all points of X' with values in unramified extensions R' of R. The same problem occurs in our presentation at the end of the construction of Néron models, where we want to prove their universal mapping property; cf. 4.4/4.

To construct the Néron model X of X_K, it is, of course, necessary to really glue translates of X''; the latter is not a standard procedure since the translates are only defined over certain unramified extensions of R. Starting with an ample invertible sheaf on X'', Néron shows that it extends to an ample invertible sheaf on the translates of X'' and, finally, on the Néron model X. So this part contains in one step the construction of X in terms of gluing translates under the birational law on X'' as well as the descent and the quasi-projectivity of the resulting model. It presents a tremendous accumulation of difficulties. In addition, explanations which are given are not very detailed and in most cases quite complicated to follow. In order to simplify things, it is possible to separate the construction into two steps. First one constructs the Néron model over an étale extension R' of R, where one has enough integral points to perform translations and where it is enough to consider the group scheme case. Then, as a second step, one goes back from R' to R by means of descent, using ample invertible sheaves and thereby proving the quasi-projectivity of the model. This is how M. Artin proceeds in [9]; the same strategy will be applied in the present book.

Finally, the universal mapping property of Néron models is established (in a rudimentary form) quite early in Néron's article, see n°4, pp. 71–73, even before Néron models are constructed. It is based on Weil's arguments [2], concerning rational maps from smooth varieties into algebraic groups.

It remains to say a few words about Néron's Chapter 3 where he constructs proper and regular minimal R-models for elliptic curves with a rational point over K. Except for a few examples, already mentioned in Section 1.5, the subject will not be touched in this book. Néron studies minimal Weierstraß equations and classifies them according to the values of their coefficients, discriminants, and j-invariants. Then he obtains the regular and proper minimal model as a successive joint of new components. His construction leads to the same diagrams as the ones obtained by Kodaira [1]. But Néron's approach of discussing minimal Weierstraß equations case by case is quite different, it does not use the existence of regular models nor does it use the intersection form. An improved version of his method was later published by Tate [2] in algorithmical form; it is known as the Tate algorithm.

Chapter 2. Some Background Material from Algebraic Geometry

In this chapter we give a review of some basic tools which are needed in later chapters for the construction of Néron models. Assuming that the reader is familiar with Grothendieck's definition of schemes and morphisms, we treat the concept of smooth and étale morphisms, of henselian rings, and of S-rational maps; moreover, we have included some facts on differential calculus and on flatness. Concerning the smoothness, we give a self-contained exposition of this notion, relating it closely to the Jacobi criterion. For the other topics we simply state results, sometimes without giving proofs. Most of the material presented in this chapter is contained in Grothendieck's treatments [EGA IV_4] and [SGA 1].

2.1 Differential Forms

In this section we define the sheaf of relative differential forms of one scheme over another. We introduce it by a purely algebraic method using derivations. So let us first review the basic facts on derivations; detailed explanations and proofs can be found in [EGA 0_{IV}], 20.5.

In the following let R be a ring, and let A be an R-algebra. An R-derivation of A into an A-module M is an R-linear map $d: A \longrightarrow M$ such that

$$d(fg) = fd(g) + gd(f) \quad \text{for all} \quad f, g \in A\ .$$

In particular, $d(r \cdot 1) = 0$ for all $r \in R$. The set $\mathrm{Der}_R(A, M)$ of all R-derivations of A into an A-module M is canonically an A-module. One defines the *module of relative differential forms* (*of degree* 1) *of A over R* as an A-module $\Omega^1_{A/R}$, together with an R-derivation $d_{A/R}: A \longrightarrow \Omega^1_{A/R}$, which is universal in the following sense: For each A-module M, the canonical map

$$\mathrm{Hom}_A(\Omega^1_{A/R}, M) \xrightarrow{\sim} \mathrm{Der}_R(A, M)\ , \qquad \varphi \longmapsto \varphi \circ d_{A/R}\ ,$$

is bijective. The map $d_{A/R}$ is called the *exterior differential.* Such a couple $(\Omega^1_{A/R}, d_{A/R})$ is uniquely determined up to canonical isomorphism. The existence can easily be verified in the following way. If A is a free R-algebra $R[T_i]_{i \in I}$ of polynomials in the variables T_i, $i \in I$, then let Ω^1 be the free A-module generated by the symbols dT_i, $i \in I$, and define $d: A \longrightarrow \Omega^1$ by the formula

$$d(P) := \sum_{i \in I} \frac{\partial P}{\partial T_i} \cdot dT_i\ ,$$

where $\partial P/\partial T_i$ is the usual partial derivative of P with respect to T_i. It is easy to see that (Ω^1, d) is the A-module of relative differential forms of A over R. In general, an R-algebra B is a residue ring $B = A/\mathfrak{a}$ of a free R-algebra of polynomials A. Then the B-module of relative differential forms of B over R is given by the B-module

$$\Omega^1_{A/R}/(\mathfrak{a}\Omega^1_{A/R} + Ad_{A/R}\mathfrak{a}) ,$$

and the exterior differential is canonically induced by $d_{A/R}$.

We give an alternate method for proving the existence of modules of differentials. Let $m : A \otimes_R A \longrightarrow A$ be the map induced by the multiplication on A, set $I = \ker(m)$ and consider the map

$$d : A \longrightarrow I/I^2 , \quad f \longmapsto 1 \otimes f - f \otimes 1 \mod I^2 .$$

The $(A \otimes_R A)$-module I/I^2 is actually an $((A \otimes_R A)/I)$-module. Using the canonical isomorphism

$$(A \otimes_R A)/I \xrightarrow{\sim} A$$

one can view I/I^2 as an A-module, and one verifies that $(I/I^2, d)$ is the A-module of relative differential forms of A over R.

The universal property of $\Omega^1_{A/R}$ implies certain functorial properties. For example, each morphism $\varphi : A \longrightarrow B$ of R-algebras induces a unique A-linear map

$$\Omega^1_{A/R} \longrightarrow \Omega^1_{B/R} , \qquad \sum_i f_i d_{A/R}(g_i) \longmapsto \sum_i \varphi(f_i) d_{B/R}(\varphi(g_i)) ,$$

and hence a B-linear map

$$\Omega^1_{A/R} \otimes_A B \longrightarrow \Omega^1_{B/R} .$$

Moreover, since each A-derivation of B is also an R-derivation, one obtains a map

$$\Omega^1_{B/R} \longrightarrow \Omega^1_{B/A} , \qquad \sum_i f_i d_{B/R}(g_i) \longmapsto \sum_i f_i d_{B/A}(g_i) .$$

Thus we have a canonical sequence

$$\Omega^1_{A/R} \otimes_A B \longrightarrow \Omega^1_{B/R} \longrightarrow \Omega^1_{B/A} \longrightarrow 0$$

which can be shown to be exact. If B is a residue algebra of A, say $B = A/\mathfrak{a}$, the R-derivation $d_{A/R}$ induces a canonical B-linear map

$$\delta : \mathfrak{a}/\mathfrak{a}^2 \longrightarrow \Omega^1_{A/R} \otimes_A B , \qquad \bar{a} \longmapsto d_{A/R}(a) \otimes 1$$

where $\bar{a}$ denotes the residue class of $a \in \mathfrak{a}$ modulo $\mathfrak{a}^2$. As a second important fact on the behavior of differentials, one shows that the sequence

$$\mathfrak{a}/\mathfrak{a}^2 \xrightarrow{\delta} \Omega^1_{A/R} \otimes_A B \longrightarrow \Omega^1_{B/R} \longrightarrow 0$$

is exact.

Next we want to globalize the notion of modules of differentials in terms of sheaves over schemes. One can either show that the formation of modules of differentials is compatible with localization or, what is more elegant, use the alternate description we have given above. Proceeding the latter way, consider a base scheme S and an S-scheme X. The diagonal morphism

$$\Delta : X \longrightarrow X \times_S X$$

yields an isomorphism of X onto its image $\Delta(X)$ which is a locally closed subscheme of $X \times_S X$; i.e., $\Delta(X)$ is a closed subscheme of an open subscheme W of $X \times_S X$. Let $\mathscr{I}$ be the sheaf of ideals defining $\Delta(X)$ as a closed subscheme of W. Then we define the *sheaf of relative differential forms* (*of degree* 1) *of* X *over* S as the sheaf

$$\Omega^1_{X/S} := \Delta^*(\mathscr{I}/\mathscr{I}^2)$$

on X. Note that $\mathscr{I}/\mathscr{I}^2$ has a natural structure of an $\mathscr{O}_{\Delta(X)}$-module; hence $\Delta^*(\mathscr{I}/\mathscr{I}^2)$ is canonically an $\mathscr{O}_X$-module. It is clear that $\Omega^1_{X/S}$ is a quasi-coherent $\mathscr{O}_X$-module, which is of finite type if X is locally of finite type over S. The canonical map

$$d_{X/S} : \mathscr{O}_X \longrightarrow \Omega^1_{X/S} ,$$

induced by the map sending a section f of $\mathscr{O}_X$ to the section $p_2^* f - p_1^* f$ of $\mathscr{I}$ (where $p_i : X \times_S X \longrightarrow X$ is the projection onto the i-th factor), is called the *exterior differential*.

Since $\Omega^1_{X/S}$ is quasi-coherent, $(\Omega^1_{X/S}, d_{X/S})$ can be described in local terms: for each open affine subset $V = \operatorname{Spec} R$ of S and for each open affine subset $U = \operatorname{Spec} A$ of X lying over V, the sheaf $\Omega^1_{X/S}|_U$ is the quasi-coherent $\mathscr{O}_X|_U$-module associated to the A-module $\Omega^1_{A/R}$, and the map $d_{X/S}|_U$ is associated to the canonical map $d_{A/R}$: $A \longrightarrow \Omega^1_{A/R}$.

The sheaf of relative differential forms has similar functorial properties as the module of relative differential forms. Given an S-morphism $f : X \longrightarrow Y$, one can pull back differential forms on Y to X. So one obtains a canonical $\mathscr{O}_X$-morphism

$$f^*\Omega^1_{Y/S} \longrightarrow \Omega^1_{X/S} .$$

Each section ω of $\Omega^1_{Y/S}$ gives rise to a section ω' of $f^*\Omega^1_{Y/S}$ and hence to a section ω'' of $\Omega^1_{X/S}$, namely to the image of ω' under the above map. It is convenient to use the notion $f^*\omega$ for both ω' and ω''; however to avoid confusion, we will always specify the module, either $f^*\Omega^1_{Y/S}$ or $\Omega^1_{X/S}$, when we talk about the section $f^*\omega$.

The canonical sequences between modules of differentials, as given above, can immediately be globalized to the case of differentials over schemes; cf. [EGA IV$_4$], 16.4:

Proposition 1. *Let* $f : X \longrightarrow Y$ *be an* S*-morphism. Then the canonical sequence of* $\mathscr{O}_X$*-modules*

$$f^*\Omega^1_{Y/S} \longrightarrow \Omega^1_{X/S} \longrightarrow \Omega^1_{X/Y} \longrightarrow 0$$

is exact.

Proposition 2. *Let* $j : Y \hookrightarrow X$ *be an immersion of* S*-schemes. Let* $\mathscr{J}$ *be the sheaf of ideals defining* Y *as a subscheme of* X. *Then the canonical sequence of* $\mathscr{O}_Y$*-modules*

$$\mathscr{J}/\mathscr{J}^2 \xrightarrow{\delta} j^*\Omega^1_{X/S} \longrightarrow \Omega^1_{Y/S} \longrightarrow 0$$

is exact.

Furthermore, we cite that the formation of sheaves of relative differentials commutes with base change and products:

Proposition 3. *Let X and S' be S-schemes. Let $X' = X \times_S S'$ be the S'-scheme obtained by base change, and let $p : X' \longrightarrow X$ be the projection. Then the canonical map*

$$p^*\Omega^1_{X/S} \xrightarrow{\sim} \Omega^1_{X'/S'}$$

is an isomorphism.

Proposition 4. *Let X_1 and X_2 be S-schemes. If $p_i : X_1 \times_S X_2 \longrightarrow X_i$ are the projections for $i = 1, 2$, the canonical map*

$$p_1^*\Omega^1_{X_1/S} \oplus p_2^*\Omega^1_{X_2/S} \xrightarrow{\sim} \Omega^1_{X_1 \times_S X_2/S}$$

is an isomorphism.

2.2 Smoothness

In this section we want to explain the basic concept of unramified, étale, and smooth morphisms from the viewpoint of differential calculus. Our approach differs from the one given in [EGA IV$_4$], 17, in so far as we have chosen the Jacobi criterion as point of departure. In the following, let S be a base scheme.

Definition 1. *A morphism of schemes $f : X \longrightarrow S$ is called unramified at a point $x \in X$ if there exist an open neighborhood U of x and an S-immersion*

$$j : U \hookrightarrow \mathbb{A}^n_S$$

of U into some linear space $\mathbb{A}^n_S$ over S such that the following conditions are satisfied:

(a) *locally at $j(x)$ (i.e., in an open neighborhood of $j(x)$), the sheaf of ideals $\mathscr{I}$ defining $j(U)$ as a subscheme of $\mathbb{A}^n_S$ is generated by finitely many sections,*

(b) *the differential forms of type dg with sections g of $\mathscr{I}$ generate $\Omega^1_{\mathbb{A}^n_S/S}$ at $j(x)$.*

The morphism $f : X \longrightarrow S$ is called unramified if it is unramified at all points of X.

Condition (a) says that unramified morphisms are locally of finite presentation. Obviously, an immersion which is locally of finite presentation is unramified. It can easily be shown that the class of unramified morphisms is stable under base change, under composition, and under the formation of products. We give some equivalent characterizations of unramified morphisms:

Proposition 2. *Let $f : X \longrightarrow S$ be locally of finite presentation, let x be a point of X, and set $s = f(x)$. Then the following conditions are equivalent:*

(a) *f is unramified at x.*

(b) *$\Omega^1_{X/S,x} = 0$*

(c) *The diagonal morphism $\Delta : X \longrightarrow X \times_S X$ is a local isomorphism at x.*

(d) *The $k(s)$-scheme $X_s = X \times_S \operatorname{Spec} k(s)$ is unramified over $k(s)$ at x.*

(e) *The maximal ideal $\mathfrak{m}_x$ of $\mathcal{O}_{X,x}$ is generated by the maximal ideal $\mathfrak{m}_s$ of $\mathcal{O}_{S,s}$, and $k(x)$ is a (finite) separable extension of $k(s)$.*

Proof. The equivalence of conditions (a) and (b) follows from the exact sequence 2.1/2. The equivalence of (b) and (c) is seen by using the identity

$$\Omega^1_{X/S} = \Delta^*(\mathscr{I}/\mathscr{I}^2)\,,$$

where $\mathscr{I}$ is the sheaf of ideals defining the diagonal in $X \times_S X$, and by applying Nakayama's lemma. Furthermore, since unramified morphisms are preserved by any base change, condition (a) implies condition (d). Conversely, if (d) is satisfied, we know already

$$\Omega^1_{X_s/k(s),x} = 0\,.$$

Let $\mathfrak{m}_s$ be the maximal ideal of $\mathcal{O}_{S,s}$. Then, since the formation of sheaves of differentials is compatible with base change, we have

$$\Omega^1_{X_s/k(s),x} = \Omega^1_{X/S,x}/\mathfrak{m}_s\Omega^1_{X/S,x}\,,$$

and Nakayama's lemma yields $\Omega^1_{X/S,x} = 0$. So condition (b) is satisfied, and we see that conditions (a) to (d) are equivalent.

In order to show that the equivalence extends to condition (e), we may assume that S is the spectrum of a field k. Then the implication (e) $\Longrightarrow$ (b) is an elementary algebraic fact, because $\Omega^1_{X/S,x} = \Omega^1_{k(x)/k}$ in this case. Conversely, let us show that condition (c) implies condition (e). We may assume that X is affine, say $X = \operatorname{Spec} A$, and that the diagonal morphism $\Delta : X \longrightarrow X \times_k X$ is an open immersion. Let $\bar{k}$ be the algebraic closure of k. It suffices to prove that $A \otimes_k \bar{k}$ is a finite direct sum of fields isomorphic to $\bar{k}$; then A will be a finite direct sum of separable field extensions of k. To do this we may assume that k is algebraically closed. For a closed point z of X, let $h_z : X \longrightarrow X$ be the constant morphism mapping X to z, and consider the morphism

$$(id_X, h_z) : X \longrightarrow X \times_k X\,.$$

Since Δ is an open immersion,

$$(id_X, h_z)^{-1}(\Delta(X)) = \{z\}$$

is open in X. Hence each closed point of X is open, and X consists of a finite number of isolated points. In particular, A is a finite-dimensional vector space over k. Shrinking X if necessary, we may assume that X consists of only one point. Then the same is true for $X \times_k X$. Since Δ is an open immersion, the corresponding morphism $\Delta^* : A \otimes_k A \longrightarrow A$ is an isomorphism and, by comparing vector space dimensions, we see $A = k$. □

If follows from condition (e) above that the relative dimension of an unramified morphism is zero. More generally, one can show that the relative dimension $\dim_x f = \dim_x f^{-1}(f(x))$ at a point x of an S-subvariety $X \subset \mathbb{A}^n_S$ with structural morphism $f : X \longrightarrow S$ is r if, locally at x, the subvariety is defined by sections g_{r+1}, $\ldots$, g_n of $\mathcal{O}_{\mathbb{A}^n_S}$ and if the differentials $dg_{r+1}(x), \ldots, dg_n(x)$ are linearly independent in $\Omega^1_{\mathbb{A}^n_S/S} \otimes k(x)$. Namely, this follows from the result above and the fact that the relative dimension decreases at most by 1 if one goes over from an S-scheme to a subscheme defined by a single equation.

Definition 3. *A morphism $f: X \longrightarrow S$ is called smooth at a point x of X (of relative dimension r) if there exist an open neighborhood U of x and an S-immersion*

$$j: U \hookrightarrow \mathbb{A}_S^n$$

of U into some linear space $\mathbb{A}_S^n$ over S such that the following conditions are satisfied:

(a) *locally at $y := j(x)$, the sheaf of ideals defining $j(U)$ as a subscheme of $\mathbb{A}_S^n$ is generated by $(n-r)$ sections $g_{r+1}, \ldots, g_n$, and*

(b) *the differentials $dg_{r+1}(y), \ldots, dg_n(y)$ are linearly independent in $\Omega^1_{\mathbb{A}_S^n/S} \otimes k(y)$.*

A morphism is called smooth if it is smooth at all points. Furthermore, a morphism is said to be étale (at a point) if it is smooth (at the point) of relative dimension 0.

Note that, as we have explained above, the integer r is indeed the relative dimension of f at x and that, due to its definition, the smooth locus of a morphism which is locally of finite presentation is open. It is an elementary task to verify that the class of smooth (resp. étale) morphisms is stable under base change, under composition, and under the formation of products. It is clear that open immersions are étale. Furthermore, étale morphisms are unramified, but the converse is not true as is seen by the following lemma.

Lemma 4. *An immersion $f: X \longrightarrow S$ is étale if and only if f is an open immersion.*

Proof. The if-part is obvious. For the only-if-part, it suffices to consider the special case where f is a closed immersion. Furthermore we may assume that, as an S-scheme, X has been realized as a closed subscheme of an affine open subscheme $V \subset \mathbb{A}_S^n$, in such a way that X is defined by n sections $g_1, \ldots, g_n$ of $\mathcal{O}_{\mathbb{A}_S^n}$ on V, where the differentials $dg_1, \ldots, dg_n$ generate $\Omega^1_{\mathbb{A}_S^n/S}|_V$. Since $f: X \longrightarrow S$ is a closed immersion, we may assume that the coordinate functions $T_1, \ldots, T_n$ of $\mathbb{A}_S^n$ vanish on X. Then we have relations

$$T_j = \sum_i a_{ij} g_i$$

with $a_{ij} \in \mathcal{O}_{\mathbb{A}_S^n}(V)$ for $i, j = 1, \ldots, n$. Taking the differentials of these equations shows that the matrix (a_{ij}) is invertible in a neighborhood of X. Due to Cramer's rule, the sheaves of ideals generated by $(T_1, \ldots, T_n)$ and $(g_1, \ldots, g_n)$ coincide in this neighborhood. This implies that f is an open immersion. □

More generally, one can show that étale morphisms are flat and, hence, open (cf. 2.4); in fact, a morphism is étale if and only if it is flat and unramified, see 2.4/8. In particular, if S is the spectrum of a field k, the notions *étale* and *unramified* coincide. In this case, each étale S-scheme X consists of isolated reduced points such that the residue field $k(x)$ of each point $x \in X$ is a finite separable extension of k.

Proposition 5. *Let $f: X \longrightarrow Y$ be a smooth morphism of schemes. Then:*

(a) *$\Omega^1_{X/Y}$ is locally free. Its rank at $x \in X$ is equal to the relative dimension of f at x.*

(b) *If f is a smooth morphism of smooth S-schemes, the canonical sequence of $\mathcal{O}_X$-modules*

$$0 \longrightarrow f^*\Omega^1_{Y/S} \longrightarrow \Omega^1_{X/S} \longrightarrow \Omega^1_{X/Y} \longrightarrow 0$$

is exact and locally split. (Actually, the assumption on X and Y to be smooth over S is unnecessary; cf. [EGA IV_4], 17.2.3.)

Proof. Since $\Omega^1_{\mathbb{A}^n_Y}$ is free of rank n, assertion (a) follows immediately from the definition of smoothness if one uses 2.1/2. In the situation (b) we know from 2.1/1 that the canonical sequence

$$f^*\Omega^1_{Y/S} \xrightarrow{\alpha} \Omega^1_{X/S} \longrightarrow \Omega^1_{X/Y} \longrightarrow 0$$

is exact. Due to (a), the three $\mathcal{O}_X$-modules are locally free of finite rank. Hence, for all $x \in X$, the $\mathcal{O}_{X,x}$-module $(f^*\Omega^1_{Y/S})_x$ is isomorphic to the direct sum of $\ker \alpha_x$ and $\operatorname{im} \alpha_x$ both of which are free. Counting the ranks, one sees $\ker \alpha = 0$. □

It is an easy consequence of (a) that, for a smooth morphism $f: X \longrightarrow S$, the map $x \longmapsto \dim_x f$ is locally constant. Next we want to characterize smoothness by the infinitesimal lifting property for morphisms.

Proposition 6. *Let $f: X \longrightarrow S$ be locally of finite presentation. The following conditions are equivalent*:

(a) *f is unramified (resp. smooth, resp. étale).*

(b) *For all S-schemes Y which are affine and for all closed subschemes Y_0 of Y defined by sheaves of ideals $\mathcal{J}$ of $\mathcal{O}_Y$ with $\mathcal{J}^2 = 0$, the canonical map*

$$\operatorname{Hom}_S(Y, X) \longrightarrow \operatorname{Hom}_S(Y_0, X)$$

is injective (resp. surjective, resp. bijective).

Proof. First we want to treat the characterization of unramified morphisms. In this situation, conditions (a) and (b) are local on X and S, so we may assume that X and S are affine, say $X = \operatorname{Spec} B$ and $S = \operatorname{Spec} R$. Let C be an R-algebra, let J be an ideal of C with $J^2 = 0$, and consider a commutative diagram

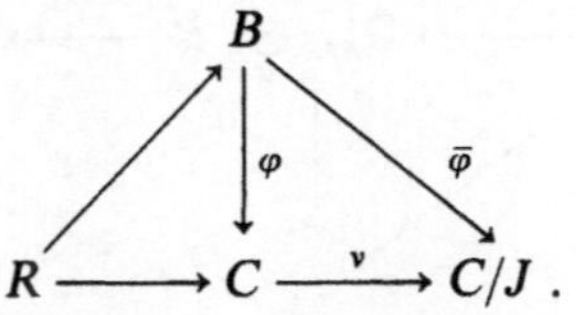

One easily shows that the map

$$\{\psi \in \operatorname{Hom}_R(B, C)\,;\, \nu \circ \psi = \bar{\varphi}\} \longrightarrow \operatorname{Der}_R(B, J)\,, \qquad \psi \longmapsto \psi - \varphi\,,$$

between the set of liftings of $\bar{\varphi}$ and the B-module of R-derivations is bijective. Notice that J is a C/J-module and, hence, a B-module via $\bar{\varphi}$.

If X is unramified over S, we know $\Omega^1_{B/R} = 0$ from Proposition 2 so that $\operatorname{Der}_R(B, J) = 0$ in this case. Thus, the implication (a) $\Longrightarrow$ (b) is clear. In order to verify the implication (b) $\Longrightarrow$ (a), set $C := (B \otimes_R B)/I^2$, where I is the kernel of the map

$$m : B \otimes_R B \longrightarrow B\,, \qquad \sum x_i \otimes y_i \longmapsto \sum x_i y_i\,.$$

Furthermore, set $J = I/I^2$. The considerations above show $\mathrm{Der}_R(B, J) = 0$. Since $J \cong \Omega^1_{B/R}$, the implication (b) $\Longrightarrow$ (a) follows.

Next we turn to the characterization of smooth morphisms. Starting with the implication (a) $\Longrightarrow$ (b), let us first consider a special case which corresponds to the local situation of a smooth morphism. So let S be affine, say $S = \operatorname{Spec} R$, and let $X = \operatorname{Spec} B$ be a closed subscheme of an affine open subscheme $V = \operatorname{Spec} A$ of $\mathbb{A}^n_S$. Let I be the ideal of A defining X. Assume that there are $g_1, \ldots, g_n \in A$ such that $dg_1, \ldots, dg_n$ form a basis of $\Omega^1_{A/R}$ and such that, for some r, the ideal $I \subset A$ of X is generated by $g_{r+1}, \ldots, g_n$. Then, since I/I^2 is generated over $B = A/I$ by the residue classes of these elements, the canonical sequence

$$(*) \qquad 0 \longrightarrow I/I^2 \longrightarrow \Omega^1_{A/R} \otimes_A B \longrightarrow \Omega^1_{B/R} \longrightarrow 0$$

is easily seen to be split exact.

Now let $Y = \operatorname{Spec} C$ be an affine S-scheme, and fix a closed subscheme $Y_0 \subset Y$ defined by an ideal J of C with $J^2 = 0$. To verify condition (b), we have to show that each R-morphism $\bar{\varphi} : B \longrightarrow C/J$ lifts to an R-morphism $\varphi : B \longrightarrow C$. Due to the universal property of a polynomial ring, we can lift $\bar{\varphi}$ to an R-morphism $\psi : A \longrightarrow C$ such that the diagram

$$\begin{array}{ccccc} & & A & \longrightarrow & B = A/I \\ & \nearrow & \downarrow{\scriptstyle \psi} & & \downarrow{\scriptstyle \bar{\varphi}} \\ R & \longrightarrow & C & \longrightarrow & C/J \end{array}$$

is commutative. Since $\psi(I) \subset J$, the map ψ gives rise to a B-linear map

$$\psi' : I/I^2 \longrightarrow J\,.$$

Since the sequence $(*)$ is split exact, the B-linear map ψ' extends to a B-linear map ψ'' as follows:

$$\begin{array}{ccccccccc} 0 & \longrightarrow & I/I^2 & \longrightarrow & \Omega^1_{A/R} \otimes_A B & \longrightarrow & \Omega^1_{B/R} & \longrightarrow & 0 \\ & & & {\scriptstyle \psi'} \searrow & \downarrow{\scriptstyle \psi''} & & & & \\ & & & & J & & & & \end{array}$$

Hence, ψ'' induces an R-derivation $\delta : A \longrightarrow J$ satisfying $\psi|_I = \delta|_I$. Then $(\psi - \delta) : A \longrightarrow C$ is an R-morphism inducing a lifting $\varphi : B \longrightarrow C$ of $\bar{\varphi}$.

It remains to reduce the general case of an arbitrary smooth morphism $f : X \longrightarrow S$ to the special case treated above. This can be done by showing that condition (b) is a local condition on X. So, as before, let $Y = \operatorname{Spec} C$ be an affine S-scheme, and let Y_0 be a closed subscheme of Y defined by a sheaf of ideals $\mathcal{J}$ of $\mathcal{O}_Y$ with $\mathcal{J}^2 = 0$. Let $\bar{\varphi} : Y_0 \longrightarrow X$ be an S-morphism. Due to the special case discussed above, there exists an open covering $\{Y_\alpha\}_\alpha$ of Y such that $\bar{\varphi}|_{Y_\alpha \cap Y_0}$ lifts to a morphism $\varphi'_\alpha : Y_\alpha \longrightarrow X$. The obstruction for (φ'_α) to define a morphism from Y to X is a cocycle with values in $\mathscr{H}om_{\mathcal{O}_{Y_0}}(\bar{\varphi}^*\Omega^1_{X/S}, \mathscr{I})$; see also [SGA 1], Exp. III, 5.1. Since this sheaf is a quasi-coherent $\mathcal{O}_{Y_0}$-module, its first cohomology group vanishes on the affine scheme Y_0. So there exist liftings $\varphi_\alpha : Y_\alpha \longrightarrow X$ of $\bar{\varphi}|_{Y_\alpha \cap Y_0}$ such that (φ_α) gives rise to a morphism $\varphi : Y \longrightarrow X$ lifting $\bar{\varphi}$. This establishes the implication (a) $\Longrightarrow$ (b) for smooth morphisms.

In order to show the converse, we may assume that X is a closed subscheme of a linear space $\mathbb{A}^n_S$ which is defined by a finitely generated sheaf of ideals $\mathscr{I} \subset \mathcal{O}_{\mathbb{A}^n_S}$. Then it suffices to show that the canonical sequence

$$0 \longrightarrow \mathscr{I}/\mathscr{I}^2 \longrightarrow \Omega^1_{\mathbb{A}^n_S/S} \otimes \mathcal{O}_X \longrightarrow \Omega^1_{X/S} \longrightarrow 0$$

is locally split exact. We will prove this in a more general situation where $\mathbb{A}^n_S$ is replaced by a smooth S-scheme Z. In order to do this, we may assume that S and Z are affine, say $S = \operatorname{Spec} R$ and $Z = \operatorname{Spec} A$, and that $X = \operatorname{Spec} B$ is defined by a finitely generated ideal $I \subset A$; in particular, we have $B = A/I$. Due to condition (b), the map

$$\bar{\varphi} = \mathrm{id} : A/I \longrightarrow A/I = (A/I^2)/(I/I^2)$$

lifts to an R-morphism $\varphi : A/I \longrightarrow A/I^2$. Then the exact sequence of R-modules

$$0 \longrightarrow I/I^2 \stackrel{\iota}{\longrightarrow} A/I^2 \stackrel{v}{\longrightarrow} A/I \longrightarrow 0$$

splits; namely, φ is a section of v, and $\mathrm{id}_{A/I^2} - \varphi \circ v$ defines an R-linear map

$$\tau : A/I^2 \longrightarrow I/I^2$$

which is a section of the inclusion ι. Since $\tau(a) \cdot \tau(b) = 0$ for all $a, b \in A/I^2$, we have

$$\tau(ab) = ab - \varphi \circ v(ab) + (a - \varphi \circ v(a))(b - \varphi \circ v(b)) = a\tau(b) + b\tau(a) .$$

Hence τ is an R-derivation giving rise to an A-homomorphism $\Omega^1_{A/R} \longrightarrow I/I^2$. Consequently, the sequence

$$0 \longrightarrow I/I^2 \longrightarrow \Omega^1_{A/R} \otimes_A B \longrightarrow \Omega^1_{B/R} \longrightarrow 0$$

is split exact.

Finally, the characterization of étale morphisms follows from what has been shown for smooth and unramified morphisms, since a morphism is étale if and only if it is smooth and unramified. □

In the definition of smoothness it is required that a smooth S-scheme X can locally be realized as a subscheme of a suitable linear space $\mathbb{A}^n_S$ such that the associated sheaf of ideals satisfies certain conditions. Now we will see that these conditions are fulfilled for each immersion of X into a smooth S-scheme.

Proposition 7. (Jacobi Criterion). *Let X and Z be S-schemes, and let $j : X \hookrightarrow Z$ be a closed immersion which is locally of finite presentation. Let $\mathscr{I}$ be the sheaf of ideals of $\mathcal{O}_Z$ which defines X as a subscheme of Z. Let x be a point of X, and set $z = j(x)$. Assume that, as an S-scheme, Z is smooth at z of relative dimension n. Then the following conditions are equivalent*:

(a) *As an S-scheme, X is smooth at x of relative dimension r.*
(b) *The canonical sequence of $\mathcal{O}_X$-modules*

$$0 \longrightarrow \mathcal{J}/\mathcal{J}^2 \longrightarrow j^*\Omega^1_{Z/S} \longrightarrow \Omega^1_{X/S} \longrightarrow 0$$

is split exact at x, and $r = \operatorname{rank}(\Omega^1_{X/S} \otimes k(x))$.

(c) *If* $dz_1, \ldots, dz_n$ *is a basis of* $(\Omega^1_{Z/S})_z$, *and if* $g_1, \ldots, g_N$ *are local sections of* $\mathcal{O}_Z$ *generating* $\mathcal{J}_z$, *there exists a re-indexing of the* $z_1, \ldots, z_n$ *and of the* $g_1, \ldots, g_N$ *such that* $g_{r+1}, \ldots, g_n$ *generate* $\mathcal{J}$ *at z and such that* $dz_1, \ldots, dz_r, dg_{r+1}, \ldots, dg_n$ *generate* $(\Omega^1_{Z/S})_z$.

(d) *There exist local sections* $g_{r+1}, \ldots, g_n$ *of* $\mathcal{O}_Z$ *generating* $\mathcal{J}_z$ *such that the differentials* $dg_{r+1}(z), \ldots, dg_n(z)$ *are linearly independent in* $\Omega^1_{Z/S} \otimes k(z)$.

Proof. The implication (a) $\Longrightarrow$ (b) follows from the preceding proposition. Namely, if condition (a) is satisfied, X has the lifting property, and, as shown in the last part of the proof of Proposition 6, the canonical exact sequence of (b) is split exact. Furthermore, $(\Omega^1_{X/S})_x$ is free of rank r by Proposition 5.

The implication (b) $\Longrightarrow$ (c) follows from Nakayama's lemma, whereas (c) $\Longrightarrow$ (d) is clear. Finally, the implication (d) $\Longrightarrow$ (a) is easily checked by using a local representation of Z at z as required for $Z \longrightarrow S$ to be smooth at z. □

Condition (d) can also be stated in terms of matrices. Namely, considering a representation

$$dg_j = \sum_{i=1}^{n} \frac{\partial g_j}{\partial z_i} dz_i$$

of the differential forms $dg_{r+1}, \ldots, dg_n$ with respect to a basis $dz_1, \ldots, dz_n$ of $(\Omega^1_{Z/S})_z$, condition (d) says that $\mathcal{J}_z$ is generated by the $(n-r)$ elements g_j and that there exists an $(n-r)$-minor of the matrix $(\partial g_j/\partial z_i)$ which does not vanish at z. So we see that Proposition 7 corresponds to the Jacobi Criterion in differential geometry. We want to derive a second version of it (see [EGA IV_4], 17.11.1 for a further generalization).

Proposition 8. *Let* $f: X \longrightarrow Y$ *be an S-morphism. Let x be a point of X, and set* $y = f(x)$. *Assume that X is smooth over S at x and that Y is smooth over S at y. Then the following conditions are equivalent:*

(a) *f is smooth at x.*
(b) *The canonical homomorphism* $(f^*\Omega^1_{Y/S})_x \longrightarrow (\Omega^1_{X/S})_x$ *is left-invertible* (*i.e., is an isomorphism onto a direct factor*).
(c) *The canonical homomorphism* $(f^*\Omega^1_{Y/S}) \otimes k(x) \longrightarrow \Omega^1_{X/S} \otimes k(x)$ *is injective.*

Proof. The implication (a) $\Longrightarrow$ (b) is a direct consequence of Proposition 5; the implication (b) $\Longrightarrow$ (c) is trivial. Concerning the implication (c) $\Longrightarrow$ (a), we will first treat the case where $Y = \mathbb{A}^s_S$. Then the morphism f is given by global sections $\bar{f}_1, \ldots, \bar{f}_s$ of $\mathcal{O}_X$, and condition (c) means that $d\bar{f}_1(x), \ldots, d\bar{f}_s(x)$ are linearly independent. Furthermore, we may assume that X is a subscheme of $\mathbb{A}^m_S$ of relative dimension r and that the sheaf of ideals defining X is generated by sections h_{r+1},

$\ldots, h_m$ such that $dh_{r+1}(x), \ldots, dh_m(x)$ are linearly independent. Let us consider the graph embedding

$$X \hookrightarrow X \times_S Y \hookrightarrow \mathbb{A}_S^m \times_S \mathbb{A}_S^s\,, \qquad x \longmapsto (x, f(x))\,.$$

We can lift the sections $\bar{f}_i$ to sections f_i defined in a neighborhood of x in $\mathbb{A}_S^m$. Then, locally at $(x, f(x))$, we have realized X as the subscheme of $\mathbb{A}_S^{m+s} = \mathbb{A}_Y^m$ which is given by

$$h_{r+1}, \ldots, h_m\,, \qquad T_1 - f_1, \ldots, T_s - f_s\,,$$

where $T_1, \ldots, T_s$ denote the coordinate functions of $\mathbb{A}_S^s = Y$. This yields a local representation of X as a subscheme of $\mathbb{A}_Y^m$ as required.

In order to handle the general case, let Y be smooth at y of relative dimension s over S. Let $g_1, \ldots, g_s$ be local sections at y of $\mathcal{O}_Y$ such that $dg_1, \ldots, dg_s$ induce a basis of $(\Omega^1_{Y/S})_y$. After shrinking X and Y, we may assume that $g_1, \ldots, g_s$ are global sections. Due to condition (c), there exist local sections $h_{s+1}, \ldots, h_r$ at x of $\mathcal{O}_X$ such that

$$f^*dg_1, \ldots, f^*dg_s\,, \qquad dh_{s+1}, \ldots, dh_r$$

is a basis of $\Omega^1_{X/S,x}$ where r is the relative dimension at x of X over S. Again, we may assume that $h_{s+1}, \ldots, h_r$ are global sections of $\mathcal{O}_X$. Setting

$$g = (g_1, \ldots, g_s) : Y \longrightarrow \mathbb{A}_S^s\,,$$
$$h = (h_{s+1}, \ldots, h_r) : X \longrightarrow \mathbb{A}_S^{r-s}\,,$$

we obtain the commutative diagram

$$\begin{array}{ccccc} X & \xrightarrow{(f,h)} & Y \times_S \mathbb{A}_S^{r-s} & \xrightarrow{p} & Y \\ & \searrow^{(g\circ f, h)} & \downarrow {\scriptstyle g \times \mathrm{id}} & & \\ & & \mathbb{A}_S^r & & \end{array}$$

By the special case above, the maps $(g \circ f, h)$ and $g \times \mathrm{id}$ are étale at x and y, respectively. Hence, due to Lemma 9 below, the morphism (f, h) is étale at x. Then, $f = p \circ (f, h)$ is a composition of smooth morphisms and, hence, smooth at x. □

Lemma 9. *Let $X \longrightarrow S$ be unramified (resp. smooth, resp. étale), and let $Y \longrightarrow S$ be unramified. Then each S-morphism $X \longrightarrow Y$ is unramified (resp. smooth, resp. étale).*

Proof. The assertion follows from Proposition 6. Namely, one verifies immediately that $X \longrightarrow Y$ satisfies the lifting property (b) of this proposition. □

Let us state the assertion of Proposition 8 for the special case of étale morphisms.

Corollary 10. *Let $f: X \longrightarrow Y$ be an S-morphism. Let x be a point of X, and set $y = f(x)$. Assume that X is smooth over S at x and that Y is smooth over S at y. Then the following conditions are equivalent:*

(a) *f is étale at x.*

(b) *The canonical homomorphism $(f^*\Omega^1_{Y/S})_x \longrightarrow (\Omega^1_{X/S})_x$ is bijective.*

Thinking of the classical inverse function theorem, the corollary suggests an analogy between the notions of étale morphisms in algebraic geometry and in differential geometry. But note that, in algebraic geometry, if one wants to view étale morphisms as local isomorphisms, the Zariski topology has to be replaced by the so-called étale topology (cf. 2.3/8). In differential geometry, the implicit function theorem shows that, locally, smooth morphisms are fibrations by open subsets of linear spaces. Up to localization by étale morphisms, the same is true in algebraic geometry:

Proposition 11. *Let $f: X \longrightarrow S$ be a morphism, and let x be a point of X. Then the following conditions are equivalent:*

(a) *f is smooth at x of relative dimension n.*

(b) *There exists an open neighborhood U of x and a commutative diagram*

$$\begin{array}{ccc} U & \xrightarrow{\ g\ } & \mathbb{A}_S^n \\ & {\scriptstyle f|_U}\searrow & \downarrow{\scriptstyle p} \\ & & S \end{array}$$

where g is étale and p is the canonical projection.

Proof. That condition (b) implies condition (a) is clear, since the composition of smooth morphisms is smooth. To show the converse, choose local sections $g_1, \ldots, g_n$ of $\mathcal{O}_X$ such that $dg_1, \ldots, dg_n$ generate $\Omega^1_{X/S}$ at x. Due to Corollary 10, the latter is equivalent to the fact that $g_1, \ldots, g_n$ define an étale map from an open neighborhood U of x to $\mathbb{A}_S^n$. □

Remark 12. If X is a smooth S-scheme and if $g_1, \ldots, g_n$ are local sections of $\mathcal{O}_X$ at a point $x \in X$, then, by Nakayama's lemma, the differentials $dg_1, \ldots, dg_n$ generate $\Omega^1_{X/S}$ at x if and only if the differentials $dg_1(x), \ldots, dg_n(x)$ form a basis of the $k(x)$-vector space $\Omega^1_{X/S,x} \otimes k(x)$. Furthermore, as we have mentioned in the preceding proof, this condition is equivalent to the fact that $g_1, \ldots, g_n$ define an étale morphism from an open neighborhood U of x to $\mathbb{A}_S^n$. If $g_1, \ldots, g_n$ satisfy these equivalent conditions, they will be called a *system of local coordinates at x* (over S). This terminology is justified since, up to an étale morphism, $g_1, \ldots, g_n$ indeed behave like a set of coordinates of the affine n-space $\mathbb{A}_S^n$.

As a consequence of Proposition 11, we obtain the following useful fact.

Corollary 13. *If X is a smooth scheme over a field k, the set of closed points x of X such that $k(x)$ is a separable extension of k is dense in X.*

Proof. For each point x_0 of X, there exists an open neighborhood U of x_0 and a factorization

$$U \xrightarrow{\ g\ } \mathbb{A}_k^n \xrightarrow{\ p\ } \operatorname{Spec} k$$

where g is étale. Then, if x is a point of U, the extension $k(x)$ of $k(g(x))$ is finite and separable. Hence it is enough to show $g(U)$ contains a closed point y such that $k(y)$

is a separable extension of k. The set of closed points y such that $k(y)$ is separable over k is dense in $\mathbb{A}_k^n$. Namely, this is clear if k is perfect. If k is not perfect, it contains infinitely many elements so that the set of k-valued points is dense in $\mathbb{A}_k^n$. Thus it suffices to show that $g(U)$ contains a non-empty open subset. However, the latter is clear by reasons of dimension, since $g(U)$ is constructible (cf. [EGA IV_1], 1.8.4). (Actually, $g(U)$ is open, because an étale map is flat and hence open.) □

Next we apply Proposition 7 in order to construct étale sections of smooth morphisms.

Proposition 14. *Let $f: X \longrightarrow S$ be a smooth morphism. Let s be a point of S, and let x be a closed point of the fibre $X_s = X \times_S \operatorname{Spec} k(s)$ such that $k(x)$ is a separable extension of $k(s)$. Then there exist an étale morphism $g: S' \longrightarrow S$ and a point $s' \in S'$ above s such that the morphism $f': X \times_S S' \longrightarrow S'$ obtained from f by the base change $S' \longrightarrow S$ admits a section $h: S' \longrightarrow X \times_S S'$, where $h(s')$ lies above x, and where $k(s') = k(x)$.*

Proof. Let n be the relative dimension of X over S at x. Let $\mathscr{J} \subset \mathcal{O}_{X_s}$ be the sheaf of ideals associated to the closed point x of X_s. Since $\operatorname{Spec} k(x) \longrightarrow \operatorname{Spec} k(s)$ is étale, the ideal $\mathscr{J}_x$ is generated by n elements $\bar{g}_1, \ldots, \bar{g}_n$ such that their differentials $d\bar{g}_1, \ldots, d\bar{g}_n$ generate $\Omega^1_{X/S} \otimes k(x)$, as seen by the Jacobi criterion (Proposition 7). Now we lift $\bar{g}_1, \ldots, \bar{g}_n$ to sections $g_1, \ldots, g_n$ of $\mathcal{O}_X$ defined on an open neighborhood of x in X. Then let S' be the subscheme of X defined by $g_1, \ldots, g_n$. Again by Proposition 7, the scheme S' is étale over S at x. After shrinking S' we may assume that $S' \longrightarrow S$ is étale. Then the tautological section $h': S' \longrightarrow X'$ is a section as required. □

Using Proposition 7, the smoothness of a scheme X over a field k can be characterized by algebraic properties of the local rings of X. A k-scheme X which is locally of finite type is called *regular* if, for each closed point x of X, the local ring $\mathcal{O}_{X,x}$ is regular. (One knows then that $\mathcal{O}_{X,x}$ is regular also for non-closed points $x \in X$; cf. [EGA 0_{IV}], 17.3.2).

Proposition 15. *Let X be locally of finite type over a field k. Let x be a point of X. Then the following conditions are equivalent:*

(a) *X is smooth over k at x.*

(b) *$(\Omega^1_{X/k})_x$ is generated by* $\dim_x X$ *elements (and hence free).*

(c) *There exist an open neighborhood U of x and a perfect field extension k' of k such that $U \otimes_k k'$ is regular.*

(d) *There exists an open neighborhood U of x such that $U \otimes_k k'$ is regular for all field extensions k' of k.*

Proof. We start with the implication (a) $\Longrightarrow$ (d). Due to Proposition 11, there exists an étale morphism $g: U \longrightarrow \mathbb{A}_k^n$, defined on an open neighborhood $U \subset X$ of x. Then Proposition 2 shows for each $y \in U$ that the maximal ideal $\mathfrak{m}_y$ is generated by $\mathfrak{m}_{g(y)}$. So $\mathfrak{m}_y$ is generated by $n = \dim U$ elements because $\mathbb{A}_k^n$ is regular; hence U is regular. Since the situation remains essentially the same after extending the field k to k', the assertion follows.

The implication (d) $\Longrightarrow$ (c) is trivial. So let us consider the implication (c) $\Longrightarrow$ (b). We may assume $k = k'$ and $X = U$. Moreover, it suffices to show for each closed point $y \in X$ that $(\Omega^1_{X/k})_y$ is generated by $\dim \mathcal{O}_{X,y}$ elements. For such a point y, the field $k(y)$ is separable over k. Hence $\Omega^1_{k(y)/k} = 0$, and the exact sequence of 2.1/2 yields an exact sequence

$$\mathfrak{m}_y/\mathfrak{m}_y^2 \longrightarrow (\Omega^1_{X/k})_y \otimes k(y) \longrightarrow 0 .$$

Since $\mathfrak{m}_y/\mathfrak{m}_y^2$ is generated by $\dim \mathcal{O}_{X,y}$ elements (due to assumption (c)), the assertion follows with the help of Nakayama's lemma.

Finally, we turn to the implication (b) $\Longrightarrow$ (a). We may assume that X is a closed subscheme of an open subscheme V of $\mathbb{A}^n_k$, via the immersion $j: X \hookrightarrow \mathbb{A}^n_k$. Let $\mathcal{J}$ be the sheaf of ideals of $\mathcal{O}_V$ defining X, and let $r = \dim_x X$. Looking at the exact sequence of 2.1/2

$$(\mathcal{J}/\mathcal{J}^2)_x \longrightarrow (j^*(\Omega^1_{\mathbb{A}^n_k/k}))_x \longrightarrow (\Omega^1_{X/k})_x \longrightarrow 0 ,$$

we see that there exist local sections $g_{r+1}, \ldots, g_n$ of $\mathcal{J}$ at x such that $dg_{r+1}, \ldots, dg_n$ generate a free direct factor of $(\Omega^1_{\mathbb{A}^n_k/k})_x$ of rank $(n-r)$. We may assume that $g_{r+1}, \ldots, g_n$ are defined on V and give rise to a smooth subscheme $X' \subset V$ of dimension r. So X is a closed subscheme of X' and has the same dimension at x as X'. Let y be a closed point of X, which is a specialization of x. Then, by what we have already seen, $\mathcal{O}_{X',y}$ is an integral domain. Since $\dim \mathcal{O}_{X,y} \geqq r$, the surjective map $\mathcal{O}_{X',y} \longrightarrow \mathcal{O}_{X,y}$ has to be injective by reasons of dimension. This shows that X and X' coincide in a neighborhood of x. □

The property (d) of the preceding proposition gives rise to the following definition. A scheme X which is locally of finite type over a field k is called *geometrically reduced* (resp. *geometrically normal*, resp. *geometrically regular*) if $X \otimes_k k'$ is reduced (resp. normal, resp. regular) for all field extensions k' of k.

Proposition 16. *Let X be locally of finite type over a field k. If X is geometrically reduced, the smooth locus of X is open and dense in X.*

Proof. It is clear that the smooth locus is open. For the proof of the density, consider a generic point x of X. For any field extension k' of k, the algebra $k(x) \otimes_k k'$ is reduced. Then it is an elementary algebraic fact that $\Omega^1_{k(x)/k}$ is generated by n elements where n is the degree of transcendency of $k(x)$ over k; cf. Bourbaki [1], Chap. V, § 16, n°7, Thm. 5. Since n equals the dimension of X at x, Proposition 15 shows x is contained in the smooth locus of X. Thus, the smooth locus contains all generic points of X. □

2.3 Henselian Rings

In the following we want to have a closer look at the local structure of étale morphisms, in particular, we want to construct the (strict) henselization of a local ring; references for this section are [EGA IV_4], 18, and Raynaud [5]. Let R be a local ring with maximal ideal $\mathfrak{m}$ and residue field k. Let S be the affine (local) scheme of R, and let s be the closed point of S. From a geometric point of view, henselian and strictly henselian rings can be introduced via schemes which satisfy certain aspects of the inverse function theorem.

Definition 1. *The local scheme S is called henselian if each étale map $X \longrightarrow S$ is a local isomorphism at all points x of X over s with trivial residue field extension $k(x) = k(s)$. If, in addition, the residue field $k(s)$ is separably closed, S is called strictly henselian.*

Notice that if S is strictly henselian, any étale morphism $X \longrightarrow S$ is a local isomorphism at all points of X over s. Usually one introduces the notion of henselian rings in terms of properties of the local ring R; namely, one requires Hensel's lemma to be true for R. As we will explain later (cf. Proposition 4), it suffices to require a seemingly weaker condition.

Definition 1'. *The local ring R is called henselian if, for each monic polynomial $P \in R[T]$, all k-rational simple zeros of the residue class $\bar{P} \in k[T]$ lift to R-rational zeros of P. If, in addition, the residue field k is separably closed, R is called strictly henselian.*

It is easily seen that the ring R is (strictly) henselian if the scheme S is (strictly) henselian. The converse is also true, but the proof is not so easy; it is mainly a consequence of Zariski's Main Theorem. For the statement of this theorem let us recall the definition of quasi-finite morphisms. Let $f: X \longrightarrow Y$ be a morphism which is locally of finite type. Then f is said to be *quasi-finite at a point x of X* if x is isolated in the fibre $X_y = X \times_Y \operatorname{Spec} k(y)$ over the image point $y := f(x)$; the latter is equivalent to the fact that the ring $\mathcal{O}_{X,x}/\mathfrak{m}_y\mathcal{O}_{X,x}$ is a finite-dimensional vector space over the field $k(y) = \mathcal{O}_{Y,y}/\mathfrak{m}_y$, cf. [EGA II], 6.2.1. For example, unramified morphisms are quasi-finite at all points. The set of points $x \in X$ such that f is quasi-finite at x is open in X, cf. [EGA IV_3], 13.1.4. The morphism f is called *quasi-finite* if f is quasi-finite at all points $x \in X$ and if f is of finite type. For example, a composition of a quasi-compact open immersion $X \hookrightarrow Z$ and a finite morphism $Z \longrightarrow Y$ is quasi-finite. Zariski's Main Theorem says that essentially every quasi-finite morphism is obtained in this way.

Theorem 2 (Zariski's Main Theorem). *Let $f: X \longrightarrow Y$ be quasi-finite and separated. Furthermore, assume that Y is quasi-compact and quasi-separated. Then there exists a factorization*

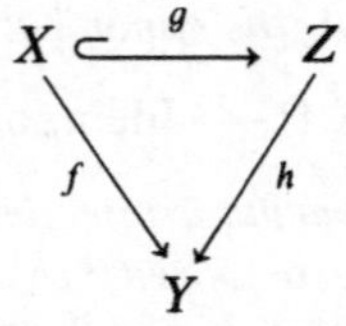

of f, where g is an open immersion and where h is finite.

For a proof see [EGA IV_4], 18.12.13; a more direct argument (for the local case) can be found in Peskine [1]. For our applications we will need a weaker version which is close to Zariski's original form of the theorem, cf. [EGA IV_3], 8.12.10.

Theorem 2′. *Let $f: X \longrightarrow Y$ be quasi-finite and separated. Assume that X is reduced, that Y is normal, and that there exist dense open subschemes $U \subset X$ and $V \subset Y$ such that $f|_U : U \longrightarrow V$ is an isomorphism. Then f is an open immersion.*

Theorem 2 can be used to investigate the local structure of étale morphisms. In terms of the corresponding extension of algebras, an étale extension is sort of a lifting of a finite separable field extension which, due to the theorem of the primitive element, is always generated by a single element.

Proposition 3. *Let $f: X \longrightarrow Y$ be a morphism of schemes, let x be a point of X, and set $y = f(x)$. Assume that f is étale at x. Then there exist an affine open neighborhood $U = \operatorname{Spec} B$ of x, an affine open neighborhood $V = \operatorname{Spec} A$ of y with $f(U) \subset V$ and a Y-immersion $U \hookrightarrow \mathbb{A}_V^1$ such that U becomes an open subscheme of a closed subscheme $Z \subset \mathbb{A}_V^1$, where Z is defined by a monic polynomial $P \in A[T]$ and where the derivative P' of P has no zeros on the image of U. Moreover, B is isomorphic to $(A[T]/(P))_Q$ for some $Q \in A[T]$.*

A detailed proof is given in Raynaud [5], Chap. V. The idea of the proof is easy to explain. Namely, we may assume that X and Y are affine, and, due to Theorem 2, that X is an open subscheme of a scheme $X' = \operatorname{Spec} B'$ which is finite over Y. Since $k(x)$ is finite and separable over $k(y)$, there exists a non-zero element $\bar{b} \in k(x)$ such that $\bar{b}$ generates $k(x)$ over $k(y)$. Let $b \in B'$ be a lifting of $\bar{b}$ which vanishes at all points of the fibre of $X' \longrightarrow Y$ over y, except at x. Now b gives rise to a morphism $X' \longrightarrow \mathbb{A}_Y^1$. Since X' is finite over Y, one can verify that this morphism induces an open immersion of a neighborhood of x into a subscheme Z of $\mathbb{A}_Y^1$ of the required type. □

It follows immediately from Proposition 3 that the notions of henselian local rings and henselian local schemes are equivalent. This equivalence can be extended by further conditions, cf. [EGA IV_4], 18.5, or Raynaud [5], Chap. I.

Proposition 4. *Let R be a local ring, and set $S = \operatorname{Spec} R$. Then the following conditions are equivalent:*

(a) *R is henselian.*

(b) *S is henselian.*

(c) *For each finite R-algebra A, the canonical map*

$$\operatorname{Idempotent}(A) \longrightarrow \operatorname{Idempotent}(A \otimes_R k)$$

between the sets of idempotent elements is bijective.

(d) *Each finite R-algebra A decomposes into a product of local rings.*

(e) *For each quasi-finite morphism $X \longrightarrow S$, and for each point x above the closed point of S, there exists an open neighborhood U of x such that $U \longrightarrow S$ is finite.*

We will only sketch the *proof*, following the ideas of Grothendieck. The implications (a) $\Longrightarrow$ (b) and (d) $\Longrightarrow$ (e) (which are the hard ones) are clear by Proposition 3 and Theorem 2. In order to show that (b) implies (c), one has to observe that it suffices to establish (c) in the case where A is a free R-module. Then one can write down formally what the idempotent elements of A must look like, and one notices that they are represented by an étale R-scheme. So it remains to show that such an étale R-scheme admits an R-section. The proof of the remaining implications is more or less trivial. □

The main reason for us to introduce strictly henselian rings is the fact that smooth schemes over strictly henselian rings admit many sections. Due to the geometric characterization of henselian rings, this property follows directly from 2.2/13 and 2.2/14.

Proposition 5. *Let R be a local henselian ring with residue field k. Let X be a smooth R-scheme. Then the canonical map $X(R) \longrightarrow X(k)$ from the set of R-valued points of X to the set of k-valued points of X is surjective. In particular, if R is strictly henselian, the set of k-valued points of $X_k = X \otimes_R k$ which lift to R-valued points of X is dense in X_k.*

Examples of henselian rings are local rings occurring in analytic geometry such as rings of germs of holomorphic functions. Furthermore, local rings which are separated and complete with respect to the maximal-adic topology are henselian. In the latter case the condition mentioned in Definition 1′ is established by Hensel's lemma; cf. Bourbaki [2], Chap. III, §4, n°3, Thm. 1. Alternatively, using the infinitesimal lifting property 2.2/6 for étale morphisms one can verify directly that such rings fulfill Definition 1. Since a noetherian local ring R is always a subring of its maximal-adic completion $\hat{R}$, these local rings R are a priori subrings of henselian rings. The "smallest" henselian ring containing R is called the henselization of R.

Definition 6. *A henselization of a local ring R is a henselian local ring R^h together with a local morphism $i: R \longrightarrow R^h$ such that the following universal property is satisfied: For any local morphism $u: R \longrightarrow A$ from R to a henselian local ring A, there exists a unique local morphism $u^h: R^h \longrightarrow A$ such that $u^h \circ i = u$.*

If the henselization exists, it is unique up to canonical isomorphism. Moreover, the residue field of R^h must be k. In view of Definition 1, the henselization of R must be the "union" of all local rings $\mathcal{O}_{X,x}$ of étale R-schemes at points x above the closed point s of $S = \operatorname{Spec} R$, whose residue fields coincide with k. That such a "union" exists in terms of inductive limits, becomes clear by the following result:

Lemma 7. *Let S' be an étale R-scheme and let s' be a point of S' above the closed point s of $S = \operatorname{Spec} R$. Let R' be the local ring $\mathcal{O}_{S',s'}$ of S' at s' and let k' be the residue field of R'. Furthermore, let A be a local R-algebra with residue field k_A. Then all R-algebra morphisms from R' to A are local. So there is a canonical map*

$$\operatorname{Hom}_R(R', A) \longrightarrow \operatorname{Hom}_k(k', k_A) .$$

This map is always injective; it is bijective if A is henselian.

Proof. Since the maximal ideal of R' is generated by the maximal ideal of R, all R-morphisms $R' \longrightarrow A$ are local. The injectivity of the map follows from the fact that the diagonal morphism $S' \longrightarrow S' \times_S S'$ is an open immersion. The surjectivity is due to the characterization of henselian local rings given in Definition 1. □

For the construction of the henselization of R, one considers the family $(R_i)_{i \in I^h}$ of all isomorphism classes of R-algebras which occur as local rings of étale R-schemes at points over the closed point of Spec R and which have the same residue field as R. Due to Proposition 3, the family I^h is a set and, due to Lemma 7, there is a natural partial order on I^h. Namely, one defines $i \leqq j$ for $i, j \in I^h$ if there exists an R-morphism $u_{ij}: R_i \longrightarrow R_j$. So $(R_i)_{i \in I^h}$ is an inductive system, which is seen to be directed and one easily proves that

$$R^h := \varinjlim_{i \in I^h} R_i$$

is a henselization of R (for details see Raynaud [5], Chap. VIII).

If one wants to introduce the smallest strictly henselian ring containing R, one has to be a little bit more careful. Namely, in view of Lemma 7, there may be different R-morphisms between two (local) étale R-algebras unless we require that the residue extension is trivial. One has to eliminate this ambiguity, and then one can proceed as in the case of the henselization.

Definition 6′. *A strict henselization of a local ring R is a strictly henselian local ring R^{sh}, whose residue field coincides with the separable algebraic closure k_s of k, together with a local morphism $i: R \longrightarrow R^{sh}$ such that the following universal property is satisfied: For any local morphism $u: R \longrightarrow A$ from R to a strictly henselian ring A, and for any k-morphism $\alpha: k_s \longrightarrow k_A$ from k_s to the residue field k_A of A, there exists a unique local morphism $u^{sh}: R^{sh} \longrightarrow A$ such that $u^{sh} \circ i = u$ and such that u^{sh} induces α on the residue fields.*

If R^{sh} exists, it is unique up to canonical isomorphism. For the construction of R^{sh}, let $(R_i)_{i \in I}$ be the family of all isomorphism classes of R-algebras which occur as local rings of étale R-schemes at points over the closed point of Spec R. Let I^{sh} be the set of all couples (R_i, α_{ij}) where R_i is a member of I and where $\alpha_{ij}: R_i \longrightarrow k_s$ varies over all R-morphisms into a fixed separable closure k_s of k. Due to Lemma 7, there exists a natural order on I^{sh}. So $((R_i, \alpha_{ij}))_{(i,j) \in I^{sh}}$ is a directed inductive system, and one easily verifies that

$$R^{sh} = \varinjlim_{(i,j) \in I^{sh}} (R_i, \alpha_{ij})$$

is the strict henselization of R; cf. Raynaud [5], Chap. VIII.

As an application of this construction, we want to mention some results on étale localizations of quasi-finite morphisms. Let us call $Y' \longrightarrow Y$ an *étale neighborhood of a point y in Y* if $Y' \longrightarrow Y$ is étale and if y is contained in the image of Y'.

Proposition 8. *Let $f: X \longrightarrow Y$ be locally of finite type. Let x be a point of X, and set $y = f(x)$.*

(a) *If f is quasi-finite at x, then there exists an étale neighborhood $Y' \longrightarrow Y$ of y such that the morphism $f': X' \longrightarrow Y'$, obtained from f by the base change $Y' \longrightarrow Y$, induces a finite morphism $f'|_{U'}: U' \longrightarrow Y'$, where U' is an open neighborhood of the fibre of $X' \longrightarrow X$ above x. If, in addition, f is separated, U' is a connected component of X'.*

(b) *If f is unramified at x (resp. étale at x), there exists an étale neighborhood $Y' \longrightarrow Y$ of y such that, locally at each point of X' above x, the morphism f' (as in (a)) is an immersion (resp. an open immersion).*

Proof. Let R be a strict henselization of the local ring $\mathcal{O}_{Y,y}$ of Y at y, and set $S = \operatorname{Spec} R$. Then R is the limit of all local rings $\mathcal{O}_{Y',y'}$ which occur as local rings of étale neighborhoods Y' of $y \in Y$ at points y' above y. Using limit arguments (cf. [EGA IV_3], 8.10.5), it suffices to prove the assertions in the case where $Y = S$. Then (a) follows from Proposition 4, and (b) is a consequence of the fact that each finite, local, and unramified R-algebra A is a quotient of R. Namely, the assumptions yield $R/\mathfrak{m} = A/\mathfrak{m}A$, where $\mathfrak{m}$ is the maximal ideal of R, and so Nakayama's lemma applies. Finally, the case of étale morphisms is deduced from the case of unramified ones by means of 2.2/4. □

The preceding proposition justifies the interpretation of unramified, resp. étale, resp. smooth morphisms given in 2.2. Namely, Proposition 8 tells us that, up to base change by étale morphisms, unramified morphisms are immersions and étale morphisms are open immersions. So, if we look at S-schemes X only up to étale base change, as it is done within the context of the étale topology or, more generally, in the theory of algebraic spaces, we may view unramified morphisms as immersions and étale morphisms as open immersions. Furthermore, Proposition 2.2/11 says that smooth morphisms may be viewed as fibrations by open subsets of linear spaces $\mathbb{A}^n_S$.

The local structure of étale morphisms $X \longrightarrow Y$ (cf. Proposition 3) can be used to study how algebraic properties are transmitted from Y to X. By a minor calculation (cf. Raynaud [5], Chap. VII), one shows that all étale schemes over a reduced (resp. normal) base are reduced (resp. normal) again. Using the elementary fact that polynomial rings inherit such properties from the base, it follows from 2.2/11 that smooth schemes over a reduced (resp. normal) base are reduced (resp. normal) again. Finally, since polynomial rings over regular rings are regular, smooth schemes over regular schemes are regular again; use 2.2/11 and 2.2/2(e). Summarizing, we can say:

Proposition 9. *Let $X \longrightarrow Y$ be a smooth morphism. If Y is reduced (resp. normal, resp. regular), then X is reduced (resp. normal, resp. regular).*

Obviously, a directed inductive limit R of reduced (resp. normal) rings R_i is reduced (resp. normal). So we have the permanence of reducedness and normality for the (strict) henselization. Moreover, since the maximal ideal $\mathfrak{m}$ of R generates the maximal ideal $\mathfrak{m}_i$ of each R_i which occurs in the construction of the (strict) henselization of R, it is clear that $\mathfrak{m}$ also generates the maximal ideal of the (strict) henselization. In particular, we see that the (strict) henselization of a discrete valuation ring is a discrete valuation ring, and that a uniformizing parameter of R yields a uniformizing parameter of the (strict) henselization. Furthermore, one can show that properties of local rings such as being noetherian or regular are preserved by the process of (strict) henselization. We state this for later reference:

Proposition 10. *If R is a reduced (resp. normal, resp. regular, resp. noetherian) local ring, the (strict) henselization is reduced (resp. normal, resp. regular, resp. noetherian) again. In particular, if R is a discrete valuation ring with uniformizing parameter π, then the (strict) henselization is a discrete valuation ring, and π gives rise to a uniformizing element there.*

Finally, we want to have a closer look at the ring extensions

$$R \longrightarrow R^h \longrightarrow R^{sh} .$$

Due to the local structure of étale morphisms (Proposition 3), these canonical homomorphisms are injective. Since R^{sh} can also be interpreted as the strict henselization of R^h, it follows from the construction of R^{sh} that the extension $R^h \hookrightarrow R^{sh}$ is integral, as can be seen by using the characterization of henselian rings mentioned in Proposition 4(e). If R is normal, the rings R^h and R^{sh} are normal and, hence, integral domains. Thus we can consider their fields of fractions

$$K \subset K^h \subset K^{sh} ,$$

which are separable algebraic over K. Moreover, K^{sh} is a Galois extension of K^h, the Galois group of K^{sh} over K^h acts on R^{sh}, and the fixed subring of R^{sh} is R^h. Due to Lemma 7, the Galois group is canonically isomorphic to the Galois group of k_s over k.

Proposition 11. *Let R be normal with field of fractions K. Let K_s be a separable closure of K, and let G be the Galois group of K_s over K. Let R_s be the integral closure of R in K_s, and let $\mathfrak{m}_s$ be a maximal ideal of R_s lying over the maximal ideal $\mathfrak{m}$ of R. Let*

$$D = \{\sigma \in G; \sigma(\mathfrak{m}_s) = \mathfrak{m}_s\}$$

be the decomposition group of $\mathfrak{m}_s$, and let

$$I = \{\sigma \in D; \sigma(\bar{x}) = \bar{x} \text{ for } \bar{x} \in R_s/\mathfrak{m}_s\}$$

be the inertia group of $\mathfrak{m}_s$. Then the following assertions hold:

(a) *The localization R' of the fixed ring R_s^D of R_s under D at the maximal ideal $\mathfrak{m}_s \cap R_s^D$ is the henselization of R.*

(b) *The localization R'' of the fixed ring R_s^I of R_s under I at the maximal ideal $\mathfrak{m}_s \cap R_s^I$ is the strict henselization of R.*

(c) *The extension $R^h \subset R^{sh}$ is Galois. Its Galois group D/I is canonically isomorphic to the Galois group of the residue field extension k_s over k.*

Proof. (a) Let $P(T) \in R'[T]$ be a monic polynomial whose reduction $\bar{P}(T)$ has a simple zero $\bar{\alpha}$ lying in the residue field of R'. Now $P(T)$ has a zero α lying in $(R_s)_{\mathfrak{m}_s}$ which induces $\bar{\alpha}$ if we regard $\bar{\alpha}$ as an element of $R_s/\mathfrak{m}_s$. Since $\bar{\alpha}$ is simple, there is only one zero α of this kind. Then it is easily seen that α is invariant under D. Hence α lies in R'. Thus we see R' is henselian. Moreover it is known that R' is a limit of étale extensions R_i of R which have the same residue fields as R; cf. Raynaud [5], Chap. X. So R' is a henselization of R.

(b) follows similarly as (a), one has only to replace the decomposition group by the inertia group. Assertion (c) follows from (a) and (b) by formal arguments. □

2.4 Flatness

Let R be a ring, and let M be an R-module. Then M is called *flat over* R (or a flat R-module) if

$$Mod_R \longrightarrow Mod_R\,, \qquad N \longmapsto N \otimes_R M\,,$$

constitutes an exact functor on the category of R-modules Mod_R. If R is a field, flatness poses no condition, and if R is a Dedekind domain, the flatness of M means that M has no torsion. Flatness is a local property; i.e., an R-module M is flat over R if and only if, for each prime ideal $\mathfrak{p}$ of R, the localization $M_{\mathfrak{p}}$ is flat over $R_{\mathfrak{p}}$. For a local ring R, a finitely generated R-module is flat if and only if it is free; cf. Bourbaki [2], Chap. I, §2, ex. 23. But, in general, flat modules do not need to be free or projective (in the sense of being a direct factor of a free module); for example, the field of fractions of a discrete valuation ring R is a flat R-module which cannot be free. Nevertheless, it can be shown that an R-module M is flat if and only if M is a direct limit of free R-modules of finite type; cf. Lazard [1], Thm. 1.2, or Bourbaki [1], Chap. X, §1, n°6, Thm. 1. A flat R-module M is called *faithfully flat* if the tensor product by M is a faithful functor; i.e., if $N \otimes_R M \neq 0$ for all R-modules $N \neq 0$. Viewing R-algebras as R-modules, one has also the notion of flatness (resp. faithful flatness) for R-algebras. For example, localizations $S^{-1}R$ are flat R-algebras and polynomial rings $R[T_1, \ldots, T_n]$ are faithfully flat R-algebras. Furthermore, we want to mention that a local flat morphism $R \longrightarrow A$ of local rings is automatically faithfully flat.

Now, turning to schemes, a morphism $f: X \longrightarrow S$ of schemes is called *flat at a point* x of X if $\mathcal{O}_{S,f(x)} \longrightarrow \mathcal{O}_{X,x}$ is flat, and f is called *flat* if it is flat at all points of X. Furthermore, a morphism $f: X \longrightarrow S$ is said to be *faithfully flat* if f is flat and surjective. If X and S are affine, say $X = \operatorname{Spec} A$ and $S = \operatorname{Spec} R$, then f is flat (resp. faithfully flat) if and only if $f^*: R \longrightarrow A$ is flat (resp. faithfully flat). Obviously, open immersions are flat, and it is easy to see that the class of flat (resp. faithfully flat) morphisms is stable under composition, base change, and formation of products; cf. [EGA IV_2], 2.1 and 2.2. In the case where S is the spectrum of a discrete valuation ring, $f: X \longrightarrow S$ is flat if and only if $\mathcal{O}_X$ has no R-torsion. So there are no irreducible and no embedded components of X which are contained in the special fibre. Since the notion of flatness is quite transparent over valuation rings, it is useful to know that there is a valuative criterion for flatness which applies to the geometric case.

Proposition 1 ([EGA IV_3], 11.8.1). *Let $f: X \longrightarrow S$ be locally of finite presentation. Let x be a point of X, and set $s = f(x)$. Assume that $\mathcal{O}_{S,s}$ is reduced and noetherian. Then f is flat at x if and only if, for each scheme S' which is the spectrum of a discrete valuation ring, and each morphism $S' \longrightarrow S$ sending the special point s' of S' to s, the morphism $f': X' \longrightarrow S'$ obtained from f by the base change $S' \longrightarrow S$ is flat at all points $x' \in X'$ lying over x.*

It is much more difficult to understand the notion of flatness in the case where the base has nilpotent elements, for example, where the base is a non-trivial artinian ring. In this case there exists no criterion to test flatness by geometric properties.

Furthermore, we want to mention a criterion which allows to test the flatness of an S-morphism between flat S-schemes on fibres.

Proposition 2 ([EGA IV_3], 11.3.11). *Let $g: X \longrightarrow S$ and $h: Y \longrightarrow S$ be locally of finite presentation. Let $f: X \longrightarrow Y$ be an S-morphism. The following conditions are equivalent:*

(a) *f is flat, and h is flat at the points of $f(X)$.*

(b) *$f_s = f \times_S k(s)$ is flat for all $s \in S$, and g is flat.*

Now let us illustrate the meaning of flatness by some geometric properties of flat morphisms of finite presentation. In the following, let $f: X \longrightarrow Y$ always be a morphism of finite presentation. There are two general facts concerning the geometry of such morphisms. First, the image $f(C)$ of a constructible subset C of X is constructible in Y if Y is quasi-compact; a subset of a topological space is called constructible if it is a union of a finite collection of locally closed subsets; cf. [EGA IV_1], 1.8.4. Second, the function of relative dimension of f

$$X \longrightarrow \mathbb{N}, \qquad x \longmapsto \dim_x f^{-1}(f(x)),$$

is upper semi-continuous; i.e., for each $n \in \mathbb{N}$ the subset where the relative dimension is $\geqq n$ is closed; cf. [EGA IV_3], 13.1.3. If we assume that, in addition, f is flat, the situation becomes much better.

Proposition 3 ([EGA IV_2], 2.4.6). *Let $f: X \longrightarrow Y$ be locally of finite presentation. If f is flat, then f is open.*

Proposition 4 ([EGA IV_3], 14.2.2). *Let $f: X \longrightarrow Y$ be locally of finite type and flat. Assume that X is irreducible and that Y is locally noetherian. Then the relative dimension of f is constant on X.*

Dropping the finiteness condition in Proposition 3, its assertion has to be weakened.

Proposition 5 ([EGA IV_2], 2.3.12). *Let $f: X \longrightarrow Y$ be faithfully flat and quasi-compact. Then the topology of Y is the quotient topology of X with respect to f; i.e., a subset $V \subset Y$ is open if and only if $f^{-1}(V)$ is open in X.*

It is impossible to characterize the flatness of an S-scheme X of finite type by geometric properties when the base S is not reduced. But under reducedness conditions on the base and on the fibres, flatness is equivalent to universal openness; cf. [EGA IV_3], 15.2.3. Moreover, if the base S is reduced and noetherian, each S-scheme X of finite type is generically flat.

Proposition 6 ([EGA IV_2], 6.9.1). *Let S be reduced and noetherian, and let X be an S-scheme of finite type. Then there exists a dense open subscheme S' of S such that $X \times_S S'$ is flat over S'.*

Anyway, the flat locus of an S-scheme which is locally of finite presentation is open.

Proposition 7 ([EGA IV_3], 11.3.1). *Let X be an S-scheme which is locally of finite presentation. Then the set of points $x \in X$ such that X is flat over S at x is open.*

Non-trivial examples of flat morphisms of finite presentation are the smooth ones; see below. Furthermore, there is a useful criterion which relates smoothness over a general base to flatness and smoothness of the fibres. The latter are schemes over fields; in this case one can apply the nice criterion 2.2/15 to test smoothness.

Proposition 8. *Let $f: X \longrightarrow S$ be locally of finite presentation. Let x be a point of X, and set $s = f(x)$. The following conditions are equivalent:*

(a) *f is smooth at x.*

(b) *f is flat at x and the fibre $X_s = X \times_S k(s)$ is smooth over $k(s)$ at x.*

In Section 2.2, we gave detailed proofs for all statements concerning smoothness. Proceeding similarly with Proposition 8, let us give its *proof.* For the implication (a) $\Longrightarrow$ (b), it is only necessary to explain that smooth morphisms are flat. Due to 2.2/11, it suffices to treat the étale case. But in this case the assertion follows easily by looking at the local structure of étale morphisms; cf. 2.3/3.

If one wants to verify this implication without using the local structure of étale morphisms (which involves Zariski's Main Theorem), one can proceed as follows. If Z is a smooth S-scheme which is flat over S, and if X is a subscheme of Z given by one equation, say $g = 0$, such that $d_{X/S}(g)$ does not vanish at a certain point $x \in X$, then X is flat over S at x. It suffices to prove this statement, since, in the general case, we can use an induction argument on the number of equations describing X locally at x as a subscheme of $\mathbb{A}^n_S$. In order to prove the assertion above, we may assume that S is noetherian. Then consider the exact sequence

$$\mathcal{O}_{Z,x} \xrightarrow{g} \mathcal{O}_{Z,x} \longrightarrow \mathcal{O}_{X,x} \longrightarrow 0$$

If S is the spectrum of a field, then $\mathcal{O}_{Z,x}$ is an integral domain and g must be a regular element, so the map on the left-hand side is injective in this case. Since smoothness is stable under any base change, we see that the map $g \otimes k(s)$ is injective, where $k(s)$ is the residue field at the image s of x. Because Z is flat over S, we get

$$\operatorname{Tor}_1^{\mathcal{O}_{S,s}}(\mathcal{O}_{X,x}, k(s)) = 0 .$$

Hence X is flat over S at x, cf. Bourbaki [2], Chap. III, § 5, n°2, Thm. 1.

For the implication (b) $\Longrightarrow$ (a), we may assume that X is a closed subscheme of a linear space $\mathbb{A}^n_S$ over an affine scheme $S = \operatorname{Spec} R$ which is defined by a finitely generated ideal $I \subset R[T_1, \ldots, T_n]$. Let r be the relative dimension of X_s at x. Since X_s is smooth over $k(s)$ at x, there exist sections $g_{r+1}, \ldots, g_n$ of I such that, locally at x, the induced functions $\bar{g}_{r+1}, \ldots, \bar{g}_n$ define X_s as a subscheme of $\mathbb{A}^n_s$ and such that $d\bar{g}_{r+1}(x), \ldots, d\bar{g}_n(x)$ are linearly independent in $\Omega^1_{\mathbb{A}^n_S/S} \otimes k(x)$; cf. 2.2/7. Now let Z be the S-scheme defined by $g_{r+1}, \ldots, g_n$. Notice that Z is smooth at x and that Z contains X as a closed subscheme. The fibres of X_s and Z_s coincide locally at x. Now let B be the algebra associated to Z, and let A be the algebra associated to X. Then A is a quotient B/J of B by a finitely generated ideal J of B. Since A is flat over R at x, the exact sequence

$$0 \longrightarrow J \longrightarrow B \longrightarrow A \longrightarrow 0$$

remains exact at x after tensoring with $k(s)$ over R. Since X_s coincides with Z_s locally at x, we see that $J \otimes_R k(s)$ vanishes at x. Nakayama's lemma yields $J_x = 0$. So X and Z coincide in a neighborhood of x and, hence, X is smooth over S at x. □

Since étale morphisms are flat, henselization and strict henselization are direct limits of flat ring extensions and, hence, they are flat extensions of the given ring.

Corollary 9. *Let R be a local ring. The ring extensions $R \longrightarrow R^h \longrightarrow R^{sh}$, where R^h is a henselization and R^{sh} a strict henselization of R, are faithfully flat.*

Apart from the nice geometric results for flat morphisms of finite presentation, the importance of flatness is expressed in the descent techniques for faithfully flat and quasi-compact morphisms. We want to mention here only the descent for properties of morphisms, the more involved program of the descent for modules or schemes will be explained in Section 6.1. Consider the following situation. Let

$$\begin{array}{ccccccc} X & \xrightarrow{f} & Y & & X' & \xrightarrow{f'} & Y' \\ & \searrow \swarrow & & & & \searrow \swarrow & \\ & S & & \longleftarrow & & S' & \end{array}$$

be a commutative diagram of morphisms, and assume that the triangle on the right-hand side is obtained from the one on the left by means of the base change $S' \longrightarrow S$. Frequently one wants to show that f enjoys a certain property provided it is known that f' has this property. So it is useful to know that quite a lot of properties descend under a faithfully flat and quasi-compact base change $S' \longrightarrow S$; for example, topological and set-theoretical properties (cf. [EGA IV_2], 2.6), finiteness properties (cf. [EGA IV_2], 2.7.1), and smoothness properties (cf. [EGA IV_4], 17.7.3). For precise statements, the reader is referred to the literature.

2.5 S-Rational Maps

A rational map $X \dashrightarrow Y$ between schemes X and Y is generally defined as an equivalence class of morphisms from dense open subschemes of X to Y; cf. [EGA I], 7. Two such morphisms $U \longrightarrow Y$ and $U' \longrightarrow Y$ are called equivalent if they coincide on a dense open part of $U \cap U'$. However, when working over a base scheme S, this notion does not behave well with respect to a base change $S' \longrightarrow S$. So we want to introduce a relative version of rational maps over a base scheme S which is compatible with base change. For our purposes, it is enough to consider S-rational maps between *smooth* S-schemes. So we will restrict ourselves to this case; for more general versions see [EGA IV_4], 20.

An open subscheme U of a smooth S-scheme X is called *S-dense* if, for each $s \in S$, the fibre $U_s = U \times_S \operatorname{Spec} k(s)$ is Zariski-dense in the fibre $X_s = X \times_S k(s)$. Clearly, finite intersections of S-dense open subschemes of X are S-dense in X again. Furthermore, if U is S-dense and open in X and if V is an open subscheme of X, then $U \cap V$ is S-dense in V. Considering a second smooth S-scheme Y, an *S-rational map* $\varphi : X \dashrightarrow Y$ is defined as an equivalence class of S-morphisms $U \longrightarrow Y$, where U is some S-dense open subscheme of X. Two such S-morphisms $U \longrightarrow Y$ and $U' \longrightarrow Y$ are called equivalent if they coincide on an S-dense open part of $U \cap U'$. We will say that $\varphi : X \dashrightarrow Y$ is *defined at a point* $x \in X$ if there is a morphism $U \longrightarrow Y$ representing φ with $x \in U$. The set of all points $x \in X$ where φ is defined constitutes an S-dense open subscheme of X. It is called the *domain of definition* of X; we denote it by $\operatorname{dom}(\varphi)$; but note that, without any further assumptions, there is no global morphism $\operatorname{dom}(\varphi) \longrightarrow Y$ defining φ. Furthermore, if $\varphi : X \dashrightarrow Y$ can be defined by an S-morphism $U \longrightarrow Y$ which induces an isomorphism from U onto an S-dense open subscheme of Y, then $\varphi : X \dashrightarrow Y$ is called *S-birational*. In this case we have an S-birational map $\varphi^{-1} : Y \dashrightarrow X$ which serves as an inverse of φ. It is clear that the notions S-dense, S-rational, and S-birational are preserved by any base change $S' \longrightarrow S$. In general, the same is not true for the domain of definition of S-rational maps. For example, set $S = \operatorname{Spec} \mathbb{Z}$, and consider the $\mathbb{Z}$-rational map $\varphi : \mathbb{A}^1_{\mathbb{Z}} \dashrightarrow \mathbb{A}^1_{\mathbb{Z}}$ given by the rational function $(T+1)/(T-1)$. Then the base change $\operatorname{Spec} \mathbb{Z}/2\mathbb{Z} \longrightarrow \operatorname{Spec} \mathbb{Z}$ transforms φ into a morphism $\mathbb{A}^1_{\mathbb{Z}/2\mathbb{Z}} \longrightarrow \mathbb{A}^1_{\mathbb{Z}/2\mathbb{Z}}$.

Let $f : X \longrightarrow Y$ be a quasi-compact and quasi-separated morphism between arbitrary schemes X and Y. Then the direct image $f_* \mathcal{O}_X$ of the structure sheaf of X is a quasi-coherent $\mathcal{O}_Y$-module, cf. [EGA I], 9.2.1, and the kernel $\mathcal{I}$ of the canonical morphism $\mathcal{O}_Y \longrightarrow f_* \mathcal{O}_X$ is a quasi-coherent sheaf of ideals in $\mathcal{O}_Y$. The *schematic image* of f is defined to be the subscheme of Y associated to $\mathcal{I}$; it is the smallest closed subscheme of Y that f factors through. If V is a subscheme of Y such that the inclusion $j : V \hookrightarrow Y$ is quasi-compact, the schematic image of j is also referred to as the *schematic closure* of V in Y. Furthermore, if the schematic closure of V in Y coincides with Y, we will say V is *schematically dense* in Y.

Lemma 1. *Let Y be a smooth S-scheme, and let V be an open quasi-compact subscheme of Y.*

(a) *If Y is of finite presentation, the set of points $s \in S$ such that V_s is not dense in Y_s is locally constructible in S (i.e. constructible if S is quasi-compact; cf.* [EGA 0_{III}], 9.1.12).

(b) *If V is S-dense in Y, it is schematically dense in Y.*

Proof. (a) We may assume that the base S is noetherian. Let A be the closed reduced subscheme $Y - V$, and denote by $p: A \longrightarrow S$ the structural morphism. Then consider the set

$$F = \{y \in A; \dim_y p^{-1}(p(y)) = \dim_y(Y/S)\} .$$

It is clear that V_s is not dense in Y_s if and only if $s \in p(F)$. Due to [EGA IV_3], 13.1.3, the set F is closed in Y and, due to [EGA IV_1], 1.8.5, the image $p(F)$ is locally constructible in S.

(b) follows from [EGA IV_3], 11.10.10. But, for the convenience of the reader, we will treat the case where the base is locally noetherian. It is enough to show that the restriction map $\mathcal{O}_Y(Y') \longrightarrow \mathcal{O}_Y(V \cap Y')$ is injective for each open subscheme Y' in a basis of the topology of Y; note that $V \cap Y'$ is S-dense in Y' for each open subscheme Y' of Y. So we may assume that S is an affine scheme $\operatorname{Spec} R$, and that Y is an affine scheme $\operatorname{Spec} A$. It suffices to show that $A \longrightarrow \mathcal{O}_Y(V)$ is injective.

Since A is flat over R, cf. 2.4/8, the associated prime ideals of A are just the associated prime ideals $\mathfrak{p}_{ij}$ of $\mathfrak{p}_i A$ where $\mathfrak{p}_1, \ldots, \mathfrak{p}_n$ are the associated prime ideals of R; cf. [EGA IV_2], 3.3.1. Since A is smooth over R, the prime ideals $\mathfrak{p}_{ij}$ are the minimal prime ideals over $\mathfrak{p}_i A$. So V meets each component $V(\mathfrak{p}_{ij})$ and, hence, the restriction map $A \longrightarrow \mathcal{O}_Y(V)$ is injective. □

For later reference we state that the schematic image is compatible with flat base change.

Proposition 2. *Let $f: X \longrightarrow Y$ be an S-morphism which is quasi-compact and quasi-separated. Let $g: S' \longrightarrow S$ be a flat morphism, and denote by $f': X' \longrightarrow Y'$ the S'-morphism obtained from f by base change. Let Z (resp. Z') be the schematic image of f (resp. of f'). Then, $Z \times_S S'$ is canonically isomorphic to Z'.*

The assertion follows immediately from the fact that the pull-back of $\mathcal{O}_Y$-modules with respect to the projection $Y' \longrightarrow Y$ gives rise to an exact functor from the category of $\mathcal{O}_Y$-modules to the category of $\mathcal{O}_{Y'}$-modules; cf. [EGA IV_2], 2.3.2.

Next we want to define the *graph of an S-rational map* $\varphi: X \dashrightarrow Y$, where X and Y are smooth S-schemes of finite type. Let U be an S-dense open subscheme of X such that φ is given by an S-morphism $U \longrightarrow Y$. We need to know that we may assume U to be quasi-compact.

Lemma 3. *Let U be an S-dense open subscheme of a smooth and quasi-compact S-scheme X. Then U contains an S-dense open subscheme which is quasi-compact.*

Proof. Let $\{U_i\}_{i \in I}$ be an affine open covering of U and, for each $i \in I$, consider the second projection $\tau_i: X \times_S U_i \longrightarrow U_i$. It admits a section $\delta_i: U_i \longrightarrow X \times_S U_i$, namely the tautological one. Denote by V_i the union of all connected components of fibres of τ_i which meet the image of δ_i. Then, τ_i being smooth, V_i is open in $X \times_S U_i$ by [EGA IV_3], 15.6.5. Let $\mathrm{Sat}(U_i)$ be the image of V_i under the first projection $X \times_S U_i \longrightarrow X$. Since U_i is smooth and, hence, flat over S, the image $\mathrm{Sat}(U_i)$ is open in X and contains U_i; it may be viewed as a saturation of U_i with respect to the structural morphism $X \longrightarrow S$. Now $\{\mathrm{Sat}(U_i)\}_{i \in I}$ is an open covering of X because U is S-dense in X, and this covering contains a finite subcover $\{\mathrm{Sat}(U_{i_1}), \ldots, \mathrm{Sat}(U_{i_n})\}$ because X is quasi-compact. Thus $U' := U_{i_1} \cup \cdots \cup U_{i_n}$ is S-dense and quasi-compact in U. □

So we have seen that $\varphi : X \dashrightarrow Y$ can be represented by an S-morphism $U \longrightarrow Y$ where U is S-dense open and quasi-compact in X. Let Γ_U be the graph of this morphism; it is a locally closed subscheme of $U \times_S Y$ (closed if Y is separated over S). Since U is quasi-compact over S, one can define the graph Γ of φ as the schematic closure of $U \cong \Gamma_U$ in $X \times_S Y$. In order to see that the definition is independent of the choice of U, it suffices to mention the fact that any quasi-compact S-dense open subscheme $V \subset U$ is schematically dense in U due to Lemma 1; hence V and U have the same schematic closure Γ in $X \times_S Y$.

Now let Ω be the largest open subscheme of X such that the projection $p : X \times_S Y \longrightarrow X$ onto the first factor induces an isomorphism

$$\Gamma \cap p^{-1}(\Omega) \xrightarrow{\sim} \Omega .$$

Then $\Omega \subset \mathrm{dom}(\varphi)$. Furthermore, if Y is *separated over* S, each graph Γ_U as above is closed in $U \times_S Y$ so that $\Gamma \cap (U \times_S Y) = \Gamma_U$. Therefore we have an isomorphism

$$\Gamma \cap p^{-1}(U) \xrightarrow{\sim} U ,$$

which shows $U \subset \Omega$. This shows $\mathrm{dom}(\varphi) \subset \Omega$ and thus $\mathrm{dom}(\varphi) = \Omega$. In particular, there is a unique S-morphism $\mathrm{dom}(\varphi) \longrightarrow Y$ corresponding to the S-rational map $\varphi : X \dashrightarrow Y$; but note that, in general, $\mathrm{dom}(\varphi)$ is not necessarily quasi-compact.

Example 4. Let $\xi = (\xi_i)_{i \in I}$ and $\eta = (\eta_j)_{j \in J}$ be systems of variables, and let k be a field with $\mathrm{char}(k) \neq 2$. Let R denote the k-algebra $k[\xi, \eta]/(\xi\eta)$ where $(\xi\eta)$ is the ideal generated by all products $\xi_i\eta_j$, $i \in I$ and $j \in J$. Set $S = \mathrm{Spec}\, R$. Then we can view $X = \mathrm{Spec}\, k[\xi]$ and $Y = \mathrm{Spec}\, k[\eta]$ as closed subschemes of S, intersecting each other at a single point, namely, at the origin of X and Y. Furthermore, the union of X and Y is S. Now fix indices $i_0 \in I$ and $j_0 \in J$, and consider the S-rational map $\varphi : \mathbb{A}^1_S \dashrightarrow \mathbb{A}^1_S$ given by the rational function

$$\frac{T^2 - 1}{(T - \xi_{i_0} + 1)(T - \eta_{j_0} - 1)} ,$$

where T is a coordinate of $\mathbb{A}^1_S$. Let D be the complement in $\mathbb{A}^1_S$ of the domain of definition $\mathrm{dom}(\varphi)$. Then $D \cap \mathbb{A}^1_X$ is the union of two closed subsets of $\mathbb{A}^1_X$; namely, of the zero set of $(T - \xi_{i_0} + 1)$ and of the closed point $(\xi, T - 1)$ which lies over the origin of X. A similar assertion is true for $D \cap \mathbb{A}^1_Y$. Since $\mathrm{char}(k) \neq 2$, both parts are disjoint. Thus, if the system ξ contains infinitely many variables, the domain of definition $\operatorname{dom}(\varphi)$ cannot be quasi-compact, since a subset of $\mathbb{A}^1_X$ consisting of a single closed point cannot be described by finitely many equations.

Proposition 5. *Let X, X', Y be smooth S-schemes of finite type, and assume that Y is separated over S. Let $\varphi : X \dashrightarrow Y$ be an S-rational map, and consider a flat S-morphism $f : X' \longrightarrow X$. Then $f^{-1}(\operatorname{dom}(\varphi))$ is an S-dense open subscheme of X', and $\varphi \circ f$ is an S-rational map from X' to Y which satisfies*

$$\operatorname{dom}(\varphi \circ f) = f^{-1}(\operatorname{dom}(\varphi)) .$$

In particular, if f is faithfully flat and if $\varphi \circ f$ is defined everywhere on X', the map φ is defined everywhere on X.

Proof. Since f is flat and locally of finite presentation, cf. [EGA IV_1], 1.4.3, the map f is open. Using this fact, one shows $f^{-1}(\operatorname{dom}(\varphi))$ is S-dense in X'. So $\varphi \circ f$ is an S-rational map and $\operatorname{dom}(\varphi \circ f)$ contains $f^{-1}(\operatorname{dom}(\varphi))$. Denote by $\Gamma \subset X \times_S Y$ the graph of φ and by $\Gamma' \subset X' \times_S Y$ the graph of $\varphi \circ f$. Then we see from Proposition 2 that

$$X' \times_X \Gamma = \Gamma' .$$

Let $p : \Gamma \longrightarrow X$ and $p' : \Gamma' \longrightarrow X'$ be the projections onto the first factors. Set $U' := \operatorname{dom}(\varphi \circ f)$, and consider its image $U := f(U')$ which is an open subscheme of X. Since $U' \longrightarrow U$ is faithfully flat, the projection p is an isomorphism over U if and only if p' is an isomorphism over U'. Therefore $U \subset \operatorname{dom}(\varphi)$, and the assertion is clear. □

Finally we want to show that the domain of definition of S-rational maps is compatible with flat base change.

Proposition 6. *Let $\varphi : X \dashrightarrow Y$ be an S-rational map between smooth S-schemes of finite type where Y is separated over S. Let $S' \longrightarrow S$ be a flat morphism, and denote by $\varphi' : X' \dashrightarrow Y'$ the S'-rational map obtained from f by base change. Then*

$$\operatorname{dom}(\varphi') = \operatorname{dom}(\varphi) \times_S S' .$$

Proof. It is clear that $\operatorname{dom}(\varphi) \times_S S' \subset \operatorname{dom}(\varphi')$. To show the opposite inclusion, denote the graph of φ by $\Gamma \subset X \times_S Y$ and the graph of φ' by $\Gamma' \subset X' \times_{S'} Y'$. Since the schematic closure commutes with flat base change, we have

$$\Gamma \times_S S' = \Gamma' .$$

Let $p : \Gamma \longrightarrow X$ and $p' : \Gamma' \longrightarrow X'$ be the projections onto the first factors. Furthermore, consider a point $x' \in \operatorname{dom}(\varphi')$, and let x be its image in X. Then we get a commutative diagram

$$\begin{array}{ccc} \operatorname{Spec} \mathcal{O}_{X',x'} & \longrightarrow & \operatorname{Spec} \mathcal{O}_{X,x} \\ \downarrow & & \downarrow \\ X' & \longrightarrow & X , \end{array}$$

where the map in the first row is faithfully flat. Therefore, the fact that p' is an isomorphism over Spec $\mathcal{O}_{X',x'}$ implies that p is an isomorphism over Spec $\mathcal{O}_{X,x}$. Since Y is of finite type over S, we see that Γ is of finite type over X. Hence, there exists an affine open neighborhood W of x such that p induces a closed immersion $p^{-1}(W) \longrightarrow W$. Let Z be the schematic image in W of this map and let U be a quasi-compact S-dense open subscheme of X where φ is defined. Then the open subscheme $U \cap W$ of W is contained in Z. Since $U \cap W$ is S-dense in W, the scheme Z coincides with W. Thus $p^{-1}(W) \longrightarrow W$ is an isomorphism, and x is contained in dom(φ). □

Chapter 3. The Smoothening Process

The smoothening process, in the form needed in the construction of Néron models, is presented in Sections 3.1 to 3.4. After we have explained the main assertion, we discuss the technique of blowing-up which is basic for obtaining smoothenings. The actual proof of the existence of smoothenings is carried out in Sections 3.3 and 3.4. As an application, we construct weak Néron models under appropriate conditions.

Our version of the smoothening process differs from the one of Néron insofar as we have added a constructibility assertion, thereby avoiding the use of provarieties; for more details see Section 1.6. A generic form of Néron's smoothening process has also been explained by M. Artin in [4].

The chapter ends with a generalization of the smoothening along a section where the base is a polynomial ring over an excellent discrete valuation ring. This kind of smoothening technique is very close to that developed by M. Artin [4] for the proof of his approximation theorem; see also Artin and Rotthaus [1].

3.1 Statement of the Theorem

In the following let R be a discrete valuation ring with field of fractions K, with residue field k, and with uniformizing element π. We denote by R^h a henselization of R and by R^{sh} a strict henselization of R. Then R^h and R^{sh} are discrete valuation rings with uniformizing element π and the residue field of R^{sh} equals the separable closure k_s of k. For any R-scheme X, let $X_K = X \otimes_R K$ be its generic fibre and $X_k = X \otimes_R k$ its special fibre.

Definition 1. *Let X be an R-scheme of finite type whose generic fibre X_K is smooth over K. A smoothening of X is an R-morphism $f: X' \longrightarrow X$ which satisfies the following conditions:*

(i) *f is proper and is an isomorphism on generic fibres.*

(ii) *For each étale R-algebra R', each R'-valued point of X lifts uniquely to an R'-valued point of X' which factors through the smooth locus X'_{smooth} of X'. More precisely, the canonical map $X'_{\text{smooth}}(R') \longrightarrow X(R')$ is bijective.*

Each étale R-algebra R' is semi-local. So in order to test condition (ii), one may restrict oneself to local extensions R' of R which are étale. In particular, such rings are discrete valuation rings; they are flat over R. Due to the valuative criterion of properness [EGA II], 7.3.8, condition (i) implies that the map $X'(R') \longrightarrow X(R')$

deduced from f is bijective for any flat R-algebra R' which is a discrete valuation ring. Hence, if condition (i) is satisfied, condition (ii) says that, for each local étale extension R' of R, the R'-valued points of X' factor through the smooth locus of X'. As seen in Section 2.3, the strict henselization R^{sh} of R is the direct limit of all local étale extensions of R. So condition (ii) is fulfilled if and only if the canonical map $X'_{\text{smooth}}(R^{sh}) \longrightarrow X(R^{sh})$ is bijective.

In general, a smoothening $X' \longrightarrow X$ is not a desingularization of X (i.e., a proper morphism $X'' \longrightarrow X$ from a regular scheme X'' to X which is an isomorphism over the regular locus of X), because the points in the complement of the smooth locus of X' do not need to be regular. However, a desingularization of X is always a smoothening, as we will see by using the following fact from commutative algebra.

Proposition 2. *Let $\iota : R \longrightarrow A$ and $\varepsilon : A \longrightarrow R$ be morphisms of regular local rings such that $\varepsilon \circ \iota = \text{id}_R$ (i.e., ε defines a section of the morphism* Spec $A \longrightarrow$ Spec R *associated to ι). Then the image of each regular system of parameters of R under ι is part of a regular system of parameters of A. If $\mathfrak{J}$ is the kernel of ε, then $\mathfrak{J}$ is generated by a part of a regular system of parameters. If $t_1, \ldots, t_n$ is a minimal system of generators of $\mathfrak{J}$, the completion of A with respect to $\mathfrak{J}$ is canonically isomorphic to $R[[t_1, \ldots, t_n]]$.*

Proof. Let $\mathfrak{m}$ be the maximal ideal of R, and let $s_1, \ldots, s_m$ be a minimal system of generators of $\mathfrak{m}$. Let $\mathfrak{m}'$ be the maximal ideal of A. As $\varepsilon \circ \iota = \text{id}_R$, the residue fields $R/\mathfrak{m}$ and $A/\mathfrak{m}'$ are canonically isomorphic, and $\mathfrak{m}/\mathfrak{m}^2$ may be viewed as a subspace of $\mathfrak{m}'/\mathfrak{m}'^2$. Hence $\iota(s_1), \ldots, \iota(s_m)$ is a part of a regular system of parameters of A. So there exist elements $t_1, \ldots, t_n$ in $\mathfrak{m}'$ such that $\iota(s_1), \ldots, \iota(s_m), t_1, \ldots, t_n$ is a regular system of parameters in A. After replacing t_i by $t_i - \iota(\varepsilon(t_i))$, we may assume that $t_1, \ldots, t_n$ are in the kernel $\mathfrak{J}$ of ε. An easy calculation shows $\mathfrak{J} = (t_1, \ldots, t_n)$ as required. The assertion concerning the $\mathfrak{J}$-adic completion of A follows immediately from the definition of a regular system of parameters. □

In order to show that a desingularization $X'' \longrightarrow X$ is a smoothening of X one has only to verify that, for any étale R-algebra R', each $a \in X''(R')$ factors through the smooth locus of X''. One knows that $X'' \otimes_R R'$ is a desingularization of $X \otimes_R R'$ (see 2.3/9) and, furthermore, that the image of $a :$ Spec $R' \longrightarrow X''$ factors through the smooth locus of X'' if the corresponding fact is true for $(a, \text{id}) :$ Spec $R' \longrightarrow X'' \otimes_R R'$ ([EGA IV_4], 17.7.4). So we may assume $R = R'$. Then it follows from Proposition 2 that X'' is smooth over R along a; cf. [EGA IV_4], 17.5.3.

Theorem 3 (Smoothening Process). *Let X be an R-scheme of finite type whose generic fibre X_K is smooth over K. Then X admits a smoothening $f : X' \longrightarrow X$.*

Moreover, one can construct f as a finite sequence of blowing-ups with centers in the special fibres. In particular, if X is quasi-projective over R, the same is true for X'.

Removing from X' the non-smooth locus, we see:

Corollary 4. *Let X be as before. Then there is an R-morphism $f : X'' \longrightarrow X$ from a smooth R-scheme X'' of finite type to X such that*

(i) *f is an isomorphism on generic fibres, and*
(ii) *the canonical map* $X''(R^{sh}) \longrightarrow X(R^{sh})$ *is bijective.*

Such schemes X'' are not unique, and they do not need to be proper over R, even if X is proper over R.

The smoothening process provides a first step towards the construction of Néron models. For example, if X_K is an abelian variety with a projective embedding $X_K \subset \mathbb{P}^n_K$, one can apply the smoothening process to the schematic closure X of X_K in $\mathbb{P}^n_R$. Restricting the resulting R-scheme to its smooth locus, we obtain a *smooth* R-model of X_K which, although it might not be proper over R, nevertheless satisfies the valuative criterion of properness for the special class of valuation rings which are étale over R.

3.2 Dilatation

We have claimed that a smoothening of X can be constructed by blowing up subschemes of the special fibre. First, let us explain what happens to the sections $X(R)$ when such a blowing-up is applied to X. Consider the following example. Set $X := \operatorname{Spec} R[T]$, where $T = (T_1, \ldots, T_n)$ is a set of variables, and let Y_k be the reduced subscheme of X which consists of the origin of the special fibre X_k of X. Then Y_k is defined by the ideal $\mathfrak{I} \subset R[T]$ which is generated by $\pi, T_1, \ldots, T_n$. Using an absolute value on K belonging to the valuation ring R, the R-valued points of X correspond bijectively to the rational points $x_K \in \mathbb{A}^n_K$ with $|T_i(x_K)| \leqq 1$, $i = 1, \ldots, n$. Furthermore, the R-valued points of X which specialize into Y_k correspond to the rational points $x_K \in \mathbb{A}^n_K$ with $|T_i(x_K)| \leqq |\pi|$. Now let $X' \longrightarrow X$ be the blowing-up of Y_k in X. Let $\mathscr{I}'$ be the sheaf of ideals of $\mathcal{O}_{X'}$ generated by $\mathfrak{I}$, and denote by X'_π the set of points of X' at which $\mathscr{I}'$ is generated by π. Then $X'_\pi = \operatorname{Spec} R[T']$, where $T' = (T'_1, \ldots, T'_n)$ is a second set of variables, and the morphism $X'_\pi \longrightarrow X$ corresponds to the morphism induced by sending T_i to $\pi T'_i$ for $i = 1, \ldots, n$. It is seen that $X'_\pi(R)$ is mapped bijectively onto the set of those R-valued points of X which specialize into Y_k; hence $X'_\pi(R)$ corresponds to the rational points $x_K \in \mathbb{A}^n_K$ which satisfy $|T_i(x_K)| \leqq |\pi|$. Furthermore, two points $x, y \in X'_\pi(R)$ have the same specialization over k if and only if $|T_i(x_K) - T_i(y_K)| \leqq |\pi^2|$ for all i. We will call X'_π the *dilatation* of Y_k in X.

In order to construct dilatations of more general type, consider an arbitrary R-scheme X of finite type and a closed subscheme Y_k of the special fibre X_k. Let $\mathscr{I}$ be the associated sheaf of ideals in $\mathcal{O}_X$; in particular, $\pi \in \mathscr{I}$. The blowing-up X' of $\mathscr{I}$ on X is defined as the homogeneous spectrum $\operatorname{Proj}(\mathscr{S})$ of the graded $\mathcal{O}_X$-algebra $\mathscr{S} = \bigoplus_{n \geqq 0} \mathscr{I}^n$ (cf. [EGA II], 3.1 and 8.1.3). Locally, it is defined as follows. If $X = \operatorname{Spec} A$, the sheaf of ideals $\mathscr{I}$ is associated to an ideal $\mathfrak{I}$ of A. Since A is noetherian, $\mathfrak{I}$ is generated by finitely many elements $g_0 = \pi, g_1, \ldots, g_n$ of A. Then X' is the closed subscheme of $\mathbb{P}^n_A$ which is given by the homogeneous ideal

$$\mathfrak{I}' = \ker(A[T_0, \ldots, T_n] \longrightarrow \bigoplus_{n \geqq 0} \mathfrak{I}^n),$$

where we consider the graded homomorphism sending the variable T_i to $g_i \in \mathfrak{J}^1$. Let U_i be the affine open subscheme of $\mathbb{P}_A^n$ where T_i does not vanish. Then $X' \cap U_i$ is affine, and the A-algebra of its global sections is given by

$$A\left[\frac{g_0}{g_i}, \ldots, \frac{g_n}{g_i}\right] \Big/ (g_i\text{-torsion})$$

where, suggestively, we have written

$$A\left[\frac{g_0}{g_i}, \ldots, \frac{g_n}{g_i}\right] := A[T_0, \ldots, T_{i-1}, T_{i+1}, \ldots, T_n]/(g_j - g_i T_j)_{j \neq i} \, .$$

That we have to divide by the g_i-torsion corresponds to the fact that the sheaf of ideals $\mathscr{I}' = \mathscr{I} \cdot \mathcal{O}_{X'}$ is invertible on X'. Furthermore, one shows that X' is R-flat (i.e., has no π-torsion) if the same is true for X.

Returning to the case of a global R-scheme X, we set

$$X'_\pi := \{x \in X'; \mathscr{I}'_x \text{ is generated by } \pi\} \, ,$$

which is an open subscheme of X'. Over an affine open part $\operatorname{Spec} A$ of X, it consists of the affine A-scheme $\operatorname{Spec} A'_{(\pi)}$, where

$$A'_{(\pi)} = A\left[\frac{g_1}{\pi}, \ldots, \frac{g_n}{\pi}\right] \Big/ (\pi\text{-torsion}) \, .$$

So X'_π is always flat over R, even if X is not. Let $u : X'_\pi \longrightarrow X$ be the canonical morphism, and denote by an index k restrictions to special fibres. The pair (X'_π, u) has the following universal property:

If Z is a flat R-scheme, and if $v : Z \longrightarrow X$ is an R-morphism such that its restriction v_k to special fibres factors through Y_k, then v factors uniquely through u.

Indeed, since the problem is local on X and Z, we may assume that both schemes are affine, say $X = \operatorname{Spec} A$ and $Z = \operatorname{Spec} B$. Using notations as before, the fact that v_k factors through Y_k implies that the ideal $\mathfrak{J} \cdot B$ is contained in πB. Hence there exist elements $h_i \in B$ with $v^* g_i = \pi h_i$; the elements h_i are unique, because B has no π-torsion. Thus, the A-morphism $A[T_0, \ldots, T_n] \longrightarrow X$ sending T_i to h_i yields a morphism $w^* : A'_{(\pi)} \longrightarrow B$ and hence a morphism $w : Z \longrightarrow X'_\pi$ such that $v = u \circ w$.

We summarize what we have shown.

Proposition 1. *Let X be an R-scheme of finite type, let Y_k be a closed subscheme of its special fibre X_k, and let $\mathscr{I}$ be the sheaf of ideals of $\mathcal{O}_X$ defining Y_k. Let $X' \longrightarrow X$ be the blowing-up of Y_k on X, and let $u : X'_\pi \longrightarrow X$ denote its restriction to the open subscheme of X' where $\mathscr{I} \cdot \mathcal{O}_{X'}$ is generated by π. Then*

(a) *X'_π is a flat R-scheme, and $u_k : (X'_\pi)_k \longrightarrow X_k$ factors through Y_k.*

(b) *For any flat R-scheme Z and for any R-morphism $v : Z \longrightarrow X$ such that $v_k : Z_k \longrightarrow X_k$ factors through Y_k, there exists a unique R-morphism $v' : Z \longrightarrow X'_\pi$ such that $v = u \circ v'$.*

Due to property (b), the couple (X'_π, u) is unique (up to canonical isomorphism) in the class of all couples (Z, v) satisfying property (a). We call X'_π the *dilatation of* Y_k *on* X. It is clear that one can construct dilatations also for locally closed subschemes of X_k. We want to mention some elementary properties of dilatations.

Proposition 2. (a) *All dilatations factor through the largest flat R-subscheme of X, which is given by the ideal of π-torsion in* $\mathcal{O}_X$.

(b) *Dilatations commute with flat base change* $R \longrightarrow R'$ *where* R' *is a discrete valuation ring such that π is also a uniformizing element of* R'.

(c) *Let X be a closed subscheme of an R-scheme Z, and let* Y_k *be a closed subscheme of* X_k. *Then the dilatation* X'_π *of* Y_k *on X can be realized as a closed subscheme of the dilatation* Z'_π *of* Y_k *on Z.*

(d) *Dilatations commute with products: Let* X^i *be R-schemes, and let* Y_k^i *be subschemes of* X_k^i *for* $i = 1, 2$. *Then the dilatation of* $Y_k^1 \times_k Y_k^2$ *on* $X^1 \times_R X^2$ *is the product* $(X^1)'_\pi \times_R (X^2)'_\pi$ *of the dilatations of* Y_k^i *on* X^i. *In particular, if X is an R-group scheme, and if* Y_k *is a subgroup scheme of* X_k, *the dilatation* X'_π *of* Y_k *on X is an R-group scheme and the canonical map* $X'_\pi \longrightarrow X$ *is a group homomorphism.*

Finally we investigate how dilatations behave with respect to smoothness.

Proposition 3. *Let X be a smooth R-scheme, and let* Y_k *be a smooth k-subscheme of* X_k. *Then the dilatation* X'_π *of* Y_k *on X is smooth over R.*

Proof. Let $u: X'_\pi \longrightarrow X$ be the dilatation of Y_k on X, let x' be a point of the special fibre of X'_π, and set $x = u(x')$. Let n be the dimension of X_k at x, and let r be the dimension of Y_k at x. Let $\mathcal{J}$ be the sheaf of ideals of $\mathcal{O}_X$ defining Y_k, and let $\bar{\mathcal{J}} = \mathcal{J}/\pi\mathcal{O}_X$ denote the sheaf of ideals of $\mathcal{O}_{X_k}$ defining Y_k in X_k. Due to the Jacobi Criterion 2.2/7 there exist $\bar{f}_1, \ldots, \bar{f}_r \in \mathcal{O}_{X_k,x}$ and $\bar{g}_{r+1}, \ldots, \bar{g}_n \in \bar{\mathcal{J}}_x$ such that $\bar{f}_1, \ldots, \bar{f}_r, \bar{g}_{r+1}, \ldots, \bar{g}_n$ form a system of local coordinates of X_k at x (cf. 2.2/12), and such that $\bar{g}_{r+1}, \ldots, \bar{g}_n$ generate $\bar{\mathcal{J}}_x$. On an affine neighborhood U of x in X there exist liftings $f_i \in \mathcal{O}_X(U)$ of $\bar{f}_i$ and $g_j \in \mathcal{J}(U)$ of $\bar{g}_j$. Then $f_1, \ldots, f_r, g_{r+1}, \ldots, g_n$ form a system of local coordinates of X over R at x, and $\pi, g_{r+1}, \ldots, g_n$ generate $\mathcal{J}$ at x. From the local construction of X'_π we see that $df_1, \ldots, df_r, dg'_{r+1}, \ldots, dg'_n$ generate $\Omega^1_{X'_\pi/R}$ at x', where $g'_j \in \mathcal{O}_{X'_\pi,x'}$ satisfies $g_j = \pi g'_j$. Hence $\Omega^1_{X'_\pi/R}$ is generated by n elements at x'. Since the relative dimension of X'_π over R is at least n at x' (cf. [EGA IV_3], 13.1.3), it follows from 2.4/8 and 2.2/15 that X'_π is smooth over R at x'. □

3.3 Néron's Measure for the Defect of Smoothness

Throughout this section, let X be an R-scheme *of finite type* whose *generic fibre* X_K is *smooth* over K. Let a be an R^{sh}-valued point of X, and let a_K (resp. a_k) denote its generic (resp. special) fibre. Consider the pull-back $a^*\Omega^1_{X/R}$ of the $\mathcal{O}_X$-module of relative differential forms from X to Spec R^{sh}. By abuse of notation, we will identify it with its module of global sections. Thereby $a^*\Omega^1_{X/R}$ becomes an R^{sh}-module of finite type. Since R^{sh} is a discrete valuation ring, this module splits into a direct sum of a free part and of a torsion part. The rank of the free part is just the rank of $\Omega^1_{X/R}$ at a_K which is the dimension of X_K at a_K (since X_K is smooth at a_K). Looking at the torsion part, we define

$$\delta(a) := \text{length of the torsion part of } a^*\Omega^1_{X/R}$$

as *Néron's measure for the defect of smoothness at a*. First we want to show that, indeed, $\delta(a)$ provides a measure of how far X is from being smooth at a.

Lemma 1. *Let a be an R^{sh}-valued point of X. Then a factors through the smooth locus of X if and only if $\delta(a) = 0$.*

Proof. If a is contained in the smooth locus of X, then $\Omega^1_{X/R}$ is locally free at a_k and, hence, $a^*\Omega^1_{X/R}$ is free. Thus we have $\delta(a) = 0$. Conversely, if $\delta(a) = 0$, then $a^*\Omega^1_{X/R}$ can be generated by d elements, where d is the dimension of X_K at a_K. In particular, $\Omega^1_{X/R}$ and, hence, $\Omega^1_{X_k/k}$ can be generated by d elements at a_k. Since the relative dimension at a_k is at least d (cf. [EGA IV_3], 13.1.3), it follows from 2.2/15 that X_k is smooth over k at a_k of relative dimension d. Then X is smooth over R at a_k. This follows from 2.4/8, if it is known that X is R-flat at a_k. Avoiding the interference of flatness, one can proceed as follows. Choose a representation of a neighborhood $U \subset X$ of a_k as a closed subscheme of some $\mathbb{A}^n_R$. Due to the Jacobi Criterion 2.2/7(c), there exist local sections $g_{d+1}, \ldots, g_n$ on a neighborhood of $a_k \in \mathbb{A}^n_R$ which vanish on U, and which have the property that U_k is defined by $(\pi, g_{d+1}, \ldots, g_n)$ at a_k and that $dg_{d+1}, \ldots, dg_n$ generate a direct factor of $\Omega^1_{\mathbb{A}^n_R/R}$ at a_k. Then, in a neighborhood of a, the subscheme Z of $\mathbb{A}^n_R$ given by $g_{d+1}, \ldots, g_n$ is smooth of relative dimension d; furthermore locally at a, the scheme Z contains U as a closed subscheme. Thus, by reasons of dimension and of smoothness, the generic fibres U_K and Z_K coincide at a_K and, hence, U and Z coincide at a_k. □

The Jacobi Criterion provides a useful method to calculate $\delta(a)$. Namely, let $U \subset X$ be a neighborhood of a which can be realized as a closed subscheme of an R-scheme Z where Z is smooth over R and has constant relative dimension n. Assume that there exist $z_1, \ldots, z_n$ on Z such that $dz_1, \ldots, dz_n$ constitute a basis of $\Omega^1_{Z/R}$, and let $g_1, \ldots, g_m$ be functions on Z which generate the sheaf of ideals of $\mathcal{O}_Z$ defining U in Z. Representing the relative differentials dg_μ with respect to the basis $dz_1, \ldots, dz_n$, say

$$dg_\mu = \sum_{\nu=1}^{n} \frac{\partial g_\mu}{\partial z_\nu} dz_\nu ,$$

we define the Jacobi matrix of $g_1, \ldots, g_m$ by

$$J = \left(\frac{\partial g_\mu}{\partial z_\nu} \right)_{\substack{\mu=1,\ldots,m \\ \nu=1,\ldots,n}} .$$

If d is the relative dimension of X at a_K, we call Λ the set of all $(n-d)$-minors Δ of J. In this situation, Néron's measure for the defect of smoothness at a can be calculated from the minors $\Delta \in \Lambda$. To give a precise statement, let $v(r)$ denote the π-order of elements $r \in R$.

Lemma 2. $\delta(a) = \min\{v(a^*\Delta); \Delta \in \Lambda\}$.

Proof. Due to the Jacobi Criterion 2.2/7, there exists a minor $\Delta \in \Lambda$ with $a^*\Delta \neq 0$; any minor Δ' of J with more than $n-d$ rows will satisfy $a^*\Delta' = 0$. Furthermore, it follows from 2.1/2 that $a^*\Omega^1_{X/R}$ is representable as a quotient F/M, where $F := a^*\Omega^1_{Z/R}$ is a free R^{sh}-module of rank n, and where M is the submodule which is generated by $a^*dg_1, \ldots, a^*dg_m$. Since the rank of M is $(n-d)$, one can find a basis e_1, $\ldots$, e_n of F such that M is generated by elements $r_{d+1}e_{d+1}, \ldots, r_ne_n$ where $r_i \in R^{sh}$ and $r_i \neq 0$; this follows from the theory of elementary divisors. Thus the length of the torsion part of F/M, which is $\delta(a)$ by definition, is given by the formula

$$\delta(a) = v(r_{d+1}) + \ldots + v(r_n) .$$

Now consider the ideal in R^{sh} which is generated by all elements $a^*\Delta$, $\Delta \in \Lambda$; it equals the ideal generated by all values which are assumed on M by alternating $(n-d)$-forms on F. Obviously, this ideal is generated by the product $r_{d+1} \ldots r_n$, and there exists a minor $\Delta \in \Lambda$ with $(a^*\Delta) = (r_{d+1} \ldots r_n)$. Thus the assertion is clear. □

The method we have just used can easily show that $\delta(a)$ is bounded when a varies over the set of R^{sh}-valued points of X.

Proposition 3. *There exists an integer c such that $\delta(a) \leqq c$ for all $a \in X(R^{sh})$.*

Proof. Since an R-scheme of finite type is quasi-compact by definition, we may assume that X is an affine R-scheme Spec A. Choose a representation

$$A = R[z_1, \ldots, z_n]/(g_1, \ldots, g_m)$$

of A as a quotient of a free polynomial ring $R[z_1, \ldots, z_n]$. For integers d, let $(X_K)_d$ be the union of all irreducible components of dimension d of X_K. Then $(X_K)_d$ is non-empty for at most finitely many d and, since X_K is smooth, X_K is the disjoint sum of the $(X_K)_d$. Let X_d be the schematic closure of $(X_K)_d$ in X; i.e., let X_d be the subscheme of X which is defined by the kernel of the homomorphism $A \longrightarrow \mathcal{O}_X((X_K)_d)$. Let A_d be its ring of global sections. Considering the Jacobi matrix

$$J = \left(\frac{\partial g_\mu}{\partial z_\nu}\right)_{\substack{\mu=1,\ldots,m \\ \nu=1,\ldots,n}},$$

let Λ be the set of all $(n-d)$-minors Δ of J. Then, due to the Jacobi Criterion 2.2/7, we see for each $x \in (X_K)_d$ that there exists a minor $\Delta \in \Lambda$ satisfying $\Delta(x) \neq 0$. Hence the family $(\Delta)_{\Delta \in \Lambda}$ generates the unit ideal in $A_d \otimes_R K$. After chasing denominators, one can find elements $f_1, \ldots, f_t \in A_d$, minors $\Delta_1, \ldots, \Delta_t \in \Lambda$, as well as an integer $c \geqq 0$ such that

$$\sum_{i=1}^{t} f_i \Delta_i |_{X_d} = \pi^c .$$

Hence, by Lemma 2, we have $\delta(a) \leqq c$ for all $a \in X(R^{sh})$ whose generic fibre a_K belongs to $(X_K)_d$. Since only finitely many of the schemes $(X_K)_d$ are non-empty, we see that δ is bounded on $X(R^{sh})$. □

It follows that the function δ assumes its maximum on $X(R^{sh})$. The maximum of δ can be viewed as a global measure of how far X is from being smooth at the points of $X(R^{sh})$. Since we want to construct a smoothening of X by blowing up subschemes of X_k, we have to define suitable centers Y_k in the special fibre such that the defect of smoothness, i.e., the maximum of δ, decreases. Smooth R^{sh}-schemes have many sections (cf. 2.3/5). So it is natural to look at subschemes $Y_k \subset X_k$ such that there exist enough R^{sh}-valued points of X whose special fibres factor through Y_k. More precisely, if k_s denotes the residue field of R^{sh}, we will consider the following property (N) for couples (X, Y_k) consisting of an R-scheme X of finite type and of a closed subscheme $Y_k \subset X_k$:

(N) *The family of those k_s-valued points of Y_k, which lift to R^{sh}-valued points of X, is schematically dense in Y_k.*

For the notion of schematic density (more precisely, of schematic dominance) see [EGA IV$_3$], 11.10.2. In our situation the condition just means that the sheaf of $\mathcal{O}_X$-ideals defining Y_k equals the intersection of all kernels of morphisms $a^* : \mathcal{O}_X \longrightarrow a_* \mathcal{O}_{\operatorname{Spec} k_s}$, where a varies over the set of R^{sh}-valued points of X whose special fibres factor through Y_k.

Since the strict henselization R^{sh} is the limit over all local étale extensions R' of R, condition (N) is equivalent to the following condition: the set of closed points of Y_k which lift to R'-valued points of X for some local étale extension R' of R is schematically dense in Y_k. For example, if X is smooth over R, and if Y_k is a geometrically reduced closed subscheme of X_k, then it follows from 2.2/16, 2.2/13, and 2.2/14 that (X, Y_k) has the property (N).

Lemma 4. *If the couple (X, Y_k) satisfies property (N), then Y_k is geometrically reduced, and the smooth locus of the k-scheme Y_k is open and dense in Y_k.*

Proof. Property (N) yields that the k_s-valued points of Y_k are schematically dense in Y_k. Since k_s is a geometrically reduced k-algebra, Y_k is also geometrically reduced (cf. [EGA IV$_3$], 11.10.7). So the assertion follows from 2.2/16. □

Next we want to establish the key tool which is needed for the proof of Theorem 3.1/3. It provides us with a means of lowering the defect of smoothness of X so that eventually X becomes smooth at the points we are interested in.

Proposition 5. *Let Y_k be a closed subscheme of X_k such that the couple (X, Y_k) satisfies property (N). Let U_k be an open subscheme of Y_k such that U_k is smooth over k and such that the pull-back $\Omega^1_{X/R}|_{U_k}$ of $\Omega^1_{X/R}$ to U_k is locally free. Let $X'_\pi \longrightarrow X$ be the dilatation of Y_k in X and, for each $a \in X(R^{sh})$ with $a_k \in Y_k$, denote by $a' \in X'_\pi(R^{sh})$ the unique lifting of a. Then if $a \in X(R^{sh})$ specializes into a point of U_k, we have*

$$\delta(a') \leqq \max\{0, \delta(a) - 1\} .$$

In particular, $\delta(a') < \delta(a)$ for all R^{sh}-valued points a of X which specialize into points of U_k and which are not contained in the smooth locus of X.

First we want to look at an example which explains how the proposition works in a special situation. Let X be the closed subscheme of $\mathbb{A}^2_R = \operatorname{Spec} R[T_1, T_2]$ which is defined by the equation $T_1 T_2 = \pi^2$. Then X is affine, and its R-algebra of global sections is

$$A = R[T_1, T_2]/(T_1 T_2 - \pi^2) .$$

Let Y_k be the closed subscheme of X_k which is defined by (π, T_1, T_2); it consists of a single k-valued point. Using the R-morphism

$$A \longrightarrow R , \qquad T_1 \longmapsto \pi , \qquad T_2 \longmapsto \pi ,$$

this point lifts to an R-valued point of X. Hence (X, Y_k) satisfies property (N). Furthermore, an easy calculation shows $\delta(a) = 1$. The dilatation X'_π of Y_k in X is an affine A-scheme with coordinate ring

$$A' = A[T'_1, T'_2]/(T_1 - \pi T'_1, T_2 - \pi T'_2) = R[T'_1, T'_2]/(T'_1 T'_2 - 1) .$$

In particular, X'_π is smooth over R, and the lifting $a' \in X'_\pi(R^{sh})$ of a, which corresponds to the R-morphism

$$A' \longrightarrow R , \qquad T'_1 \longmapsto 1 , \qquad T'_2 \longmapsto 1 ,$$

fulfills $\delta(a') = 0$.

Proof of Proposition 5. Since the problem is local on X, it is enough to work in a neighborhood of a point $u \in U_k$. So we may assume that X is affine, say $X = \operatorname{Spec} A$, that U_k coincides with Y_k, and that the latter is irreducible. Let r be the dimension of Y_k. Then the sheaves $\Omega^1_{Y_k/k}$ and $\Omega^1_{X/R}|_{Y_k}$ are locally free and the first is obtained from the second one by dividing through the submodule which is generated by all differentials dg of functions $g \in A$ vanishing on Y_k (cf. 2.1/2). Shrinking X if necessary, we can assume that both sheaves are free and that there exist elements $\bar{y}_1, \ldots, \bar{y}_r$, $\bar{z}_1, \ldots, \bar{z}_n \in A$ having the following properties:

The differentials $d\bar{y}_1, \ldots, d\bar{y}_r$ give rise to a basis of $\Omega^1_{Y_k/k}$, the functions $\bar{z}_1, \ldots, \bar{z}_n$ vanish on Y_k, and $d\bar{y}_1, \ldots, d\bar{y}_r, d\bar{z}_1, \ldots, d\bar{z}_n$ give rise to a basis of $\Omega^1_{X/R}|_{Y_k}$.

It follows then from Nakayama's lemma that $\Omega^1_{X/R}$ is generated by $d\bar{y}_1, \ldots, d\bar{y}_r$, $d\bar{z}_1, \ldots, d\bar{z}_n$ at all points of Y_k. However, in general we will not have a basis, because $\Omega^1_{X/R}$ does not need to be locally free. Therefore we want to construct a closed embedding $X \hookrightarrow Z$ into a smooth R-scheme Z such that the above generators of $\Omega^1_{X/R}$ lift to a basis of $\Omega^1_{Z/R}$. This is possible after shrinking X.

Namely, represent A as a quotient of a free polynomial ring $R[T_1,\ldots,T_{r+n+m}]$ with respect to an ideal H and require that T_i is a lifting of $\bar{y}_i$ for $i = 1, \ldots, r$ and that T_{r+j} is a lifting of $\bar{z}_j$ for $j = 1, \ldots, n$. Since $\Omega^1_{X/R}|_{Y_k}$ is free of rank $r + n$, we know that $\Omega^1_{X/R} \otimes k(u)$ is of dimension $r + n$ over $k(u)$ where u is the point in Y_k around which we want to work. Hence there exist $h_1, \ldots, h_m \in H$ such that the Jacobi matrix

$$\left(\frac{\partial h_i}{\partial T_j}(u)\right)_{\substack{i=1,\ldots,m\\ j=1,\ldots,r+n+m}}$$

at u is of rank m. Writing Z for the closed subscheme of $\mathbb{A}_R^{r+n+m}$ which is defined by $h_1, \ldots, h_m$, we have closed immersions

$$Y_k \hookrightarrow X \hookrightarrow Z\ ,$$

where Z is smooth at u of relative dimension $r + n$. Let C be the R-algebra of global sections of $\mathcal{O}_Z$, and represent the algebras of global sections on Y_k and X as quotients of C; say $A = C/I$ with $I = \operatorname{Id}(X)$ and $B = C/J$ with $J = \operatorname{Id}(Y_k)$. So we know $I \subset J$. Furthermore, let $y_i \in C$ be the image of T_i for $i = 1, \ldots, r$, and $z_j \in C$ the image of T_{r+j} for $j = 1, \ldots, n$. Then y_i is a lifting of $\bar{y}_i \in A$, and the same is true for z_j and $\bar{z}_j$. Replacing Z by an affine open neighborhood of u, we may assume that Z is smooth over R of relative dimension $r + n$ and that $dy_1, \ldots, dy_r, dz_1, \ldots, dz_n$ form a basis of $\Omega^1_{Z/R}$. Also we may assume that Y_k, as a subscheme of Z, is defined by π, $z_1, \ldots, z_n$; i.e., that $J = (\pi, z_1, \ldots, z_n)$. Namely, these functions define a smooth k-subscheme Y_k' of Z of dimension r. Since Y_k is contained in Y_k' and since Y_k is smooth of dimension r, we have $Y_k = Y_k'$ locally at u.

Now we come to the key point of the proof. We claim $I \subset J^2$. This relation will enable us to give the desired estimate for the function δ, when X is replaced by the dilatation X'_π. So consider an element $f \in I$. Since $I \subset J$, we can write

$$f = g\pi + \sum_{i=1}^{n} g_i z_i$$

where $g, g_i \in C$. The differential df vanishes on X and hence on Y_k. Therefore we have

$$\sum_{i=1}^{n} g_i dz_i|_{Y_k} = df|_{Y_k} = 0\ .$$

Then $g_i|_{Y_k} = 0$, i.e., $g_1, \ldots, g_n \in J$, since $z_1, \ldots, z_n$ have been chosen in such a way that their differentials form part of a basis of $\Omega^1_{X/R}|_{Y_k}$. In particular, we can write f as

$$(*) \qquad f = g\pi + h$$

with

$$h = g_1 z_1 + \ldots + g_n z_n \in J^2$$

since $z_1, \ldots, z_n \in J$. For any $a \in X(R^{sh})$ with $a_k \in Y_k$, we know $h'(a) \equiv 0 \pmod{\pi}$ for all $h' \in J$. Therefore $h(a) \equiv 0 \pmod{\pi^2}$. On the other hand, we have $f(a) = 0$ for all $a \in X(R^{sh})$. Thus the equation $(*)$ implies $g(a) \equiv 0 \pmod{\pi}$ for all $a \in X(R^{sh})$ such that $a_k \in Y_k$. Since the couple (X, Y_k) satisfies property (N), this yields $g|_{Y_k} = 0$ and, hence, $g \in J$. So $I \subset J^2$ as claimed.

Next consider the dilatation X'_π of Y_k in X. It can be realized as a closed subscheme of the dilatation Z'_π of Y_k in Z. Giving a more precise description of these dilatations, we have $Z'_\pi = \operatorname{Spec} C'$ where

$$C' = C\left[\frac{z_1}{\pi}, \ldots, \frac{z_n}{\pi}\right],$$

and Z'_π is smooth over R, since Z is smooth over R (cf. 3.2/3). Writing $z'_j := \frac{z_j}{\pi}$, the differentials $dy_1, \ldots, dy_r, dz'_1, \ldots, dz'_n$ form a basis of $\Omega^1_{Z'_\pi/R}$. Then $X'_\pi = \operatorname{Spec} A'$ with $A' = C'/I'$, and the ideal $I' \subset C'$ is the smallest one such that I' contains the image of I and such that C'/I' has no π-torsion; i.e., I' consists of those elements $c' \in C'$ such that $\pi^\nu c' \in IC'$ for some $\nu \in \mathbb{N}$. Since $I \subset J^2$, any element $f \in I$ can be written as

$$(\dagger) \qquad f = \pi^2 f'$$

with $f' \in C'$; hence $f' \in I'$. The differential of f has a representation

$$df = \sum_{i=1}^{r} b_i dy_i + \sum_{j=1}^{n} c_j dz_j$$

in $\Omega^1_{Z/R}$, where $b_i, c_j \in C$. It implies the representation

$$df = \sum_{i=1}^{r} b_i dy_i + \sum_{j=1}^{n} \pi c_j dz'_j$$

in $\Omega^1_{Z'_\pi/R}$. Furthermore, we have a representation

$$df' = \sum_{i=1}^{r} b'_i dy_i + \sum_{j=1}^{n} c'_j dz'_j$$

in $\Omega^1_{Z'_\pi/R}$, where $b'_i, c'_j \in C'$. Then the relation $(\dagger)$ implies

$$(\dagger\dagger) \qquad b_i = \pi^2 b'_i \,, \qquad c_j = \pi c'_j \,,$$

since the dy_i, dz'_j form a basis of $\Omega^1_{Z'_\pi/R}$. Now choose a point $a \in X(R^{sh})$ with $a_k \in U_k = Y_k$ and let $a' \in X'_\pi(R^{sh})$ be the lifting of a. Let d be the dimension of X_K at a_K. In order to relate $\delta(a')$ to $\delta(a)$, we want to apply Lemma 2. So let $f_1, \ldots, f_l$ be generators of I. There exists an $(r + n - d)$-minor Δ of the Jacobi matrix

$$\left(\frac{\partial f_\lambda}{\partial y_i}, \frac{\partial f_\lambda}{\partial z_j}\right)_{\substack{\lambda=1,\ldots,l \\ i=1,\ldots,r;\, j=1,\ldots,n}}$$

such that $\delta(a) = v(\Delta(a_K))$. Then, using the equation $(\dagger)$, we can define elements $f'_\lambda \in I'$ by $f'_\lambda := \pi^{-2} f_\lambda$. Let Δ' be the minor of the Jacobi matrix

$$\left(\frac{\partial f'_\lambda}{\partial y_i}, \frac{\partial f'_\lambda}{\partial z'_j}\right)_{\substack{\lambda=1,\ldots,l \\ i=1,\ldots,r;\, j=1,\ldots,n}}$$

which corresponds to Δ. Then the relations $(\dagger\dagger)$ show that Δ' is obtained from Δ by multiplying each column of Δ with a factor π^{-1} or π^{-2}. Thus

$$v(\Delta'(a_K)) \leqq v(\Delta(a_K)) - (n + r - d)$$

and, hence,

$$\delta(a') \leqq \delta(a) - (n + r - d) .$$

If $n + r - d > 0$, the assertion of the proposition is clear. If $n + r = d$, the smooth R-scheme Z has relative dimension d, and this is just the dimension of X_K at a_K. So Z_K and X_K coincide on an open neighborhood of a_K. Since X is a closed subscheme of Z, and since Z_K is schematically dense in Z, we see that X coincides with Z locally at a. So a factors through the smooth locus of X, and $\delta(a) = 0$ in this case. □

We mention here that, as we have seen, the proof actually yields a better estimate for the defect of smoothness than the one stated in Proposition 5. For example, if X_K is equidimensional of dimension d, if Y_K is equidimensional of dimension r, and if $\Omega^1_{X/R}|_{Y_k}$ is locally free of rank $r + n$, then

$$\delta(a') \leqq \delta(a) - (n + r - d) .$$

3.4 Proof of the Theorem

In order to prove Theorem 3.1/3, let us fix the notation we will use. As in the preceding section, X is an R-scheme of finite type whose generic fibre X_K is smooth over K. Let E be a subset of $X(R^{sh})$. A closed subscheme Y_k of X_k is called *E-permissible* if the following conditions are satisfied:

(i) *The set of k_s-valued points of Y_k which lift to R^{sh}-valued points in E is schematically dense in Y_k; in particular, the couple (X, Y_k) has the property* (N).

(ii) *Let U_k be the largest open subscheme of Y_k which is smooth over k and where $\Omega^1_{X/R}|_{Y_k}$ is locally free. Then there is no k_s-valued point in $Y_k - U_k$ which lifts to a point in E.*

Note that the subscheme $U_k \subset Y_k$ of (ii) is always Zariski-dense in Y_k due to Lemma 3.3/4. Using the notion of E-permissible subschemes, we can formulate Proposition 3.3/5 in a more precise form.

Lemma 1. *Let Y_k be an E-permissible subscheme of X_k, and let $X' \longrightarrow X$ be the blowing-up of Y_k on X. For a point $a \in E$, denote by $a' \in X'(R^{sh})$ its (unique) lifting.*

(a) *If a does not specialize into a point of Y_k, then $\delta(a) = \delta(a')$.*

(b) *If a specializes into a point of Y_k, then $\delta(a') \leqq \max\{0, \delta(a) - 1\}$.*

Proof. If $a_k \notin Y_k$, there exists an open neighborhood of a over which the blowing-up is an isomorphism; hence $\delta(a) = \delta(a')$. If $a_k \in Y_k$, Proposition 3.2/1 shows that the point a' is necessarily contained in the dilatation X'_π of Y_k in X. Since X'_π is an open subscheme of X' and since Y_k is E-permissible in X, Proposition 3.3/5 yields the desired estimate for $\delta(a')$. □

If Y_k is E-permissible in X, the blowing-up $X' \longrightarrow X$ of Y_k on X is said to be *E-permissible*. For any blowing-up $X' \longrightarrow X$ of a subscheme of the special fibre X_k, one has a canonical bijection $X'(R^{sh}) \xrightarrow{\sim} X(R^{sh})$. So we may identify $E \subset X(R^{sh})$ with the corresponding subset of $X'(R^{sh})$. Hence we get the notion of E-permissible blowing-ups for X' again. This allows us to formulate a more precise version of Theorem 3.1/3.

Theorem 2. *Let X be an R-scheme of finite type with a smooth generic fibre X_K, and let E be a subset of $X(R^{sh})$. Then there exists a proper morphism $X' \longrightarrow X$ which consists of a finite sequence of E-permissible blowing-ups with centers contained in the non-smooth parts of the corresponding schemes, such that each R^{sh}-valued point $a \in E$ factors through the smooth locus of X'. In particular, if X is quasi-projective over R, so is X'.*

Proof. For a subset $E \subset X(R^{sh})$, we introduce the defect of smoothness of X along E by

$$\delta(X, E) := \max\{\delta(a); a \in E\} .$$

Due to Proposition 3.3/3, we know $\delta(X, E)$ is finite. So we can proceed by induction on $\delta(X, E)$. If $\delta(X, E) = 0$, then each $a \in E$ factors through the smooth locus of X (cf. Lemma 3.3/1), and the assertion is trivial. So let $\delta(X, E) > 0$. Since we consider only blowing-ups with centers in the non-smooth locus, we can remove from E all points which factor through the smooth locus of X, and thereby we may assume $\delta(a) > 0$ for all $a \in E$.

For the induction step, we have to arrange things in such a way that Lemma 1 can be applied. We do this by introducing a canonical partition of the set $E \subset X(R^{sh})$. First let us fix some notations. For a subset $F \subset X(R^{sh})$, we denote by F_k the subset of $X(k_s)$ which is induced from F by specialization. Identifying points in F_k with their associated closed points in X_k, let $\overline{F_k}$ be the Zariski closure of F_k in X_k, provided with the canonical reduced structure. Then $(X, \overline{F_k})$ satisfies property (N).

In order to obtain the desired partition of E, set $F^1 := E$ and $Y_k^1 := \overline{F_k^1}$. Let U_k^1 be the largest open subscheme of Y_k^1 which is smooth over k and where $\Omega^1_{X/R}|_{Y_k^1}$ is locally free, and define

$$E^1 := \{a \in F^1; a_k \in U_k^1\} .$$

Proceeding in the same way with $F^2 := F^1 - E^1$, and so on, we obtain

(i) a decreasing sequence $F^1 \supset F^2 \supset \ldots$ in $X(R^{sh})$,

(ii) subsets $E^1, E^2, \ldots \subset X(R^{sh})$ such that E decomposes into a disjoint union

$$E = E^1 \,\dot{\cup}\, \ldots \,\dot{\cup}\, E^i \,\dot{\cup}\, F^{i+1} ,$$

(iii) dense open subschemes $U_k^i \subset Y_k^i := \overline{F_k^i}$ such that $E_k^i \subset U_k^i$ and, moreover, $Y_k^{i+1} \subset Y_k^i - U_k^i$; in particular, $\dim Y_k^{i+1} < \dim Y_k^i$ if $Y_k^i \neq \emptyset$.

So we see that necessarily $Y_k^{t+1} = \emptyset$ for some $t \in \mathbb{N}$ big enough and, consequently, that $F^{t+1} = \emptyset$. Hence we have the partition

$$E = E^1 \,\dot{\cup}\, \ldots \,\dot{\cup}\, E^t .$$

Since each U_k^i is smooth over k, and since $\Omega^1_{X/R}|_{Y_k^i}$ is locally free on U_k^i, it follows that each Y_k^i is E^i-permissible, and that Y_k^t is, in fact, E-permissible. Furthermore, note that, in terms of subsets of X, each Y_k^i is disjoint from the smooth locus of X, since E_k and, hence, all F_k^i are disjoint from it, and since the non-smooth locus of X is a closed subset of X_k.

Now we can carry out the induction step. Let $X' \longrightarrow X$ be the blowing-up of Y_k^t on X. Then

$$\delta(X', E^t) < \delta(X, E^t)$$

by Lemma 1, because Y_k^t is E^t-permissible. Furthermore, due to the induction hypothesis, there exists a morphism $X'' \longrightarrow X'$ which consists of a sequence of E^t-permissible blowing-ups with centers contained in the non-smooth loci of the corresponding schemes, such that each $a \in E^t$, when viewed as an R^{sh}-valued point of X'', factors through the smooth locus of X''. Considering the composition $X'' \longrightarrow X' \longrightarrow X$, this modification does not affect the set $E - E^t$. So it is a sequence of E-permissible blowing-ups.

Writing $(E'')^i$ for the lifting of E^i to $X''(R^{sh})$, let us consider the partition

$$E'' = (E'')^1 \dot{\cup} \ldots \dot{\cup} (E'')^{t-1} ,$$

where E'' is obtained from the lifting of E by removing $(E'')^t$; i.e., by removing the set of points which factor through the smooth locus of X''. Then, obviously, this partition equals the canonical partition of E''. Since $\delta(X'', E'') \leqq \delta(X, E)$, a second induction on the length of such a partition yields a sequence of E''-permissible blowing-ups $X''' \longrightarrow X''$ with centers in non-smooth loci such that all points of E'', when viewed as R^{sh}-valued points of X''', factor through the smooth locus of X'''. Then

$$X''' \longrightarrow X'' \longrightarrow X' \longrightarrow X$$

is a sequence of E-permissible blowing-ups as desired. □

Remark 3. If in the situation of Theorem 2 it is not known that the generic fibre X_K is smooth, the assertion nevertheless remains true if the generic fibres of the points in E factor through the smooth locus of X_K and have a bounded defect of smoothness. Namely, these are the properties of E and X_K which are used in the proof.

3.5 Weak Néron Models

In the following let X_K be a smooth and separated K-scheme of finite type, and let K^{sh} be the field of fractions of a strict henselization R^{sh} of R. As a first step towards the construction of a Néron model of X_K, we want to look for a smooth and separated R-model of finite type, say X, such that each K^{sh}-valued point of X_K extends to an R^{sh}-valued point of X. We will see that such R-models X of X_K even satisfy certain aspects of the universal mapping property characterizing Néron models.

If X_K admits a separated R-model X of finite type such that the canonical map $X(R^{sh}) \longrightarrow X_K(K^{sh})$ is bijective, we can apply Corollary 3.1/4 to get a smooth R-model of the type we are looking for. For example, in the case of an abelian variety X_K we can proceed in this way, since there is a closed immersion $X_K \hookrightarrow \mathbb{P}^n_K$ into a projective space; we can take X to be the schematic closure of X_K in $\mathbb{P}^n_R$.

If it is only known that $X_K(K^{sh})$ is bounded in X_K, and if no separated R-model X of finite type such that $X(R^{sh}) \longrightarrow X_K(K^{sh})$ is bijective is given in an obvious way, we will consider a finite collection of separated R-models instead of a single one as before. Using the flattening techniques of Raynaud and Gruson [1], one can actually show that there exists a single separated R-model X of finite type such that each K^{sh}-valued point of X_K extends to an R^{sh}-valued point of X; we will give a sketch of proof in Proposition 6 below. But, for our purpose, it is not necessary to make use of this result, since we are mainly interested in group schemes X_K. Namely, in this case, it makes no difference if we start with a finite collection of R-models, since group arguments will help us later to reduce to a single R-model. As the second method is much more elementary, we will use it for our construction. We begin with a definition characterizing the collections of R-models of X_K we want to work with.

Definition 1. *A weak Néron model of X_K is a finite collection $(X_i)_{i \in I}$ of smooth and separated R-models of finite type such that each K^{sh}-valued point of X_K extends to an R^{sh}-valued point in at least one of these R-models.*

Theorem 2. *Let X_K be a smooth and separated K-scheme of finite type. If $X_K(K^{sh})$ is bounded in X_K, there exists a weak Néron model of X_K.*

Proof. Since $X_K(K^{sh})$ is bounded in X_K, it follows from 1.1/7 that there exists a finite family $(X_i)_{i \in I}$ of separated R-models of finite type such that each K^{sh}-valued point of X_K extends to an R^{sh}-valued point in at least one of these R-models. Applying Corollary 3.1/4 to each X_i, we obtain smooth and separated R-models X_i' of finite type such that the R^{sh}-valued points of X_i' and X_i correspond bijectively to each other. Hence $(X_i')_{i \in I}$ is a weak Néron model of X_K. □

Weak Néron models satisfy a certain mapping property which later leads to the universal mapping property characterizing Néron models.

Proposition 3 (Weak Néron Property). *Let $(X_i)_{i \in I}$ be a weak Néron model of X_K, and let Z be a smooth R-scheme with irreducible special fibre Z_k. Furthermore, let $u_K : Z_K \dashrightarrow X_K$ be a K-rational map. Then there exists an $i \in I$ such that u_K extends to an R-rational map $u : Z \dashrightarrow X_i$.*

Proof. There is an open dense subscheme $V_K \subset Z_K$ such that u_K is defined on V_K. Let F be the schematic closure of $F_K := Z_K - V_K$ in Z. Since we are working over a discrete valuation ring, F_k is nowhere dense in Z_k, and we may replace Z by $V := Z - F$ which is R-dense in Z. Thereby we may assume that u_K is defined on all of Z_K and thus is a K-morphism $Z_K \longrightarrow X_K$. Moreover, we may assume that Z is of finite type.

Consider the graph of u_K and denote its schematic closure in $Z \times_R X_i$ by Γ^i. Let $p_i : \Gamma^i \longrightarrow Z$ and $q_i : \Gamma^i \longrightarrow X_i$ be the projections. It is only necessary to show that, for some $i \in I$, the projection p_i is invertible on an R-dense open part of Z. Then $u := q_i \circ p_i^{-1} : Z \dashrightarrow X_i$ is a solution of our problem. One knows from Chevalley's theorem ([EGA IV_1], 1.8.4) that T_k^i, the image of Γ_k^i under p_i, is a constructible subset of Z_k, and we claim that, for some $i \in I$, the set T_k^i must contain a non-empty open part of Z_k. To verify this, we may assume $R = R^{sh}$, and hence, that k coincides with its separable algebraic closure. Then, by 2.2/13, the set of k-rational points is Zariski-dense in Z_k, and each $z_k \in Z_k(k)$ lifts to a point $z \in Z(R)$. Let $z_K \in Z(K)$ be the associated generic fibre, and set $x_K := u_K(z_K)$. By the definition of weak Néron models, there is an index $i \in I$ such that x_K extends to a point $x \in X_i(R)$. Consequently, we must have $(z, x) \in \Gamma^i(R)$ and thus $z_k \in T^i(k)$. This consideration shows that $\bigcup_{i \in I} T^i(k)$ is Zariski-dense in Z_k, and, since all T_k^i are constructible and I is finite, that there is some T_k^i containing a non-empty open part of Z_k.

Fixing such an index $i \in I$, let us consider the projection $p_i : \Gamma^i \longrightarrow Z$. The local ring $\mathcal{O}_{Z,\eta}$ at the generic point η of Z_k is a discrete valuation ring. Furthermore, as we have seen, there is a point $\xi \in \Gamma^i$ above η. Thus $\mathcal{O}_{\Gamma^i,\xi}$ dominates $\mathcal{O}_{Z,\eta}$. Since p_i is an isomorphism on generic fibres and since Γ^i is flat over R, both local rings give rise to the same field as total ring of fractions so that $\mathcal{O}_{Z,\eta} \longrightarrow \mathcal{O}_{\Gamma^i,\xi}$ is an isomorphism. Since Z and Γ^i are of finite type over R, there exist open neighborhoods U of η in Z and V of ξ in Γ^i such that p_i induces an isomorphism between U and V. Hence p_i is invertible over an R-dense open part of Z. □

Corollary 4. *Let Z be a smooth R-scheme, and let ζ be a generic point of the special fibre of Z. Denote by R' the local ring $\mathcal{O}_{Z,\zeta}$ of Z at ζ and by K' the field of fractions of R'. If $(X_i)_{i \in I}$ is a weak Néron model of X_K, then $(X_i \otimes_R R')_{i \in I}$ is a weak Néron model of $X_K \otimes_K K'$.*

Proof. Since the strict henselization of R' is a direct limit of étale extensions of R', it suffices to show that, for any étale Z-scheme Z', for any point ζ' of Z' above ζ, and for any K'-rational map $u'_{K'}$ from $Z'_{K'}$ to X_K, there exists an index $i \in I$ such that $u'_{K'}$ extends to a rational map $u' : Z' \dashrightarrow X_i$ which is defined at ζ'. Since ζ' is a generic point of the special fibre of Z', the assertion follows from Proposition 3. □

In the situation of Proposition 3, one cannot expect, in general, that the R-rational map $Z \dashrightarrow X$ is a morphism if $Z_K \dashrightarrow X_K$ is a morphism, even if the weak Néron model $(X_i)_{i \in I}$ of X_K consists of a single proper R-model of X_K. In particular, weak Néron models fail to be unique, even if one restricts to the class of weak Néron models consisting of a single R-model of X_K.

Example 5. Set $Z = X = \mathbb{P}_R^r$, the r-dimensional projective space over R, and consider a K-isomorphism $u_K : Z_K \xrightarrow{\sim} X_K$; i.e., a K-automorphism $u_K : \mathbb{P}_K^r \xrightarrow{\sim} \mathbb{P}_K^r$. Using a set of homogeneous coordinates $x_0, \ldots, x_r$ of $\mathbb{P}_R^r$, we can describe u_K by

$$x_i \longmapsto \sum_{j=0}^{r} a_{ij} x_j\,, \qquad i = 0, \ldots, r\,,$$

where $A := (a_{ij})$ is a matrix in $\mathrm{Gl}_{r+1}(K)$. We may assume that all coefficients a_{ij} belong to R. Then, by the theory of elementary divisors, there are matrices $S, T \in \mathrm{Gl}_{r+1}(R)$ and integers $0 \le n_0 \le \ldots \le n_r$ such that

$$\mathrm{SAT} = \begin{pmatrix} \pi^{n_0} & & 0 \\ & \ddots & \\ 0 & & \pi^{n_r} \end{pmatrix}.$$

Hence there exist sets of homogeneous coordinates $x_0, \ldots, x_r$ and $x'_0, \ldots, x'_r$ of $\mathbb{P}^r_R$ such that u_K is described by

$$x_i \longmapsto \pi^{n_i} x'_i\,, \qquad i = 0, \ldots, r\,,$$

where we may assume $n_0 = 0$.

If $n_0 = \ldots = n_r = 0$, it is clear that $u_K : \mathbb{P}^r_K \xrightarrow{\sim} \mathbb{P}^r_K$ extends to an automorphism $u : \mathbb{P}^r_R \xrightarrow{\sim} \mathbb{P}^r_R$. However, if $n_0 = \ldots = n_s = 0$ and $n_{s+1}, \ldots, n_r > 0$ for some $s < r$, then u_K extends only to an R-rational map $u : \mathbb{P}^r_R \dashrightarrow \mathbb{P}^r_R$. Namely, u is defined on the R-dense open subscheme $V \subset \mathbb{P}^r_R$ which consists of the generic fibre $\mathbb{P}^r_K$ and of the open part $V_k \subset \mathbb{P}^r_k$ complementary to the linear subspace Q_k where $x_0, \ldots, x_s$ vanish. In fact, if Q'_k is the linear subspace in $\mathbb{P}^r_k$ where $x'_{s+1}, \ldots, x'_r$ vanish, we can view u_k as a projection of $\mathbb{P}^r_k$ onto Q'_k with center Q_k.

Finally, as indicated at the beginning of this section, we want to show how, for a separated K-scheme X_K of finite type, one can always find a single separated R-model X of finite type such that $X(R^{sh}) \longrightarrow X_K(K^{sh})$ is bijective. The key fact which has to be established is the following result:

Proposition 6. *Let X_K be a separated (not necessarily smooth) K-scheme of finite type. Let $X_1, \ldots, X_n$ be separated R-models of X_K which are of finite type. Then there exist a separated R-model X of finite type of X_K and proper morphisms $X'_i \longrightarrow X_i$, $i = 1, \ldots, n$, consisting of finite sequences of blowing-ups with centers in the special fibres such that the given isomorphisms*

$$X'_i \otimes K \xrightarrow{\sim} X \otimes K$$

extend to open immersions $X'_i \hookrightarrow X$.

Thus, using the valuative criterion of properness, we obtain the desired characterization of boundedness.

Corollary 7. *$X_K(K^{sh})$ is bounded in X_K if and only if X_K admits a separated R-model X of finite type such that each K^{sh}-valued point of X_K extends to an R^{sh}-valued point of X.*

Before starting the proof, let us list some elementary facts we will need. Let U, U', V, V' be separated and flat R-schemes of finite type and, for shortness, let us refer here to an R-morphism $W \longrightarrow U$ as a *blowing-up* if it is a finite sequence of blowing-ups with centers in the special fibres; note that W is separated, flat, and of finite type if U is.

(a) Let $U' \longrightarrow U$ be a blowing-up, and let $U \hookrightarrow V$ be an open immersion. Then there exists a blowing-up $V' \longrightarrow V$ such that $U' \longrightarrow U$ is obtained from $V' \longrightarrow V$ by the base change $U \hookrightarrow V$.

Just extend the center of the blowing-up $U' \longrightarrow U$ to a subscheme of V and define V' by blowing up this subscheme in V.

(b) If $U'_i \longrightarrow U$, $i = 1, 2$, are blowing-ups, then there exists a commutative diagram of blowing-ups

$$\begin{array}{ccc} U' & \longrightarrow & U'_1 \\ \downarrow & & \downarrow \\ U'_2 & \longrightarrow & U\,. \end{array}$$

Namely, if $U'_i \longrightarrow U$ is the blowing-up of the ideal $\mathscr{I}_i$ of $\mathcal{O}_U$, $i = 1, 2$, then define U' as the blowing-up of $\mathscr{I}_1 \cdot \mathscr{I}_2$ on U. Note that U' is isomorphic to the blowing-up on U'_2 of the pull-back of $\mathscr{I}_1$ under $U'_2 \longrightarrow U$ and to the blowing-up on U'_1 of the pull-back of $\mathscr{I}_2$ under $U'_1 \longrightarrow U$.

(c) Let $f: U \longrightarrow V$ be a flat R-morphism such that f_K is an open immersion. Then f is an open immersion.

Let us justify the latter statement. Since f is open, we may assume f faithfully flat. Furthermore, it is enough to show that f is an open immersion after faithfully flat base change. So we may perform the base change $U \longrightarrow V$ and thereby assume that f has a section ε. Then it is to verify that ε is an isomorphism. We know already that ε is a closed immersion, since f is separated. Thus we have the canonical surjective map

$$\alpha : \mathcal{O}_U \longrightarrow \varepsilon_* \mathcal{O}_V .$$

Since f_K is an isomorphism, the kernel of $\alpha \otimes_R K$ vanishes. But $\mathcal{O}_U$ is flat over R, so the kernel of α must vanish identically. Then α is an isomorphism and, hence, ε is an isomorphism.

Finally we mention the technique of flattening by blowing up which will serve as a key point in the proof of Proposition 6; cf. Raynaud and Gruson [1], Thm. 5.2.2.

Let $f: U \longrightarrow V$ be an R-morphism such that f_K is flat. Then there exists a blowing-up $V' \longrightarrow V$ such that the strict transform $f' : U' \longrightarrow V'$ of f is flat.

Here U' is the schematic closure of U_K in $U \times_V V'$ (the strict transform of U), and f' is the restriction of $f \times_V \mathrm{id}_{V'}$ to U'.

Now let us give the *proof of Proposition* 6. By an induction argument, one reduces to the case where only two R-models X_1 and X_2 are given. Denote by Γ the schematic closure of the graph of the isomorphism $X_1 \otimes K \xrightarrow{\sim} X_2 \otimes K$ in $X_1 \times_R X_2$. Applying the flattening by blowing up, there exist blowing-ups $X'_i \longrightarrow X_i$, $i = 1, 2$, such that the strict transform $p'_i : \Gamma'_i \longrightarrow X'_i$ of the i-th projection $p_i : \Gamma \longrightarrow X_i$ is flat. Notice that the canonical map $\Gamma'_i \longrightarrow \Gamma$ is a blowing-up, too. Then, by (c), the map p'_i is an open immersion and, by (b), there is a commutative diagram of blowing-ups

$$\begin{array}{ccc} \Gamma'' & \longrightarrow & \Gamma'_1 \\ \downarrow & & \downarrow \\ \Gamma'_2 & \longrightarrow & \Gamma . \end{array}$$

Furthermore, since $p'_i : \Gamma'_i \longrightarrow X'_i$ is an open immersion, there exists a blowing-up $X''_i \longrightarrow X'_i$ such that $\Gamma'' \longrightarrow \Gamma'_i$ is obtained from $X''_i \longrightarrow X'_i$ by restriction to Γ'_i; see (a). Then $\Gamma'' \longrightarrow X''_i$ is an open immersion, and we can glue X''_1 and X''_2 along Γ''. Thereby we obtain an R-model X of X_K which is of finite type, and which contains X''_1 and X''_2 as open subschemes. Moreover, X is separated. Namely, let Γ^* be the schematic closure of the graph of the isomorphism $X''_1 \otimes K \xrightarrow{\sim} X''_2 \otimes K$ in $X''_1 \times_R X''_2$. Since Γ'' is flat over R, the canonical isomorphism $\Gamma'' \otimes K \longrightarrow \Gamma^* \otimes K$ extends by continuity to a morphism $\Gamma'' \longrightarrow \Gamma^*$. Similar arguments show that the canonical morphism $\Gamma^* \otimes K \longrightarrow \Gamma \otimes K$ extends to a morphism $\Gamma^* \longrightarrow \Gamma$. Then, due to its construction, the morphism $\Gamma'' \longrightarrow \Gamma$ is proper, and it follows from [EGA II], 5.4.3, that $\Gamma'' \longrightarrow \Gamma^*$ is proper. Thus Γ'' is closed in Γ^* and hence closed in $X''_1 \times_R X''_2$. Thereby it is seen that X is separated over R. □

3.6 Algebraic Approximation of Formal Points

Apart from its importance for the construction of Néron models, the smoothening process is also a necessary tool for the proof of M. Artin's approximation theorem, which will be the subject of this section. As a first step, we have to show that a smoothening $X' \longrightarrow X$ of an R-scheme X satisfies the lifting property not only for R'-valued points, where R' is étale over R, but even for a larger class of extensions R'/R. For example, we are concerned with the case where R' is the π-adic completion $\hat{R}$ of R.

Definition 1. *A flat local extension $R \longrightarrow R'$ of discrete valuation rings is said to have ramification index* 1 *if a uniformizing element π of R induces a uniformizing element of R', and if the extension of the residue fields $k' = R'/\pi R'$ over $k = R/\pi R$ is separable.*

Recall that an extension of fields k'/k is separable if and only if $k' \otimes_k l$ is reduced for all fields l over k; cf. Bourbaki [1], Chap. VIII, § 7, n°3.

To illustrate the definition, we mention that the π-adic completion $\hat{R}$ of R has ramification index 1 over R. Furthermore, if R' is essentially of finite type over R, it has ramification index 1 over R if and only if R' is a local ring of a smooth R-scheme at a generic point of the special fibre. In this case, $R \longrightarrow R'$ or, better, the morphism $\operatorname{Spec} R' \longrightarrow \operatorname{Spec} R$ is regular in the sense of [EGA IV_2], 6.8.1. The class of ring extensions of ramification index 1 is stable under the formation of direct limits and completions.

If $R \longrightarrow R'$ has ramification index 1 and if, in addition, the extension of fields of fractions K'/K is separable, the extension R'/R is regular. For example, the extension $\hat{R}/R$ is regular or, equivalently, the extension of fields of fractions $Q(\hat{R})/Q(R)$ is separable, if and only if R is excellent (cf. [EGA IV_2], 7.8.2).

Lemma 2. *Let R be an excellent discrete valuation ring. If $R \longrightarrow R'$ has ramification index* 1*, then $R \longrightarrow R'$ is regular. In particular, since the completion of R' is of ramification index* 1 *over R, it follows that R' is excellent.*

Proof. Let K (resp. K') be the field of fractions of R (resp. R'). We have only to prove that K' is separable over K. So we may assume $p = \operatorname{char} K > 0$. It suffices to show that $L \otimes_K K'$ is reduced for each finite radicial extension L of K; cf. [EGA IV_2], 6.7.7. Let us first consider the case where the extension L/K is radicial of degree p. Since R is excellent, the integral closure $\tilde{R}$ of R in L is an R-module of finite type (cf. [EGA IV_2], 7.8.3) and, hence, a free R-module of rank p. Moreover, $\tilde{R}$ is a discrete valuation ring. So let $\tilde{k}$ be the residue field of $\tilde{R}$. If the degree of $\tilde{k}$ over k is p, then π is a uniformizing element of $\tilde{R}$, and $\tilde{R} \otimes_R R'/(\pi)$ is isomorphic to $\tilde{k} \otimes_k k'$. The latter is a field, since k' is separable over k and since $\tilde{k}$ is radicial over k; hence $\tilde{R} \otimes_R R'$ is a discrete valuation ring with uniformizing element π. If $\tilde{k} = k$, the p-th power of a uniformizing element $\tilde{\pi}$ of $\tilde{R}$ gives rise to a uniformizing element of R, and $\tilde{R} \otimes_R R'$ is a discrete valuation ring with uniformizing element $\tilde{\pi} \otimes 1$. In both cases, $\tilde{R} \otimes_R R'$ is a discrete valuation ring. Considering its field of fractions, it follows that $L \otimes_K K'$ is reduced. Since a finite radicial extension can be broken up into radicial subextensions of degree p, the same assertion remains true for arbitrary radicial extensions L of K. □

We mention that the ring of integers $\mathbb{Z}$ as well as all fields are excellent and that any R-algebra which is essentially of finite type over an excellent ring R is excellent; see [EGA IV_2], 7.8.3 and 7.8.6.

We want to show that smoothenings are compatible with ring extensions R'/R of ramification index 1. In order to do this, certain parts of the smoothening process have to be generalized. So let X be an R-scheme of finite type, and let R'/R be a ring extension of ramification index 1. Let a be an R'-valued point of X such that its generic fibre a_K factors through the smooth locus of the generic fibre X_K. Then, as in 3.3, we set

$$\delta(a) := \text{length of the torsion part of } a^*\Omega^1_{X/R} .$$

Without changes, the proof of 3.3/1 shows that $\delta(a) = 0$ if and only if a factors through the smooth locus of X. Furthermore, the key proposition of the smoothening process remains valid:

Proposition 3. *Let Y_k be the schematic closure of a_k in X_k. Let $X'_\pi \longrightarrow X$ be the dilatation of Y_k in X, and denote by a' the (unique) lifting of a to an R'-valued point of X'_π. Then $\delta(a') \leq \max\{0, \delta(a) - 1\}$.*

Literally the same proof as the one of 3.3/5 works in this case; namely, one has only to observe the fact that a_k factors through the smooth locus of the k-scheme Y_k. Since Y_k is geometrically reduced, the generic point of Y_k, which is a_k, is contained in the smooth locus of the k-scheme Y_k; cf. 2.2/16. Applying Proposition 3 finitely many times, one obtains an analogue of 3.1/3.

Proposition 4. *Let X be an R-scheme of finite type, and consider an extension R'/R of ramification index 1. Let a be an R'-valued point of X such that a_K factors through the smooth locus of X_K. Then there exists an R-morphism $X' \longrightarrow X$, which consists of a finite sequence of dilatations with centers in special fibres, such that a lifts to an R'-valued point of X' which factors through the smooth locus of X'.*

Proposition 4 enables us to show that smoothenings are compatible with ring extensions R'/R of ramification index 1. One has only to justify the following fact.

Lemma 5. *Let X be an R-scheme of finite type with smooth generic fibre, let $X' \longrightarrow X$ be a smoothening of X, and consider an extension R'/R of ramification index 1. Then each R'-valued point a of X lifts to an R'-valued point a' of X' which factors through the smooth locus of X'.*

Proof. Due to the properness of $X' \longrightarrow X$, the point $a \in X(R')$ lifts to a point $a' \in X'(R')$. Due to Proposition 4, there exists a finite sequence of dilatations $\sigma : X'' \longrightarrow X'$ such that σ is an isomorphism on generic fibres and such that the (unique) lifting a'' of a' factors through the smooth locus of X''. Since the schematic closure A''_k of a''_k in X''_k is geometrically reduced and, hence, generically smooth over k by 2.2/16, the set of those closed points $x \in A''_k \cap X''_{\text{smooth}}$ which have a separable residue field $k(x)$ over k is dense in A''_k; cf. 2.2/13. Since all these points lift to R^{sh}-valued points of X'', the image of a''_k in X', which equals a'_k, is contained in the smooth locus of X' (because X' is a smoothening of X). □

Corollary 6. *Let X be an R-scheme of finite type with a smooth generic fibre, let $X' \longrightarrow X$ be a smoothening of the R-scheme X, and consider an extension R'/R of ramification index 1. Then $X' \otimes_R R' \longrightarrow X \otimes_R R'$ is a smoothening of the R'-scheme $X \otimes_R R'$.*

Proof. Since $R \longrightarrow (R')^{sh}$ has ramification index 1, the assertion follows from Lemma 5. □

Using the preceding result and the existence of Nagata compactifications (Nagata [1] and [2]) for separated schemes of finite type over R, we can generalize 3.5/4 and show that weak Néron models are stable under extensions R'/R of ramification index 1. As usual, fields of fractions are denoted by K, residue fields by k, and strict henselizations by an index "*sh*".

Proposition 7. *Let X_K be a smooth K-scheme of finite type admitting a weak Néron model $(X_i)_{i \in I}$ over R. Let R'/R be of ramification index 1. Then $(X_i \otimes_R R')_{i \in I}$ is a weak Néron model of $X_{K'}$ over R'.*

Proof. Using 3.5/6, one easily reduces to the case where the index set I consists of a single element. So let X be a smooth and separated R-model of finite type of X_K such that the canonical map $X(R^{sh}) \longrightarrow X(K^{sh})$ is bijective, and consider a K'-valued point of X_K; i.e., a K-morphism $a_K : \operatorname{Spec} K' \longrightarrow X_K$. We have to show that a_K extends to an R-morphism $a : \operatorname{Spec} R' \longrightarrow X$. In order to do this, let $\bar{X}$ be a Nagata compactification of X. The latter is a proper R-scheme containing X as a dense open subscheme. Since X is flat over R, we see that X_K is dense in X and, hence, that X_K is dense in $\bar{X}_K$.

By the properness of $\bar{X}$, the morphism a_K extends to an R-morphism $\bar{a} : \operatorname{Spec} R' \longrightarrow \bar{X}$ such that the image of the generic point of $\operatorname{Spec} R'$ is contained in X_K and, thus, in the smooth locus of $\bar{X}_K$. So we can apply Proposition 4 and thereby find a finite sequence of dilatations $\bar{X}' \longrightarrow \bar{X}$ with centers in special fibres such that $\bar{a}$ lifts to an R'-valued point $\bar{a}'$ of the smooth locus of $\bar{X}'$. Similarly as in the proof of Lemma 5, let A_k be the schematic image of the special fibre of $\bar{a}'$ in the special fibre of $\bar{X}'$. Since A_k is generically smooth over k, the set E_k of its closed points x_k which have separable residue field $k(x_k)$ and which belong to the smooth part of $\bar{X}'$ is dense in A_k.

All points $x_k \in E_k$ lift to R^{sh}-valued points of $\bar{X}'$ by 2.2/14, and we claim that the liftings can be chosen in such a way that their generic fibres factor through X_K. Namely, as in the proof of 2.2/14, one uses the Jacobi Criterion 2.2/7 in order to construct local coordinates $g_1, \ldots, g_n$ in a neighborhood $U \subset \bar{X}'$ of x_k which, on the special fibre, generate the ideal of x_k. The g_i give rise to an étale morphism $g : U \longrightarrow \mathbb{A}^n_K$. Since the image of $\bar{X}_K - X_K$ under g is thin in $\mathbb{A}^n_K$, it follows that x_k can be lifted to a point $x \in \bar{X}'(R^{sh})$ whose generic fibre belongs to $X_K(K^{sh})$ as claimed.

Now, composing each such $x \in \bar{X}'(R^{sh})$ with the morphism $\bar{X}' \longrightarrow \bar{X}$, we obtain a set of points $F \subset \bar{X}(R^{sh})$ whose generic fibres belong to X_K and whose special fibres are dense in A_k. But then, since X is a weak Néron model of X_K, we must have $F \subset X(R^{sh})$, and it follows that the generic point of A_k belongs to X. Consequently, the R-morphism $\bar{a} : \operatorname{Spec} R' \longrightarrow \bar{X}$ factors through X giving rise to the desired extension of $a_K : \operatorname{Spec} K' \longrightarrow X_K$. □

For the remainder of this section, we will be concerned with approximation theory. Let A be a local noetherian ring with maximal ideal $\mathfrak{m}$, and denote by $\hat{A}$ its $\mathfrak{m}$-adic completion. We say A satisfies the *approximation property* if, for each

A-scheme X of finite type and for each $\hat{A}$-valued point $\hat{a}$ of X, there exists an A-valued point a of X such that the diagram

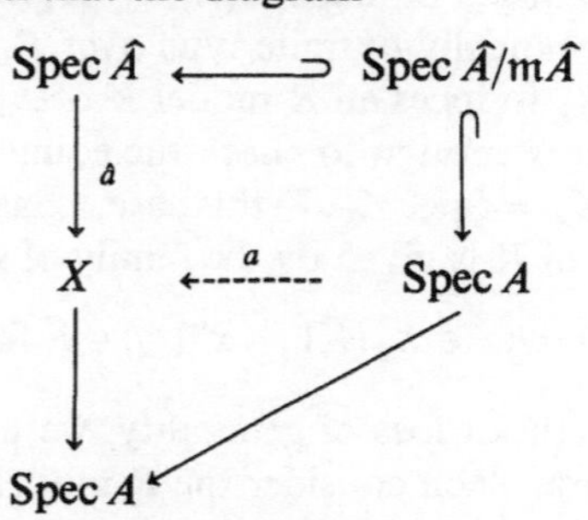

is commutative. Since $\hat{A}$ is henselian, it is clear by Definition 2.3/1′ that A is henselian if it satisfies the approximation property. Morever, if A is henselian, we see from 2.3/5 that, for each $\hat{A}$-valued point $\hat{a}$ of X which factors through the smooth locus of X, there exists an A-valued point of X which coincides with $\hat{a}$ on Spec $\hat{A}/m\hat{A}$.

Using the smoothening process, it is easy to verify the approximation property for discrete valuation rings which are henselian and excellent, as can be seen from the following proposition.

Proposition 8. *Let R be an excellent discrete valuation ring, and let $\hat{R}$ be its completion. Furthermore, let X be an R-scheme of finite type, and let σ be an $\hat{R}$-valued point of X. Then there exists a commutative diagram of R-morphism*

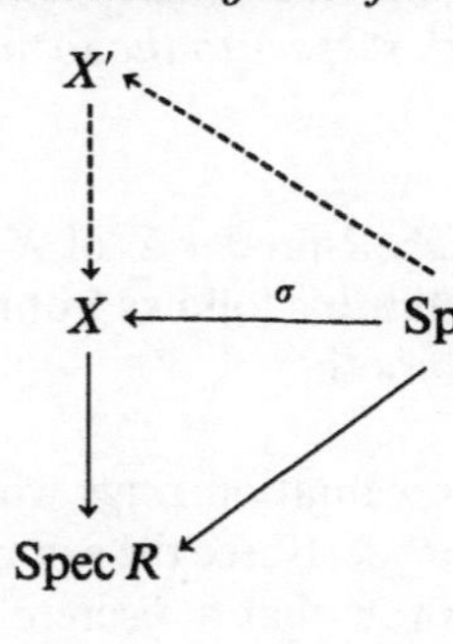

where X' is smooth over R.

Proof. We may assume that σ is schematically dense in X. Since R is excellent, the generic fibre X_K is geometrically reduced and, hence, smooth at the generic point; cf. 2.2/16. So σ_K factors through the smooth locus of X_K and the assertion follows from Proposition 4. □

Corollary 9. *Let R be a discrete valuation ring which is henselian and excellent. Then R satisfies the approximation property.*

In the following we denote by $\hat{K}$ the field of fractions of $\hat{R}$. If X_K is a K-scheme which is locally of finite type, we can provide $X_K(\hat{K})$ with the canonical topology, which is induced by the valuation on K. We claim that this topology coincides with the one generated by all images of maps $X(\hat{R}) \longrightarrow X_K(\hat{K})$, where X varies over all R-models of X_K which are locally of finite type over R. Namely, each R-model U of an open subset $U_K \subset X_K$ induces an R-model X of X_K by gluing U and X_K over U_K. Since $X(\hat{R}) = U(\hat{R})$, it is enough to check the equality of the topologies for an affine K-scheme X_K, say $X_K = \operatorname{Spec} A_K$. In this case, a basis of the topology of $X_K(\hat{K})$ induced by the valuation of K is given by the family of subsets of type

$$U(g_1,\ldots,g_r) = \{x \in X_K(\hat{K}) \,;\quad x^*(g_i) \in \hat{R} \text{ for } i = 1,\ldots,r\}$$

where $g_1, \ldots, g_r \in A_K$. Without loss of generality, we may assume that $g_1, \ldots, g_r$ generate A_K as a K-algebra. Then consider the R-model $X = \operatorname{Spec} A$ of X_K, where A is the image of the R-morphism

$$R[T_1,\ldots,T_r] \longrightarrow A_K\,, \qquad T_i \longmapsto g_i\,.$$

It follows that $U(g_1,\ldots,g_r)$ is the image of $X(\hat{R}) \longrightarrow X_K(\hat{K})$. Conversely, let X be an R-model of locally finite type of X_K. It remains to show that the image of $X(\hat{R}) \longrightarrow X_K(\hat{K})$ is open in $X_K(\hat{K})$. We may assume that X is affine, say $X = \operatorname{Spec} A$. Let $h_1, \ldots, h_r$ generate A as an R-algebra and denote by g_i the pull-back of h_i to X_K. Then the image of $X(\hat{R}) \longrightarrow X_K(\hat{K})$ coincides with the set $U(g_1,\ldots,g_r)$ (as defined above) and, hence, is open in $X_K(\hat{K})$.

Corollary 10. *Let R be a henselian discrete valuation ring and let X_K be a K-scheme which is locally of finite type. Assume either that R is excellent or that X_K is smooth. Then $X_K(K)$ is dense in $X_K(\hat{K})$ with respect to the topology induced by the valuation of K.*

Proof. It suffices to show that each R-model X of X_K which admits an $\hat{R}$-valued point admits an R-valued point. But this follows from Corollary 9 if R is excellent, and from Proposition 4 if X_K is smooth. □

There are examples of discrete valuation rings which are henselian, but which do not satisfy the approximation property; see the example below. Such rings cannot be excellent. In fact, it is easy to show that a discrete valuation ring R is excellent if it satisfies the approximation property. Thus, the approximation property for R is equivalent to the fact that R is henselian and excellent.

Example 11. Let $k = \mathbb{F}_p$ be the prime field of characteristic $p > 0$, and let Λ be the localization of the polynomial ring $k[T]$ at the maximal ideal generated by T. The completion $\hat{\Lambda}$ of Λ with respect to T is the ring $k[[T]]$ of formal power series. Looking at the cardinality of $k[[T]]$ (resp. of $k[T]$), it is clear that the extension $k((T))/k(T)$ of the fields of fractions is not algebraic. So pick an element $\xi \in \hat{\Lambda}$ which is not algebraic over $k(T)$. Set $U = \xi^p$, and let L be the field generated by T and U over k. Now define R as the intersection of L with $\hat{\Lambda}$. Then R is a discrete valuation ring whose completion $\hat{R}$ coincides with $k[[T]]$. Furthermore, $\hat{K} = Q(\hat{R})$ is not separable over $K = Q(R)$ since $\xi \in \hat{R} - K$. So R is not excellent. The henselization R^h of R can be viewed as the set of all elements of $k[[T]]$ which are separably algebraic over K. In particular, ξ is not contained in R^h, and it is easily verified that R^h does not satisfy the approximation property.

Next we want to generalize Proposition 8 to the case where the base consists of a polynomial ring over an excellent discrete valuation ring. The resulting assertion will be crucial in the proof of M. Artin's approximation theorem.

Theorem 12. *Let R be an excellent discrete valuation ring, and denote by $\hat{R}$ its π-adic completion. Let $T_1, \ldots, T_n$ be variables, and set*

$$S = \operatorname{Spec} R[T_1, \ldots, T_n]\,,$$

$$\hat{S} = \operatorname{Spec} \hat{R}[[T_1, \ldots, T_n]]\,.$$

Let X be an S-scheme of finite type, and let σ be an $\hat{S}$-valued point of X. Then there exists a commutative diagram of S-morphisms

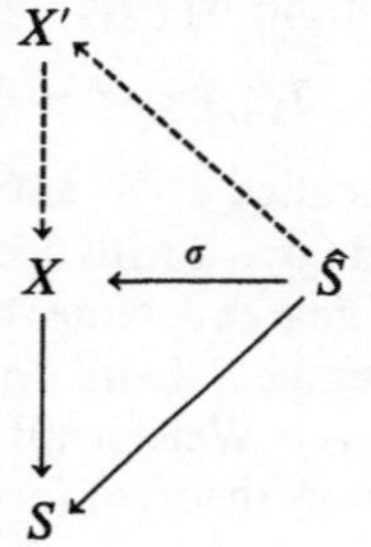

where X' is smooth over S.

The *proof* is done by induction on the number n of variables $T_1, \ldots, T_n$. The case $n = 0$ is settled by Proposition 8. So let $n > 0$. We may assume that X is a closed subscheme of $\mathbb{A}_S^N$ and that X is defined by global sections of $\mathcal{O}_{\mathbb{A}_S^N}$, say

$$X = V(f_1, \ldots, f_r) \subset \mathbb{A}_S^N\,;$$

the coordinate functions of $\mathbb{A}_S^N$ will be denoted by $Y_1, \ldots, Y_N$. Let η (resp. $\hat{\eta}$) be the generic point of the special fibre of S (resp. $\hat{S}$), let $\hat{s}$ be the closed point of $\hat{S}$, and let s be its image in S.

In order to carry out the induction step, we will establish three lemmata, the first and the third one under the assumption of the induction hypothesis; i.e., under the assumption that Theorem 12 is true for less than n variables.

Lemma 13. *Let f_0 be a global section of $\mathcal{O}_{\mathbb{A}_S^N}$ such that $\sigma^* f_0$ does not vanish at $\hat{\eta}$. Then there exists a commutative diagram of S-morphisms*

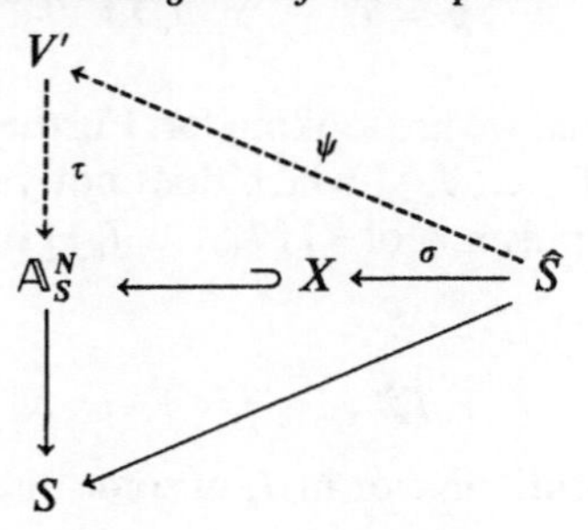

such that V' is smooth over S and such that τ^*f_0 divides each τ^*f_i, $i = 1, \ldots, r$, in $\Gamma(V', \mathcal{O}_{V'})$.

In the proof of the lemma, we will use Weierstraß division for the formal power series ring $\hat{R}[[T_1, \ldots, T_n]]$; cf. Bourbaki [2], Chap. VII, § 3, n°8. Let us first recall some basic facts of this theory. An element $f \in \hat{R}[[T_1, \ldots, T_n]]$ is called a *Weierstraß divisor in T_n of degree $d \geqq 0$* if the coefficients $a_\nu \in \hat{R}[[T_1, \ldots, T_{n-1}]]$ of the power series expansion

$$f = \sum_{\nu=0}^{\infty} a_\nu T_n^\nu$$

satisfy the conditions

(1) $$a_d \text{ is a unit in } \hat{R}[[T_1, \ldots, T_{n-1}]] ,$$

(2) $$a_\delta \in (\pi, T_1, \ldots, T_{n-1}) \text{ for } \delta = 0, \ldots, d-1 .$$

An element of $\hat{R}[[T_1, \ldots, T_n]]$ is called a *Weierstraß polynomial in T_n of degree d* if it is a monic polynomial in T_n of degree d with coefficients in $\hat{R}[[T_1, \ldots, T_{n-1}]]$ and if it is a Weierstraß divisor in T_n of degree d. Note that an element $f \in \hat{R}[[T_1, \ldots, T_n]]$ is a Weierstraß divisor in T_n of degree d if and only if the reduction of f modulo π, as an element of $k[[T_1, \ldots, T_n]]$, is a Weierstraß divisor in T_n of degree d. Since $\hat{R}$ is complete, the Weierstraß division theorem for $k[[T_1, \ldots, T_n]]$ lifts to a division theorem for $\hat{R}[[T_1, \ldots, T_n]]$:

If $f \in \hat{R}[[T_1, \ldots, T_n]]$ is a Weierstraß divisor in T_n of degree d, then $\hat{R}[[T_1, \ldots, T_n]]$ decomposes into a direct sum

(*) $$\hat{R}[[T_1, \ldots, T_n]] = \bigoplus_{\delta=0}^{d-1} \hat{R}[[T_1, \ldots, T_{n-1}]] T_n^\delta \oplus \hat{R}[[T_1, \ldots, T_n]] \cdot f$$

of $\hat{R}[[T_1, \ldots, T_{n-1}]]$-modules. Furthermore, f can be written as a product of a unit in $\hat{R}[[T_1, \ldots, T_n]]$ and a Weierstraß polynomial of degree d.

The last assertion follows easily if one applies the decomposition (*) to the element T_n^d, say

$$T_n^d = \sum_{\delta=0}^{d-1} a_\delta T_n^\delta + u \cdot f .$$

Then u is a unit, and

$$p = T_n^d - \sum_{\delta=0}^{d-1} a_\delta T_n^\delta$$

is the Weierstraß polynomial we are looking for. Further, we want to mention that, for each element $f \in \hat{R}[[T_1, \ldots, T_n]]$ which does not vanish identically modulo π, there exists an $\hat{R}$-automorphism φ of $\hat{R}[[T_1, \ldots, T_n]]$ of type

$$T_n \longmapsto T_n$$

$$T_i \longmapsto T_i + T_n^{b_i} , \qquad i = 1, \ldots, n-1 ,$$

such that $\varphi(f)$ is a Weierstraß divisor in T_n of some degree $d \geq 0$.

Proof of Lemma 13. If $\sigma^* f_0$ is a unit, then f_0 is invertible in a neighborhood of $\sigma(\hat{s})$ and, hence, the assertion is obvious. So we may assume that $\sigma^* f_0$ is not a unit. Since $\sigma^* f_0$ does not vanish at $\hat{\eta}$, there exists an $\hat{R}$-automorphism of $\hat{R}[[T_1,\ldots,T_n]]$ of type

$$T_n \longmapsto T_n\,, \qquad T_i \longmapsto T_i + T_n^{b_i}\,, \qquad i = 1,\ldots,n-1\,,$$

such that $\sigma^* f_0$ will be transformed by this automorphism into a Weierstraß divisor of degree $d \geq 1$. So we may assume that $\sigma^* f_0$ is a Weierstraß divisor of degree $d \geq 1$. Then $\sigma^* f_0$ can uniquely be written as

$$\sigma^* f_0 = \hat{u} \cdot \hat{p}$$

with a Weierstraß polynomial

$$\hat{p} = T_n^d + a'_{d-1} T_n^{d-1} + \ldots + a'_0 \in \hat{R}[[T_1,\ldots,T_{n-1}]][T_n]$$

of degree d and a unit $\hat{u}$ in $\hat{R}[[T_1,\ldots,T_n]]$. The Weierstraß division theorem yields a decomposition of $\hat{R}[[T_1,\ldots,T_n]]$ into a direct sum

$$(*) \qquad \hat{R}[[T_1,\ldots,T_n]] = \bigoplus_{\delta=0}^{d-1} \hat{R}[[T_1,\ldots,T_{n-1}]]\,T_n^\delta \oplus \hat{R}[[T_1,\ldots,T_n]] \cdot \hat{p}$$

of $\hat{R}[[T_1,\ldots,T_{n-1}]]$-modules. We will use the decomposition $(*)$ in order to make the application of the induction hypothesis possible. First we want to construct an auxiliary S-scheme V as a subscheme of $\mathbb{A}_S^{N'}$, where

$$N' = N \cdot d + d + N\,.$$

Let

$$\begin{aligned} &Y_{\nu\delta}\,; \qquad \nu = 1,\ldots,N\,, \qquad \delta = 0,\ldots,d-1\,,\\ &A_\delta\,; \qquad \delta = 0,\ldots,d-1\,,\\ &Z_\nu\,; \qquad \nu = 1,\ldots,N\,, \end{aligned}$$

be the coordinate functions of $\mathbb{A}_S^{N'}$ so that $\mathbb{A}_S^{N'} = \operatorname{Spec} R[T_\mu, Y_{\nu\delta}, A_\delta, Z_\nu]$. Consider the polynomial

$$p = T_n^d + A_{d-1} T_n^{d-1} + \ldots + A_0$$

and define an S-morphism $\tau : \mathbb{A}_S^{N'} \longrightarrow \mathbb{A}_S^N$ by setting

$$\tau^* Y_\nu = \sum_{\delta=0}^{d-1} Y_{\nu\delta} T_n^\delta + Z_\nu p$$

for $\nu = 1,\ldots,N$. Then Euclid's division yields unique decompositions

$$(**) \qquad \tau^* f_i = \sum_{\delta=0}^{d-1} f_{i\delta} T_n^\delta + q_i \cdot p\,, \qquad i = 0,\ldots,r\,,$$

in $\mathcal{O}_{\mathbb{A}_S^{N'}}$ where $f_{i\delta}$ is independent of T_n for all i and δ. Furthermore, each $f_{i\delta}$ is independent of $Z_1,\ldots,Z_N$ by the definition of τ. Thus we have

$$f_{i\delta} \in R[T_\mu, Y_{\nu\delta'}, A_{\delta'}]_{\mu=1,\ldots,n-1;\,\nu=1,\ldots,N;\,\delta'=0,\ldots,d-1}\,.$$

Denote by S' (resp. $\hat{S}'$) the spectrum of $R[T_1,\ldots,T_{n-1}]$ (resp. $\hat{R}[[T_1,\ldots,T_{n-1}]]$), set

$$N'' = d \cdot N + N ,$$

and regard the above ring $R[T_\mu, Y_{\nu\delta'}, A_{\delta'}]$ as the ring of global sections of $\mathcal{O}_{\mathbb{A}^{N''}_{S'}}$. Then the inclusion

$$R[T_\mu, Y_{\nu\delta'}, A_{\delta'}] \hookrightarrow R[T_\mu, Y_{\nu\delta'}, A_{\delta'}, Z_\nu] ,$$

where on the left-hand side μ runs from 1 to $n - 1$ and on the right-hand side from 1 to n, defines a projection

$$\rho : \mathbb{A}^{N'}_S \longrightarrow \mathbb{A}^{N''}_{S'} .$$

Consider now the closed subschemes

$$W = V(f_{i\delta})_{\substack{i=0,\dots,r\\ \delta=0,\dots,d-1}} \subset \mathbb{A}^{N''}_{S'} , \qquad \text{and}$$

$$V = V(f_{i\delta})_{\substack{i=0,\dots,r\\ \delta=0,\dots,d-1}} \subset \mathbb{A}^{N'}_{S} .$$

Then V is the pull-back of W by the map ρ. So V is isomorphic to $\mathbb{A}^{N+1}_W$, and T_n, $Z_1, \dots, Z_N$ can be viewed as coordinate functions of $\mathbb{A}^{N+1}_W$. Due to the decomposition (*), for each ν we obtain a representation

$$\hat{y}_\nu := \sigma^* Y_\nu = y'_\nu + \hat{z}_\nu \cdot \hat{p} ,$$

where

$$y'_\nu = \sum_{\delta=0}^{d-1} y'_{\nu\delta} T_n^\delta$$

with $y'_{\nu\delta} \in \hat{R}[[T_1, \dots, T_{n-1}]]$ and $\hat{z}_\nu \in \hat{R}[[T_1, \dots, T_n]]$. Then define an S'-morphism

$$\varphi' : \hat{S}' \longrightarrow \mathbb{A}^{N''}_{S'}$$

by setting

$$(\varphi')^* Y_{\nu\delta} = y'_{\nu\delta} \quad \text{for} \quad \nu = 1, \dots, N , \quad \delta = 0, \dots, d-1 ,$$

$$(\varphi')^* A_\delta = a'_\delta \quad \text{for} \quad \delta = 0, \dots, d-1 .$$

Furthermore, consider the S-morphism

$$\varphi : \hat{S} \longrightarrow \mathbb{A}^{N'}_S$$

defined by

$$\varphi^* Y_{\nu\delta} = y'_{\nu\delta} ; \qquad \nu = 1, \dots, N , \qquad \delta = 0, \dots, d-1 ,$$

$$\varphi^* A_\delta = a'_\delta ; \qquad \delta = 0, \dots, d-1 ,$$

$$\varphi^* Z_\nu = \hat{z}_\nu ; \qquad \nu = 1, \dots, N .$$

Then we have $\sigma = \tau \circ \varphi$, $\varphi^* p = \hat{p}$, and $\varphi^* f_{i\delta} = (\varphi')^* f_{i\delta}$ for all i and δ. In order to see that φ' factors through W, one considers Taylor expansions of

$$\sigma^* f_i = f_i(\hat{y}) = f_i(y' + \hat{z} \cdot \hat{p}) ,$$

thereby obtaining

$$\sigma^* f_i \equiv f_i(y') \mod \hat{p} \cdot \hat{R}[[T_1, \ldots, T_n]] , \qquad i = 0, \ldots, r .$$

Since $\sigma^* f_i = 0$ for $i = 1, \ldots, r$, it follows

$$f_i(y') \equiv 0 \mod \hat{p} \cdot \hat{R}[[T_1, \ldots, T_n]]$$

for $i > 0$. Moreover, since $\hat{p}$ and $\sigma^* f_0$ differ by a unit in $\hat{R}[[T_1, \ldots, T_n]]$, we have

$$f_i(y') \equiv 0 \mod \hat{p} \cdot \hat{R}[[T_1, \ldots, T_n]]$$

for $i = 0$, too. On the other hand, using (**) we get relations

$$\sigma^* f_i = \varphi^* \tau^* f_i = \varphi^* \left(\sum_{\delta=0}^{d-1} f_{i\delta} \cdot T_n^\delta + q_i \cdot p \right) = \sum_{\delta=0}^{d-1} (\varphi')^* f_{i\delta} \cdot T_n^\delta + \hat{q}_i \cdot \hat{p}$$

for $i = 0, \ldots, r$, where $\hat{q}_i \in \hat{R}[[T_1, \ldots, T_n]]$. Then, since $\sigma^* f_i \equiv 0 \mod \hat{p}$, the direct sum decomposition (*) implies $(\varphi')^* f_{i\delta} = 0$ for all i and all δ. So φ' factors through W, and the induction hypothesis can be applied. Thus there exists a factorization of φ' into S'-morphisms

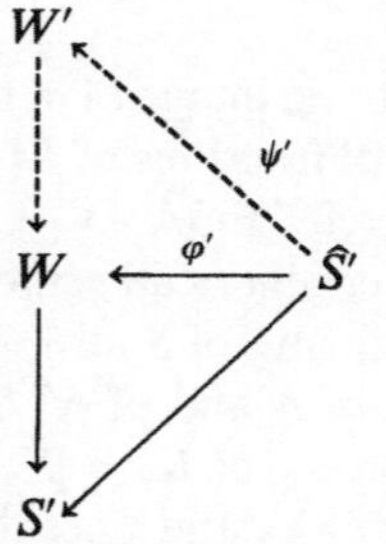

where W' is a smooth S'-scheme. By base change we obtain from W' the smooth S-scheme $W'' = W' \times_{S'} S$ and, hence, the smooth S-scheme

$$V' = \mathbb{A}^N_{W''} = \mathbb{A}^{N+1}_{W'} ,$$

where $Z_1, \ldots, Z_N$ give rise to a set of coordinates of $\mathbb{A}^N_{W''}$. Furthermore, we can define an S-morphism

$$\psi : \hat{S} \longrightarrow V'$$

(over $\hat{S}' \longrightarrow W'$) by setting

$$\psi^* Z_\nu = \hat{z}_\nu \qquad \text{for} \qquad \nu = 1, \ldots, N .$$

Then there is a commutative diagram of S-morphisms

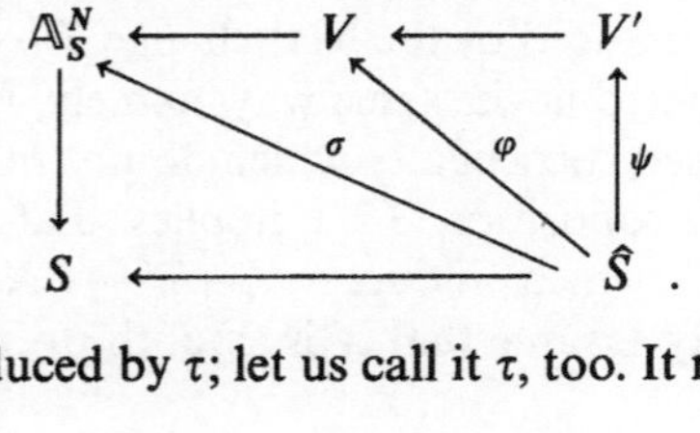

The map $V \longrightarrow \mathbb{A}^N_S$ is induced by τ; let us call it τ, too. It remains to show that $\tau^* f_0$

divides $\tau^* f_i$, $i = 1, \ldots, r$, at least locally at $\varphi(\hat{s})$. Due to the definition of V, it suffices to know that the factor q_0 defined by the relation (**) is invertible at $\varphi(\hat{s})$. But this is clear. Namely, the equation

$$\hat{u} \cdot \hat{p} = \sigma^* f_0 = \varphi^* \tau^* f_0 = \varphi^*(q_0) \cdot \hat{p}$$

shows that $\varphi^*(q_0) = \hat{u}$ is a unit in $\hat{R}[[T_1, \ldots, T_n]]$. □

We will apply the preceding lemma in the situation where f_0 is the square of a maximal minor of the Jacobi matrix

$$J = \left(\frac{\partial f_i}{\partial Y_\nu} \right)_{\substack{i=1,\ldots,r \\ \nu=1,\ldots,N}} .$$

Before this can be done, however, we have to justify the following reduction step.

Lemma 14. *It suffices to prove Theorem 12 in the case where X, at the point $\sigma(\hat{\eta})$, is smooth over S of relative dimension $N - m$ and where X, as a closed subscheme of $\mathbb{A}_S^N$, is defined by m global sections $f_1, \ldots, f_m$ of $\mathcal{O}_{\mathbb{A}_S^N}$.*

Proof. Replacing X by the schematic image of σ, one may assume σ to be schematically dense in X. Since the fields of fractions of $\hat{R}[[T_1, \ldots, T_n]]$ and of $R[T_1, \ldots, T_n]$ are separable over each other (cf. [EGA IV_2], 7.8.3), the generic fibre of X is geometrically reduced and, hence, generically smooth over S. Denote by Λ the local ring of S at η and by Λ' the local ring of $\hat{S}$ at $\hat{\eta}$. The extension $\Lambda \longrightarrow \Lambda'$ is regular, and π is a uniformizing element of Λ and of Λ'. Set $T = \operatorname{Spec} \Lambda$ and $T' = \operatorname{Spec} \Lambda'$. Then σ induces a T'-valued point $\sigma_{T'}$ of $X_T = X \times_S T$. Since the generic point t' of T' is mapped to the generic point of X_T and since the generic fibre of X_T is generically smooth over T, Proposition 4 shows the existence of a commutative diagram

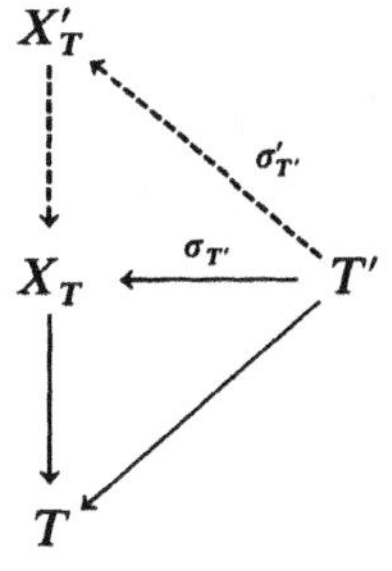

where X'_T is smooth over T and where $X'_T \longrightarrow X_T$ is constructed as a finite sequence of dilatations with centers in the special fibres. Using a limit argument, we may assume that $X'_T \longrightarrow X_T$ is induced by the base change $T \longrightarrow S$ from an S-morphism $X' \longrightarrow X$ which is constructed in the same way; namely, we can extend the centers of the blowing-ups to closed subschemes which do not meet generic fibres. Due to the construction of X', Proposition 3.2/1 implies that σ lifts (uniquely) to an R-morphism $\sigma' : \hat{S} \longrightarrow X'$ which induces $\sigma'_{T'} : T' \longrightarrow X'_T$. Obviously, σ' is an S-morphism. Thus we may assume that X is smooth over S at $\sigma(\hat{\eta})$, say of relative dimension $N - m$. Due to 2.2/7, we may assume that $f_1, \ldots, f_m$ define X as a subscheme of $\mathbb{A}_S^N$ at $\sigma(\hat{\eta})$. Now consider the closed subscheme $V \subset \mathbb{A}_S^N$ given by $f_1, \ldots, f_m$. Then $X \subset V$, and both coincide in a neighborhood of $\sigma(\hat{\eta})$. In particular, the morphism $\hat{S} \longrightarrow X$ factors through V. Since smooth S-schemes are locally integral, we may replace X by V. Namely, if $V' \longrightarrow V$ is an S-morphism from a smooth S-scheme V' to V such that $\hat{S} \longrightarrow V$ factors through $V' \longrightarrow V$, we can assume that V' is integral. Then there is an open dense subscheme $V'' \subset V'$ which is mapped into X, and it follows that the map $V' \longrightarrow V$ must factor through X because V' is integral and because X is closed in V. □

Thus we may assume that X, as a closed subscheme of $\mathbb{A}_S^N$, is defined by m global sections, say

$$X = V(f_1, \ldots, f_m) \subset \mathbb{A}_S^N ,$$

and that the determinant

$$\Delta = \det\left(\frac{\partial f_i}{\partial Y_j}\right)_{\substack{i=1,\ldots,m \\ j=1,\ldots,m}}$$

does not vanish at $\sigma(\hat{\eta})$; cf. 2.2/7. We will now finish the proof of Theorem 12 by establishing a third lemma; see Bourbaki [2], Chap. III, §4, n°5, for a similar statement.

Lemma 15. *Consider a situation as in Lemma* 13. *Assume that X is as above and that $f_0 = \Delta^2$. Then there exists a diagram*

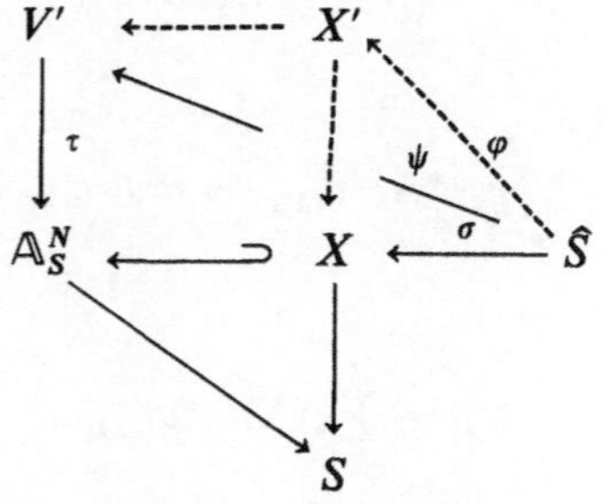

where $X' \longrightarrow V'$ is étale; in particular, X' is smooth over S. Except for the square in the upper left corner, the diagram is commutative.

Proof. In the following, we write f for the column vector $(f_1, \ldots, f_m)^t$; the index t indicates the transpose. On V' we have a relation

$$(*) \qquad \tau^* f = \tau^* \Delta^2 \cdot a'$$

with a column vector $a' = (a'_1, \ldots, a'_m)^t$ of global sections of $\mathcal{O}_{V'}$. Denote by $\Delta_1 = \Delta$, $\Delta_2, \ldots, \Delta_l$ the $(m \times m)$-minors of

$$J = \left(\frac{\partial f_i}{\partial Y_\nu}\right)_{\substack{i=1,\ldots,m \\ \nu=1,\ldots,N}} .$$

Due to Cramer's rule, there exist $(N \times m)$-matrices M_λ, $\lambda = 1, \ldots, l$, with global sections of $\mathcal{O}_{\mathbb{A}_S^N}$ as entries such that

$$(**) \qquad J \cdot M_\lambda = \Delta_\lambda \cdot I_m \, .$$

I_m is the $(m \times m)$-unit matrix. We will construct X' as a subscheme of $\mathbb{A}_{V'}^{l \cdot N}$. So denote by $Z_{\lambda\nu}$, $\lambda = 1, \ldots, l$, $\nu = 1, \ldots, N$, the coordinate functions of $\mathbb{A}_S^{l \cdot N}$. Let $Z_{(\lambda)}$ be the column vector $(Z_{\lambda 1}, \ldots, Z_{\lambda N})^t$, $\lambda = 1, \ldots, l$. Now consider the S-morphism

$$\rho : \mathbb{A}_{V'}^{l \cdot N} \longrightarrow \mathbb{A}_S^N$$

given by

$$\rho^* Y = \sum_{\lambda=1}^{l} \tau^* \Delta_\lambda \cdot Z_{(\lambda)} + \tau^* Y$$

where Y is the column vector $(Y_1, \ldots, Y_N)^t$. By Taylor expansion we get an equation

$$\rho^* f = \tau^* f + \sum_{\lambda=1}^{l} \tau^* \Delta_\lambda \cdot \tau^* J \cdot Z_{(\lambda)} + \sum_{\lambda, \mu = 1}^{l} \tau^* \Delta_\lambda \cdot \tau^* \Delta_\mu \cdot q_{(\lambda, \mu)}$$

with certain column vectors $q_{(\lambda,\mu)} = (q_{\lambda\mu 1}, \ldots, q_{\lambda\mu m})^t$. Each $q_{\lambda\mu i}$ is a polynomial in the variables $Z_{\lambda\mu}$ with global sections of $\mathcal{O}_{V'}$ as coefficients, and each monomial of $q_{\lambda\mu i}$ has degree ≥ 2. Using $(*)$ and $(**)$, we can write

$$\tau^* f = \tau^* \Delta \cdot (\tau^* \Delta \cdot I_m) \cdot a' = \tau^* \Delta \cdot \tau^* J \cdot a'_{(1)}$$

with

$$a'_{(1)} = \tau^* M_1 \cdot a' \, .$$

Furthermore, we have

$$\sum_{\mu=1}^{l} \tau^* \Delta_\mu \cdot q_{(\lambda,\mu)} = \tau^* J \cdot q_{(\lambda)}$$

with

$$q_{(\lambda)} = \sum_{\mu=1}^{l} \tau^* M_\mu \cdot q_{(\lambda,\mu)} \, .$$

Setting $a'_{(\lambda)} = 0$ for $\lambda = 2, \ldots, l$, we see

$$\rho^* f = \sum_{\lambda=1}^{l} \tau^* \Delta_\lambda \cdot \tau^* J \cdot [a'_{(\lambda)} + Z_{(\lambda)} + q_{(\lambda)}] \, .$$

Then let X' be the closed subscheme of $\mathbb{A}_{V'}^{l \cdot N}$ which is defined by the global sections

$$a'_{(\lambda)} + Z_{(\lambda)} + q_{(\lambda)}, \quad \lambda = 1, \ldots, l \, .$$

Due to 2.2/10, the projection $X' \longrightarrow V'$ is étale along the zero section of $\mathbb{A}_{V'}^{l \cdot N} \longrightarrow V'$. Obviously, the morphism $X' \longrightarrow \mathbb{A}_S^N$ induced by ρ factors through X. Since $\sigma^* \Delta$ is not a zero divisor, the relation

$$0 = \sigma^* f = \sigma^* \Delta^2 \cdot \psi^* a'$$

implies $\psi^* a' = 0$ and, hence, $\psi^* a'_{(\lambda)} = 0$ for $\lambda = 1, \ldots, l$. Thus, the zero section of

$\mathbb{A}_{V'}^{l;N}$ induces a lifting φ of ψ. Replacing X' by the étale locus of $X' \longrightarrow V'$, the assertion of the lemma is clear. □

Thereby we have finished the proof of Theorem 12. The statement of Theorem 12 was announced by M. Artin in [8]. Its proof, given in terms of commutative algebra, has been published recently by M. Artin and C. Rotthaus; cf. Artin and Rotthaus [1]. The method of proof is similar to the one used in Artin [4], where it is shown that the henselization of $R[T_1,\ldots,T_n]$ at $(\pi, T_1,\ldots,T_n)$ satisfies the approximation property. In fact, the latter result can be obtained as a consequence of Theorem 12.

Theorem 16. (M. Artin). *Let R be a field or an excellent discrete valuation ring, and let A be a henselization of a local R-algebra A_0 which is essentially of finite type over R. Let $\mathfrak{m}$ be a proper ideal of A, and let $\hat{A}$ be the $\mathfrak{m}$-adic completion of A. Then, given a system of polynomial equations*

$$f(Y) = 0$$

where $Y = (Y_1,\ldots, Y_N)$ are variables and $f = (f_1,\ldots,f_r)$ are polynomials in Y with coefficients in A, given a solution $\hat{y} = (\hat{y}_1,\ldots,\hat{y}_N) \in \hat{A}^N$ and an integer c, there exists a solution $y = (y_1,\ldots,y_N) \in A^N$ such that

$$y_\nu \equiv \hat{y}_\nu \mod \mathfrak{m}^c \cdot \hat{A}$$

for $\nu = 1,\ldots, N$.

Proof. Following M. Artin, we will reduce the assertion to the special case where A_0 is the localization of $R[T_1,\ldots,T_n]$ at the point $(\pi, T_1,\ldots,T_n)$ of $\operatorname{Spec} R[T_1,\ldots,T_n]$, where the integer c is 1, and where the ideal $\mathfrak{m}$ is the maximal ideal of A. In this case, the assertion is an easy consequence of Theorem 12. So let us start with the reductions.

One may assume that $\mathfrak{m}$ is the maximal ideal of A and that the integer c is 1. Namely, there exist elements $a_\nu \in A$ such that

$$\hat{y}_\nu \equiv a_\nu \mod \mathfrak{m}^c \cdot \hat{A}$$

for $\nu = 1,\ldots, N$. Let $m_1,\ldots, m_t$ be a system of generators of $\mathfrak{m}^c$. Then there exist elements $\hat{y}_{\nu j}$ of $\hat{A}$ such that

$$\hat{y}_\nu - a_\nu - \sum_{j=1}^{t} \hat{y}_{\nu j} m_j = 0 .$$

Let

$$g_\nu = Y_\nu - a_\nu - \sum_{j=1}^{t} Y_{\nu j} m_j \in A[Y_{\nu'}, Y_{\nu' j'}]_{\substack{\nu'=1,\ldots,N \\ j'=1,\ldots,t}}$$

and consider the system of polynomial equations given by $f_1,\ldots, f_r, g_1,\ldots, g_N$ in the variables $(Y_{\nu'})$ and $(Y_{\nu' j'})$. This system has the solution $((\hat{y}_\nu),(\hat{y}_{\nu j}))$ over $\hat{A}$, and any solution of this system lying in A gives rise to a solution of the required type of the system we started with.

We may assume that R is a discrete valuation ring and that the maximal ideal $\mathfrak{m}$ of A lies over the closed point of Spec R. Namely, if R is a discrete valuation ring and if $\mathfrak{m}$ lies over the generic point of R, we can replace R by its field of fractions. If R is a field, we can replace it by the power series ring $R[[T]]$, and view A as an $R[[T]]$-algebra by sending T to zero. Since $R[[T]]$ is excellent, this reduction is permissible.

We may assume that the residue field $k' = A/\mathfrak{m}$ is finite over $k = R/\pi R$. Since A_0 is essentially of finite type over R, the field k' is finitely generated over k. Let d be its degree of transcendence. Then there exist elements $z_1, \ldots, z_d \in A_0$ such that k' is finite over $k(\bar{z}_1,\ldots,\bar{z}_d)$, where $\bar{z}_\delta$ denotes the residue class of z_δ mod $\mathfrak{m}$. Let R' be the localization of $R[Z_1,\ldots,Z_d]$ at the prime ideal (π). The R-morphism

$$R[Z_1,\ldots,Z_d] \longrightarrow A_0$$

sending Z_δ to z_δ for $\delta = 1,\ldots,d$ factors through R', since $\bar{z}_1,\ldots,\bar{z}_d$ are transcendental over k. Furthermore, R' is an excellent discrete valuation ring, see [EGA IV_2], 7.8.3, and A_0 is essentially of finite type over R'.

We may assume that A is a finite S-algebra where S is a henselization of the localization S_0 of a polynomial ring $R[T_1,\ldots,T_n]$ at $(\pi, T_1,\ldots,T_n)$. Namely, let t_1, $\ldots$, t_n be a system of generators of the maximal ideal of A_0. The R-morphism

$$\varphi: R[T_1,\ldots,T_n] \longrightarrow A_0$$

sending T_i to t_i for $i = 1,\ldots,n$ induces a morphism $S_0 \longrightarrow A_0$. Since A_0 is essentially of finite type and since the residue field $A/\mathfrak{m}$ is finite over k, it is easily seen that Spec $A_0 \longrightarrow$ Spec S_0 is quasi-finite at the maximal ideal of A_0. Then the extension $S \longrightarrow A_0 \otimes_{S_0} S$ is finite (cf. 2.3/4); so $A_0 \otimes_{S_0} S$ is a direct sum of local henselian rings. Since A is among them, the extension $S \longrightarrow A$ is finite.

It suffices to prove the theorem for a henselization S of the localization S_0 of a polynomial ring $R[T_1,\ldots,T_n]$ at $(\pi, T_1,\ldots,T_n)$. Since we may assume that A is finite over S, the $\mathfrak{m}$-adic completion $\hat{A}$ of A is isomorphic to $A \otimes_S \hat{S}$. Write A as a quotient of a polynomial ring over S, say

$$0 \longrightarrow \mathfrak{a} \longrightarrow S[X_1,\ldots,X_m] \longrightarrow A \longrightarrow 0\,.$$

Then let $a_1, \ldots, a_l$ be a finite system of generators of $\mathfrak{a}$. Lifting the system $f(Y)$ over A to a system $g(Y)$ over $S[X]$ and lifting the given solution $\hat{y}$ of $f(Y)$ to $\hat{y}' = (\hat{y}'_1,\ldots,\hat{y}'_N)$ with $\hat{y}'_1, \ldots, \hat{y}'_N \in S[X] \otimes_S \hat{S}$, we get a relation

$$(*) \qquad g(\hat{y}') = \sum_{\lambda=1}^{l} a_\lambda \hat{z}_{(\lambda)}$$

where $\hat{z}_{(\lambda)} = (\hat{z}_{\lambda 1},\ldots,\hat{z}_{\lambda r})^t$ is a column vector of elements of $S[X] \otimes_S \hat{S}$. Then consider the system of equations

$$(**) \qquad g(Y) - \sum_{\lambda=1}^{l} a_\lambda Z_{(\lambda)} = 0$$

over $S[X]$, where $Y = (Y_1,\ldots,Y_N)$ and $Z = (Z_{\lambda i})$, for $\lambda = 1, \ldots, l$, $i = 1, \ldots, r$, are variables. Due to $(*)$, the system $(**)$ has a solution in $\hat{S}[X]$. Looking at the coefficients of the polynomials in $X_1, \ldots, X_m$ appearing in $(*)$, we can rewrite $(**)$

as a finite system of polynomial equations over S which has a solution over $\hat{S}$. Clearly, a solution over S of this system induces a solution over A of the system we started with.

Now let us show how, in this situation, the proof of the theorem follows from Theorem 12. The polynomials $f_1, \ldots, f_r \in S[Y_1, \ldots, Y_N]$ define a closed subscheme X of $\mathbb{A}_S^N$. Since only finitely many coefficients occur in $f_1, \ldots, f_r$, the scheme X is actually defined over an $R[T_1, \ldots, T_n]$-algebra of finite type. So we may view X as an $R[T_1, \ldots, T_n]$-scheme of finite type. The solution $\hat{y} = (\hat{y}_1, \ldots, \hat{y}_N) \in \hat{S}^N$ gives rise to an $\hat{R}[[T_1, \ldots, T_n]]$-valued point σ of X. Then Theorem 12 yields a commutative diagram

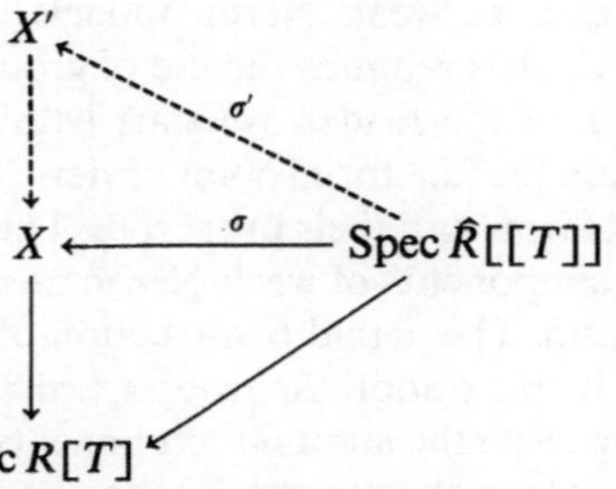

where X' is smooth over $R[T]$. The closed point $\hat{s}$ of $\operatorname{Spec} \hat{R}[[T]]$ induces a k-rational point $x' = \sigma'(\hat{s})$ of X'. Due to 2.3/5, the k-valued point x' lifts to an S-valued point of X' and, hence, to an S-valued point x of X. Then, x gives rise to a solution y over S of $f(Y) = 0$, the one we are looking for. □

Let us conclude with some remarks on the history of the approximation property. Corollary 9 was first established in Greenberg [2], where the author actually proves a much stronger result, the so-called strong approximation property for discrete valuation rings. Theorem 16 is due to M. Artin, cf. Artin [4]; he even shows the strong approximation property for polynomial rings $k[T_1, \ldots, T_n]$, where k is a field. By methods of model theory, it can also be seen from Artin's result (Theorem 16) that all rings $R[T_1, \ldots, T_n]$ satisfy that property whenever R is an excellent discrete valuation ring; cf. Becker, Denef, Lipshitz, van den Dries [1]. Artin's conjecture that the approximation property holds for every excellent ring A was recently verified by C. Rotthaus for the case where A contains the rational numbers; see Rotthaus [1].

The importance of the approximation theorem is based on the applications to moduli problems; there it is a powerful tool to show that certain functors are representable by algebraic spaces; cf. Artin [5] and [6]. We will come back to this point in Section 8.3.

Chapter 4. Construction of Birational Group Laws

In the previous chapter, we discussed the smoothening process and, as an application, proved the existence of weak Néron models. The next step towards the construction of Néron models requires the use of group arguments.

For the convenience of the reader, we start with a general section on group schemes where we explain the functorial point of view. Then we discuss the existence of invariant differential forms and their properties. They are used in order to define the so-called minimal components of weak Néron models, which are unique up to R-birational isomorphism. The actual construction of Néron models is continued in Section 4.3. Starting with a smooth K-group scheme X_K of finite type and a weak Néron model $(X_i)_{i \in I}$, we select the minimal components from the X_i. After a possible shrinking, we glue them along the generic fibre to produce a smooth and separated R-model X of X_K and we show that the group structure on X_K extends to an R-birational group law on X. Admitting the fact (to be obtained in Chapters 5 and 6) that X with its R-birational group law can uniquely be enlarged to an R-group scheme $\bar{X}$, we show in Section 4.4 that $\bar{X}$ will be a Néron model of X_K. This is done by employing an argument of A. Weil, saying that a rational map from a smooth scheme to a separated group scheme is defined everywhere if it is defined in codimension 1.

4.1 Group Schemes

Let C be a category; for example, let C be the category (Sch/S) of schemes over a fixed scheme S. Each object $X \in C$ gives rise to its functor of points

$$h_X : C \longrightarrow (\text{Sets})$$

which associates to any $T \in C$ the set

$$h_X(T) := X(T) := \operatorname{Hom}(T, X)$$

of T-valued points of X. Each morphism $X \longrightarrow X'$ in C induces a morphism $h_X \longrightarrow h_{X'}$ of functors by the composition of morphisms in C. In this way one gets a covariant functor

$$h : C \longrightarrow \operatorname{Hom}(C^0, (\text{Sets}))$$

of C to the category of covariant functors from C^0 (the dual of C) to the category of sets; the category $\operatorname{Hom}(C^0, (\text{Sets}))$ is denoted by $\hat{C}$; it is called the category of contravariant functors from C to (Sets).

Proposition 1. *The functor* $h: C \longrightarrow \hat{C}$ *is fully faithful; i.e., for any two objects* X, $X' \in C$, *the canonical map*

$$\mathrm{Hom}_C(X, X') \longrightarrow \mathrm{Hom}_{\hat{C}}(h_X, h_{X'})$$

is bijective. More generally, for all objects $X \in C$ *and* $\mathscr{F} \in \hat{C}$, *there is a canonical bijection*

$$\mathscr{F}(X) \xrightarrow{\sim} \mathrm{Hom}_{\hat{C}}(h_X, \mathscr{F})$$

mapping $u \in \mathscr{F}(X)$ *to the morphism* $h_X \longrightarrow \mathscr{F}$ *which to a T-valued point* $g \in h_X(T)$, *where* T *is an object of* C, *associates the element* $\mathscr{F}(g)(u) \in \mathscr{F}(T)$. *The bijection coincides with the above one if* $\mathscr{F} = h_{X'}$ *and is functorial in* X *and* $\mathscr{F}$ *in the sense that* $\mathscr{F} \longmapsto \mathrm{Hom}_{\hat{C}}(h(\cdot), \mathscr{F})$ *defines an isomorphism* $\hat{C} \longrightarrow \hat{C}$.

Proof. Consider an element $u \in \mathscr{F}(X)$. We have only to show that there is a unique functorial morphism $h_X \longrightarrow \mathscr{F}$ mapping the so-called *universal point* $\mathrm{id}_X \in h_X(X)$ onto $u \in \mathscr{F}(X)$ and that it is as stated. Then all assertions of the proposition are immediately clear. So let us show how to justify this claim. For any object $T \in C$, each T-valued point $g: T \longrightarrow X$ factors through the universal point of X. Thus, if $h_X \longrightarrow \mathscr{F}$ exists as claimed, the image of g under $h_X(T) \longrightarrow \mathscr{F}(T)$ must coincide with the image of u under $\mathscr{F}(g): \mathscr{F}(X) \longrightarrow \mathscr{F}(T)$. Conversely, taking the latter as a definition, we see that $h_X \longrightarrow \mathscr{F}$ can be constructed as required. □

In particular, if a functor $\mathscr{F} \in \mathrm{Hom}(C^0, (\mathrm{Sets}))$ is isomorphic to a functor h_X, then X is uniquely determined by $\mathscr{F}$ up to an isomorphism in the category C. In this case, the functor $\mathscr{F}$ is said to be *representable*. Thus Proposition 1 says that the functor h defines an equivalence between the category C and the full subcategory of $\mathrm{Hom}(C^0, (\mathrm{Sets}))$ consisting of all representable functors.

In order to define group objects in the category C, it is necessary to introduce the notion of a law of composition on an object X of C. By the latter we mean a functorial morphism

$$\gamma: h_X \times h_X \longrightarrow h_X .$$

Thus, a law of composition on X consists of a collection of maps

$$\gamma_T: h_X(T) \times h_X(T) \longrightarrow h_X(T)$$

(laws of composition on the sets of T-valued points of X) where T varies over the objects in C. The functoriality of γ means that all maps γ_T are compatible with canonical maps between points of X, i.e., for any morphism $u: T' \longrightarrow T$ in C, the diagram

$$\begin{array}{ccc} h_X(T) \times h_X(T) & \xrightarrow{\gamma_T} & h_X(T) \\ \Big\downarrow{\scriptstyle h_X(u) \times h_X(u)} & & \Big\downarrow{\scriptstyle h_X(u)} \\ h_X(T') \times h_X(T') & \xrightarrow{\gamma_{T'}} & h_X(T') \end{array}$$

is commutative. If the law of composition has the property that $h_X(T)$ is a group under γ_T for all T, then γ defines on h_X the structure of a group functor, i.e., of a contravariant functor from C to the category of groups. In this case, γ is called a group law on X.

Definition 2. *A group object in C is an object X together with a law of composition $\gamma : h_X \times h_X \longrightarrow h_X$ which is a group law.*

It follows that a group object in C is equivalent to a group functor which, as a functor to the category of sets, is representable.

When dealing with group objects, it is convenient to know that the category in question contains direct products and a final object, say S. The latter means that, for each object T of C, there is a uniqe morphism $T \longrightarrow S$. So, in the following, assume that C is of this type, and consider a group object X of C with group law γ. Then, since the product $X \times X$ exists in C and since the functor $h : C \longrightarrow \mathrm{Hom}(C^0, (\mathrm{Sets}))$ commutes with direct products, the law of composition $\gamma : h_X \times h_X \longrightarrow h_X$ corresponds to a morphism $m : X \times X \longrightarrow X$, as is seen by using Proposition 1. Furthermore, the injection of the unit element into each group $h_X(T)$ yields a natural transformation from h_S to h_X, hence it corresponds to a morphism

$$\varepsilon : S \longrightarrow X ,$$

called the *unit section of* X, which is a section of the unique morphism $X \longrightarrow S$. Finally, the formation of the inverse in each $h_X(T)$ defines a natural transformation $h_X \longrightarrow h_X$ and hence a morphism

$$\iota : X \longrightarrow X ,$$

called the *inverse map on* X. The group axioms which are satisfied by the groups $h_X(T)$, and hence by the functor h_X, correspond to certain properties of the maps m, ε and ι. Namely, the following diagrams are commutative:

(a) *associativity*

$$\begin{array}{ccc} X \times X \times X & \xrightarrow{m \times \mathrm{id}_X} & X \times X \\ \Big\downarrow{\scriptstyle \mathrm{id}_X \times m} & & \Big\downarrow{\scriptstyle m} \\ X \times X & \xrightarrow{\quad m \quad} & X \end{array}$$

(b) *existence of a left-identity*

$$\begin{array}{ccccc} X & \xrightarrow{(p, \mathrm{id}_X)} & S \times X & \xrightarrow{\varepsilon \times \mathrm{id}_X} & X \times X \\ & \searrow^{\mathrm{id}_X} & & & \Big\downarrow{\scriptstyle m} \\ & & & & X \end{array} ,$$

where $p : X \longrightarrow S$ is the morphism from X to the final object S.

(c) *existence of a left-inverse*

$$\begin{array}{ccc} X & \xrightarrow{(\iota, \mathrm{id}_X)} & X \times X \\ \downarrow{\scriptstyle p} & & \downarrow{\scriptstyle m} \\ S & \xrightarrow{\varepsilon} & X \end{array}$$

(d) *commutativity* (only if all groups $h_X(T)$ are commutative)

$$\begin{array}{ccc} X \times X & \xrightarrow{\tau} & X \times X \\ & \searrow{\scriptstyle m} & \downarrow{\scriptstyle m} \\ & & X \end{array}$$

where τ commutes the factors.

Note that a left-identity is also a right-identity and that a left-inverse is also a right-inverse. It is clear that once we have an object X and morphisms m, ε, and ι with the above properties, we can construct a group object in the given category from these data, and furthermore, that group objects in C and data $(X, m, \varepsilon, \iota)$ correspond bijectively to each other.

Proposition 3. *The group objects in a category C correspond one-to-one to data $(X, m, \varepsilon, \iota)$ where X is an object of C and where*

$$m : X \times X \longrightarrow X, \qquad \varepsilon : S \longrightarrow X, \qquad \iota : X \longrightarrow X$$

are morphisms in C such that the diagrams (a), (b), (c) *above are commutative. Furthermore, a group object in C is commutative if and only if, in addition, the corresponding diagram* (d) *is commutative.*

In the following we restrict ourselves to the category (Sch/S) of S-schemes where S is a fixed base scheme. Then the direct product in (Sch/S) is given by the fibred product of schemes over S, and the S-scheme S is a final object in (Sch/S).

Definition 4. *An S-group scheme is a group object in the category of S-schemes* (Sch/S).

Due to Proposition 3, an S-group scheme G can be viewed as an S-scheme X together with appropriate morphisms m, ε, and ι. When no confusion about the group structure is possible, we will not mention these morphisms explicitly. In particular, in our notation we will make no difference between the group object G and the associated representing scheme X. Also we want to point out that there exist group functors on (Sch/S) which are not representable and thus do not correspond to S-group schemes. For example, let X be a smooth S-scheme and, for any S-scheme T, let $\mathcal{R}_{X/S}(T)$ be the set of all T-birational automorphisms of $X_T = X \times_S T$. Then, in general, the group functor $\mathcal{R}_{X/S}$ is not representable by a scheme, even if X is the projective line over a field.

It follows immediately from Definition 4 that the technique of base change can be applied to group schemes. Thus, for any base change $S' \longrightarrow S$, one obtains from an S-group scheme G an S'-group scheme $G_{S'} := G \times_S S'$. If $S = \operatorname{Spec} R$ for some ring R, we talk also about R-group schemes instead of S-group schemes. Furthermore, if $K = R$ is a field, an algebraic K-group is meant to be a K-group scheme of finite type (not necessarily smooth).

There are many notions for ordinary groups which have a natural analogue for group functors and thus for group schemes. For example, a homomorphism of group functors $\mathscr{G}' \longrightarrow \mathscr{G}$ is a natural transformation between $\mathscr{G}'$ and $\mathscr{G}$ (viewed as functors from (Sch/S) to (Groups)). If $\mathscr{G}'$ and $\mathscr{G}$ are represented by S-schemes G' and G, respectively, such a homomorphism corresponds to a morphism $G' \longrightarrow G$ which is compatible with the group law on G' and on G. We also have the notions of subgroup, kernel of a homorphism, monomorphism, etc., for group functors. However, when dealing with S-group schemes G, we reserve the notion of *subgroup schemes* to such representable subgroup functors which are represented by *subschemes* of G (the latter is not automatic). A subscheme Y of G defines a subgroup scheme of G if and only if the following conditions are satisfied:

(i) the unit-section $\varepsilon : S \longrightarrow G$ factors through Y,

(ii) the group law $m : G \times_S G \longrightarrow G$ restricts to a morphism $Y \times_S Y$, and

(iii) the inverse map $\iota : G \longrightarrow G$ restricts to a morphism $Y \longrightarrow Y$.

Let us look at some examples of S-group schemes. We start with the classical groups $\mathbb{G}_a$ (the additive group), $\mathbb{G}_m$ (the multiplicative group), GL_n (the general linear group), and PGL_n (the projective general linear group). In terms of group functors, these groups are defined as follows. For any S-scheme T set

$$\begin{aligned} \mathbb{G}_a(T) &:= \text{the additive group } \mathcal{O}_T(T) \\ \mathbb{G}_m(T) &:= \text{the group of units in } \mathcal{O}_T(T) \\ \mathrm{GL}_n(T) &:= \text{the group of } \mathcal{O}_T(T)\text{-linear automorphisms of } (\mathcal{O}_T(T))^n \\ \mathrm{PGL}_n(T) &:= \mathrm{Aut}_T\,(\mathbb{P}(\mathcal{O}_T^n))\,. \end{aligned}$$

All these group functors are representable by affine schemes over $\mathbb{Z}$. Working over $S := \operatorname{Spec} \mathbb{Z}$, the additive group is represented by the scheme

$$X := \operatorname{Spec} \mathbb{Z}[\zeta]$$

(ζ is an indeterminate), where the group law $m : X \times X \longrightarrow X$ corresponds to the algebra homomorphism

$$\mathbb{Z}[\zeta] \longrightarrow \mathbb{Z}[\zeta] \otimes_{\mathbb{Z}} \mathbb{Z}[\zeta]\,, \qquad \zeta \longmapsto \zeta \otimes 1 + 1 \otimes \zeta\,.$$

Similarly, for $\mathbb{G}_m$, the representing object is $\operatorname{Spec} \mathbb{Z}[\zeta, \zeta^{-1}]$ with the group law given by $\zeta \longmapsto \zeta \otimes \zeta$. In the case of GL_n we consider a set ζ_{ij} of n^2 indeterminates. Then

$$X := \operatorname{Spec} \mathbb{Z}[\zeta_{ij}, \det(\zeta_{ij})^{-1}]$$

is a representing object; the group law is defined by the multiplication of matrices. Finally, PGL_n is represented by the open subscheme

$$X \subset \operatorname{Proj} \mathbb{Z}[\zeta_{ij}]$$

where $\det \zeta_{ij}$ does not vanish. For a general base S, the representing objects are obtained from the ones over $\operatorname{Spec} \mathbb{Z}$ by base extension. It is clear that the above procedure works as well for further classical groups ($\mathrm{SL}_n, \mathrm{Sp}_n, \mathrm{O}_n, \ldots$). Also it should be mentioned that one can define $\mathrm{GL}_V, \mathrm{PGL}_V, \ldots$ for any vector bundle V over S. Just replace $\mathcal{O}_T^n$ in the above definitions by the pull-back of V with respect to $T \longrightarrow S$.

All the above group schemes are *affine*, i.e., the representing schemes are affine over the base S. Another important class of group schemes consists of the so-called *abelian schemes* over S. Thereby we mean smooth proper S-group schemes with connected fibres (the latter are abelian varieties in the usual sense). They are always commutative. As examples one may consider elliptic curves over fields which have a rational point or, more generally, Jacobians of smooth complete curves.

4.2 Invariant Differential Forms

Throughout this section, let G be a group scheme over a fixed scheme S. First we want to introduce the notion of translations on G. In order to do this, consider a T-valued point

$$g : T \longrightarrow G$$

of G; i.e., an S-morphism from an S-scheme T to G. Then g gives rise to the T-valued point

$$g_T := (g, \mathrm{id}_T) : T \longrightarrow G_T := G \times_S T$$

of the T-scheme $G_T := G \times_S T$. If $p_1 : G_T \longrightarrow G$ denotes the first projection, we have $g = p_1 \circ g_T$. In the special case where $T := G$ and $g := \mathrm{id}_G$ is the universal point of G, the morphism g_T equals the diagonal morphism Δ of G. For any other T-valued point g of G, the morphism g_T is obtained from Δ by performing the base change $g : T \longrightarrow G$.

As usual, let $m : G \times_S G \longrightarrow G$ be the group law of G and write m_T for its extension when a base change $T \longrightarrow S$ is applied to G. Then, for any T-valued point g of G, we define the left translation by

$$\tau_g : G_T \xrightarrow{\sim} T \times_T G_T \xrightarrow{g_T \times \mathrm{id}} G_T \times_T G_T \xrightarrow{m_T} G_T$$

and the right translation by

$$\tau'_g : G_T \xrightarrow{\sim} G_T \times_T T \xrightarrow{\mathrm{id} \times g_T} G_T \times_T G_T \xrightarrow{m_T} G_T .$$

Both morphisms are isomorphisms. Quite often we will drop the index T and characterize the map τ_g by writing

$$\tau_g : G \longrightarrow G , \qquad x \longmapsto gx ;$$

the same procedure will be applied for τ'_g and for similar morphisms. In the special case where $T := G$ and $g := \mathrm{id}_G$ is the universal point, τ_g is the so-called *universal left translation*, namely the morphism

$$\Phi : T \times_S G \longrightarrow T \times_S G\,, \qquad (x, y) \longmapsto (x, xy).$$

Similarly, τ'_g gives rise to the *universal right translation*

$$\Psi : G \times_S T \longrightarrow G \times_S T\,, \qquad (x, y) \longmapsto (xy, y)\,.$$

Each left translation by a T-valued point $g : T \longrightarrow G$ is obtained from the universal left translation Φ by performing the base change $g : T \longrightarrow G$; in a similar way one can proceed with right translations.

Now let us consider the sheaf $\Omega^i_{G/S}$ of relative differential forms of some degree $i \geqq 0$ on G; it is defined as the i-th exterior power of $\Omega^1_{G/S}$. For any S-scheme T and any T-valued point $g \in G(T)$, the left translation $\tau_g : G_T \longrightarrow G_T$ gives rise to an isomorphism

$$\tau_g^* \Omega^i_{G_T/T} \xrightarrow{\sim} \Omega^i_{G_T/T}\,.$$

A global section ω in $\Omega^i_{G/S}$ is called *left-invariant* if $\tau_g^* \omega_T = \omega_T$ in $\Omega^i_{G_T/T}$ for all $g \in G(T)$ and all T, where ω_T is the pull-back of ω with respect to the projection $p_1 : G_T \longrightarrow G$ (see 2.1/3 for the canonical isomorphism $p_1^* \Omega^i_{G/S} \xrightarrow{\sim} \Omega^i_{G_T/T}$; see also Section 2.1 for our notational convention on the pull-back of differential forms). Using right translations τ'_g, one defines *right-invariant* differential forms in the same way. Since each translation on the group scheme G_T is obtained by base change from the universal translation, it is clear that one has to check the invariance under translations only for the universal translation. Generally, in connection with translations, we will drop the index T and write ω instead of ω_T if no confusion is possible.

In the following we will frequently use the fact that two global sections ω and ω' of a sheaf $\mathscr{F}$ on G are equal provided they coincide on every T-valued point g of G; i.e., provided $g_T^* \omega_T = g_T^* \omega'_T$ in $g_T^* \mathscr{F}_T$, where $\mathscr{F}_T$ is the pull-back of $\mathscr{F}$ to G_T. This is easily verified by using the universal point $g := \mathrm{id}_G$ of G; namely, for $T = G$, we have the commutative diagram

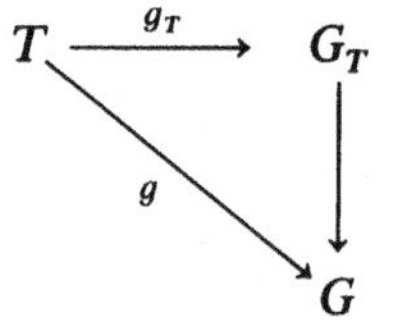

where $G_T \longrightarrow G$ is the projection. Similarly, one shows that two sheaves $\mathscr{F}$ and $\mathscr{G}$ are isomorphic if their restrictions to each T-valued point of G are isomorphic.

Proposition 1. *Let G be an S-group scheme with unit section $\varepsilon : S \longrightarrow G$. Then, for each $\omega_0 \in \Gamma(S, \varepsilon^* \Omega^i_{G/S})$, there exists a unique left-invariant differential form $\omega \in \Gamma(G, \Omega^i_{G/S})$ such that $\varepsilon^* \omega = \omega_0$ in $\varepsilon^* \Omega^i_{G/S}$. The same assertion is true for right-invariant differential forms.*

Proof. It is only necessary to consider left-invariant differential forms since the inverse map $G \longrightarrow G, x \longmapsto x^{-1}$, transforms left-invariant forms into right-invariant ones.

The uniqueness assertion is easy to obtain. Consider two global left-invariant sections ω, ω' of $\Omega^i_{G/S}$ such that $\varepsilon^*\omega = \varepsilon^*\omega' = \omega_0$ in $\varepsilon^*\Omega^i_{G/S}$. Then we have $g^*\omega = g^*\omega'$ in $g^*\Omega^i_{G/S}$ for each point $g \in G(S)$, since $g = \tau_g \circ \varepsilon$. Hence ω and ω' coincide at all points of $G(S)$. This fact remains true after base change. So ω and ω' coincide at the universal point of G and we have $\omega = \omega'$.

In order to settle the existence part, it is only necessary to consider the case where $i = 1$. Furthermore, the problem is local on S; so we may assume that ω_0 lifts to a section ω' of $\Omega^1_{G/S}$ which is defined over a neighborhood U of the unit section $\varepsilon : S \longrightarrow G$. Then the decomposition

$$(*) \qquad \Omega^1_{G \times_S G/S} \cong p_1^*\Omega^1_{G/S} \oplus p_2^*\Omega^1_{G/S}$$

of 2.1/4 gives a decomposition

$$m^*\omega' = \omega_1 \oplus \omega_2$$

over $m^{-1}(U)$, where $m : G \times_S G \longrightarrow G$ is the group law. If

$$\delta : G \longrightarrow G \times_S G\,, \qquad x \longmapsto (x^{-1}, x)$$

denotes the twisted diagonal morphism, $m^*\omega'$ is defined in a neighborhood of the image of δ so that $\delta^*\omega_2$ gives rise to a global section ω of $\Omega^1_{G/S}$. We claim that ω is left-invariant and satisfies $\varepsilon^*\omega = \omega_0$ in $\varepsilon^*\Omega^1_{G/S}$.

For an arbitrary T-valued point $g \in G(T)$, the commutative diagram

$$\begin{array}{ccc} G & \xrightarrow{\ \tau_g\ } & G \\ \downarrow{\scriptstyle \delta} & & \downarrow{\scriptstyle \delta} \\ G \times G & \xrightarrow{\ \tau'_{g^{-1}} \times \tau_g\ } & G \times G \end{array}$$

gives $\tau_g^*\delta^*\omega_2 = \delta^*(\tau'_{g^{-1}} \times \tau_g)^*\omega_2$ in $\Omega^1_{G_T/T}$. So ω will be left-invariant if we can show $(\tau'_{g^{-1}} \times \tau_g)^*\omega_2 = \omega_2$. Since the product map $\tau'_{g^{-1}} \times \tau_g$ respects the decomposition $(*)$ over $m^{-1}(U)$, we see

$$\tilde{\omega}_j := (\tau'_{g^{-1}} \times \tau_g)^*\omega_j \in \Gamma(m^{-1}(U), p_j^*(\Omega^1_{G_S/S}))\,, \qquad j = 1, 2\,.$$

However $m \circ (\tau'_{g^{-1}} \times \tau_g) = m$ so that

$$m^*\omega' = \omega_1 \oplus \omega_2 = \tilde{\omega}_1 \oplus \tilde{\omega}_2\,.$$

The two decompositions must coincide. Hence $\tilde{\omega}_2 = \omega_2$, and ω is left-invariant.

It remains to show $\varepsilon^*\omega = \omega_0$ in $\varepsilon^*\Omega^1_{G/S}$. Consider the morphism

$$\varepsilon_T : T := G \longrightarrow G \times_S T = G \times_S G$$

obtained from the unit section $\varepsilon : S \longrightarrow G$ by the base change $T \longrightarrow S$. Since $\varepsilon_T^* p_1^* \Omega^1_{G/S}$ vanishes in $\Omega^1_{G/S}$ and since $m \circ \varepsilon_T = \mathrm{id}_G$, we have

$$\varepsilon_T^*\omega_2 = \varepsilon_T^*(\omega_1 + \omega_2) = \varepsilon_T^* m^*\omega' = \omega' \quad \text{in} \quad \Omega^1_{G/S}\,.$$

Since $p_2 \circ \varepsilon_T = id_G = p_2 \circ \delta$, there is a canonical identification

$$\varepsilon_T^* p_2^* \Omega^1_{G/S} = \Omega^1_{G/S} = \delta^* p_2^* \Omega^1_{G/S} .$$

Then $\delta \circ \varepsilon = \varepsilon_T \circ \varepsilon$ implies

$$\varepsilon^* \delta^* \omega_2 = \varepsilon^* \varepsilon_T^* \omega_2 \quad \text{in} \quad \varepsilon^* \Omega^1_{G/S} .$$

Furthermore, we know $\delta^* \omega_2 = \omega$. So we get

$$\varepsilon^* \omega = \varepsilon^* \delta^* \omega_2 = \varepsilon^* \varepsilon_T^* \omega_2 = \varepsilon^* \omega' = \omega_0 \quad \text{in} \quad \varepsilon^* \Omega^1_{G/S} .$$

Thus ω is as desired. □

Using the structural morphism $p : G \longrightarrow S$, we can state the result of Proposition 1 more elegantly in the following form:

Proposition 2. *There are canonical isomorphisms*

$$p^* \varepsilon^* \Omega^i_{G/S} \xrightarrow{\sim} \Omega^i_{G/S} , \qquad i \in \mathbb{N} ,$$

which are obtained by extending sections in $\varepsilon^* \Omega^i_{G/S}$ to left-invariant sections in $\Omega^i_{G/S}$. Similar isomorphisms are obtained by using right-invariant differential forms.

Actually, Proposition 1 provides only an $\mathcal{O}_G$-module homomorphism $p^* \varepsilon^* \Omega^i_{G/S} \longrightarrow \Omega^i_{G/S}$ which, under the pull-back by ε, becomes an isomorphism. However, applying translations, the same assertion is true for any S-valued point of G. In particular, after base change $T := G \longrightarrow S$, the above homomorphism is an isomorphism at the point $g_T \in G_T(T)$ which is induced by the universal point g of G. Hence, the above homomorphism is an isomorphism already over G. □

We are specially interested in the case where G is a smooth group scheme over a local scheme S. Then each $\mathcal{O}_G$-module $\Omega^i_{G/S}$ is locally free, and $\varepsilon^* \Omega^i_{G/S}$ is a free $\mathcal{O}_S$-module. Thus we see:

Corollary 3. *Let G be a smooth group scheme of relative dimension d over a local scheme S. Then each $\Omega^i_{G/S}$, $0 \le i \le d$, is a free $\mathcal{O}_G$-module generated by $\binom{d}{i}$ left-invariant differential forms of degree i. The same is true for right-invariant differential forms.*

For the rest of this section, let us assume that G is a *smooth S-group scheme of relative dimension d*, and that there is a *left-invariant differential form $\omega \in \Omega^d_{G/S}(G)$ generating $\Omega^d_{G/S}$* as an $\mathcal{O}_G$-module. For an arbitrary T-valued point g of G we can consider the interior automorphism

$$\text{int}_g = \tau_g \circ \tau'_{g^{-1}} : G \longrightarrow G , \qquad x \longmapsto g x g^{-1} ,$$

given by g.

Proposition 4. *There exists a unique group homomorphism $\chi : G \longrightarrow \mathbb{G}_m$ (a character on G) such that*

$$\mathrm{int}_g^* \omega = \tau_{g^{-1}}'^* \omega = \chi(g)\omega$$

for each T-valued point g of G.

Proof. Since left translations commute with right translations, we see immediately that

$$\mathrm{int}_g^* \omega = \tau_{g^{-1}}'^* \tau_g^* \omega = \tau_{g^{-1}}'^* \omega$$

is left-invariant (on G_T) for any T-valued point g of G. Hence, since ω and $\mathrm{int}_g^* \omega$ generate $\Omega^d_{G_T/T}$, there exists a well-defined unit $\chi(g) \in \mathcal{O}_T(T)^*$ such that

$$\mathrm{int}_g^* \omega = \chi(g)\omega ;$$

recalling the functorial definition of the multiplicative group $\mathbb{G}_m$ and of group homomorphisms, one easily shows that $g \longmapsto \chi(g)$ defines a group homomorphism $\chi : G \longrightarrow \mathbb{G}_m$. □

Now let us consider the case where $S = \mathrm{Spec}\, K$ and where K is the field of fractions of a discrete valuation ring R. As usual, let R^{sh} denote a strict henselization of R and K^{sh} the field of fractions of R^{sh}. Let $|\ |$ be an absolute value on K and K^{sh}, which corresponds to R and R^{sh}. We want to look a little bit closer at the character χ occurring in the above lemma.

Proposition 5. *Let G be a smooth K-group scheme of relative dimension d, and assume that $G(K)$ (resp. $G(K^{sh})$) is bounded in G. Then the character χ of Proposition 4 satisfies $|\chi(g)| = 1$ for each $g \in G(K)$ (resp. each $g \in G(K^{sh})$).*

Proof. The character χ is bounded on $G(K)$; hence we may view $\chi(G(K))$ as a bounded subgroup of K^*. Such a subgroup consists of units in R. □

Remark 6. If, in the situation of Proposition 5, the group G is connected, one can actually show that the character χ is trivial. To see this, one uses the fact that G contains a maximal torus T defined over K, [SGA 3_{II}], Exp. XIV, 1.1. If χ is non-trivial, it induces a surjective map $T \longrightarrow \mathbb{G}_m$, and T must contain a subtorus isogenous to $\mathbb{G}_m$. But then $G(K)$ cannot be bounded.

4.3 R-Extensions of K-Group Laws

Let R be a discrete valuation ring with uniformizing element π, with field of fractions K, and with residue field k. As usual, R^{sh} denotes a strict henselization of R, and K^{sh} denotes the field of fractions of R^{sh}. Let X_K be a *smooth* K-group scheme *of dimension* d, assume that X_K is *of finite type*, and that $X_K(K^{sh})$ is *bounded* in X_K. As a group scheme over a field, X_K is automatically separated. The purpose of this section is to construct a smooth and separated R-scheme X of finite type with generic fibre X_K such that the group law of X_K extends to an R-birational group law on X and such that each translation on X_K by a K^{sh}-valued point extends to an R^{sh}-birational morphism of X. Later, it will turn out that X is already an R-dense open subscheme of the Néron model of X_K.

We start our construction by choosing a weak Néron model $(X_i)_{i\in I}$ of X_K; for the existence see Theorem 3.5/2. There is no restriction in assuming that the special fibres $X_i \otimes_R k$ are (non-empty and) irreducible for all $i \in I$. We will pick certain "minimal components" of this collection and glue them along the generic fibre to obtain the R-model X of X_K we are looking for.

In order to define minimal components, consider a left-invariant differential form ω of degree d on X_K which generates $\Omega^d_{X_K/K}$; for the existence see 4.2/1 and 4.2/3. It follows that ω is unique up to a constant in K^*. We want to define the order of ω on smooth R-models X of X_K which have an irreducible special fibre X_k, always assuming that X is separated and of finite type over R.

To do this, consider a general situation where $\mathscr{L}$ is a line bundle on a smooth R-scheme Z and where ζ is a generic point of the special fibre Z_k. Then the local ring $\mathcal{O}_{Z,\zeta}$ is a discrete valuation ring with uniformizing element π and, for any section f of $\mathscr{L}$ over the generic fibre Z_K which does not vanish at the generic point of Z_K lying over ζ, there is a unique integer n such that $\pi^{-n}f$ extends to a generator of $\mathscr{L}$ at ζ. The integer n is called the order of f at ζ, denoted by $\operatorname{ord}_\zeta f$.

Going back to the situation where we considered the section ω over the generic fibre of X, there is a unique generic point ξ of the special fibre X_k, since the latter has been assumed to be irreducible. We call $\operatorname{ord}_\xi \omega$ the order of ω at X and we denote it by $\operatorname{ord}_X \omega$. If $n = \operatorname{ord}_X \omega$, then $\pi^{-n}\omega$ generates $\Omega^d_{X/R}$ over X. Namely, $\pi^{-n}\omega$ is defined on X up to points of codimension $\geqq 2$, and X being normal, $\pi^{-n}\omega$ extends to a global section of X. Furthermore, since the zero set of a non-zero section in a line bundle is of pure codimension 1 on an irreducible normal scheme, it is seen that $\pi^{-n}\omega$ extends to a generator of $\Omega^d_{X/R}$ over X. Similarly, for sections $a \in \Gamma(X_K, \mathcal{O}_{X_K})$ (provided a is non-zero at the generic point of X_K lying over X_k), the order $\operatorname{ord}_X a$ can be defined. If $m = \operatorname{ord}_X a$, it follows that $\pi^{-m}a$ extends to a global section of $\mathcal{O}_X$. The latter is invertible if a is invertible over X_K. In this case, we have $|a(x)| = |\pi^m|$ for each K^{sh}-valued point x of X which extends to an R^{sh}-valued point of X.

Lemma 1. *Let X' and X'' be smooth and separated R-models of X_K which as above have irreducible special fibre each. Consider an R-rational map $u: X' \dashrightarrow X''$ which is an isomorphism on generic fibres; in particular, there is a unit $a \in \Gamma(X_K, \mathcal{O}^*_{X_K})$ satisfying $u^*_K\omega = a\omega$. Assume that $|a(x)| = 1$ for some $x \in X_K(K^{sh})$ such that x extends to a point in $X'(R^{sh})$. Then:*

(i) $n' := \operatorname{ord}_{X'}\omega \geqq n'' := \operatorname{ord}_{X''}\omega$.

(ii) *If U is the domain of definition of u, the morphism $u: U \longrightarrow X''$ is an open immersion if and only if $n' = n''$.*

Proof. Since $\pi^{-n'}\omega$ (resp. $\pi^{-n''}\omega$) generates $\Omega^d_{X'/R}$ (resp. $\Omega^d_{X''/R}$), there is a section $b \in \Gamma(X', \mathcal{O}_{X'})$ such that

$$u^*(\pi^{-n''}\omega) = b\pi^{-n'}\omega$$

over X'. Actually, b is only defined over U; however $X' - U$ is of codimension $\geqq 2$ in X' so that b extends to a section over X'. The preceding equation gives $a = \pi^{n''-n'}b$ over X_K. Since $\text{ord}_{X'}a = 0$ by our assumption on a, we see

$$n' - n'' = \text{ord}_{X'}b \geqq 0 .$$

This verifies the first assertion.

To obtain the second one, we see from 2.2/10 that u is étale on U if and only if $u^*\Omega^d_{X''/R} \longrightarrow \Omega^d_{U/R}$ is bijective; i.e., if and only if b is invertible over U and hence over X'. The latter is equivalent to $n' - n'' = 0$. Furthermore, since u_K is an isomorphism, Zariski's Main Theorem 2.3/2' implies that u is étale on U if and only if it is an open immersion. □

Let X' and X'' be smooth, separated R-models of X_K which are of finite type over R and which have irreducible special fibres. Then X' and X'' are called *equivalent* if the identity on X_K extends to an R-birational map $X' \dashrightarrow X''$.

Proposition 2. *Let X_K be a smooth K-group scheme of finite type such that $X_K(K^{sh})$ is bounded in X_K.*

(i) *There exists a largest integer n_0 such that $\text{ord}_X\omega \geqq n_0$ for all R-models X of X_K which are smooth, separated, and of finite type over R, and which have an irreducible special fibre X_k. All such R-models X with $\text{ord}_X\omega = n_0$ are called ω-minimal.*

(ii) *Up to equivalence there exist only finitely many R-models $X_1, \ldots, X_n$ of X_K which are ω-minimal.*

Proof. (i) Let $(X_i)_{i\in I}$ be a weak Néron model of X_K; for the existence see 3.5/2. We may assume that the special fibre of each X_i is irreducible. So the order of ω is defined with respect to each X_i. Let n_0 be the minimum of the finite set $\{\text{ord}_{X_i}\omega; i \in I\}$. We claim that n_0 satisfies assertion (i). Namely, consider a smooth R-model X of X_K which is separated and of finite type over R and which has an irreducible special fibre. Due to the weak Néron property 3.5/3, the identity on X_K extends to an R-rational map $u : X \dashrightarrow X_i$ for some $i \in I$. Then $\text{ord}_X\omega \geqq n_0$ by Lemma 1. In a similar way, assertion (ii) is deduced from Lemma 1 (ii). □

Since ω, as a left-invariant differential form of degree d, is unique up to a constant in K^*, it is clear that the notion of ω-minimality does not depend on the choice of ω. One has to interpret the ω-minimal R-models as those smooth R-models with irreducible special fibre, which are of "biggest" size, just as can be seen from the two R-models

$$\text{Spec}\, R[\zeta, \zeta^{-1}] \quad \text{and} \quad \text{Spec}\, R[\zeta, \zeta^{-1}, (\zeta - 1)/\pi]$$

of the multiplicative group $\mathbb{G}_m$ over K, and from the left-invariant differential form $\omega := \zeta^{-1}d\zeta$. Furthermore, we leave it to the reader to verify that, for the additive group $\mathbb{G}_a$ over K and for the left-invariant differential form $\omega := d\zeta$, there does not exist any ω-minimal R-model.

Lemma 3. *Let Z be a smooth R-scheme, and let η be a generic point of the special fibre of Z. Denote by R' the local ring $\mathcal{O}_{Z,\eta}$ of Z at η, and by K' the field of fractions of R'. If $X_1, \ldots, X_n$ is a set of representatives of the ω-minimal R-models of X_K, then, up to a splitting of special fibres into connected components, $X_1 \otimes_R R', \ldots, X_n \otimes_R R'$ represent the ω'-minimal R'-models of $X_K \otimes_K K'$, where ω' is the pull-back of ω to $X_K \otimes K'$.*

Proof. Due to 3.5/4, weak Néron models are compatible with the base change $R \longrightarrow R'$. Furthermore, each generic point ξ' of the special fibre of $X_i \otimes_R R'$ lies over a generic point ξ of the special fibre of X_i. Thus, we have $\mathrm{ord}_\xi \omega = \mathrm{ord}_{\xi'} \omega'$. Hence the R'-extension of an ω-minimal R-model of X_K decomposes into a union of ω'-minimal R'-models of $X_{K'}$. □

Now we are able to construct the R-model X of X_K we are looking for.

Proposition 4. *Let X_K be a smooth K-group scheme of finite type such that the set of K^{sh}-valued points of X_K is bounded in X_K. Then there exists an R-model X of X_K which is smooth, separated, faithfully flat, and of finite type over R and which satisfies the following conditions:*

(i) *Each open subscheme of X which is an R-model of X_K with irreducible special fibre is ω-minimal.*

(ii) *For each ω-minimal R-model X' of X_K, the identity on X_K extends to an R-rational map $X' \dashrightarrow X$ which is an open immersion on its domain of definition.*

(iii) *Let R' be the local ring $\mathcal{O}_{Z,\zeta}$ of a smooth R-scheme Z at a generic point ζ of the special fibre, and let K' be the field of fractions of R'. Then each translation on $X_{K'}$ by a K'-valued point of $X_{K'}$ extends to an R'-birational morphism of $X \otimes_R R'$, which is an open immersion on its domain of definition.*

Proof. Let $X_1, \ldots, X_n$ be a set of representatives of the ω-minimal R-models of X_K. By shrinking the special fibre of each X_i, we may assume that the following condition is satisfied:

(∗) For each pair of indices $i, j \in \{1, \ldots, n\}$ with $i \neq j$, the diagonal of $X_K \times_K X_K$ constitutes a Zariski-closed subset in $X_i \times_R X_j$.

Namely, let Δ_K be the diagonal in $X_K \times_K X_K$, and consider its schematic closure Δ in $X_i \times_R X_j$. Let $p_h : \Delta \longrightarrow X_h$ for $h = i$ or j be the projection onto the first or second factor. It is enough to know that the image of Δ_k under p_i is nowhere dense in $(X_i)_k$. Assume the contrary. Then the image of Δ_k contains a non-empty open part of $(X_i)_k$ and, hence, there is a point $\eta \in \Delta$ above the generic point ξ of the special fibre of X_i. Thus the local ring $\mathcal{O}_{\Delta,\eta}$ dominates $\mathcal{O}_{X_i,\xi}$. Since p_i is an isomorphism on generic fibres and since Δ is flat over R, both local rings give rise to the same field of fractions. But then, $\mathcal{O}_{X_i,\xi}$ being a discrete valuation ring, the map $\mathcal{O}_{X_i,\xi} \longrightarrow \mathcal{O}_{\Delta,\eta}$ is an isomorphism. Since Δ is of finite type over X_i, there exist open neighborhoods U of ξ in X_i and V of η in Δ such that p_i induces an isomorphism between V and U; cf. [EGA I], 6.5.4. Hence p_i is invertible over an R-dense open part of X_i, and

$$p_j \circ (p_i|_\Delta)^{-1} : X_i \dashrightarrow X_j$$

constitutes an R-birational map, as is seen by Lemma 1. However, this contradicts the choice of $X_1, \ldots, X_n$.

Now we can construct the desired R-model X of X_K by gluing all models $X_1, \ldots, X_n$ along generic fibres. Then X is separated due to condition (*), and it satisfies conditions (i) and (ii) by construction.

To verify condition (iii), assume first $R = R'$, and consider a translation $\tau_K : X_K \longrightarrow X_K$ on X_K by a K-valued point. Fix an open subscheme U of X consisting of the generic fibre X_K and of an irreducible component of the special fibre X_k. Furthermore, let $(X_i)_{i \in I}$ be a weak Néron model of X_K, where the special fibre of X_i is irreducible for each $i \in I$. Then, due to the weak Néron property 3.5/3, there exists an index $i \in I$ such that τ_K extends to an R-rational map $\tau : U \dashrightarrow X_i$. Since U is ω-minimal, the map τ is R-birational; it is an open immersion on its domain of definition by Lemma 1 (note that, for right translations, the assumption of Lemma 1 is satisfied by 4.2/5). Moreover, X_i is ω-minimal. Then it is clear that τ_K extends to an R-rational map

$$\tau : X \dashrightarrow X.$$

Likewise, one can construct an R-rational extension

$$\tau' : X \dashrightarrow X$$

of the inverse translation $(\tau_K)^{-1}$ on X_K. Since τ and τ' are composable with each other in terms of R-rational maps, it is easily seen that they are, in fact, R-birational. Finally, Lemma 1 shows that τ is an open immersion on its domain of definition. So, if $R = R'$, condition (iii) is satisfied. In the general case, we can perform the base change $R \longrightarrow R'$, and thereby reduce to the above special case by using 3.5/4 and Lemma 3. □

Applying assertion (iii) of the preceding proposition, we want to show next that we can extend the K-group law on X_K to an R-birational group law on the R-scheme X we have just constructed.

Proposition 5. *Let X_K be a smooth K-group scheme of finite type such that the set of K^{sh}-valued points of X_K is bounded in X_K. Let X be the R-model obtained in Proposition 4 by gluing a set of representatives of ω-minimal R-models. Then the group law m_K on X_K extends to an R-birational group law on X.*

More precisely, m_K extends to an R-rational map

$$m : X \times_R X \dashrightarrow X$$

such that the universal translations

$$\Phi : X \times_R X \dashrightarrow X \times_R X , \qquad (x, y) \longmapsto (x, m(x, y))$$

$$\Psi : X \times_R X \dashrightarrow X \times_R X , \qquad (x, y) \longmapsto (m(x, y), y)$$

are R-birational. Furthermore, m is associative; i.e., the usual diagram for testing associativity is commutative as far as the occurring rational maps are defined.

Proof. Let ξ be a generic point of the special fibre X_k *of* X, and denote by R' the local ring $\mathcal{O}_{X,\xi}$ of X at ξ. Let S' be the spectrum of R'; it can be viewed as an X-scheme and as an R-scheme. Due to Proposition 4, the translation τ_K obtained from Φ_K by the base change $S'_K \longrightarrow X_K$ extends to an S'-birational map

$$\tau_\xi : S' \times_R X \dashrightarrow S' \times_R X .$$

Now consider the commutative diagram of rational maps

$$\begin{array}{ccc} \tau_\xi : S' \times_R X & \dashrightarrow & S' \times_R X \\ \downarrow & & \downarrow \\ \Phi : X \times_R X & \dashrightarrow & X \times_R X . \end{array}$$

It follows from 2.5/5 or by a simple direct argument that Φ is defined at all generic points of the special fibre of $X \times_R X$ which project to ξ under the first projection. As we can apply this reasoning to any generic point of the special fibre X_k, we see that Φ is R-rational. Since each τ_ξ is S'-birational, it follows that Φ is R-birational.

Dealing with Ψ_K in the same way as with Φ_K yields an R-birational extension Ψ of Ψ_K. Choose an R-dense open part $W \subset X \times_R X$ containing the generic fibre such that Φ and Ψ are defined on W. Then, composing Φ with the projection onto the second factor of $X \times_R X$, and Ψ with the projection onto the first factor, we obtain two extensions $W \longrightarrow X$ of the group law m_K of X_K, which must coincide. Thus, m_K extends to an R-rational map

$$m : X \times_R X \dashrightarrow X ,$$

and we see that Φ and Ψ can be described by m as stated. In particular, the associativity is a consequence of the uniqueness of R-rational extensions of K-morphisms. □

It is a general fact that an R-birational group law on X, as obtained in the preceding proposition, always determines an R-group scheme $\bar{X}$; cf. 5.1/5.

Theorem 6. *Let X_K be a smooth K-group scheme of finite type. Let X be a smooth and separated R-model of X_K which is of finite type, and assume that the group law m_K of X_K extends to an R-birational group law $m : X \times_R X \dashrightarrow X$. Then there is a smooth and separated R-group scheme $\bar{X}$ of finite type, containing X as an R-dense open subscheme, and having X_K as generic fibre such that the group law on $\bar{X}$ extends the R-birational group law m on X. Up to canonical isomorphism, $\bar{X}$ is unique.*

This result which, for the case of birational group laws over a field, goes back to A. Weil [2], § V, n°33, Thm. 15, will be proved in Chapter 5 for a strictly henselian base ring R. The generalization for an arbitrary discrete valuation ring will follow in Chapter 6 by means of descent theory. That $\bar{X}$ satisfies the Néron mapping property will be shown in the next section by using an extension theorem for morphisms into group schemes; cf. 4.4/4.

4.4 Rational Maps into Group Schemes

In order to establish the Néron mapping property for the R-group scheme $\bar{X}$ which has been introduced in the last section, we want to make use of an extension argument of Weil for rational maps into group schemes; cf. Weil [2], § II, n°15, Prop. 1.

Theorem 1. *Let S be a normal noetherian base scheme, and let $u: Z \dashrightarrow G$ be an S-rational map from a smooth S-scheme Z to a smooth and separated S-group scheme G. Then, if u is defined in codimension $\leqq 1$, it is defined everywhere.*

As in Weil's proof, which deals with the case where the base consists of a field, we will proceed by reducing to the diagonal; the following basic fact is needed:

Lemma 2. *Let $u: Z \dashrightarrow \operatorname{Spec} A$ be a rational map from a normal noetherian scheme Z into an affine scheme* Spec A*. Then the set of points in Z, where u is not defined, is of pure codimension* 1*. In particular, if u is defined in codimension $\leqq 1$, it is defined everywhere.*

The assertion (cf. [EGA IV_4], 20.4.12) is due to the fact that a rational function on Z, which is defined in codimension $\leqq 1$, is defined everywhere or, equivalently, that any noetherian normal integral domain equals the intersection over all its localizations at prime ideals of height 1.

Now let us start the *proof of Theorem* 1. Consider the rational map

$$v: Z \times_S Z \dashrightarrow G, \qquad (z_1, z_2) \longmapsto u(z_1)u(z_2)^{-1},$$

and let U (resp. V) denote the domain of definition of u (resp. v). Then $U \times_S U$ is contained in V. First we want to show that V contains the diagonal Δ of $Z \times_S Z$. Since

$$V \cap \Delta \supset (U \times_R U) \cap \Delta = U$$

(where we have identified Z with Δ), we see that $v|_{V \cap \Delta}$ factors through the unit section $\varepsilon: S \longrightarrow G$. Set $F := (Z \times_S Z) - V$. We have to show $F \cap \Delta = \emptyset$. Consider a point x of $F \cap \Delta$, and let $s \in S$ be the image of x in S. Let H be an affine open neighborhood of $\varepsilon(s)$ in G. Then there exists an open neighborhood W of x in $Z \times_S Z$ such that v induces a rational map

$$v' := v|_W : W \dashrightarrow H .$$

Let V' be the domain of definition of v'; we have $V' \subset V$. Since $v|_{V \cap \Delta}$ factors through H, we see $V' \cap \Delta = V \cap \Delta$. Furthermore, set $F' := W - V'$. Since H is affine and $Z \times_S Z$ is normal (cf. 2.3/9), it follows from Lemma 2 that F' is of pure codimension 1 in W. Since

$$F' \cap \Delta = F \cap \Delta \subset Z - U$$

(where we have identified Z with Δ again), we know that $F' \cap \Delta$ is of codimension $\geqq 2$ in Δ if u is defined in codimension $\leqq 1$. Let d be the relative dimension of Z over S at x. Since F' is of pure codimension 1 in $W \subset Z \times_S Z$, and, since $\Delta \subset Z \times_S Z$ is defined locally by d equations, due to the smoothness of Z, we get

$$\dim_x(F' \cap \Delta) \geqq \dim_x F' - d = \dim_x(Z \times_S Z) - 1 - d = \dim_x \Delta - 1 .$$

However, this contradicts the fact that $F' \cap \Delta$ is of codimension $\geqq 2$ in Δ. Thus V must contain the diagonal Δ as claimed.

It remains to show that this fact implies $U = Z$. Due to 2.5/5 it is enough to show that there exists a faithfully flat S-morphism $f: Z' \longrightarrow Z$ from a smooth S-scheme Z' of finite type to Z such that $u \circ f$ is defined everywhere. So, let Z' be the intersection of V with $Z \times_S U$ in $Z \times_S Z$. Then the first projection from $Z \times_S Z$ to Z gives rise to a faithfully flat morphism $f: Z' \longrightarrow Z$. Namely, since smooth morphisms are flat, it only remains to show that f is surjective. So, let $z: T \longrightarrow Z$ be a geometric point of Z; i.e., T is the spectrum of a field. Viewing V as a Z-scheme with respect to the first projection, the scheme $T \times_Z V$ is non-empty since V contains the diagonal Δ of $Z \times_S Z$. Furthermore, the domain of definition U of u is S-dense open in Z. Hence the intersection of $T \times_Z V$ with $T \times_S U$ in $T \times_S Z$ is not empty. Thus we see that the morphism f is surjective and, hence, faithfully flat. Now look at the morphism

$$V \cap (Z \times_S U) \longrightarrow G , \qquad (z_1, z_2) \longmapsto v(z_1, z_2) u(z_2).$$

It is clear that this map coincides with $u \circ f$, in terms of S-rational maps. Thus, the S-rational map u is defined everywhere, and we have finished the proof of Theorem 1. □

Remark 3. The method of reduction to the diagonal which was used in the proof of Theorem 1 works also within the context of formal schemes or rigid analytic spaces. So, if R is a complete discrete valuation ring, the assertion of Theorem 1 remains true if Z is of type $\operatorname{Spec} R[[t]]$ or $\operatorname{Spec} R\{t\}$ (formal or strictly convergent power series in a finite number of variables $t_1, \ldots, t_n$).

Now it is easy to show that the R-group scheme $\bar{X}$ we have introduced in Section 4.3 satisfies the Néron mapping property and thereby to end the proof of the existence theorem 1.3/1 for Néron models over a discrete valuation ring R (modulo the proof of Theorem 4.3/6). Recall the situation of 4.3. Starting with a smooth K-group scheme of finite type X_K such that the set of its K^{sh}-valued points is bounded in X_K, we have constructed in 4.3/4 a smooth and separated R-model of finite type X such that the group law on X_K extends to an R-birational group law on X; cf. 4.3/5. In 4.3/6 we have claimed that there is a unique extension of X to a smooth and separated R-group scheme of finite type $\bar{X}$ containing X as an R-dense open subscheme.

Corollary 4. *Let X be the R-model of X_K as considered in* 4.3/4 *and* 4.3/5, *and let $\bar{X}$ be the associated R-group scheme in the sense of* 4.3/6. *Then $\bar{X}$ is a Néron model of X_K over the ring R.*

Furthermore, for each ω-minimal R-model X' of X_K, the identity on X_K extends to an open immersion $X' \hookrightarrow \bar{X}$ over R.

Proof. In order to show that $\bar{X}$ satisfies the Néron mapping property let Z be a smooth R-scheme and let $u_K : Z_K \longrightarrow X_K$ be a K-morphism. We have to show that u_K extends to an R-morphism $u : Z \longrightarrow \bar{X}$.

It is enough to consider the case where Z has an irreducible special fibre. Let ζ be the generic point of the special fibre of Z, and let $R' = \mathcal{O}_{Z,\zeta}$ be the local ring of Z at ζ.

Look first at the rational map

$$Z \times_R X \dashrightarrow Z \times_R X \,, \qquad (z, x) \longrightarrow (z, u_K(z)x) \,,$$

which is defined on the generic fibre. Applying the base change $\operatorname{Spec} R' \longrightarrow Z$, this map is turned into an R'-rational map; cf. 4.3/4. Then it follows from 2.5/5 that the map

$$\tau : Z \times_R \bar{X} \dashrightarrow \bar{X} \,, \qquad (z, x) \longrightarrow u_K(z)x \,,$$

is defined at all generic points of the special fibre of $Z \times_R \bar{X}$ which project to ζ under the first projection. So τ is an R-rational map. Since it is defined at the generic fibre, it is defined everywhere by Theorem 1. Therefore, if we denote by p the structural morphism of Z, and by ε the unit section of $\bar{X}$, the composition of the morphism

$$(\mathrm{id}_Z, \varepsilon \circ p) : Z \longrightarrow Z \times_R \bar{X}$$

with τ yields an R-morphism $u : Z \longrightarrow \bar{X}$ extending u_K. The uniqueness of u follows from the separatedness of $\bar{X}$.

If X' is an ω-minimal R-model of X_K, the identity on X_K extends to an R-rational map from X' to X by 4.3/4. Hence it extends to an R-morphism from X' to $\bar{X}$ by Theorem 1. Then it is an open immersion, due to 4.3/1. □

Chapter 5. From Birational Group Laws to Group Schemes

For the construction of Néron models, we need the fact that an S-birational group law on a smooth S-scheme with non-empty fibres can be birationally enlarged to a smooth S-group scheme; see 4.3/6. The purpose of the present section is to prove this result in the case where S is strictly henselian. In Chapter 6, the result will be extended to a more general base.

The technique of constructing group schemes from birational group laws is originally due to A. Weil [2], § V, n°33, Thm. 15; he considered birational group laws over fields. More general situations were dealt with by M. Artin in [SGA 3_{II}], Exp. XVIII, among them birational group laws over strictly henselian rings. The proof we give in this chapter, essentially follows the exposition of M. Artin [9]. Finally, in Chapter 6, descent techniques can be used to handle the case where the base is of a more general type.

5.1 Statement of the Theorem

In the following, let S be a scheme, and let X be a smooth separated S-scheme of finite type. Furthermore, we will assume that X has non-empty fibres over S or, which amounts to the same, that X is faithfully flat over S.

Definition 1. *An S-birational group law on X is an S-rational map*

$$m : X \times_S X \dashrightarrow X \,, \qquad (x, y) \longmapsto xy \,,$$

such that

(a) *the S-rational maps*

$$\Phi : X \times_S X \dashrightarrow X \times_S X \,, \qquad (x, y) \longmapsto (x, xy) \,,$$

$$\Psi : X \times_S X \dashrightarrow X \times_S X \,, \qquad (x, y) \longmapsto (xy, y) \,,$$

are S-birational, and

(b) *m is associative; i.e., $(xy)z = x(yz)$ whenever both sides are defined.*

Just as in the case of group schemes, the maps Φ and Ψ will be referred to as universal left or right translations.

Note that, in place of (a), it is enough to require Φ and Ψ to be open immersions on an S-dense open subscheme U of $X \times_S X$. To see this, one has only to verify that the images $V = \Phi(U)$ and $W = \Psi(U)$ are S-dense in $X \times_S X$. Since each fibre of U over S has the same number of components as the corresponding fibre of $X \times_S X$ over S, the same is true for the fibres of V and W over S. Hence V and W are S-dense in $X \times_S X$ if Φ and Ψ are open immersions on U.

The notion of S-birational group law is compatible with base change. Furthermore, an S-birational group law on X induces an S-birational group law on each S-dense open subscheme of X. In particular, if $\bar{X}$ is an S-group scheme and if X is an S-dense open subscheme of $\bar{X}$, the group law of $\bar{X}$ induces an S-birational group law on X. But there are S-birational group laws which do not occur in this way. Namely, even if the base consists of a field, one can blow up a subscheme of a group scheme $\bar{X}$ and consider the induced birational group law on the blowing-up. So it is natural to shrink X in order to expect that an S-birational group law on X extends to a group law on an S-scheme $\bar{X}$ containing X.

Definition 2. *Let m be an S-birational group law on a separated and smooth S-scheme X which is faithfully flat and of finite type over S. A solution of m is a separated and smooth S-group scheme $\bar{X}$ of finite type over S with group law $\bar{m}$, together with an S-dense open subscheme $X' \subset X$ and an open immersion $X' \hookrightarrow \bar{X}$ such that*

(a) *the image of X' is S-dense in $\bar{X}$, and*

(b) *$\bar{m}$ restricts to m on X'.*

First we want to show that solutions of S-birational group laws are unique.

Proposition 3. *Let m be an S-birational group law on a separated and smooth S-scheme X which is faithfully flat and of finite type over S. If there exists a solution of m, it is uniquely determined up to canonical isomorphism.*

For the proof we need the following well-known lemma.

Lemma 4. *Let G be a smooth S-group scheme, and let U be an S-dense open subscheme of G. Then the morphism*

$$U \times_S U \longrightarrow G, \qquad (x, y) \longmapsto xy$$

is smooth and surjective.

Proof of Proposition 3. Let

$$\sigma_1 : X_1' \hookrightarrow \bar{X}_1 \qquad \text{and} \qquad \sigma_2 : X_2' \hookrightarrow \bar{X}_2$$

be solutions of the S-birational group law m on X, and set $Y := X_1' \cap X_2'$. Then Y is an S-dense open subscheme of X, and each $\sigma_i(Y)$ is S-dense open in $\bar{X}_i$, $i = 1, 2$. The group laws $\bar{m}_i$ of $\bar{X}_i$ give rise to morphisms

$$\bar{m}_i \circ (\sigma_i \times \sigma_i) : Y \times_S Y \longrightarrow \bar{X}_i, \qquad i = 1, 2,$$

which are faithfully flat by Lemma 4. Furthermore, the morphisms σ_1 and σ_2 yield an S-birational map

$$\alpha = \sigma_2 \circ \sigma_1^{-1} : \bar{X}_1 \dashrightarrow \bar{X}_2$$

which is compatible with the group laws; i.e.,

$$\bar{m}_2 \circ (\sigma_2 \times \sigma_2) = \alpha \circ \bar{m}_1 \circ (\sigma_1 \times \sigma_1).$$

So, due to 2.5/5, the map α is defined everywhere. Since α is compatible with the group laws, it is clear that α is a group homomorphism. Similarly, $\beta = \sigma_1 \circ \sigma_2^{-1}$ is a group homomorphism which is defined everywhere. Since α and β are inverse to each other, they yield S-isomorphisms between $\bar{X}_1$ and $\bar{X}_2$. □

Next we want to look at the existence of solutions of S-birational group laws. In the present chapter we will consider the case where the base consists of a discrete valuation ring; see 6.6/1 for the case where the base is more general.

Theorem 5. *Let S be the spectrum of a field or of a discrete valuation ring, and let m be an S-birational group law on a smooth separated S-scheme X which is faithfully flat and of finite type over S. Then there exists a solution of m, i.e., a smooth separated S-group scheme $\bar{X}$ of finite type with a group law $\bar{m}$, together with an S-dense open subscheme $X' \subset X$ and an S-dense open immersion $X' \hookrightarrow \bar{X}$ such that $\bar{m}$ restricts to m on X'.*

The group scheme $\bar{X}$ is unique, up to canonical isomorphism. If (in the case where the base S consists of a valuation ring) the generic fibre X_K of X is a group scheme under the law m_K, the above assertion is true for $X' = X$. So, in this case, it is not necessary to shrink X.

The proof of the existence will follow in the subsequent sections (cf. 5.2/2, 5.2/3, and 6.5/2), whereas the uniqueness has already been proved. So, accepting the existence of $\bar{X}$, let us concentrate on the additional assertion on the domain X' where the group laws on X and $\bar{X}$ coincide. Assume that the base S consists of a discrete valuation ring and that the generic fibre X_K is a group scheme. By the uniqueness assertion, the S-rational map

$$\iota : X \dashrightarrow \bar{X}$$

induced by $X' \hookrightarrow X$ restricts to a K-isomorphism

$$\iota_K : X_K \longrightarrow \bar{X}_K .$$

Hence ι is defined in codimension ≤ 1 so that, by 4.4/1, the rational map ι is defined everywhere. Now let ω be a differential form generating $\Omega^d_{\bar{X}/S}$, where d is the relative dimension of $\bar{X}$ over S; cf. 4.2/3. Pulling back ω, we get a differential form $\iota^*\omega$ on X which generates $\Omega^d_{X/S}$ over $X' \cup X_K$; hence $\iota^*\omega$ generates $\Omega^d_{X/S}$ in codimension ≤ 1. Since on a normal scheme, the zero set of a non-vanishing section of a line bundle is empty or of pure codimension ≤ 1, we see that $\iota^*\omega$ has no zeros. Thus ι is étale by 2.2/10. Since ι is birational, Zariski's Main Theorem 2.3/2′ implies that ι is an open immersion. □

5.2 Strict Birational Group Laws

In the following, let S be a scheme, and let X be a smooth separated S-scheme of finite type. Furthermore, we assume that X is faithfully flat over S.

If X is an S-dense open subscheme of an S-group scheme $\bar{X}$, then, for each T-valued point $x : T \longrightarrow X$, the set of points $y \in T \times_S X$ which is characterized symbolically by the conditions

$$xy \in T \times_S X\,, \qquad x^{-1}y \in T \times_S X\,, \qquad \text{and} \qquad xy^{-1} \in T \times_S X$$

is T-dense and open in $T \times_S X$. Thus, we see that the group law of X induces an S-birational group law on X which is of a special type. Namely, there is an open subscheme U of $X \times_S X$ which is X-dense in $X \times_S X$ (with respect to both projections $p_i : X \times_S X \longrightarrow X$, $i = 1, 2$; i.e., X-dense when $X \times_S X$ is viewed as an X-scheme via each p_i), such that the universal translations

$$\Phi : X \times_S X \dashrightarrow X \times_S X\,, \qquad (x, y) \longmapsto (x, xy),$$

$$\Psi : X \times_S X \dashrightarrow X \times_S X\,, \qquad (x, y) \longmapsto (xy, y)\,,$$

are defined and open immersions on U, and their images $V := \Phi(U)$ and $W := \Psi(U)$ are X-dense in $X \times_S X$. Just take for U the intersection of $X \times_S X$ with the inverse images of $X \times_S X$ under the group law and both universal translations on X. So it is natural to introduce the following terminology:

Definition 1. *An S-birational group law on X is called a strict (S-birational) group law if it satisfies the following condition: There is an X-dense open subscheme U of $X \times_S X$, on which m is defined, such that the universal translations*

$$\Phi : X \times_S X \dashrightarrow X \times_S X\,, \qquad (x, y) \longmapsto (x, xy)\,,$$

$$\Psi : X \times_S X \dashrightarrow X \times_S X\,, \qquad (x, y) \longmapsto (xy, y)\,,$$

are isomorphisms from U onto X-dense open subschemes $V := \Phi(U)$ and $W := \Psi(U)$ in $X \times_S X$. (As before, X-density is meant with respect to both projections from $X \times_S X$ onto its factors.)

Note that X-density implies S-density. So the subschemes U, V, and W above are S-dense in $X \times_S X$. The first step in the existence proof of 5.1/5 consists in showing that each S-birational group law on X induces a strict group law on an S-dense open subscheme of X if S consists of a field or of a discrete valuation ring.

Proposition 2. *Let S consist of a field or of a discrete valuation ring. Let X be a smooth separated S-scheme of finite type, and consider an S-birational group law m on X. Then there exists an S-dense open subscheme X' of X such that m restricts to a strict group law on X'.*

Proof. Let U be the S-dense open subscheme of $X \times_S X$ such that m is defined at U and such that the universal translations Φ and Ψ are open immersions on U. Set $V = \Phi(U)$ and $W = \Psi(U)$. Since U, V, and W are S-dense in $X \times_S X$, the set

$$Z = U \cap V \cap W$$

is again S-dense open in $X \times_S X$. We want to show that there exists an S-dense open subscheme Ω_1 of X such that $Z \cap (\Omega_1 \times_S X)$ is Ω_1-dense in $\Omega_1 \times_S X$ with respect to the first projection p_1. Due to 2.5/1, the set

$$T_1 = \{x \in X;\, Z \cap (x \times_S X) \text{ is not dense in } x \times_S X\}$$

is constructible in X. Since Z is S-dense in $X \times_S X$, the generic points of the fibres of X over S do not belong to T_1. Hence the closure $\overline{T}_1$ of T_1 in X cannot be dense in any fibre of X if S consists of a discrete valuation ring. So the open subscheme $\Omega_1 = X - \overline{T}_1$ is S-dense in X and has the required property. Similarly, one defines a subscheme Ω_2 of X by considering the second projection. Then the subscheme

$$X' = \Omega_1 \cap \Omega_2$$

is S-dense open in X, and $Z \cap (X' \times_S X')$ is X'-dense in $X' \times_S X'$ (with respect to both projections).

Setting

$$\begin{aligned} U' &:= U \cap (X' \times_S X') \cap (m|_U)^{-1}(X')\,, \\ V' &:= \Phi(U')\,, \\ W' &:= \Psi(U'), \end{aligned}$$

it remains to show that these open subschemes are X'-dense in $X' \times_S X'$. As a general argument, we will use the fact that U, V, and W give rise to X'-dense open subschemes in $X' \times_S X'$, because $Z = U \cap V \cap W$. Now consider a point $a \in X'$. We may assume that the base S is a field and that a is an S-valued point of X'. First we will show that U' is X'-dense in $X' \times_S X'$ with respect to the first projection p_1. If we view $X \times_S X$ as an X-scheme via p_1, the base change $a \longrightarrow X$ transforms Φ into

$$\Phi(a, \cdot) : U \cap (a \times_S X) \xrightarrow{\sim} V \cap (a \times_S X) \subset a \times_S X\,,$$

which is an open immersion with dense image. Then the open subscheme

$$\Phi(a, \cdot)^{-1}(V \cap (a \times_S X')) = (m|_U)^{-1}(X') \cap (a \times_S X)$$

is also dense in $a \times_S X$. This shows that U' is p_1-dense, i.e., X'-dense with respect to p_1. In a similar way, using Ψ, one shows U' is p_2-dense. Next, it is clear that V' is p_1-dense, since $V' \cap (a \times_S X')$ is the image of the dense open subscheme $U' \cap (a \times_S X')$ of $a \times_S X$ under the open immersion $\Phi(a, \cdot)$; the latter has a dense image in $a \times_S X$. By the same argument, using $\Psi(\cdot, a)$, we see that W' is p_2-dense. In order to show that W' is p_1-dense, set $U_a := m^{-1}(a)$, and consider the diagram of isomorphisms

$$\begin{array}{ccccc}
U_a & & \xrightarrow{\quad\sim\quad} & & W \cap (a \times_S X) =: W_a \\
\Big\downarrow\wr & \searrow & & & \Big\downarrow\cap \\
 & & U & \xrightarrow{\sim\Psi} & W \\
 & & \Big\downarrow{\wr\,\Phi} & & \\
V_a := V \cap (X \times_S a) & \subset & V & &
\end{array}$$

Since a belongs to X', the set V_a is dense in $X \times_S a$, and W_a is dense in $a \times_S X$. The same is true if we replace V_a by its restriction to $X' \times_S a$ and W_a by its restriction to $a \times_S X'$. Taking inverse images with respect to Φ and Ψ, the set

$$U_a \cap U' = \Phi^{-1}(V_a \cap (X' \times_S a)) \cap \Psi^{-1}(W_a \cap (a \times_S X'))$$

is open and dense in U_a. Hence its image under Ψ, which is $W' \cap (a \times_S X)$, is open and dense in $a \times_S X$. Thereby we see that W' is p_1-dense. Similarly, one shows that V' is p_2-dense. □

The proposition reduces the proof of Theorem 5.1/5 to the problem of enlarging a strict group law on X to a group law on a group scheme $\bar{X}$. If the base scheme S is normal and strictly henselian (of any dimension), we will construct the group scheme $\bar{X}$ in a direct way. The case where S consists of a field or of a discrete valuation ring, without assuming that the latter is strictly henselian, will be reduced to the preceding one by means of descent theory, cf. 6.5/2. For further generalizations see Section 6.6.

Theorem 3. *Let S be the spectrum of a strictly henselian local ring which is noetherian and normal, and let m be a strict group law on a separated smooth S-scheme X which is faithfully flat and of finite type over S. Then there exists an open immersion $X \hookrightarrow \bar{X}$ with S-dense image into a smooth separated S-group scheme $\bar{X}$ of finite type such that the group law $\bar{m}$ of $\bar{X}$ restricts to m on X.*

The S-group scheme $\bar{X}$ is unique, up to canonical isomorphism.

The uniqueness assertion of Theorem 5.1/5, which has already been proved in Section 5.1, yields the uniqueness assertion of the present theorem. A proof of the existence part will be given in Section 5.3, assuming that the base S is strictly henselian. The idea is easy to describe, although a rigorous proof requires the consideration of quite a lot of unpleasant technical details. Namely, a smooth scheme X over a strictly henselian base S admits many sections in the sense that the points of the special fibre X_k which lift to S-valued points of X are schematically dense in X_k; cf. 2.3/5. So the idea is to construct $\bar{X}$ by gluing "translates" of X. More precisely, consider an S-valued point a of X and a copy $X(a)$ of X, thought of as a left translate of X by a. Then one can glue X and $X(a)$ along the correspondence given by the left translation by a

$$\Phi(a, \cdot) : X \dashrightarrow X .$$

The result is a new S-scheme $X' = X \cup X(a)$, and it has to be verified that the strict group law m on X extends to a strict group law m' on X'. The left translation by a

$$\Phi'(a, \cdot) : X' \dashrightarrow X'$$

is now defined on the open subscheme X of X'. Repeating such a step finitely many times with suitable S-valued points $a_1, \ldots, a_n \in X(S)$, and applying a noetherian argument, one ends up with an S-scheme $\bar{X} = X^{(n)}$ such that the strict group law m on X extends to a strict group law $\bar{m}$ on $\bar{X}$, such that the S-rational map

$$\bar{m} : \bar{X} \times_S \bar{X} \dashrightarrow \bar{X}$$

is defined on the open subscheme $X \times_S X \subset \bar{X} \times_S \bar{X}$. Then it is not difficult to show that $\bar{m}$ defines a group law on $\bar{X}$, and that $\bar{X}$ is the S-group scheme we are looking for.

The technical problems in the proof of Theorem 3 are due to the fact that, for a point $a \in X$, the product ax is only defined for "generic" $x \in X$. This drawback disappears, when we look at the situation from the point of view of group functors. Let m be a strict group law on X, as in Theorem 3, and consider the group functor

$$\mathscr{R}_{X/S} : (\mathrm{Sch}/S) \longrightarrow (\mathrm{Sets})$$

which associates to each S-scheme T the set of T-birational maps from $X_T = X \times_S T$ onto itself. Identifying X with its functor of points $h_X = \mathrm{Hom}(\cdot, X)$, cf. 4.1, we claim that there is a monomorphism $X \hookrightarrow \mathscr{R}_{X/S}$ respecting the laws of composition on X and $\mathscr{R}_{X/S}$. Namely, due to the definition of strict group laws, one knows that the universal left translation

$$\Phi : X \times_S X \dashrightarrow X \times_S X , \qquad (x, y) \longmapsto (x, m(x, y))$$

is X-birational if $X \times_S X$ is viewed as an X-scheme via the first projection. So, for any S-scheme T and any T-valued point $a \in X(T)$, the map

$$\tau_a : T \times_S X \dashrightarrow T \times_S X ,$$

the "left translation" by a obtained from Φ by means of the base change $a : T \longrightarrow X$, is T-birational and thus belongs to $\mathscr{R}_{X/S}(T)$. It is clear that the maps

$$X(T) \longrightarrow \mathscr{R}_{X/S}(T) , \qquad a \longmapsto \tau_a ,$$

constitute a morphism of functors $X \longrightarrow \mathscr{R}_{X/S}$.

Lemma 4. *The morphism $X \longrightarrow \mathscr{R}_{X/S}$ is a monomorphism which respects the laws of composition on X and on $\mathscr{R}_{X/S}$; i.e., for any S-scheme T and all T-valued points a, $b, c \in X(T)$ satisfying $m(a, b) = c$, one has $\tau_a \circ \tau_b = \tau_c$.*

Proof. We have to show that all maps $X(T) \longrightarrow \mathscr{R}_{X/S}(T)$ are injective. So consider $a, b \in X(T)$ with $\tau_a = \tau_b$. Applying the base change $T \longrightarrow S$ to our situation, we may consider T as the new base, writing S instead of T. Let U be the X-dense open subscheme of $X \times_S X$ required by Definition 1 (on which the universal translations are open immersions). Using the X-density of U with respect to the first projection, we see that the compositions

$$\Psi_a : S \times_S X \xrightarrow{a \times \mathrm{id}_X} X \times_S X \overset{\Psi}{\dashrightarrow} X \times_S X ,$$
$$\Psi_b : S \times_S X \xrightarrow{b \times \mathrm{id}_X} X \times_S X \overset{\Psi}{\dashrightarrow} X \times_S X$$

are defined as S-rational maps. Since $\Psi_a = (\tau_a, \mathrm{id}_X)$ and $\Psi_b = (\tau_b, \mathrm{id}_X)$ when $S \times_S X$ is identified with X, we see that $\tau_a = \tau_b$ yields $\Psi_a = \Psi_b$. Now Ψ is an open immersion on U, so $a \times \mathrm{id}_X$ and $b \times \mathrm{id}_X$ must coincide on the S-dense open subscheme

$$X' := (a \times \mathrm{id}_X)^{-1}(U) \cap (b \times \mathrm{id}_X)^{-1}(U)$$

of $S \times_S X$, hence on all of $S \times_S X$. In particular, their first components agree, i.e., $a = b$. Thus we see that $X \longrightarrow \mathscr{R}_{X/S}$ is a monomorphism. That this transformation respects the laws of composition follows immediately from the associativity of m. □

If X has been expanded into an S-group scheme $\bar{X}$ such that X is S-dense and open in $\bar{X}$ and such that the group law on $\bar{X}$ restricts to the strict group law m on X, then there is a canonical commutative diagram of natural transformations

$$\begin{array}{ccc} X & \longrightarrow & \mathcal{R}_{X/S} \\ \downarrow & & \downarrow \wr \\ \bar{X} & \longrightarrow & \mathcal{R}_{\bar{X}/S} \end{array}$$

where the vertical arrow on the right-hand side is an isomorphism, since X is S-dense in $\bar{X}$. Although it is not in general true that the group functor $\bar{X}$ is generated by X, i.e., that $X(T)$ generates the group $\bar{X}(T)$ for all S-schemes T, the latter is nevertheless correct if T is a strictly henselian local S-scheme. Namely the group law on $\bar{X}$ induces a surjective and smooth S-morphism

$$X \times_S X \longrightarrow \bar{X}\,,$$

c.f. 5.1/4, so that, by 2.3/5, each T-valued point of $\bar{X}$ lifts to a T-valued point of $X \times_S X$.

5.3 Proof of the Theorem for a Strictly Henselian Base

We have already seen in 5.2/2 that Theorem 5.2/3 implies Theorem 5.1/5 if the base is strictly henselian. So we may restrict ourselves to strict group laws and give only a proof of 5.2/3. In this section we assume that *the base* S consists of a *strictly henselian local ring* which is *noetherian* and *normal*. Furthermore, let X be a smooth and separated S-scheme which is faithfully flat and of finite type over S, and let m be a strict group law on X; the symbols Φ, Ψ, and U, V, W will be used in the sense of 5.2/1.

Introducing further notational conventions, let X^n be the n-fold fibred product of X over S, and, for integers $1 \leqq i_1 < \ldots < i_r \leqq n$, let

$$p_{i_1 \ldots i_r} : X^n \longrightarrow X^r$$

be the projection of X^n onto the product of the factors with indices $i_1, \ldots, i_r$. In such a situation, we can view X^n as an X^r-scheme with respect to the morphism $p_{i_1 \ldots i_r}$. So we have the notion of X^r-density in X^n; to be more precise, we will speak of $p_{i_1 \ldots i_r}$-density. Sometimes, we will write $x = (x_1, \ldots, x_n)$ for points in X^n and $(x_{i_1}, \ldots, x_{i_r})$ instead of $p_{i_1 \ldots i_r}(x)$ for their projections onto X^r. As usual, the S-rational map $m : X^2 \dashrightarrow X$ will be characterized by $(x_1, x_2) \longmapsto x_1 x_2$.

Lemma 1. *Let Ω be the set of points $(x, y, z, w) \in X^4$ such that*

$$(z, w) \in U\ , \qquad (y, w) \in U\ , \qquad \textit{and} \qquad (x, yw) \in U\ .$$

Then Ω is p_{123}-dense in X^4.

Proof. Recall that the intersection of finitely many p_{123}-dense open subschemes of X^4 is p_{123}-dense and open again. Since U is p_1-dense in X^2, the first two conditions pose no problem. So it remains to show that the set Ω' of all points $(x, y, w) \in X^3$, satisfying $(y, w) \in U$ and $(x, yw) \in U$, is p_{12}-dense and open in X^3. We can describe Ω' as the inverse image of U with respect to the following morphism:

$$\begin{array}{ccccc} X \times_S U & \xrightarrow{\mathrm{id}_X \times \Phi} & X^3 & \xrightarrow{p_{13}} & X^2\ , \\ (x, y, w) & \longmapsto & (x, y, yw) & \longmapsto & (x, yw)\ . \end{array}$$

Since U is p_1-dense in X^2, and since Φ leaves the first component fixed and is an open immersion on U with a p_1-dense image in X^2, we see that Ω' is p_{12}-dense and open in X^3. □

The assertion of Lemma 1 is only an example for similar assertions of this type. Roughly speaking, it says that, fixing x, y, and z, the stated conditions form open conditions on w; these are satisfied if w is generic.

Lemma 2. *Let Γ be the schematic closure in X^3 of the graph of $m: U \longrightarrow X$. Let T be an S-scheme. If (a, b, c) is a T-valued point in $\Gamma(T) \subset X^3(T)$, then, using the functor $\mathscr{R}_{X/S}$ of 5.2, the T-birational maps τ_a, τ_b, and τ_c of X_T satisfy $\tau_a \circ \tau_b = \tau_c$ in $\mathscr{R}_{X/S}(T)$.*

Proof. Let Ω be the p_{123}-dense open subscheme of X^4 which was considered in Lemma 1. Then the S-rational maps

$$\lambda: X^4 \dashrightarrow X^4\ , \qquad (x, y, z, w) \longmapsto (x, y, x(yw), w),$$
$$\mu: X^4 \dashrightarrow X^4\ , \qquad (x, y, z, w) \longmapsto (x, y, zw, w)\ ,$$

are defined on Ω. Next, let $\Omega' := \Omega \cap p_{12}{}^{-1}(U)$. We claim that $\Omega' \cap (\Gamma \times_S X)$ is schematically dense in $\Omega \cap (\Gamma \times_S X)$. Namely, $p_{12}{}^{-1}(U) \cap (\Gamma \times_S X)$ is schematically dense in $\Gamma \times_S X$ by the definition of Γ (since X is flat over S), and this density is not destroyed when we intersect both sets with an open subscheme of X^4 such as Ω. Since the law m is associative, the morphism $\mu|_{(\Gamma \times_S X) \cap \Omega'}$ factors through Λ, the schematic image of $\lambda|_\Omega$. By continuity, also $\mu|_{(\Gamma \times_S X) \cap \Omega}$ factors through Λ, and thus yields a morphism

$$\mu: (\Gamma \times_S X) \cap \Omega \longrightarrow \Lambda\ .$$

Now set

$$\varphi = (a, b, c) \times \mathrm{id}_X : T \times_S X \longrightarrow X^4\ ,$$

and $\Omega_{a,b,c} := \varphi^{-1}(\Omega)$. Then $\Omega_{a,b,c}$ is T-dense and open in X_T. Let $\varphi_T : X_T \longrightarrow X_T^4$ be the T-morphism derived from φ, and let μ_T be the T-morphism obtained from μ by means of the base change $T \longrightarrow S$. Then $p_3 \circ \mu_T \circ \varphi_T$ coincides with τ_c on $\Omega_{a,b,c}$, but also with $\tau_a \circ \tau_b$ since $\mu \circ \varphi$ factors through Λ. Hence, we have $\tau_a \circ \tau_b = \tau_c$ in $\mathscr{R}_{X/S}(T)$. □

We state an important consequence of Lemma 2.

Lemma 3. *Let Γ be the schematic closure in X^3 of the graph of $m: U \longrightarrow X$, and let $q_{ij}: \Gamma \longrightarrow X^2$ be the morphisms induced from the projections $p_{ij}: X^3 \longrightarrow X^2$. Then each q_{ij} is an open immersion and has an image which is p_1-dense and p_2-dense in X^2*

Proof. First we want to show that each q_{ij} is injective as a map of sets. If (a, b, c) is a T-valued point in $\Gamma(T)$ for some S-scheme T, then $\tau_a \circ \tau_b = \tau_c$ by Lemma 2. Since this is an identity in the group $\mathscr{R}_{X/S}(T)$, any two of the maps τ_a, τ_b, τ_c determine the third one. As stated in 5.2/4, the natural transformation $X \longrightarrow \mathscr{R}_{X/S}$ is a monomorphism. Hence a point of Γ is known if two of its components are given. This implies that q_{ij} is injective as a map of sets and, hence, that q_{ij} is quasi-finite. We claim that the maps q_{ij} are, in fact, S-birational. Namely, using the notation of 5.2/1, the projection q_{12} gives rise to an isomorphism $q_{12}^{-1}(U) \xrightarrow{\sim} U$ because m is defined on U. Furthermore, q_{13} defines an isomorphism $q_{13}^{-1}(V) \xrightarrow{\sim} V$ because q_{13} is injective and because $\Phi_{|U}$ is an isomorphism $U \xrightarrow{\sim} V$. Likewise, q_{23} defines an isomorphism $q_{23}^{-1}(W) \xrightarrow{\sim} W$ because q_{23} is injective and because $\Psi_{|U}$ is an isomorphism $U \xrightarrow{\sim} W$. Thus, by Zariski's Main Theorem 2.3/2′ (recall that S is normal), each q_{ij} is an open immersion and, due to the X-density of U, V, and W in X^2, the image of each q_{ij} is X-dense in X^2 (with respect to p_1 and p_2). □

Fixing points $a, b, c \in X(T)$ for some S-scheme T, we see from the preceding lemma that there exists at most one point $x \in X(T)$ such that $ax = c$ and at most one point $y \in X(T)$ such that $yb = c$. Suggestively, we will write $a^{-1}c$ for x and cb^{-1} for y. With this notation the assertion of Lemma 3 can be interpreted as follows: The maps

$$q_{13} \circ q_{12}^{-1}: X^2 \dashrightarrow X^2, \qquad (a, b) \longmapsto (a, ab),$$

$$q_{23} \circ q_{12}^{-1}: X^2 \dashrightarrow X^2, \qquad (a, b) \longmapsto (b, ab),$$

$$q_{23} \circ q_{13}^{-1}: X^2 \dashrightarrow X^2, \qquad (a, c) \longmapsto (a^{-1}c, c),$$

$$q_{12} \circ q_{13}^{-1}: X^2 \dashrightarrow X^2, \qquad (a, c) \longmapsto (a, a^{-1}c),$$

$$q_{13} \circ q_{23}^{-1}: X^2 \dashrightarrow X^2, \qquad (b, c) \longmapsto (cb^{-1}, c),$$

$$q_{12} \circ q_{23}^{-1}: X^2 \dashrightarrow X^2, \qquad (b, c) \longmapsto (cb^{-1}, b),$$

are S-birational. They are open immersions on their domains of definition; the latter as well as the corresponding images are X-dense in X^2 (with respect to both projections). In addition, the lemma shows that the law $m: X^2 \dashrightarrow X$ is defined at a point $(x, y) \in X^2$ as soon as the fibre $q_{12}^{-1}((x, y))$ is non-empty. This fact will be needed in the next lemma.

Lemma 4. *Let a be an S-valued point of X, and consider another point $b \in X$. Then $a \times_S b$ can be viewed as a point in X^2, and the law $m : X^2 \dashrightarrow X$ is defined at $a \times_S b$ if and only if the birational map $\tau_a : X \dashrightarrow X$ is defined at b.*

Proof. It is only necessary to verify the if-part of the assertion. Considering the S-dense open subscheme $U_a := U \cap (a \times_S X)$ of $a \times_S X \cong X$, we know that τ_a is at least defined on U_a. Let Γ_a be the schematic closure in X^2 of the graph of $\tau_{a|U_a}$. Then we have

$$(a \times_S \Gamma_a) \cap (a \times_S U_a \times_S X) \subset \Gamma$$

and, by continuity, also $a \times_S \Gamma_a \subset \Gamma$. Since the image of the morphism

$$a \times_S \Gamma_a \hookrightarrow \Gamma \xrightarrow{q_{12}} X^2$$

contains the point $a \times_S b$, the fibre over it with respect to q_{12} is non-empty. Thus, the assertion follows from Lemma 3. □

The preceding lemma is very useful if one wants to expand the domain of definition of $m : X^2 \dashrightarrow X$ by means of enlarging X. Namely, one has only to enlarge the domain of definition of $\tau_a : X \dashrightarrow X$ for suitable sections $a \in X(S)$. This can be done by introducing sort of a translate of X by a and by gluing it to X.

Therefore, fix a section $a \in X(S)$ and, as in the proof of Lemma 4, consider the schematic closure Γ_a in X^2 of the graph of the S-birational map τ_a. Then $a \times_S \Gamma_a \subset \Gamma$ and, by Lemma 3, both projections $p_i : \Gamma_a \longrightarrow X$ are injective as maps of sets. Since τ_a is S-birational, Zariski's Main Theorem implies that p_1 and p_2 are open immersions; furthermore, p_1 and p_2 have S-dense images in X. So these projections define gluing data, and we obtain an S-scheme

$$X' = X \cup_{\Gamma_a} X ,$$

which is smooth and of finite type over S, and which is covered by two S-dense open subschemes isomorphic to X. Due to its definition, Γ_a is closed in X^2, hence X' is separated over S.

We need to distinguish between the two copies of X which cover X'. So let us write more precisely

$$p_1 : \Gamma_a \longrightarrow X(a) ,$$

$$p_2 : \Gamma_a \longrightarrow X$$

for the gluing data, where $X(a)$ is another copy of X. This way we have fixed one of the two canonical embeddings of our original S-scheme X into X'. We want to show that $X(a)$ can be interpreted as a "left translate" (in X') of X by a. Namely, consider the S-birational map $\tau_a : X \dashrightarrow X$. It is defined at least on U_a so that we have the following factorization:

$$\begin{array}{ccccc} & & \Gamma_a & & \\ & \nearrow & & \searrow{\scriptstyle p_2} & \\ U_a & & \xrightarrow{\tau_a} & & X . \end{array}$$

Working in X', we can write this diagram also in the form

$$\begin{array}{ccc} & \Gamma_a & \\ \nearrow & & \searrow^{p_1} \\ U_a & \longrightarrow & X(a)\,. \end{array}$$

Since the horizontal map is the restriction to U_a of the canonical isomorphism $X \xrightarrow{\sim} X(a)$, we see that $\tau_a : X \dashrightarrow X$ extends to an isomorphism $\tau_a : X \xrightarrow{\sim} X(a)$, namely the canonical one. In particular, τ_a extends to an S-birational map $X' \dashrightarrow X'$ which is defined on X.

Lemma 5. *As before, let X' be the S-scheme obtained by gluing a left translate $X(a) = \tau_a(X)$ for some point $a \in X(S)$ to X. Then X' contains X as an S-dense open subscheme, and the strict group law m on X extends to a strict group law m' on X'.*

Proof. We have already seen that X is S-dense in X'. So it is clear that m extends to an S-birational group law m' on X', and we have only to show that m' is strict, i.e., that there exists an X'-dense (with respect to both projections) open subscheme $U' \subset X' \times_S X'$ satisfying the following conditions:

(a) m' is defined on U',

(b) the universal translations

$$\Phi' : X' \times_S X' \dashrightarrow X' \times_S X' , \qquad (x, y) \longmapsto (x, xy) ,$$

$$\Psi' : X' \times_S X' \dashrightarrow X' \times_S X' , \qquad (x, y) \longmapsto (xy, y) ,$$

are open immersions on U', and the images $V' := \Phi(U')$ and $W' := \Psi(U')$ are X'-dense in $X' \times_S X'$ (with respect to both projections).

The product $X' \times_S X'$ is the union of the open subschemes

$$X \times_S X , \qquad X(a) \times_S X , \qquad X \times_S X(a) , \qquad \text{and} \qquad X(a) \times_S X(a) .$$

In order to define U', let U, as before, be the open subscheme of $X \times_S X$ whose existence is required in Definition 5.2/1 for the strict group law m on X. Furthermore, let U_1 be the image of U under the isomorphism

$$\tau_a \times \mathrm{id}_X : X \times_S X \xrightarrow{\sim} X(a) \times_S X .$$

Then m' is defined on U since m is defined on U, and the isomorphism $\tau_a : X \xrightarrow{\sim} X(a)$ can be used in order to obtain the morphism

$$U_1 \longrightarrow X(a) , \qquad (\tau_a(x), y) \longmapsto \tau_a(xy) ,$$

from $m : U \longrightarrow X$. Both morphisms coincide on an S-dense open part of U, due to the associativity of m. Thus m' is defined on the open subscheme $U \cup U_1$ of $X' \times_S X'$; the latter is X'-dense with respect to the first projection.

Next consider the open subscheme

$$\{(x, y, z) \in X^3 \, ; (x, y) \in U \, , (xy, z) \in U\}$$

of X^3. Similarly as in the proof of Lemma 1, one shows that it is p_{23}-dense in X^3. Hence, intersecting it with $X \times_S a \times_S X$ and applying the isomorphism

$$X \times_S a \times_S X \xrightarrow{p_{13}} X^2 \xrightarrow{\mathrm{id}_X \times \tau_a} X \times_S X(a) ,$$

we obtain an open subscheme U_2 of $X \times_S X(a)$ which is $X(a)$-dense with respect to the second projection. Then the morphism

$$(*) \qquad U_2 \longrightarrow X , \qquad (x, \tau_a(y)) \longmapsto (xa)y$$

is defined and, using the associativity of m, it coincides with the multiplication $m : U \longrightarrow X$ on an S-dense open part of U. Thus, writing U' for the X'-dense (with respect to both projections) open subscheme $U \cup U_1 \cup U_2$ of $X' \times_S X'$, we see that m' is defined on U' and, hence, that U' satisfies condition (a).

In order to verify condition (b), notice that the universal translations Φ' and Ψ' corresponding to m' extend the universal translations Φ and Ψ corresponding to m. Thus, since Φ and Ψ are open immersions on U, we see that Φ' and Ψ' are open immersions on each one of the schemes U, U_1, and U_2. In particular, Φ' and Ψ' are quasi-finite on U'. Since these are S-birational maps on $X' \times_S X'$, Zariski's Main Theorem 2.3/2′ implies that they are open immersions on U'.

As in 5.2/1, set $V := \Phi(U)$. Furthermore, let V_1 be the image of V under the isomorphism

$$\tau_a \times \tau_a : X \times_S X \xrightarrow{\sim} X(a) \times_S X(a) .$$

Then $V' := \Phi'(U')$ contains $V \cup V_1$, and the latter is X'-dense in $X' \times_S X'$ (with respect to both projections); in particular, V' is X'-dense in $X' \times_S X'$.

Similarly, one shows that $W' := \Psi'(U')$ is X'-dense in $X' \times_S X'$ with respect to the first projection. In order to see that the same is true for the second projection, notice that $W_1 := \Psi'(U')$ is X-dense in $X' \times_S X$ with respect to the second projection. Furthermore, consider the open subscheme

$$W_2 := \Psi'(U_2) \subset X \times_S X(a)$$

and look at the description $(*)$ of m' on U_2 which was discussed above. Then W_2 is seen to be $X(a)$-dense in $X' \times_S X(a)$ with respect to the second projection since, for any T-valued point z of X, the right translation

$$X_T \dashrightarrow X_T , \qquad x \longmapsto xz ,$$

is T-birational. Hence $W' = \Psi'(U')$ is X'-dense in $X' \times_S X'$ with respect to both projections. The latter finishes the verification of condition (b). □

Now consider a sequence $a_1, a_2, \ldots$ of S-valued points of X. Iterating the construction of X' by using these points, we obtain a sequence of S-schemes

$$X = X^{(0)} \subset X^{(1)} \subset X^{(2)} \subset \ldots ,$$

where $X^{(i)} = X^{(i-1)} \cup X^{(i-1)}(a_i)$. Each $X^{(i)}$ contains X as an S-dense open subscheme, and $X^{(i)}$ is separated, smooth, and of finite type over S. Furthermore, Lemma 5 shows that the strict group law m on X extends to a strict group law $m^{(i)}$ on each $X^{(i)}$. Using a noetherian argument, we want to show that the sequence $X^{(0)} \subset X^{(1)} \subset X^{(2)} \subset \ldots$ becomes stationary at a certain $X^{(n)}$. Then, for a suitable choice of the a_i, we will see that $X^{(n)}$ is the S-group scheme we are looking for.

Lemma 6. *There exist finitely many S-valued points $a_1, \ldots, a_n \in X(S)$ such that, for $X^{(n)}$ as above, the S-rational map $m: X \times_S X \dashrightarrow X$ extends to an S-morphism $X \times_S X \longrightarrow X^{(n)}$.*

Proof. First we show that we can find $a_1, \ldots, a_n \in X(S)$ in such a way that, for each $a \in X(S)$, the S-birational map $\tau_a: X \dashrightarrow X$ extends to an S-morphism $X \longrightarrow X^{(n)}$. Proceeding indirectly, consider a sequence $a_1, a_2, \ldots$ in $X(S)$ such that

$$\tau_{a_{i+1}}: X \dashrightarrow X^{(i)}, \qquad i = 1, 2, \ldots,$$

is not defined everywhere on X. Let $\Gamma^{(i)}$ be the schematic closure in $(X^{(i)})^3$ of the graph of $m: U \longrightarrow X$. It coincides with the schematic closure of the graph of the induced strict group law $m^{(i)}$ on $X^{(i)}$; so we know from Lemma 3 that

$$p_{12}: \Gamma^{(i)} \longrightarrow X^{(i)} \times_S X^{(i)}$$

is an open immersion. Setting

$$Q^{(i)} := p_{12}(\Gamma^{(i)}) \cap (X \times_S X),$$

the $Q^{(i)}$ form an increasing sequence of open subschemes of $X \times_S X$, since the $\Gamma^{(i)}$ form an increasing sequence. However, the base S consists of a noetherian ring, which implies that the topological space $X \times_S X$ is noetherian. Thus the $Q^{(i)}$ must become stationary at a certain index $n \in \mathbb{N}$, and we claim that, for $a = a_{n+1}$, the map $\tau_a: X \dashrightarrow X^{(n)}$ is defined everywhere. Namely, consider a point $b \in X$. By the definition of $X^{(n+1)}$, the birational map $\tau_a: X \dashrightarrow X^{(n+1)}$ is defined everywhere. So we see from Lemma 4 that the law $m^{(n+1)}$ on $X^{(n+1)}$ is defined at $a \times_S b$. Hence the fibre over $a \times_S b$ of

$$p_{12}: \Gamma^{(n+1)} \longrightarrow X^{(n+1)} \times_S X^{(n+1)}$$

is non-empty, and $a \times_S b \in Q^{(n+1)}$. But, since $Q^{(n+1)} = Q^{(n)}$, the fibre over $a \times_S b$ of

$$p_{12}: \Gamma^{(n)} \longrightarrow X^{(n)} \times_S X^{(n)}$$

cannot be empty, and we see from Lemma 3 that the law $m^{(n)}$ on $X^{(n)}$ is defined at $a \times_S b$. In particular, $\tau_{a_{n+1}} = \tau_a: X \dashrightarrow X^{(n)}$ is defined at b. This contradicts our assumption on the sequence $a_1, a_2, \ldots$; so there must exist $a_1, \ldots, a_n \in X(S)$ such that $\tau_a: X \dashrightarrow X^{(n)}$ is defined everywhere for each $a \in X(S)$.

It remains to show that, in this situation, the S-rational map $m: X \times_S X \dashrightarrow X^{(n)}$ is defined everywhere. We know already from Lemma 4 that m is defined on $a \times_S X$ for each S-valued point a of X. However, this is not enough, and we now have to use the fact that our assumption on X to be a faithfully flat and smooth scheme over a *strictly henselian base S* yields the following property:

Let t be a point of S, and let C_t be the reduced subscheme of $X \times_S t$ whose underlying topological space is the closure in $X \times_S t$ of the set of points $\{a(t); a \in X(S)\}$. Then there exists a component X_t^0 of X_t contained in C_t; cf. Lemma 7 below.

Moreover, let k' be an extension field of $k(t)$, and let t' be the scheme of k'. Then $C_t \times_t t'$ coincides with the reduced subscheme of $X \times_S t'$ whose underlying topological space is the closure of the points $\{a(t'); a \in X(S)\}$; cf. [EGA IV$_3$], 11.10.7.

In particular, if $Z_{t'}$ is a dense open subscheme of $X \times_S t'$, there exists a point $a \in X(S)$ such that $a \times_S t'$ gives rise to a point of $Z_{t'}$.

Now let us continue the proof of Lemma 6. Using the notation of Lemma 3, we know that

$$q_{23} \circ q_{13}^{-1} : X \times_S X \dashrightarrow X \times_S X \,, \qquad (w,x) \longmapsto (w^{-1}x, x) \,,$$

is an S-birational map. It is an open immersion on its domain of definition D, and this domain as well as its image are X-dense in X^2 with respect to both projections. Now consider a point $t \in X^2$. It follows that the set

$$Z := \{(w,x,y) \in X^3 \,;\, (w,x) \in D \text{ and } (w^{-1}x, y) \in U\} \,,$$

where U is as in 5.2/1, is open and p_{23}-dense in X^3 and, hence, open and dense in $X \times_S t$. So, applying the base change $t \longrightarrow X^2$ to $X \times_S X^2$, the assumption on X as explained above implies the existence of a point $a \in X(S)$ such that $a \times_S t \in Z$. Then the S-rational map

$$X \times_S X \dashrightarrow X \,, \qquad (x,y) \longmapsto (a^{-1}x)y \,,$$

is defined at t. Furthermore, since the left translation

$$\tau_a : X \dashrightarrow X^{(n)}$$

is defined everywhere, we see that

$$X \times_S X \dashrightarrow X^{(n)} \,, \qquad (x,y) \longmapsto a((a^{-1}x)y) \,,$$

is defined at t. However, this map coincides on $X \times_S X$ with the strict group law m, since m is associative. So we see that m extends to an S-rational map

$$X \times_S X \dashrightarrow X^{(n)}$$

which is defined at all points of X^2. □

Lemma 7. *Let T be a noetherian scheme, let $Y \longrightarrow T$ be a morphism of finite type, and let $\{a_i, i \in I\}$ be a family of sections of Y. Let t_1 and t_0 be points of T such that t_0 is a specialization of t_1. Let C_j be the closure of the set of points $\{a_i(t_j), i \in I\}$ in the fibre Y_{t_j}, $j = 0, 1$. Then* $\dim C_1 \geqq \dim C_0$.

In particular, if T is strictly henselian and noetherian, and if $Y \longrightarrow T$ is smooth and surjective, then, for each point $t \in T$, there exists a connected component Y_t^0 of the fibre Y_t such that the set of the points $\{a(t), a \in Y(T)\}$ is dense in Y_t^0.

Proof. It suffices to show the first assertion after a base change $\varphi : T' \longrightarrow T$ such that the points t_0, t_1 belong to the image of φ. So, due to [EGA II], 7.1.4, we may assume that T consists of a discrete valuation ring with generic point t_1 and closed point t_0. Denote by V the schematic closure of C_1 in Y; so V is flat over T, since T consists of a discrete valuation ring. Then it is clear that

$$\dim V_{t_1} \geqq \dim V_{t_0} \,;$$

cf. [EGA IV$_3$], 14.3.10. Since $C_0 \subset V$, the first assertion is clear.

For the second, we may assume that the relative dimension of Y over T is constant on Y. Due to 2.3/5 the closure of the set of points $\{a(t_0), a \in Y(T)\}$ is Y_{t_0}

for the closed point t_0 of T. Hence the second assertion follows from the first one. □

Now the proof of Theorem 5.2/3 is quite easy. Namely, let $\bar{X}$ be the S-scheme $X^{(n)}$ constructed in Lemma 6. Then $\bar{X}$ is separated, smooth, of finite type, and contains X as an S-dense open subscheme. Furthermore, by Lemmata 5 and 6, the strict group law m on X extends to a strict group law $\bar{m}$ on $\bar{X}$, and the S-rational map $\bar{m}: \bar{X}^2 \dashrightarrow \bar{X}$ is defined on X^2. It is a general fact that $\bar{X}$ is an S-group scheme in this situation; so we can end the proof of 5.2/3 by establishing the following result:

Lemma 8. *Let $\bar{X}$ be a smooth and separated S-scheme of finite type which is equipped with a strict group law $\bar{m}$. Assume that $\bar{X}(S)$ is non-empty and that there exists an S-dense open subscheme X of $\bar{X}$ such that $\bar{m}$ is defined on the open subscheme X^2 of $\bar{X}^2$. Then $\bar{X}$ is an S-group scheme with respect to the law $\bar{m}$.*

Proof. First we want to show that

$$\bar{m}: \bar{X} \times_S \bar{X} \dashrightarrow \bar{X}, \qquad (x, y) \longmapsto xy,$$

is defined everywhere. Since the domain of definition is compatible with faithfully flat base change (2.5/6), it suffices to show that, for each point $(b, c) \in \bar{X}^2$, the map

$$\bar{m}_X = \mathrm{id}_X \times \bar{m}: X \times_S \bar{X}^2 \dashrightarrow X \times_S \bar{X}$$

is defined at some point $(a, b, c) \in X \times_S \bar{X}^2$ above (b, c). For example, let (a, b, c) be a generic point of the fibre over (b, c). Then $(a, b) \in X \times_S \bar{X}$ is a generic point in the fibre over b and the map

$$X \times_S \bar{X} \dashrightarrow X, \qquad (w, x) \longmapsto xw,$$

is defined at (a, b), since $\bar{m}$ is a strict group law on $\bar{X}$. Likewise, using Lemma 3, the map

$$X \times_S \bar{X} \dashrightarrow X, \qquad (w, y) \longmapsto w^{-1}y,$$

is defined at (a, c) which is a generic point in the fibre over c. Since $\bar{m}$ is defined on X^2, the map

$$m': X \times_S \bar{X} \times_S \bar{X} \dashrightarrow X \times_S \bar{X}, \qquad (w, x, y) \longmapsto (w, (xw)(w^{-1}y)),$$

is defined at (a, b, c), and the associativity of $\bar{m}$ shows that m' coincides with $\bar{m}_X$. Thus $\bar{m}$ is defined on all of $\bar{X}^2$.

Similar arguments show that the map

$$\bar{X} \times_S \bar{X} \longrightarrow \bar{X}, \qquad (x, y) \longmapsto x^{-1}y,$$

is defined everywhere. But then $\bar{m}$ defines on $\bar{X}$ the structure of an S-group scheme. Namely, returning to the functorial point of view, consider the monomorphism

$$\bar{X} \hookrightarrow \mathscr{R}_{\bar{X}/S}$$

of 5.2/4. The group law on $\mathscr{R}_{\bar{X}/S}$ restricts to the law $\bar{m}$ on $\bar{X}$, and $\bar{X}(T) \neq \emptyset$ for $T = S$ and, hence, for all S-schemes T. Thus, since the map $(x, y) \longmapsto (x^{-1}y)$ is defined on $\bar{X} \times_S \bar{X}$, we see that each $\bar{X}(T)$ is a subgroup of $\mathscr{R}_{\bar{X}/S}(T)$. So $\bar{X}$ is a subgroup functor of $\mathscr{R}_{\bar{X}/S}$ and in fact, the representability being granted, an S-group scheme with group law $\bar{m}$. □

So we have finished the proof of Lemma 8 and thereby also the proofs of 5.2/3 and of 5.1/5 for the case where the base S consists of a strictly henselian valuation ring or of a separably closed field.

Chapter 6. Descent

During the years 1959 to 1962, Grothendieck gave a series of six lectures at the Séminaire Bourbaki, entitled "Technique de descente et théorèmes d'existence en géométrie algébrique". In the first lecture [FGA], n°190, the general technique of faithfully flat descent is introduced. It is an invaluable tool in algebraic geometry. Quite often it happens that a certain construction can be carried out only after faithfully flat base change. Then one can try to use descent theory in order to go back to the original situation one started with. Before Grothendieck, descent was certainly known in the form of Galois descent.

We begin by describing the basic facts of Grothendieck's formalism and by discussing some general criteria for effective descent, including several examples. Then, working over a Dedekind scheme, our main objective is to study the descent of torsors under smooth group schemes; see Raynaud [4]. As a preparation, we discuss the theorem of the square and use it to show the quasi-projectivity of torsors. Relying on the latter fact, effective descent of torsors can be described in a very convenient form; we do this in Section 6.5. As an application, we look at existence and descent of Néron models for torsors. Also, working over a more general base, we are able to extend the technique of associating group schemes to birational group laws as discussed in Chapter 5. The chapter ends with an example of non-effective descent.

6.1 The General Problem

Let $p: S' \longrightarrow S$ be a morphism of schemes and consider the functor $\mathscr{F} \longrightarrow p^*\mathscr{F}$, which associates to each quasi-coherent S-module $\mathscr{F}$ its pull-back under p. Then, in its simplest form, the problem of descent relative to $p: S' \longrightarrow S$ is to characterize the image of this functor. The procedure of solution is as follows. Set $S'' := S' \times_S S'$, and let $p_i: S'' \longrightarrow S'$ be the projection onto the i-th factor ($i = 1, 2$). For any quasi-coherent S'-module $\mathscr{F}'$, call an S''-isomorphism $\varphi: p_1^*\mathscr{F}' \longrightarrow p_2^*\mathscr{F}'$ a *covering datum* of $\mathscr{F}'$. Then the pairs $(\mathscr{F}', \varphi)$ of quasi-coherent S'-modules with covering data form a category in a natural way. A morphism between two such objects $(\mathscr{F}', \varphi)$ and $(\mathscr{G}', \psi)$ consists of an S'-morphism $f: \mathscr{F}' \longrightarrow \mathscr{G}'$ which is compatible with the covering data φ and ψ; thereby we mean that the diagram

$$\begin{array}{ccc} p_1^*\mathcal{F}' & \xrightarrow{\ \varphi\ } & p_2^*\mathcal{F}' \\ \Big\downarrow {\scriptstyle p_1^*f} & & \Big\downarrow {\scriptstyle p_2^*f} \\ p_1^*\mathcal{G}' & \xrightarrow{\ \psi\ } & p_2^*\mathcal{G}' \end{array}$$

is commutative.

Starting with a quasi-coherent S-module $\mathcal{F}$, we have a natural covering datum on $p^*\mathcal{F}$, which consists of the canonical isomorphism

$$p_1^*(p^*\mathcal{F}) \cong (p \circ p_1)^*\mathcal{F} = (p \circ p_2)^*\mathcal{F} \cong p_2^*(p^*\mathcal{F}) .$$

So we can interpret the functor $\mathcal{F} \longmapsto p^*\mathcal{F}$ as a functor into the category of quasi-coherent S'-modules with covering data. It is this functor which will be of interest in the following. We will show that it is fully faithful if $p : S' \longrightarrow S$ is faithfully flat and quasi-compact, and that, furthermore, it is an equivalence of categories if, instead of covering data, we consider descent data; i.e., special covering data which satisfy a certain cocycle condition. The problem of descent can be viewed as a natural generalization of a patching problem; cf. Example 6.2/A.

As usual we will call a diagram

$$A \xrightarrow{\ \alpha\ } B \overset{\beta}{\underset{\gamma}{\rightrightarrows}} C$$

of maps between sets *exact* if α is injective and if $\operatorname{im} \alpha = \ker(\beta, \gamma)$, where $\ker(\beta, \gamma)$ consists of all elements $b \in B$ such that $\beta(b) = \gamma(b)$. Working in the category of abelian groups, the exactness of such a diagram is equivalent to the exactness of the sequence

$$0 \longrightarrow A \xrightarrow{\ \alpha\ } B \xrightarrow{\ \beta - \gamma\ } C .$$

Proposition 1. *Assume that $p : S' \longrightarrow S$ is faithfully flat and quasi-compact. Let $\mathcal{F}$ and $\mathcal{G}$ be quasi-coherent S-modules, and set $q := p \circ p_1 = p \circ p_2$. Then, identifying $q^*\mathcal{F}$ canonically with $p_i^*(p^*\mathcal{F})$ for $i = 1, 2$, likewise for $q^*\mathcal{G}$, the diagram*

$$\operatorname{Hom}_S(\mathcal{F}, \mathcal{G}) \xrightarrow{\ p^*\ } \operatorname{Hom}_{S'}(p^*\mathcal{F}, p^*\mathcal{G}) \overset{p_1^*}{\underset{p_2^*}{\rightrightarrows}} \operatorname{Hom}_{S''}(q^*\mathcal{F}, q^*\mathcal{G})$$

is exact. In other words, the functor $\mathcal{F} \longmapsto p^*\mathcal{F}$ from quasi-coherent S-modules to quasi-coherent S'-modules with covering data is fully faithful.

Proof. The assertion is local on S, so we can assume that S is affine. Then S' is quasi-compact, and it is covered by finitely many affine open subschemes $S_i' \subset S'$, $i \in I$. Consider the disjoint union $\overline{S}' := \coprod_{i \in I} S_i'$ of these schemes.

Let $u : \overline{S}' \longrightarrow S'$ be the canonical morphism, $\bar{p} : \overline{S}' \longrightarrow S$ its composition with $p : S' \longrightarrow S$, and let $\bar{p}_1, \bar{p}_2$ denote the projections of $\overline{S}'' := \overline{S}' \times_S \overline{S}'$ onto its factors. Then we obtain a diagram

$$\begin{array}{ccccc} \operatorname{Hom}_S(\mathcal{F}, \mathcal{G}) & \xrightarrow{\ p^*\ } & \operatorname{Hom}_{S'}(p^*\mathcal{F}, p^*\mathcal{G}) & \overset{p_1^*}{\underset{p_2^*}{\rightrightarrows}} & \operatorname{Hom}_{S''}(q^*\mathcal{F}, q^*\mathcal{G}) \\ \Big\| & & \Big\downarrow {\scriptstyle u^*} & & \Big\downarrow {\scriptstyle (u \times u)^*} \\ \operatorname{Hom}_S(\mathcal{F}, \mathcal{G}) & \xrightarrow{\ \bar{p}^*\ } & \operatorname{Hom}_{\overline{S}'}(\bar{p}^*\mathcal{F}, \bar{p}^*\mathcal{G}) & \overset{\bar{p}_1^*}{\underset{\bar{p}_2^*}{\rightrightarrows}} & \operatorname{Hom}_{\overline{S}''}(\bar{q}^*\mathcal{F}, \bar{q}^*\mathcal{G}) \end{array}$$

where $\bar{q} := \bar{p} \circ \bar{p}_1 = \bar{p} \circ \bar{p}_2$. The diagram is commutative if, in the right-hand square, we consider single horizontal arrows, either p_1^* and $\bar{p}_1^*$ or p_2^* and $\bar{p}_2^*$. Furthermore, u being faithfully flat, the vertical maps are injective. Using this fact, it is easily checked that the upper row is exact if the lower row has this property. In other words, we may replace $p : S' \longrightarrow S$ by $\bar{p} : \bar{S}' \longrightarrow S$ and thereby assume that S and S' are affine, say $S = \operatorname{Spec} R$ and $S' = \operatorname{Spec} R'$. Then the problem becomes a problem on R-modules.

Let

$$(*) \qquad R \longrightarrow R' \rightrightarrows R' \otimes_R R'$$

be the diagram which corresponds to the projections $S'' \rightrightarrows S' \longrightarrow S$. We claim that the assertion of the proposition follows if we can show that the tensor product of $(*)$ with any R-module M yields an exact diagram. Namely, consider R-modules M and N such that $\mathscr{F}$ (resp. $\mathscr{G}$) is associated to M (resp. N), and assume that we have exact diagrams

$$M \longrightarrow M \otimes_R R' \rightrightarrows M \otimes_R R' \otimes_R R' \,,$$

$$N \longrightarrow N \otimes_R R' \rightrightarrows N \otimes_R R' \otimes_R R' \,.$$

Then the injectivity of $N \longrightarrow N \otimes_R R'$ implies the injectivity of the map p^* in the assertion. Similarly, it is seen that any R'-homomorphism $M \otimes_R R' \longrightarrow N \otimes_R R'$, which corresponds to an element in $\ker(p_1^*, p_2^*)$, restricts to an R-homomorphism $M \longrightarrow N$. This yields $\operatorname{im} p^* \supset \ker(p_1^*, p_2^*)$. Since the opposite inclusion is trivial, our claim is justified. So, in order to finish the proof of the proposition, it remains to establish the following result:

Lemma 2. *Let $R \longrightarrow R'$ be a faithfully flat morphism of rings. Then, for any R-module M, the canonical diagram*

$$M \longrightarrow M \otimes_R R' \rightrightarrows M \otimes_R R' \otimes_R R'$$

is exact.

Proof. We may apply a faithfully flat base change over R, say with R'. Thereby we can assume that $R \longrightarrow R'$ admits a section $R' \longrightarrow R$. So all the maps in the above diagram have sections, and the exactness is obvious. □

Next we want to introduce descent data and the cocycle condition characterizing them. Set $S''' := S' \times_S S' \times_S S'$, and let $p_{ij} : S''' \longrightarrow S''$ be the projections onto the factors with indices i and j for $i < j$; $i, j = 1, 2, 3$. In order that a quasi-coherent S'-module $\mathscr{F}'$ with covering datum $\varphi : p_1^* \mathscr{F}' \longrightarrow p_2^* \mathscr{F}'$ belongs to the essential image of the functor $\mathscr{F} \longmapsto p^* \mathscr{F}$, it is necessary that the diagram

$$\begin{array}{ccccc}
p_{12}^* p_1^* \mathscr{F}' & \xrightarrow{p_{12}^* \varphi} & p_{12}^* p_2^* \mathscr{F}' = p_{23}^* p_1^* \mathscr{F}' & \xrightarrow{p_{23}^* \varphi} & p_{23}^* p_2^* \mathscr{F}' \\
\| & & & & \| \\
p_{13}^* p_1^* \mathscr{F}' & & \xrightarrow{\qquad p_{13}^* \varphi \qquad} & & p_{13}^* p_2^* \mathscr{F}'
\end{array}$$

is commutative; the unspecified identities are the canonical ones. Namely, if $\mathscr{F}'$ is the pull-back under p of a quasi-coherent S-module and if φ is the natural covering datum on $\mathscr{F}'$, then the diagram is commutative, because all occurring isomorphisms are the canonical ones. The commutativity of the above diagram is referred to as the *cocycle condition* for φ; in short, we can write it as

$$p_{13}^*\varphi = p_{23}^*\varphi \circ p_{12}^*\varphi \ .$$

It corresponds to the usual cocycle condition on triple overlaps when a global object is to be constructed by gluing local parts. A covering datum φ on $\mathscr{F}'$ which satisfies the cocycle condition is called a *descent datum* on $\mathscr{F}'$. The descent datum is called *effective* if the pair $(\mathscr{F}', \varphi)$ is isomorphic to the pull-back $p^*\mathscr{F}$ of a quasi-coherent S-module $\mathscr{F}$ where, on $p^*\mathscr{F}$, we consider the canonical descent datum. Also we want to mention that the notions of covering and descent data are compatible with base change over S.

In the case where S and S' are affine, covering and descent data can be described in terms of modules over rings. Namely, let $S = \operatorname{Spec} R$, $S' = \operatorname{Spec} R'$, and consider a quasi-coherent S'-module $\mathscr{F}'$ with a covering datum $\varphi : p_1^*\mathscr{F}' \longrightarrow p_2^*\mathscr{F}'$, where $\mathscr{F}'$ is associated to the R'-module M'. Then $p_1^*\mathscr{F}'$ and $p_2^*\mathscr{F}'$ are associated to $M' \otimes_R R'$ and $R' \otimes_R M'$, both of which are viewed as $R' \otimes_R R'$-modules. Thus the covering datum φ on $\mathscr{F}'$ corresponds to an $R' \otimes_R R'$-isomorphism

$$M' \otimes_R R' \xrightarrow{\sim} R' \otimes_R M'$$

which, again, will be denoted by φ. Using the canonical map $M' \longrightarrow M' \otimes_R R'$ as well as the composition of the canonical map $M' \longrightarrow R' \otimes_R M'$ with φ^{-1}, we arrive at a *co-cartesian* diagram $M' \rightrightarrows M' \otimes_R R'$ over the canonical diagram $R' \rightrightarrows R' \otimes_R R'$. This means that, considering associated arrows in both diagrams, $M' \otimes_R R$ is obtained from M' by tensoring with $R' \otimes_R R'$ over R'. Conversely, any such co-cartesian diagram determines a covering datum on M' and, hence, on $\mathscr{F}'$.

If φ is a descent datum on $\mathscr{F}'$, we can pull it back with respect to the projections $p_{ij} : S''' \longrightarrow S''$. Due to the cocycle condition, the various pull-backs of $\mathscr{F}'$ to S''' can be identified via the pull-backs of φ. Thereby we obtain in a canonical way homomorphisms (depending on φ)

$$M' \otimes_R R' \;\substack{\rightarrow\\\rightarrow\\\rightarrow}\; M' \otimes_R R' \otimes_R R'$$

such that the diagram

$$(*) \qquad M' \rightrightarrows M' \otimes_R R' \;\substack{\rightarrow\\\rightarrow\\\rightarrow}\; M' \otimes_R R' \otimes_R R'$$

is co-cartesian over the canonical diagram

$$(**) \qquad R' \rightrightarrows R' \otimes_R R' \;\substack{\rightarrow\\\rightarrow\\\rightarrow}\; R' \otimes_R R' \otimes_R R' .$$

Furthermore, $(*)$ satisfies certain natural commutativity conditions just as we have them for $(**)$ or for the associated diagram

$$S''' \;\substack{\rightarrow\\\rightarrow\\\rightarrow}\; S'' \rightrightarrows S' ,$$

where $p_1 \circ p_{12} = p_1 \circ p_{13}$, etc. Conversely, one can show that each co-cartesian diagram $(*)$ over $(**)$, which satisfies the commutativity conditions, determines a descent datum on M', and hence on $\mathscr{F}'$. It is clear that a descent datum φ on $\mathscr{F}'$ is effective if and only if the associated co-cartesian diagram $M' \rightrightarrows M' \otimes_R R'$ can be enlarged into a commutative co-cartesian diagram

$$M \longrightarrow M' \rightrightarrows M' \otimes_R R'$$

over the canonical diagram

$$R \longrightarrow R' \rightrightarrows R' \otimes_R R' \, .$$

Returning to the case where S and S' are arbitrary schemes, it is sometimes convenient to formulate the cocycle condition within the context of T-valued points of S', where T is an arbitrary S-scheme. So consider a quasi-coherent S'-module $\mathscr{F}'$ with a covering datum $\varphi : p_1^*\mathscr{F}' \longrightarrow p_2^*\mathscr{F}'$. For $t_1, t_2 \in S'(T)$, denote by

$$\varphi_{t_1,t_2} : t_1^*\mathscr{F}' \longrightarrow t_2^*\mathscr{F}'$$

the pull-back of φ under the morphism $(t_1, t_2): T \longrightarrow S''$. Adding a third point $t_3 \in S'(T)$, we can consider the morphism

$$(t_1, t_2, t_3): T \longrightarrow S'''$$

and compose it with each one of the projections $S''' \rightrightarrows S''$. Then, pulling back φ to T, we see that φ satisfies the cocycle condition if and only if

$$\varphi_{t_1,t_3} = \varphi_{t_2,t_3} \circ \varphi_{t_1,t_2}$$

for all t_1, t_2, $t_3 \in S'(T)$ and all T. In particular, for $t = t_1 = t_2 = t_3$, the cocycle condition implies $\varphi_{t,t} = \varphi_{t,t}^2$ and, hence, $\varphi_{t,t} = \mathrm{id}$. For example, if $t : S' \longrightarrow S'$ is the universal point of S', we see that the pull-back of a descent datum $\varphi : p_1^*\mathscr{F}' \longrightarrow p_2^*\mathscr{F}'$ with respect to the diagonal morphism $\Delta: S' \longrightarrow S''$ yields the identity on $\mathscr{F}'$.

Lemma 3. *Assume that the morphism $p : S' \longrightarrow S$ admits a section. Then any descent datum φ on a quasi-coherent S'-module $\mathscr{F}'$ is effective. More precisely, the choice of a section $s : S \longrightarrow S'$ of p determines an S-module $\mathscr{F}$, namely $\mathscr{F} := s^*\mathscr{F}'$, such that $p^*\mathscr{F}$ is isomorphic to the pair $(\mathscr{F}', \varphi)$.*

Proof. Writing $T := S'$, let us consider the points $t := \mathrm{id}_{S'}$ and $\tilde{t} := s \circ p$ of $S'(T)$. Then $t^*\mathscr{F}' = \mathscr{F}'$ and $\tilde{t}^*\mathscr{F}' = p^*\mathscr{F}$, and we can consider the isomorphism

$$f = \varphi_{t,\tilde{t}} : \mathscr{F}' \xrightarrow{\sim} p^*\mathscr{F} \, .$$

It is enough to show that f is compatible with the descent datum on $p^*\mathscr{F}$; i.e., we have to show that the diagram

$$\begin{array}{ccc} p_1^*\mathscr{F}' & \xrightarrow{\ \varphi\ } & p_2^*\mathscr{F}' \\ \downarrow{\scriptstyle p_1^*f} & & \downarrow{\scriptstyle p_2^*f} \\ p_1^*p^*\mathscr{F} & = & p_2^*p^*\mathscr{F} \end{array}$$

is commutative. In order to do this, consider the following S''-valued points of S':

$$p_1, \quad p_2, \quad \text{and} \quad t_3 := s \circ p \circ p_1 = s \circ p \circ p_2 \, .$$

Then $\varphi = \varphi_{p_1, p_2}$ since $(p_1, p_2) : S'' \longrightarrow S''$ is the identity, and we have

$$p_i^* f = p_i^* \varphi_{t, \tilde{t}} = \varphi_{p_i, t_3} \qquad \text{for } i = 1, 2 \, ,$$

since the diagram

$$\begin{array}{ccc} S'' & \xrightarrow{(p_i, t_3)} & S' \times_S S' \\ & {\scriptstyle p_i} \searrow \quad \nearrow {\scriptstyle (t, \tilde{t}) = (\mathrm{id}, s \circ p)} & \\ & S' & \end{array}$$

is commutative. Now the cocycle condition for φ yields

$$\varphi_{p_1, t_3} = \varphi_{p_2, t_3} \circ \varphi_{p_1, p_2}$$

and thus

$$p_1^* f = p_2^* f \circ \varphi \, . \qquad \square$$

Now we are ready to prove the desired result on the descent of quasi-coherent S'-modules.

Theorem 4 (Grothendieck). *Let $p : S' \longrightarrow S$ be faithfully flat and quasi-compact. Then the functor $\mathscr{F} \longmapsto p^* \mathscr{F}$, which goes from quasi-coherent S-modules to quasi-coherent S'-modules with descent data, is an equivalence of categories.*

Proof. We know already from Proposition 1 that the functor in question is fully faithful. So it is enough to show that each descent datum on a quasi-coherent S'-module is effective. The latter is clear by Lemma 3 if $p : S' \longrightarrow S$ admits a section. We will reduce to this case.

First observe that we may replace the morphism $p : S' \longrightarrow S$ by a composition $\bar{p} : \bar{S}' \xrightarrow{u} S' \xrightarrow{p} S$, where $u : \bar{S}' \longrightarrow S'$ is faithfully flat and quasi-compact. This is true since the functor $\mathscr{F}' \longmapsto u^* \mathscr{F}'$ is fully faithful (see Proposition 1) and since descent data on $\mathscr{F}'$ (with respect to p) can easily be pulled back to descent data on $u^* \mathscr{F}'$ (with respect to $\bar{p}$). So, proceeding as in the proof of Proposition 1, we may assume that S and S' are affine, say $S = \operatorname{Spec} R$ and $S' = \operatorname{Spec} R'$.

Let M' be an R'-module with descent datum $\varphi : M' \otimes_R R' \xrightarrow{\sim} R' \otimes_R M'$. Then φ determines a co-cartesian diagram $M' \rightrightarrows M' \otimes_R R'$ over $R' \rightrightarrows R' \otimes_R R'$. If M' descends to an R-module, we know from Lemma 2 that it must descend to the R-module

$$K := \ker(M' \rightrightarrows M' \otimes_R R') \, .$$

So let us work with this module. We claim that the diagram

$$K \longrightarrow M' \rightrightarrows M' \otimes_R R'$$

is commutative and co-cartesian over

$$R \longrightarrow R' \rightrightarrows R' \otimes_R R'$$

and, hence, that φ is effective. In order to verify this, we may apply a faithfully flat base change and thereby assume that $R \longrightarrow R'$ admits a section. Then it follows from Lemma 3 that (M', φ) descends to an R-module M. More precisely, $M' \rightrightarrows M' \otimes_R R'$ extends to a commutative co-cartesian diagram

$$M \longrightarrow M' \rightrightarrows M' \otimes_R R'$$

over

$$R \longrightarrow R' \rightrightarrows R' \otimes_R R' .$$

Since M is mapped bijectively onto K by Lemma 2, our claim is justified. □

Keeping the morphism $p: S' \longrightarrow S$, we want to study the problem of when an S'-scheme X' descends to an S-scheme X. The general setting will be the same as in the case of quasi-coherent modules, and the definitions we have given can easily be adapted to the new situation. For example, a *descent datum on an S'-scheme X'* is an S''-isomorphism

$$\varphi : p_1^* X' \longrightarrow p_2^* X'$$

which satisfies the cocycle condition; $p_i^* X'$ is the scheme obtained from X' by applying the base change $p_i: S'' \longrightarrow S'$. Again there is a canonical functor $X \longmapsto p^* X$ from S-schemes to S'-schemes with descent data. If $p: S' \longrightarrow S$ is faithfully flat and quasi-compact, we see from Theorem 4 that this functor gives an equivalence between affine S-schemes and affine S'-schemes with descent data. More generally, the same assertion is true with affine replaced by quasi-affine (use Theorem 6(b) below). Thus, in this case, descent data on affine or quasi-affine S'-schemes are always effective. Recall that an S'-scheme X' is called affine (resp. quasi-affine) over S' if, for each affine open subscheme $S'_0 \subset S'$, the open subscheme $S'_0 \times_{S'} X'$ of X' is affine (resp. quasi-affine). To be precise, one has, of course, to mention the fact that one can easily generalize Theorem 4 from quasi-coherent modules to quasi-coherent algebras, so that it can be applied to structure sheaves of schemes over S or S'. Working with an additional structure such as a multiplication on a quasi-coherent S'-module, this structure descends if it is compatible with the descent datum.

It is not true that descent data on schemes are always effective, even if $p: S' \longrightarrow S$ is faithfully flat and quasi-compact; see Section 6.7. So one needs criteria for effectiveness. First we mention that Lemma 3 carries over to the scheme situation. Since the proof was given by formal arguments, no changes are necessary.

Lemma 5. *Assume that $p: S' \longrightarrow S$ has a section. Then all descent data on S'-schemes are effective.*

In order to formulate another criterion, consider an S'-scheme X' with a descent datum $\varphi: p_1^* X' \longrightarrow p_2^* X'$, and let U' be an open subscheme of X'. Then U' is called *stable under* φ if φ induces a descent datum on U'; i.e., if φ restricts to an isomorphism $p_1^* U' \xrightarrow{\sim} p_2^* U'$.

Theorem 6. *Let $p: S' \longrightarrow S$ be faithfully flat and quasi-compact.*

(a) *The functor $X \longmapsto p^* X$ from S-schemes to S'-schemes with descent data is fully faithful.*

(b) *To simplify, assume S and S′ affine. Then a descent datum φ on an S′-scheme X′ is effective if and only if X′ can be covered by quasi-affine (or, alternatively, by affine) open subschemes which are stable under φ.*

Proof. Assertion (a) is an immediate consequence of Proposition 1. Namely, consider S-schemes X and Y, and write X', Y' for the schemes obtained by the base change $p: S' \longrightarrow S$. Then it is to show that the sequence

$$\mathrm{Hom}_S(X, Y) \xrightarrow{p^*} \mathrm{Hom}_{S'}(X', Y') \overset{p_1^*}{\underset{p_2^*}{\rightrightarrows}} \mathrm{Hom}_{S''}(X'', Y'')$$

is exact. The problem is local on S and Y. So we may assume that S and Y are affine. Furthermore, replacing S' by a finite disjoint sum of affine open parts of S', we may assume that S' is affine. Then, up to a local consideration on X, we can pose the problem in terms of quasi-coherent algebras so that Proposition 1 can be applied.

In order to verify the if-part of assertion (b), we may use (a) and assume that X' is quasi-affine. This means that X' is quasi-compact and can be realized as an open subscheme of an affine scheme or, equivalently, that the canonical map

$$X' \longrightarrow \mathrm{Spec}\,\Gamma(X', \mathcal{O}_{X'}) =: Z'$$

is a quasi-compact open immersion; cf. [EGA II], 5.1.2. Let $S = \mathrm{Spec}\,R$ and $S' = \mathrm{Spec}\,R'$. Then, using the fact that, for quasi-compact R'-schemes, the functor of global sections commutes with flat extensions of R', the descent datum on X' gives a descent datum on the R'-module $\Gamma(X', \mathcal{O}_{X'})$ and hence on the affine S'-scheme Z'. Thus it follows from Theorem 4 that Z' descends to an affine S-scheme Z. Considering the canonical projections

$$Z'' \overset{q_1}{\underset{q_2}{\rightrightarrows}} Z' \xrightarrow{q} Z\,,$$

where Z'' is obtained from Z by the base change $S'' \longrightarrow S$, we see $q_1^{-1}(X') = q_2^{-1}(X')$ since the descent datum of Z' is stable on X'. However, this implies $q^{-1}(q(X')) = X'$; in particular, the inverse image of $q(X')$ with respect to q is open. Using the fact that $q: Z' \longrightarrow Z$ is faithfully flat and quasi-compact and that therefore the Zariski topology on Z is the quotient of the Zariski topology on Z' (cf. [EGA IV_2], 2.3.12), we see that $q(X')$ is open. So X' descends to the quasi-affine piece $q(X')$ of Z. The only-if-part of assertion (b) is trivial. □

We want to add a criterion for the effectiveness of descent data on schemes which uses ample line bundles. Let us recall the definition of ampleness, cf. [EGA II], 4.5 and 4.6. An invertible sheaf $\mathscr{L}$ on a scheme X is called *ample* on X if X is quasi-compact and quasi-separated, and if for some $n > 0$ there are global sections $l_1, \ldots, l_r$ generating $\mathscr{L}^{\otimes n}$ such that X_{l_i}, the domain where the section l_i generates $\mathscr{L}^{\otimes n}$, is quasi-affine for each i. In fact, if $\mathscr{L}$ is ample on X, then, for any $n > 0$ and any global section l of $\mathscr{L}^{\otimes n}$, the open subscheme $X_l \subset X$ is quasi-affine as will follow from arguments given below. An invertible sheaf $\mathscr{L}$ on an S-scheme X is called *S-ample on X* (or *relatively ample over S*) if there exists an affine open covering $\{S_j\}$ of S such that the restriction of $\mathscr{L}$ onto $X \times_S S_j$ is ample for all j. The definition of S-ampleness is independent of the choice of the particular covering $\{S_j\}$, see [EGA

II], 4.6.4 and 4.6.6. If X admits an S-ample sheaf, then, by definition, it is automatically quasi-separated over S.

Consider now a quasi-compact and quasi-separated morphism $f: X \longrightarrow S$ and an invertible sheaf $\mathscr{L}$ on X. For each $n \in \mathbb{N}$, the direct image $f_* \mathscr{L}^{\otimes n}$ is a quasi-coherent sheaf on S, see [EGA I], 9.2.1. Let U_n be the open set of all points $x \in X$ such that the canonical morphism

$$(f^* f_*(\mathscr{L}^{\otimes n}))_x \longrightarrow \mathscr{L}_x^{\otimes n}$$

is surjective. Then U_n consists of all points $x \in X$ such that there is a section of $\mathscr{L}^{\otimes n}$ which is defined over the f-inverse of a neighborhood of $f(x)$ in S and which generates $\mathscr{L}^{\otimes n}$ at x. Denote by U the union of all U_n for $n \geq 1$. Let

$$\mathscr{M} = \bigoplus_{n \geq 0} f_*(\mathscr{L}^{\otimes n})$$

be the quasi-coherent graded S-algebra associated to $\mathscr{L}$, and set $P = \operatorname{Proj} \mathscr{M}$; see [EGA II], § 2. There is always a canonical S-morphism $r: U \longrightarrow P$. Namely, assuming S affine, one shows for each global section l of $\mathscr{L}^{\otimes n}$ with $n > 0$ that there is a canonical isomorphism

$$\Gamma(P_l, \mathcal{O}_P) \xrightarrow{\sim} \Gamma(X_l, \mathcal{O}_X),$$

use [EGA I], 9.3.1, and hence a morphism

$$X_l \longrightarrow P_l \hookrightarrow P .$$

The morphism is an open immersion if and only if X_l is quasi-affine over S. Thereby it is seen that *the sheaf $\mathscr{L}$ is S-ample on X if and only if $U = X$ and the canonical morphism $r: U \longrightarrow P$ is an open immersion.*

Returning to the problem of descent relative to a morphism $p: S' \longrightarrow S$, the notion of descent data generalizes naturally to pairs $(X', \mathscr{L}')$ where X' is an S'-scheme and $\mathscr{L}'$ is an invertible sheaf on X'. Namely, a descent datum on such a pair consists of a descent datum

$$\varphi: p_1^* X' \longrightarrow p_2^* X'$$

on X' and of an isomorphism

$$\lambda: \mathscr{L}_1' \longrightarrow \varphi^* \mathscr{L}_2'$$

where $\mathscr{L}_i'$ is the pull-back of $\mathscr{L}'$ with respect to the projection $p_i^* X' \longrightarrow X'$. Of course, λ must satisfy the cocycle condition, which is a cocycle condition over the cocycle condition for φ. More precisely, introducing the total space L' associated to $\mathscr{L}'$, we can say that a descent datum on $(X', \mathscr{L}')$ is a commutative diagram

$$\begin{array}{ccc} p_1^* L' & \xrightarrow{\lambda} & p_2^* L' \\ \downarrow & & \downarrow \\ p_1^* X' & \xrightarrow{\varphi} & p_2^* X' , \end{array}$$

where the vertical maps are the projections of the linear fibre spaces $p_i^* L'$ onto their bases $p_i^* X'$, where φ and λ are descent data for schemes, and where λ is an isomorphism of linear fibre spaces over φ. Another possibility is to view the descent datum φ as a cartesian diagram

$$\begin{array}{ccccc} X' \times_S S' \times_S S' & \Rrightarrow & X' \times_S S' & \overset{q_1}{\underset{q_2}{\rightrightarrows}} & X' \\ \downarrow & & \downarrow & & \downarrow \\ S''' & \Rrightarrow & S'' & \rightrightarrows & S' \end{array}$$

with natural commutativity conditions (similar to what we have explained for S'-modules), and to view λ as an isomorphism

$$\lambda : q_1^* \mathscr{L}' \longrightarrow q_2^* \mathscr{L}' \; .$$

The cocycle condition for λ can then be formulated as usual by using pull-backs with respect to the projections $X' \times_S S' \times_S S' \Rrightarrow X' \times_S S'$.

Theorem 7 (Grothendieck). *Let $p : S' \longrightarrow S$ be faithfully flat and quasi-compact. Let X' be a quasi-compact S'-scheme, and consider an invertible sheaf $\mathscr{L}'$ which is S'-ample on X'. Then, if there is a descent datum on $(X', \mathscr{L}')$, the descent is effective on X', and the pair $(X', \mathscr{L}')$ descends to a pair $(X, \mathscr{L})$ with an S-ample invertible sheaf $\mathscr{L}$ on X.*

We give only a sketch of proof for the case where S and S' are affine. First, using Theorem 4, the graded S'-algebra $\mathscr{M}' = \bigoplus_{n \geq 0} f'_*(\mathscr{L}'^{\otimes n})$, where $f' : X' \longrightarrow S'$ is the structural morphism, descends to a graded S-algebra $\mathscr{M} = \bigoplus_{n \geq 0} \mathscr{M}_n$. Next, let l' be a global section in some $\mathscr{L}'^{\otimes n}$. Then we can write

$$l' = \sum_{i=1}^{m} a_i \otimes l_i$$

with global sections a_i of $\mathcal{O}_{S'}$ and global sections l_i of $\mathscr{M}_n$. If l' generates $\mathscr{L}'^{\otimes n}$ at a certain point $x \in X'$, at least one of the global sections $1 \otimes l_i$ must generate $\mathscr{L}'^{\otimes n}$ at this point. Thereby it is seen that X' can be covered by quasi-affine open pieces X'_l where l is a global section in some $\mathscr{L}'^{\otimes n}$ which descends to a global section in $\mathscr{M}$. Then the descent datum is stable on the X'_l, and X' descends to X by Theorem 6. Finally, $\mathscr{L}'$ descends to $\mathscr{L}$ with respect to the canonical projection $X' \longrightarrow X$ since one can use Theorem 4 again.

6.2 Some Standard Examples of Descent

We start with an example which shows that the problem of descent occurs as a natural generalization of a patching problem.

Example A (*Zariski coverings*). Consider a quasi-separated scheme S and a finite affine open covering $(S_i)_{i \in I}$ of S. Let $S' := \coprod_{i \in I} S_i$ be the disjoint union of the S_i, and let $p : S' \longrightarrow S$ be the canonical projection. Note that p is faithfully flat and quasi-compact. A quasi-coherent S'-module $\mathscr{F}'$ may be thought of as a family of S_i-modules $\mathscr{F}_i$. Under what conditions does $\mathscr{F}'$ descend to a quasi-coherent S-module $\mathscr{F}$; i.e., under what conditions can one glue the $\mathscr{F}_i$ in order to obtain a quasi-coherent S-module $\mathscr{F}$ from them? By Theorem 6.1/4 we need a descent datum for $\mathscr{F}'$ with respect to $p : S' \longrightarrow S$. Such a datum consists of an isomorphism $\varphi : p_1^* \mathscr{F}' \xrightarrow{\sim} p_2^* \mathscr{F}'$ satisfying the cocycle condition, where p_1 and p_2 are the projections from S'' onto S'. In our case, we have

$$S'' = S' \times_S S' = \coprod_{i,j \in I} S_i \times_S S_j = \coprod_{i,j \in I} S_i \cap S_j \,,$$

and on $S_i \times_S S_j = S_i \cap S_j$, the first projection p_1 is the inclusion of $S_i \cap S_j$ into S_i whereas p_2 is the inclusion of $S_i \cap S_j$ into S_j. Thus the isomorphism φ consists of a family of isomorphisms

$$\varphi_{ij} : \mathscr{F}_i|_{S_i \cap S_j} \xrightarrow{\sim} \mathscr{F}_j|_{S_i \cap S_j}$$

satisfying the cocycle condition, namely, the condition that

$$\varphi_{ik}|_{S_i \cap S_j \cap S_k} = \varphi_{jk}|_{S_i \cap S_j \cap S_k} \circ \varphi_{ij}|_{S_i \cap S_j \cap S_k}$$

for all $i, j, k \in I$. So the descent datum φ is equivalent to patching data for the S_i-modules $\mathscr{F}_i$, and the cocycle condition assures that the patching data are compatible on triple overlaps.

Example B (*Galois descent*). Let $p : S' \longrightarrow S$ be a finite and faithfully flat morphism of schemes, and assume that p is a *Galois covering*; i.e., there is a finite group Γ of S-automorphisms of S' such that the morphism

$$\Gamma \times S' \longrightarrow S'' \,, \qquad (\sigma, x) \longmapsto (\sigma x, x) \,,$$

is an isomorphism; $\Gamma \times S'$ is the disjoint union of copies of S', parametrized by Γ. For example, if K'/K is a finite Galois extension of fields, the morphism $p : \operatorname{Spec} K' \longrightarrow \operatorname{Spec} K$ is such a Galois covering. Similarly, for a pair of discrete valuation rings $R \subset R'$, the morphism $p : \operatorname{Spec} R' \longrightarrow \operatorname{Spec} R$ is a Galois covering if R is henselian, R' is (finite) étale over R, and the residue extension of R'/R is Galois; use 2.3/7 and the fact that R' is henselian. We want to describe the descent of schemes with respect to $p : S' \longrightarrow S$.

Consider an S'-scheme X' with an action $\Gamma \times X' \longrightarrow X'$ which is compatible with the action of Γ on S'; i.e., we require that, for each $\sigma \in \Gamma$, the diagram

(∗)
$$\begin{array}{ccc} X' & \xrightarrow{\ \sigma\ } & X' \\ \downarrow & & \downarrow \\ S' & \xrightarrow{\ \sigma\ } & S' \end{array}$$

is commutative (for simplicity, automorphisms given by σ are again denoted by σ). Notice that the diagram is cartesian. We claim that an action on X' of the type just described is equivalent to a descent datum on X'.

Namely, from the isomorphism

$$\Gamma \times S' \xrightarrow{\sim} S'', \qquad (\sigma, x) \longmapsto (\sigma x, x),$$

we obtain an isomorphism

$$\Gamma \times \Gamma \times S' \xrightarrow{\sim} S''', \qquad (\sigma, \tau, x) \longmapsto ((\sigma \circ \tau)x, \tau x, x).$$

Taking these isomorphisms as identifications, the projections $p_{ij}: S''' \longrightarrow S''$ and $p_i: S'' \longrightarrow S'$ define projections

$$\Gamma \times \Gamma \times S' \overrightarrow{\overrightarrow{\longrightarrow}} \Gamma \times S' \rightrightarrows S'$$

which are described by

$$(0) \qquad (\sigma, \tau, x) \begin{cases} \stackrel{p_{12}}{\longmapsto} (\sigma, \tau x) \\ \stackrel{p_{13}}{\longmapsto} (\sigma \circ \tau, x), \\ \stackrel{p_{23}}{\longmapsto} (\tau, x) \end{cases} \qquad (\sigma, x) \begin{cases} \stackrel{p_1}{\longmapsto} \sigma x \\ \stackrel{p_2}{\longmapsto} x \end{cases}$$

Now assume that we have an action of Γ on X' which is compatible with the action of Γ on S'. Then we can use the same definitions (0) in order to define "projections" from $\Gamma \times \Gamma \times X'$ to $\Gamma \times X'$ and from the latter to X'. Thereby we obtain a diagram

$$(**) \qquad \begin{array}{ccccc} \Gamma \times \Gamma \times X' & \overrightarrow{\overrightarrow{\longrightarrow}} & \Gamma \times X' & \rightrightarrows & X' \\ \downarrow & & \downarrow & & \downarrow \\ \Gamma \times \Gamma \times S' & \overrightarrow{\overrightarrow{\longrightarrow}} & \Gamma \times S' & \rightrightarrows & S' \end{array}$$

where the vertical maps are the canonical ones. Since the diagram (*) is cartesian, all squares above are cartesian if in the first and second rows maps are considered which correspond to each other. Furthermore, in the last row we have the usual commutativity relations

$$\text{(i)} \quad p_1 \circ p_{12} = p_1 \circ p_{13},$$

$$\text{(ii)} \quad p_1 \circ p_{23} = p_2 \circ p_{12},$$

$$\text{(iii)} \quad p_2 \circ p_{23} = p_2 \circ p_{13}.$$

The same relations hold for the first row. Indeed, (ii) and (iii) are trivial whereas (i) is equivalent to the associativity condition

$$\sigma(\tau x) = (\sigma \circ \tau)x; \qquad \sigma, \tau \in \Gamma, \qquad x \in X'.$$

So it is clear that (**) yields a descent datum on X', the associativity of the action accounting for the cocycle condition.

Conversely, start with a descent datum φ on X'. Then, applying the base change $X' \longrightarrow S'$ to the morphism

$$\Gamma \times \Gamma \times S' \xrightarrow{p_{23}} \Gamma \times S' \xrightarrow{p_2} S',$$

we obtain the following canonical diagram

$$(***)\qquad \begin{array}{ccccc} \Gamma \times \Gamma \times X' & \longrightarrow & \Gamma \times X' & \longrightarrow & X' \\ \downarrow & & \downarrow & & \downarrow \\ \Gamma \times \Gamma \times S' & \longrightarrow & \Gamma \times S' & \longrightarrow & S' \end{array}$$

which has cartesian squares. In particular, we have canonical isomorphisms

$$\Gamma \times X' \xrightarrow{\sim} S' \times_S X' ,$$

and

$$\Gamma \times \Gamma \times X' \xrightarrow{\sim} S' \times_S S' \times_S X' .$$

Therefore we can write the descent datum φ in the form of a diagram (**). Furthermore, we may assume that (***) forms a part of (**), the one, where in both rows of (**) only the lower morphisms are considered. We claim that the morphism $\Gamma \times X' \longrightarrow X'$ over $p_1 : \Gamma \times S' \longrightarrow S'$ defines the desired action on X'. To justify this, note first that each $\sigma \in \Gamma$ acts as an automorphism on X'. Next, the commutativity conditions (ii) and (iii) imply that the morphisms

$$\Gamma \times \Gamma \times X' \rightrightarrows \Gamma \times X'$$

are defined as in (0) and, finally, as before, condition (i) accounts for the associativity of the action of Γ on X'.

As for the effectiveness of the descent, one may look at the condition given in Theorem 6.1/6. Assuming S and, hence, S' affine, as well as X' quasi-separated, a necessary and sufficient condition is that the Γ-orbit of each point $x \in X'$ is contained in a quasi-affine open subscheme of X'. Namely, considering translates of such subschemes under elements $\sigma \in \Gamma$ and taking their intersections, we can cover X' by quasi-affine open pieces which are Γ-invariant and hence stable under the descent datum. For example, if $X' \longrightarrow S'$ is quasi-projective, the condition is fulfilled, and the descent is always effective.

Example C (*Descent from R' to R, where $R \subset R'$ is an étale extension of discrete valuation rings with same residue field*). Let K (resp. K') be the field of fractions of R (resp. R'). We want to show the following result on the descent from R' to R, which will be further generalized in Example D.

Proposition C.1. *The functor which associates to an R-scheme X the triple (X_K, X', τ), consisting of the K-scheme $X_K := X \otimes_R K$, the R'-scheme $X' := X \otimes_R R'$, and the canonical isomorphism $\tau : X_K \otimes_K K' \xrightarrow{\sim} X' \otimes_{R'} K'$, is fully faithful. Its essential image consists of all triples (X_K, X', τ) which admit a quasi-affine open covering.*

The notion of an open covering of a triple (X_K, X', τ) is meant in the obvious way. Such a covering consists of a family of triples $(U_{K,i}, U'_i, \tau_i)$, where the $U_{K,i}$ (resp. the U'_i) form an open covering of X_K (resp. X'), and where τ restricts to an isomorphism $\tau_i: U_{K,i} \otimes_K K' \longrightarrow U_i' \otimes_{R'} K'$. The covering is called quasi-affine if all $U_{K,i}$ and all U_i' are quasi-affine.

Starting with a triple (X_K, X', τ), we have the canonical descent datum on $X_K \otimes_K K'$. Transporting it with τ, we obtain a descent datum on the generic fibre $X' \otimes_{R'} K'$ of X', and by the lemma below, this descent datum extends canonically to a descent datum on X'. Then the assertion of Proposition C.1 is a consequence of 6.1/6. So it is enough to show:

Lemma C.2. *For each R'-scheme X', any descent datum with respect to $K \longrightarrow K'$ on the generic fibre of X' extends canonically to a descent datum with respect to $R \longrightarrow R'$ on X'.*

Proof. Let us use the notations R'' and R''' for $R' \otimes_R R'$ and $R' \otimes_R R' \otimes_R R'$. Since R' is étale over R, the diagonal embedding $\operatorname{Spec} R' \longrightarrow \operatorname{Spec} R''$ is open (cf. 2.2/2). Thus its image, the diagonal Δ'' of $\operatorname{Spec} R''$, is a connected component of $\operatorname{Spec} R''$. Furthermore, since the residue extension of R'/R is trivial, the special fibre of Δ'' coincides with the special fibre of $\operatorname{Spec} R''$; i.e., $\operatorname{Spec} R'' = \Delta'' \,\dot{\cup}\, T''$ where the special fibre of T'' is empty. A similar assertion is true for the diagonal Δ''' in $\operatorname{Spec} R'''$.

Write K'' and K''' for the two- and threefold tensor products of K' over K. Furthermore, consider an R'-scheme X' and a descent datum with respect to $K \longrightarrow K'$ on its generic fibre. Indicating generic fibres by an index K, the descent datum on X_K' corresponds to a diagram

$$(*) \qquad \begin{array}{ccccc} X_K''' & \substack{\longrightarrow\\ \longrightarrow\\ \longrightarrow} & X_K'' & \substack{\longrightarrow\\ \longrightarrow} & X_K' \\ \downarrow{\scriptstyle p'''} & & \downarrow{\scriptstyle p''} & & \downarrow{\scriptstyle p'} \\ \operatorname{Spec} K''' & \substack{\longrightarrow\\ \longrightarrow\\ \longrightarrow} & \operatorname{Spec} K'' & \substack{\longrightarrow\\ \longrightarrow} & \operatorname{Spec} K' \end{array}$$

with cartesian squares such that the rows satisfy the usual commutativity conditions. In order to extend the descent datum to a descent datum on X', it is enough to extend the diagram $(*)$ to a diagram

$$(**) \qquad \begin{array}{ccccc} X''' & \substack{\longrightarrow\\ \longrightarrow\\ \longrightarrow} & X'' & \substack{\longrightarrow\\ \longrightarrow} & X' \\ \downarrow & & \downarrow & & \downarrow \\ \operatorname{Spec} R''' & \substack{\longrightarrow\\ \longrightarrow\\ \longrightarrow} & \operatorname{Spec} R'' & \substack{\longrightarrow\\ \longrightarrow} & \operatorname{Spec} R' \end{array}$$

of the same type. In order to do this, we have to realize that, by restriction, the lower row in $(*)$ gives rise to unique isomorphisms

$$(***) \qquad \Delta_K''' \xrightarrow{\sim} \Delta_K'' \xrightarrow{\sim} \operatorname{Spec} K',$$

and that the upper row in $(*)$ gives rise to unique isomorphisms

$$(p''')^{-1}(\Delta_K''') \xrightarrow{\sim} (p'')^{-1}(\Delta_K'') \xrightarrow{\sim} X_K'.$$

That the maps $X''_K \rightrightarrows X'_K$ coincide on the p''-inverse of Δ''_K follows from the fact that the pull-back of descent data with respect to diagonal maps always yields the identity map (cf. 6.1). A similar reasoning applies to the maps $X'''_K \substack{\longrightarrow\\ \longrightarrow\\ \longrightarrow} X''_K$.

Now it is easy to extend (∗) into (∗∗). Since the special fibre of $\operatorname{Spec} R''$ is concentrated at the open and closed subscheme Δ'', similarly for $\operatorname{Spec} R'''$ and its diagonal Δ''', we have just to extend the part of (∗) which lies over (∗∗∗). However this is trivial by the above isomorphisms.

Example D (*Descent from R' to R where $R \subset R'$ is a pair of discrete valuation rings with same uniformizing element π and with same residue field*). The situation is more general than the one considered in Example C. For example, R' can be the maximal-adic completion of the discrete valuation ring R. But we will see that, nevertheless, the results C.1 and C.2 remain valid in this case.

Consider a pair of discrete valuation rings $R \subset R'$ as required, and denote their fields of fractions by K and K'. By an index K we will indicate tensor products with K over R. Let $\delta : \operatorname{Spec} R' \longrightarrow \operatorname{Spec} R''$ be the diagonal embedding where, as usual, $R'' = R' \otimes_R R'$.

Lemma D.1. *Let M'' be an R''-module and denote by M' its pull-back with respect to δ. Assume that the quotient M''/T'' is flat over R'' where T'' is the kernel of the canonical map $M'' \longrightarrow M''_K$. Then the canonical diagram*

$$\begin{array}{ccc} M'' & \longrightarrow & M' \\ \big\downarrow & & \big\downarrow \\ M''_K & \longrightarrow & M'_K \end{array}$$

is cartesian; i.e., M'' is a fibred product of M''_K and M' over M'_K (in the category of sets, resp. R-modules, resp. R''-modules).

For example, the flatness condition on M''/T'' is satisfied if we start with an R'-module M' and take for M'' the pull-back of M' with respect to a projection $p_i : \operatorname{Spec} R'' \longrightarrow \operatorname{Spec} R'$.

Proof. Since the horizontal maps are surjective, we may extend the diagram to a commutative diagram of exact sequences

$$(*)\qquad \begin{array}{ccccccccc} 0 & \longrightarrow & L & \longrightarrow & M'' & \longrightarrow & M' & \longrightarrow & 0 \\ & & \big\downarrow & & \big\downarrow & & \big\downarrow & & \\ 0 & \longrightarrow & L_K & \longrightarrow & M''_K & \longrightarrow & M'_K & \longrightarrow & 0 \end{array}$$

The second row can be thought of as being obtained from the first one by taking tensor products over R with K. We claim that the map $L \longrightarrow L_K$ is an isomorphism; i.e., that L is already a K-vector space. Then it is immediately clear that M'' is the fibred product of M''_K and M' over M'_K; the universal property is checked by means of diagram chasing in $(*)$.

So it remains to show that L is already a K-vector space. Let us consider the first row of $(*)$ for the special case where $M'' = R''$. Thereby we obtain the exact sequence

$$(**) \qquad 0 \longrightarrow \mathfrak{I}'' \longrightarrow R'' \longrightarrow R' \longrightarrow 0$$

of R-modules (or, alternatively, R''-modules). In terms of R-modules, the sequence is split exact, since $R'' \longrightarrow R'$ admits a section. In particular, taking the tensor product of $(**)$ over R with $R/\pi^n R$ for any $n > 0$ gives a split exact sequence

$$0 \longrightarrow \mathfrak{I}''/\pi^n\mathfrak{I}'' \longrightarrow R''/\pi^n R'' \longrightarrow R'/\pi^n R' \longrightarrow 0 .$$

By the assumptions on R and R', we see that the map

$$R''/\pi^n R'' \longrightarrow R'/\pi^n R'$$

is bijective. Thus, for $n = 1$, we have $\mathfrak{I}''/\pi\mathfrak{I}'' = 0$ and, therefore, $\mathfrak{I}'' = \pi\mathfrak{I}''$. So $\mathfrak{I}''$ is a K-vector space since R'' and, hence, $\mathfrak{I}''$ have no π-torsion. Now, tensoring $(**)$ over R'' with M'' and using the fact that M' is the pull-back of M'' with respect to the diagonal morphism $\operatorname{Spec} R' \longrightarrow \operatorname{Spec} R''$, we get the exact sequence $\mathfrak{I}'' \otimes_{R''} M'' \longrightarrow M'' \longrightarrow M' \longrightarrow 0$. Comparing it with the first row in $(*)$, we have a surjective R-homomorphism $\mathfrak{I}'' \otimes_{R''} M'' \longrightarrow L$. Therefore, since $\mathfrak{I}''$ is a K-vector space, the same must be true for L, provided L has no π-torsion.

Thus it remains to show that the π-torsion of L is trivial. To do this we consider first the case where $M'' = T''$. Using a limit argument, we may assume $\pi^n M'' = 0$ for some integer n. But then the isomorphism $R''/\pi^n R'' \xrightarrow{\sim} R'/\pi^n R'$ yields an isomorphism

$$M'' = M''/\pi^n M'' \xrightarrow{\sim} M'/\pi^n M' = M'$$

so that L is trivial in this case. In the general case we tensor the exact sequence

$$0 \longrightarrow T'' \longrightarrow M'' \longrightarrow M''/T'' \longrightarrow 0$$

over R'' with R', thereby obtaining the sequence

$$0 \longrightarrow T'' \otimes_{R''} R' \longrightarrow M' \longrightarrow (M''/T'') \otimes_{R''} R' \longrightarrow 0 .$$

The latter is exact because M''/T'' is flat over R''. By the same reason, $(M''/T'') \otimes_{R''} R'$ is flat over R' and, thus, $T' := T'' \otimes_{R''} R'$ is the torsion-submodule of M'. Since the canonical homomorphism $M'' \longrightarrow M'$ maps T'' surjectively onto T', the first row of the diagram $(*)$ yields an exact sequence

$$0 \longrightarrow L \cap T'' \longrightarrow T'' \longrightarrow T' \longrightarrow 0$$

and it follows from the special case considered above that $L \cap T''$ must be trivial. So the π-torsion of L is trivial and we see that L is a K-vector space. □

Reversing arrows in the definition of cartesian diagrams and fibred products, one arrives at the notions of *co-cartesian diagrams* and *amalgamated sums*. We want to translate the assertion of the above lemma into a statement on amalgamated sums of schemes. First note that Lemma D.1 remains true if we work in the category of R-algebras or R''-algebras. So it yields a statement on amalgamated sums in the category of affine R-schemes or R''-schemes. We want to generalize it to the case of not necessarily affine schemes. Set $S = \operatorname{Spec} R$, $S' = \operatorname{Spec} R'$, $S'' = \operatorname{Spec} R''$, and let $\delta : S' \longrightarrow S''$ be the diagonal embedding. For any R-scheme X, let $X_K = X \otimes_R K$ be its generic fibre.

Proposition D.2. *Let X' be an S'-scheme and let X'' be its pull-back with respect to one of the projections $p_i : S'' \longrightarrow S'$. Then the canonical diagram*

$$\begin{array}{ccc} X'_K = \delta^* X''_K & \longrightarrow & X''_K \\ \downarrow & & \downarrow \\ X' = \delta^* X'' & \longrightarrow & X'' \end{array}$$

is co-cartesian in the category of R-schemes (resp. R''-schemes); i.e., in this category, X'' is the amalgamated sum of X' and X''_K under X'_K.

Proof. In order to reduce the assertion of the proposition to Lemma D.1, we need to know that a subset $F \subset X''$ is closed if and only if $F \cap X'$ is closed in X' and $F \cap X''_K$ is closed in X''_K; note that, in terms of sets, the above diagram consists of injections and that $X'' = X' \cup X''_K$, due to the assumption on R and R'. The necessity of the condition is clear. In order to show that it is sufficient, we may assume that X' is affine, say $X' = \operatorname{Spec} A'$. Then the above diagram of schemes corresponds to a diagram of R''-algebras

$$\begin{array}{ccc} A'' & \longrightarrow & A' \\ \downarrow & & \downarrow \\ A''_K & \longrightarrow & A'_K , \end{array}$$

which is cartesian in the category of sets. Now assume that $F \cap X'$ is closed in X' and that $F \cap X''_K$ is closed in X''_K. Let $\mathfrak{I}' \subset A'$ and $\mathfrak{I}''_K \subset A''_K$ be the corresponding reduced ideals. Since $F \cap X'$ coincides with $F \cap X''_K$ on X'_K, we have

$$\operatorname{rad}(A'_K \mathfrak{I}') = \operatorname{rad}(A'_K \mathfrak{I}''_K) .$$

The fibred product of $\mathfrak{I}'$ and $\mathfrak{I}''_K$ over A'_K exists in the category of sets. Denoting it by $\mathfrak{I}''$, we see that we have a canonical inclusion $\mathfrak{I}'' \hookrightarrow A''$; furthermore, it is easily verified that $\mathfrak{I}''$ is an ideal in A''. We claim

$$\operatorname{rad}(\mathfrak{I}'' A') = \mathfrak{I}' \qquad \text{and} \qquad \operatorname{rad}(\mathfrak{I}'' A''_K) = \mathfrak{I}''_K .$$

The inclusion "$\subset$" is trivial in both cases. To justify the opposite inclusions, consider an element $f \in \mathfrak{I}'$. Using the equation between radicals above, it is seen that a power of f has an inverse image in $\mathfrak{I}''$; so $f \in \operatorname{rad}(\mathfrak{I}'' A')$. Similarly, if $f \in \mathfrak{I}''_K$, a power of π times a power of f has an inverse image in $\mathfrak{I}''$ and, hence, $f \in \operatorname{rad} \mathfrak{I}'' A''_K$. This justifies the above description of $\mathfrak{I}'$ and $\mathfrak{I}''_K$, and it follows that the closed subset of X'' given by $\mathfrak{I}''$ coincides with $F \cap X'$ on X' and with $F \cap X''_K$ on X''_K. Hence F is closed in

X'', since $X'' = X' \cup X''_K$. Thereby we have proved the desired topological characterization of closed sets in X''. Looking at complements of closed sets, we see that a subset of X'' is open if and only if its intersection with X' is open in X' and its intersection with X''_K is open in X''_K.

Now it is easy to verify the assertion of the proposition. Consider a scheme Z and a commutative diagram

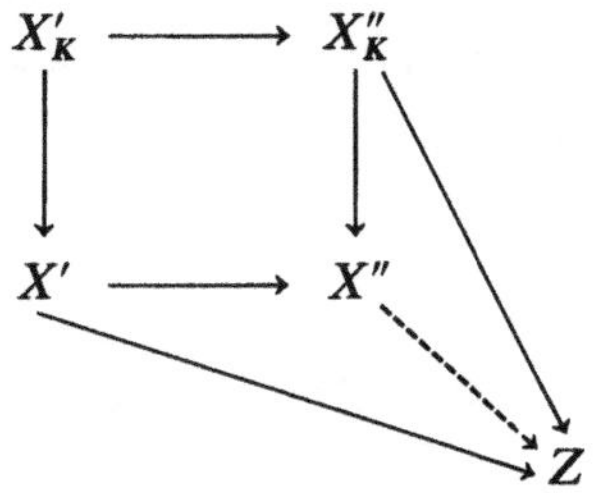

where the solid arrows are given and where the square is the canonical diagram. It has to be shown that the diagram can be supplemented by a unique morphism $X'' \dashrightarrow Z$. Let W be an open affine subscheme of Z, let U' be its inverse image in X' and U''_K its inverse image in X''_K. Then by the above topological characterization, $U'' := U' \cup U''_K$ is an open subscheme of X'' which extends U''_K and whose pull-back with respect to the diagonal embedding $\delta : X' \longrightarrow X''$ yields U'. So we can look at the problem

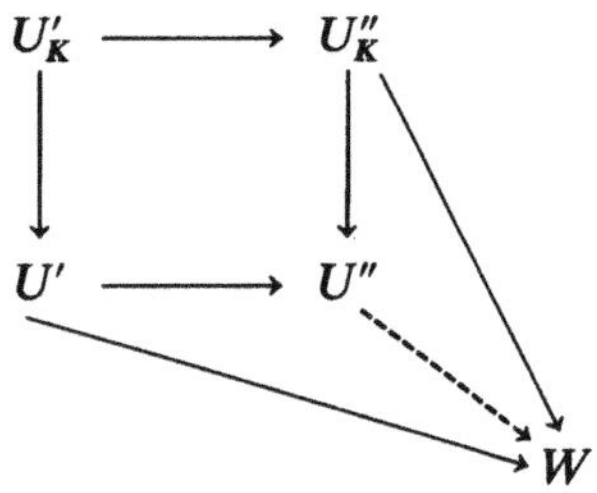

Working locally on U'' and applying Lemma D.1, we want to show that it has a unique solution. To do this, it is enough to verify the flatness condition of Lemma D.1 or, equivalently, the fact that the schematic closure $\bar{X}''$ of X''_K in X'' is flat over R''. Since the projection p_i we are considering is flat, we see that $\bar{X}''$ can be interpreted as the pull-back under p_i of the schematic closure $\bar{X}'$ of X'_K in X'; cf. 2.5/2. However, $\bar{X}'$ is flat over R' by its definition. So $\bar{X}''$ is flat over R'' and Lemma D.1 is applicable. It follows that the above local problem has a unique solution $U'' \longrightarrow W$ and, by working with respect to an affine open covering of Z, that the above global problem has a unique solution $X'' \longrightarrow Z$. □

Now we want to explain how the results D.1 and D.2 imply that descent data with respect to $\operatorname{Spec} K' \longrightarrow \operatorname{Spec} K$ extend to descent data with respect to $S' \longrightarrow S$.

Lemma D.3. *Consider an R'-module M' (resp. an R'-scheme X') and a descent datum φ_K with respect to $K \longrightarrow K'$ on M'_K (resp. on X'_K). Then φ_K extends uniquely to a descent datum with respect to $R \longrightarrow R'$ on M' (resp. on X').*

Proof. A descent datum with respect to $R \longrightarrow R'$ on M' may be viewed as a commutative diagram

$$\begin{array}{ccc} M' \otimes_R R' & \xrightarrow{\ \varphi\ } & R' \otimes_R M' \\ \big\downarrow & & \big\downarrow \\ M' & = & M' \end{array} ,$$

where φ is an isomorphism satisfying the cocycle condition and where the vertical maps are the canonical ones obtained from the diagonal map $\delta : S' \longrightarrow S''$. Similarly, for the descent datum φ_K on the generic fibre of M', we get the upper square of the following commutative diagram

$$\begin{array}{ccc} (M' \otimes_R R')_K & \xrightarrow{\ \varphi_K\ } & (R' \otimes_R M')_K \\ \big\downarrow & & \big\downarrow \\ M'_K & = & M'_K \\ \big\uparrow & & \big\uparrow \\ M' & = & M' \end{array} .$$

Then, taking the fibred product of the first and third rows over the second row, Lemma D.1 shows that φ_K extends uniquely to an R''-isomorphism

$$\varphi : M' \otimes_R R' \longrightarrow R' \otimes_R M' ,$$

whose pull-back with respect to the diagonal map $\delta : S'' \longrightarrow S'$ yields the identity on M'. That φ satisfies the cocycle condition follows in a similar way from Lemma D.1. Thus φ is a descent datum on M' which extends φ_K; it is unique. For the case of schemes, the assertion is deduced in formally the same way from Proposition D.2. □

Now, applying Theorems 6.1/4 and 6.1/6, we can derive from the above lemma the desired generalization of Proposition C.1.

Proposition D.4. (a) *The functor which associates to each R-module M the triple* (M_K, M', τ), *where* $M_K := M \otimes_R K$, $M' := M \otimes_R R'$, *and* $\tau : M_K \otimes_K K' \xrightarrow{\sim} M' \otimes_R K'$ *is the canonical isomorphism, is an equivalence of categories.*

(b) *The functor which associates to each R-scheme X the triple* (X_K, X', τ) *consisting of the K-scheme* $X_K := X \otimes_R K$, *the R′-scheme* $X' = X \otimes_R R'$, *and the canonical isomorphism* $\tau : X_K \otimes_K K' \xrightarrow{\sim} X' \otimes_{R'} K'$, *is fully faithful. Its essential image consists of all triples* (X_K, X', τ) *which admit a quasi-affine open covering.*

Finally, we want to mention that it is an easy exercise to verify assertion (a) of the proposition by a direct argument. Applying a limit argument, one reduces to the case of finitely generated R- or R'-modules, where it is possible to treat the case of torsion and of free modules separately. However, for the purpose of assertion (b), it was necessary to prove more precise results also in the module case.

6.3 The Theorem of the Square

Let S be a scheme, let X be an S-scheme, and consider an S-group scheme G which acts on X. Using the notion of T-valued points for arbitrary S-schemes T, such an action corresponds to an S-morphism

$$G \times_S X \longrightarrow X \,, \qquad (g, x) \longmapsto gx \,,$$

where

$$g(g'x) = (gg')x \qquad \text{and} \qquad 1_T x = x$$

for arbitrary points $g, g' \in G(T)$, $x \in X(T)$, and for the unit element $1_T \in G(T)$. Alternatively, interpreting G (resp. X) as a functor from the category of S-schemes to the category of groups (resp. sets), we can say that the group functor G acts on X; i.e., that, for each S-scheme T, we have an action of $G(T)$ on $X(T)$ which is compatible with S-morphisms $T' \longrightarrow T$ in the usual way. Similarly as in the case of group schemes, one defines for any $g \in G(T)$ the translation

$$\tau_g : X \longrightarrow X \,, \qquad x \longmapsto gx \,,$$

where, more precisely, τ_g has to be interpreted as a T-morphism from X_T to X_T.

Now let us fix an invertible sheaf $\mathscr{L}$ on X. Its pull-back to X_T will again be denoted by $\mathscr{L}$. So we can talk about the pull-back of $\mathscr{L}$ with respect to a translation τ_g, $g \in G(T)$, thus obtaining the invertible sheaf

$$\mathscr{L}_g := \tau_g^* \mathscr{L}$$

on X_T. Let $P_{X/S}$ be the functor which associates to any S-scheme T the group

$$\operatorname{Pic}(T \times_S X)/p^* \operatorname{Pic}(T) \,;$$

i.e., the group of invertible sheaves on $X \times_S T$ modulo the pull-back under the projection $p : T \times_S X \longrightarrow T$ of invertible sheaves on T. Then $P_{X/S}$ is a commutative group functor, and we can consider the morphism

$$\varphi_{\mathscr{L}} : G \longrightarrow P_{X/S} \,, \qquad g \longmapsto \text{class of} \quad \mathscr{L}_g \otimes \mathscr{L}^{-1} \,,$$

which is a functorial morphism between functors from the category of S-schemes to the category of sets. We will say that $\mathscr{L}$ *satisfies the theorem of the square* if $\varphi_{\mathscr{L}}$ respects group structures and, thus, is a functorial morphism between group functors. We do this in analogy to the classical case, where X is an abelian variety over a field K, and where the action of G on X is given by translation. In this case, the functor $P_{X/S}$ coincides with the relative Picard functor $\operatorname{Pic}_{X/S}$ (see 8.1/4), and the classical theorem of the square asserts that, for each invertible sheaf $\mathscr{L}$ on X, the morphism $\varphi_{\mathscr{L}}$ is a morphism of group functors. For proofs see Weil [2], § VIII, n°57, Thm. 30, Cor. 2, as well as Lang [1], Chap. III, § 3, Cor. 4, and Mumford [3], Chap. II, § 6, Cor. 4.

The purpose of the present section is to extend the classical theorem of the square to a more general situation. For the applications we have in mind, it is enough to know that, for each invertible sheaf $\mathscr{L}$ on X, a power $\mathscr{L}^{\otimes n}$ satisfies the theorem of the square.

Theorem 1. *Let S be a Dedekind scheme and let G be a smooth S-group scheme with connected fibres which acts on an S-scheme X, where $X \longrightarrow S$ is smooth, of finite type, and has geometrically connected generic fibres. Then, for any invertible sheaf $\mathscr{L}$ on X, there is an integer $n > 0$ such that $\mathscr{L}^{\otimes n}$ satisfies the theorem of the square.*

If the generic fibres of X are proper or if the local rings $\mathcal{O}_{S,\xi}$ at generic points $\xi \in S$ are perfect fields, the assertion holds for $n = 1$.

We will reduce the theorem to the classical situation where S consists of a field. In fact, we will show that $\mathscr{L}$ satisfies the theorem of the square if and only if this is the case over each generic point of S; see Lemma 2. In order to carry out this reduction step, it is necessary to write down somewhat more explicitly the condition of $\varphi_{\mathscr{L}} : G \longrightarrow P_{X/S}$ being a morphism of group functors. Let m be the group law on G. Set $T := G \times_S G$, and consider the projections $p_1, p_2 : G \times_S G \rightrightarrows G$ as T-valued points of G. Furthermore, let

$$f : G \times_S G \times_S X \longrightarrow G \times_S G$$

be the projection onto the first two factors. Then we claim that $\varphi_{\mathscr{L}}$ is a morphism of group functors if and only if

$$\mathscr{M} := \mathscr{L}_{m(p_1,p_2)} \otimes \mathscr{L}_{p_1}^{-1} \otimes \mathscr{L}_{p_2}^{-1} \otimes \mathscr{L} ,$$

as an invertible sheaf on $G \times_S G \times_S X$, is isomorphic to the pull-back $f^*\mathscr{N}$ of an invertible sheaf $\mathscr{N}$ on $G \times_S G$.

In fact, the class of $\mathscr{M}$ in $P_{X/S}(G \times_S G)$ is given by

$$\varphi_{\mathscr{L}}(m(p_1,p_2)) - \varphi_{\mathscr{L}}(p_1) - \varphi_{\mathscr{L}}(p_2) .$$

Thus it is trivial if $\varphi_{\mathscr{L}}$ is a morphism of group functors. In order to show the converse, we mention the following fact:

For an arbitrary S-scheme T and two points $g, g' \in G(T)$, the invertible sheaf $\mathscr{L}_{m(g,g')} \otimes \mathscr{L}_g^{-1} \otimes \mathscr{L}_{g'}^{-1} \otimes \mathscr{L}$ is the pull-back of $\mathscr{M}$ with respect to the morphism

$$(g, g') \times_S \mathrm{id}_X : T \times_S X \longrightarrow G \times_S G \times_S X .$$

So if $\mathscr{M} \cong f^*\mathscr{N}$ for some invertible sheaf $\mathscr{N}$ on $G \times_S G$, the commutative diagram

$$\begin{array}{ccc} T \times_S X & \xrightarrow{(g,g') \times_S \mathrm{id}_X} & G \times_S G \times_S X \\ \downarrow p & & \downarrow f \\ T & \xrightarrow{(g,g')} & G \times_S G \end{array} ,$$

where p is the projection onto the first factor, yields

$$\mathscr{L}_{m(g,g')} \otimes \mathscr{L}_g^{-1} \otimes \mathscr{L}_{g'}^{-1} \otimes \mathscr{L} \cong p^*((g,g')^*(\mathscr{N}))$$

and, hence,

$$\varphi_{\mathscr{L}}(m(g,g')) = \varphi_{\mathscr{L}}(g) + \varphi_{\mathscr{L}}(g') .$$

This justifies our claim. We will now reduce the theorem of the square to generic fibres.

Lemma 2. *Let S, G, X and $\mathscr{L}$ be as in Theorem 1, and let $\mathscr{M}$ be the invertible sheaf on $G \times_S G \times_S X$ which has been defined above. Then the following conditions are equivalent:*

(a) *There exists an invertible sheaf $\mathscr{N}$ on $G \times_S G$ such that $\mathscr{M}$ is isomorphic to the pull-back $f^*\mathscr{N}$ of $\mathscr{N}$ with respect to the projection $f: G \times_S G \times_S X \longrightarrow G \times_S G$; i.e., $\mathscr{L}$ satisfies the theorem of the square.*

(b) *For each generic point ξ of S, the invertible sheaf $\mathscr{L}$ satisfies the theorem of the square after performing the base change* Spec $k(\xi) \longrightarrow S$.

Proof. The fact that an invertible sheaf on X satisfies the theorem of the square is preserved by any base change. Thus the implication (a) $\Longrightarrow$ (b) is obvious.

In order to show the converse, we may assume that S is irreducible with generic point ξ. If condition (b) is given, there is an invertible sheaf $\mathscr{N}_\xi$ on $(G \times_S G)_\xi$ satisfying

$$\mathscr{M}_\xi \cong f_\xi^*(\mathscr{N}_\xi) ,$$

where the index ξ indicates restrictions to generic fibres. We can extend $\mathscr{N}_\xi$ to an invertible sheaf $\mathscr{N}$ on $G \times_S G$ because $G \times_S G$ is regular. For example, this can be done by considering a divisor on $(G \times_S G)_\xi$ which corresponds to $\mathscr{N}_\xi$. Taking its schematic closure in $G \times_S G$, the associated invertible sheaf on $G \times_S G$ may be viewed as an extension of $\mathscr{N}_\xi$.

Now consider the invertible sheaf $\mathscr{M}' := \mathscr{M} \otimes (f^*(\mathscr{N}))^{-1}$ on $G \times_S G \times_S X$. Using the projection $p: G \times_S G \times_S X \longrightarrow S$, we claim there is a divisor Δ on S such that

$$\mathscr{M}' \cong p^*(\mathscr{O}_S(\Delta)) .$$

Namely, $\mathscr{M}'_\xi$ is trivial. So we can choose a global generator and view it as a meromorphic section of $\mathscr{M}'$. Then it generates $\mathscr{M}'$ over an open subset of $G \times_S G \times_S X$ whose complement consists of at most finitely many fibres over closed points in S. Thus there is a divisor D on $G \times_S G \times_S X$ whose support meets only finitely many fibres of p over closed points of S such that

$$\mathscr{M}' \cong \mathscr{O}_{G \times_S G \times_S X}(D) .$$

Now look at the projection

$$p_3: G \times_S G \times_S X \longrightarrow X .$$

Since the structural morphism $G \times_S G \longrightarrow S$ is smooth and has geometrically irreducible fibres, the same is true for p_3 and it is easily seen that the pull-back of a prime divisor on X yields a prime divisor on $G \times_S G \times_S X$. Hence, the Weil divisor

D, whose support is not dominant over X, is of the tye $p_3^*(\Delta')$ with a Weil divisor Δ' of X. So we have

$$(*) \qquad \mathscr{M}' = \mathscr{M} \otimes (f^*(\mathscr{N}))^{-1} \cong p_3^*(\mathcal{O}_X(\Delta')) ,$$

and it remains to show that $\mathcal{O}_X(\Delta')$ is the pull-back of an invertible sheaf on S. If X has irreducible fibres over S, a similar argument as above shows that Δ' is pull-back of a divisor on S. In the general case, consider the morphism

$$q = (\varepsilon, \varepsilon, \mathrm{id}_X) : X \longrightarrow G \times_S G \times_S X ,$$

where ε is the composition of the structural morphism $X \longrightarrow S$ with the unit section $S \longrightarrow G$. Pulling back $(*)$ with respect to q, we get on the right-hand side $\mathcal{O}_X(\Delta')$. On the left-hand side, the pull-back of $\mathscr{M}$ is trivial; it is the evaluation of $\mathscr{M}$ at the unit section of $G \times_S G$. Furthermore, since $f \circ q : X \longrightarrow G \times_S G$ factors through S, we see that $q^*(f^*(\mathscr{N}))$ is the pull-back of an invertible sheaf on S. So $\mathcal{O}_X(\Delta')$ is the pull-back of an invertible sheaf on S as claimed; we can write it in the form $\mathcal{O}_S(\Delta)$ with a divisor Δ on S.

Now, looking at the isomorphism

$$\mathscr{M} \cong f^*(\mathscr{N}) \otimes p^*(\mathcal{O}_S(\Delta))$$

obtained from $(*)$, we can replace $\mathscr{N}$ by its tensor product with the pull-back of $\mathcal{O}_S(\Delta)$ to $G \times_S G$. Then the resulting invertible sheaf, again denoted by $\mathscr{N}$, satisfies $\mathscr{M} \cong f^*(\mathscr{N})$. Thus $\mathscr{M}$ is as required in condition (a). □

The essence of the lemma consists in the fact that an invertible sheaf $\mathscr{L}$ on X satisfies the theorem of the square as soon as the pull-back of $\mathscr{L}$ to each generic fibre of X satisfies this theorem. So, in order to establish Theorem 1, it can be assumed that S is the spectrum of a field.

In the main case where $G = X$ is an abelian variety we are done by the classical theorem of the square. For the general case, we refer to Raynaud [4], Thm. IV. 3.3, in order to see that a power of $\mathscr{L}$ satisfies the theorem of the square. In fact, one shows that $\mathscr{L}$ itself satisfies the theorem of the square if the field K is replaced by a finite radicial extension; cf. Raynaud [4], Thm. IV. 2.6.

We want to add two possibilities of obtaining the theorem of the square in special situations, always assuming that the base is a field. First, let us consider the case where X is proper. In order to show that

$$\varphi_{\mathscr{L}} : G \longrightarrow P_{X/K}$$

is a morphism of group functors, look at the relative Picard functor $\mathrm{Pic}_{X/K}$ (cf. Section 8.1). Since X is proper, smooth, and geometrically connected over K, the canonical morphism

$$P_{X/K} \longrightarrow \mathrm{Pic}_{X/K}$$

is injective (cf. 8.1/4). So it is enough to show that $\varphi_{\mathscr{L}}$ defines a morphism of group functors

$$\varphi'_{\mathscr{L}} : G \longrightarrow \mathrm{Pic}_{X/K} .$$

Now we use the fact that $\mathrm{Pic}_{X/K}$ is representable by a group scheme over K (cf. 8.2/3) and that $(\mathrm{Pic}^0_{X/K})_{\mathrm{red}}$ is an abelian variety A over K; cf. [FGA], n°236, Cor. 3.2. Since $\varphi'_{\mathscr{L}}$ maps unit sections onto each other, it must factor through A. Then the rigitiy lemma (cf. Lang [1], Chap. II, § 1, Thm. 4) shows that the resulting morphism

$$G \longrightarrow A$$

is a morphism of group functors. Hence, it follows that $\mathscr{L}$ satisfies the theorem of the square.

The second method we want to mention applies to the case where X is a torsor under G. The applications of Theorem 1 we have in mind refer to this situation. Still considering the case where S consists of a field K and replacing K by its algebraic closure, we may assume that X coincides with G and, thus, is an algebraic group over an algebraically closed field. Then, by the theorem of Chevalley 9.2/1, there is an exact sequence of algebraic groups over K

$$1 \longrightarrow G_{\mathrm{aff}} \longrightarrow G \longrightarrow A \longrightarrow 1$$

where G_{aff} is smooth, connected, and affine, and where A is an abelian variety. Since the Picard group of the affine group G_{aff} consists only of torsion, one can show that a power of $\mathscr{L}$ is the pull-back of an invertible sheaf on A. So one is essentially reduced to the case where G is an abelian variety.

6.4 The Quasi-Projectivity of Torsors

We want to introduce the notion of torsors, a notion which is closely related to the concept of group schemes. Consider a base scheme S, an S-scheme X, and an S-group scheme G which acts on X by means of a morphism

$$G \times_S X \longrightarrow X , \qquad (g,x) \longmapsto gx .$$

Assume that G is (faithfully) flat and locally of finite presentation over S. Then X is called a *torsor* (with respect to the fppf-topology), more precisely, an *S-torsor under G* if

(i) the structural morphism $X \longrightarrow S$ is faithfully flat and locally of finite presentation, and

(ii) the morphism $G \times_S X \longrightarrow X \times_S X$, $(g,x) \longmapsto (gx,x)$, is an isomorphism.

Viewing $G \times_S X$ and $X \times_S X$ as X-schemes with respect to the second projections, we see that the isomorphism in (ii) is, in fact, an X-isomorphism. In other words, applying the base change $X \longrightarrow S$ to X and G, both schemes become isomorphic. The same is, of course, true for any base change $Y \longrightarrow S$ which factors through X. In particular, if $X(S) \neq \varnothing$, the choice of an S-valued point of X gives rise to an S-isomorphism from G to X, and there is no essential difference between G and the torsor X. We say that the torsor X is *trivial* in this case. Furthermore, $X \longrightarrow S$ satisfies any of the conditions listed in [EGA IV_2], 2.7.1 and [EGA IV_4], 17.7.4, for example, being smooth, separated, or of finite type, provided these conditions are satisfied by $G \longrightarrow S$. Namely, in order to apply the cited results, it is enough to consider the case where S is affine. Then, since $X \longrightarrow S$ is open, there exists a quasi-compact open subscheme Y of X such that $Y \longrightarrow S$ is surjective and, hence, faithfully flat as well as locally of finite presentation. So, what we have claimed follows from the isomorphism $G \times_S Y \xrightarrow{\sim} X \times_S Y$ by faithfully flat and quasi-compact descent. In particular, if G is smooth, X is smooth and it can be trivialized after a surjective étale base change $S' \longrightarrow S$ because, after performing a suitable base change of this type, X will have sections by 2.2/14.

Examples of torsors are easy to describe. Consider a finite Galois extension L/K of fields. Then Spec L is a (Spec K)-torsor under the constant group Gal(L/K). Or, consider an invertible sheaf $\mathscr{L}$ on a scheme X and remove the zero section from its associated total space. The resulting scheme is an X-torsor under the multiplicative group $(\mathbb{G}_m)_X$. It is trivial if and only if $\mathscr{L}$ is trivial. We want to formulate now the main result to be proved in this section.

Theorem 1. *Let S be a Dedekind scheme, and let X be a torsor under an S-group scheme G. Assume that G is smooth, separated, and of finite type over S. Then X is quasi-projective over S. In particular, G itself is quasi-projective over S.*

For the proof we have to construct an S-ample invertible sheaf $\mathscr{L}$ on X. In order to do so, we use the theorem of the square.

First we show that, for any effective divisor D on X, the associated invertible sheaf $\mathscr{L} := \mathscr{O}_X(D)$ is S-ample if $X - \operatorname{supp}(D)$ satisfies certain properties.

Proposition 2. *Let S be a Dedekind scheme and let G be a smooth S-group scheme with connected fibres which acts on an S-scheme X, where X is smooth and of finite type over S. Assume there exists an open subscheme $U \subset X$ such that U is affine over S and such that U meets all G-orbits of points in X; i.e., such that the action of G induces a surjective morphism $G \times_S U \longrightarrow X$. Then, for any effective divisor D on X with support $X - U$, the invertible sheaf $\mathscr{L} = \mathscr{O}_X(D)$ is S-ample.*

For example, $X - U$ provided with its reduced structure gives rise to such a divisor D; cf. [EGA IV$_4$], 21.12.7.

Proof. In a first step we want to reduce to the case where S is local. So assume $\mathscr{L}$ is an invertible sheaf on X such that, for each $s \in S$, the pull-back $\mathscr{L}(s)$ of $\mathscr{L}$ to $X(s) := X \times_S \operatorname{Spec} \mathscr{O}_{S,s}$ is ample. Then there exist global sections $l_1, \ldots, l_r$ generating a certain power $\mathscr{L}(s)^{\otimes n}$ such that the open subscheme $X(s)_{l_i} \subset X(s)$ where l_i generates $\mathscr{L}(s)^{\otimes n}$ is *affine*; use [EGA II], 4.5.2, or the characterization of ample invertible sheaves given in Section 6.1. By a limit argument, the l_i extend to sections l'_i of $\mathscr{L}^{\otimes n}$ over some neighborhood S' of $s \in S$ and, by [EGA IV$_3$], 8.10.5, we may assume that the l'_i generate $\mathscr{L}^{\otimes n}$ over S', that the projection $X \times_S S' \longrightarrow S'$ is separated, and that the open subscheme $X'_{l'_i} \subset X \times_S S'$ where l'_i generates $\mathscr{L}^{\otimes n}$ is affine. Thereby we see that $\mathscr{L}$ is ample over a neighborhood of each point $s \in S$ and, thus, that $\mathscr{L}$ is S-ample on X.

Let us assume now that S is the spectrum of a local ring R. Since ampleness can be checked after faithfully flat and quasi-compact base change, as follows from [EGA IV$_2$], 2.7.2, it is enough to treat the case where R is strictly henselian. Using the fact that G has geometrically connected fibres, we see that G operates on the connected components of X. So we can assume that X is connected. We claim that it is enough to consider the case where the structural morphism $X \longrightarrow S$ is surjective. In fact, $X \longrightarrow S$ is open and, if $X \longrightarrow S$ is not surjective, we replace S by the image of X. However, doing so, we may loose the property of S being local and strictly henselian. In this case we have to go back to the beginning and to start the proof anew. Therefore, by induction on the dimension of S, we are reduced to the case where S is local and strictly henselian, where $X \longrightarrow S$ is surjective, and where X is connected. Then X has sections by 2.3/5 and, thus, its generic fibre is *geometrically connected* by [EGA IV_2], 4.5.13.1.

In this situation, we want to establish the assertion of the proposition. Replacing the divisor D by a multiple of itself, we can assume that the invertible sheaf $\mathscr{L} = \mathcal{O}_X(D)$ satisfies the theorem of the square; see 6.3/1. Then the divisor $D_g + D_{g^{-1}}$ is linearly equivalent to $2D$, where we have written D_g for the translate of D under g. Hence there is a section $l \in \Gamma(X, \mathscr{L}^{\otimes 2})$ such that

$$X_l = X - \operatorname{supp}(D_g + D_{g^{-1}}) = gU \cap g^{-1}U .$$

As the intersection of two affine open subschemes of a noetherian scheme, X_l is quasi-affine. Furthermore, it follows that $\mathscr{L}$ is ample, provided we can show that the open subschemes $gU \cap g^{-1}U$ cover X if g varies over $G(S)$.

So it remains to verify the latter fact. Fix a point $x \in X$. Write s for its image in S and set $k = k(s)$. We claim that there is a dense open subscheme $Z_s \subset G_s$ such that

$$x \in gU_s \cap g^{-1}U_s$$

for each $g \in Z_s(k)$. To see this, we may assume that x is a closed point of X_s. Then we apply the base change $k \longrightarrow k'$ to fibres over s, where $k' = k(x)$ is finite over k. Let W be the inverse of $U_s \otimes_k k'$ under the morphism

$$G_s \otimes_k k' \longrightarrow X_s \otimes_k k' , \qquad a \longmapsto ax ,$$

and write W^{-1} for its inverse under the group law on $G_s \otimes_k k'$. Then, since U meets all G-orbits of points in X and since G has geometrically connected fibres, $W \cap W^{-1}$ is a dense open subscheme of $G_s \otimes_k k'$. Furthermore, the relation $x \in g(U_s \otimes_k k') \cap g^{-1}(U_s \otimes_k k')$ is equivalent to $g^{-1}x \in U_s \otimes_k k'$ and $gx \in U_s \otimes_k k'$. Thus $x \in g(U_s \otimes_k k') \cap g^{-1}(U_s \otimes_k k')$ for all $g \in (W \cap W^{-1})(k')$. Then, using methods of descent, we find a dense open subscheme of $W \cap W^{-1}$ descending to a dense open subscheme Z_s of G_s such that $x \in gU_s \cap g^{-1}U_s$ for all $g \in Z_s(k)$.

Now it is easy to see that the open subschemes $gU \cap g^{-1}U$ cover X if g varies over $G(S)$. Namely, we have only to realize that, for each dense open subscheme $Z_s \subset G_s$ of a fibre over a point $s \in S$, there exists a section in $G(S)$ which, by restriction to G_s, yields a section of Z_s. If s is the closed point of S, this follows from 2.3/5. If s belongs to the generic fibre of S, we can consider the schematic closure of $G_s - Z_s$ in G. Its special fibre is nowhere dense in the special fibre of G so that an argument as the one given before will finish the proof of Proposition 2.

Later, in 6.6/1, we will use the same idea of proof again without the restriction that the base S is of dimension ≤ 1. In this case, one can apply the assertion of 5.3/7 in order to end the proof. □

In order to derive the assertion of Theorem 1 from Proposition 2, we need some further preparations. Let G^0 be the identity component of G; i.e., G^0 is the open subscheme of G which is the union of all identity components of the fibres G_s over points $s \in S$ (cf. [EGA IV_3], 15.6.5). Then G^0 has geometrically connected fibres, and it acts on X. Therefore we can apply Proposition 2 if we can find an open subscheme $U \subset X$ such that U is affine over S and such that U meets all G^0-orbits of points in X. As is easily checked on geometric fibres, the latter condition is equivalent to the fact that U is S-dense in X:

Lemma 3. *Let X be a torsor under a smooth S-group scheme G which is of finite type over S, and consider an open subscheme $U \subset X$. Then U meets all G^0-orbits of points in X if and only if U is S-dense in X.*

In order to really construct an open subscheme $U \subset X$ as required in Proposition 2, we have to derive some information on the existence of affine open subschemes of X.

Lemma 4. *Let $S = \operatorname{Spec} R$ be an affine scheme which is noetherian, and consider an S-scheme X of finite type which is normal and separated. Let $(x_i)_{i \in I}$ be a finite family of points of codimension ≤ 1 in X. Then there exists an affine open subscheme $U \subset X$ containing all points x_i.*

Proof. We may assume that X is connected with field of rational functions L and, furthermore, that all x_i are of codimension 1. Then the local rings $\mathcal{O}_{X,x_i}$ are discrete valuation rings contained in L, and they are pairwise different since X is separated. So we can use the approximation theorem for inequivalent valuations (cf. Bourbaki [2], Chap. VI, § 7, n°1, Prop. 1) and see that

$$A := \bigcap_{i \in I} \mathcal{O}_{X,x_i}$$

is a semi-local ring with local components $\mathcal{O}_{X,x_i}$. We can write A as a direct limit of R-algebras A_j of finite type. Interpreting the elements of each A_j as rational functions on X, we obtain for each j a rational map

$$u_j : X \dashrightarrow \operatorname{Spec} A_j$$

which is an S-morphism in a neighborhood of each x_i. Since X and A_j are of finite type over R, our construction shows that u_j is an open immersion at each x_i if j is big enough; cf. [EGA IV_2], 8.10.5. Thus we have reduced the assertion of the lemma to the case where X is quasi-affine and where it is easily verified. □

Now we are able to *prove the assertion of Theorem* 1. As explained before, we have only to construct an S-dense open subscheme $U \subset X$ which is affine over S. In order to do this, fix a closed point $s \in S$. Working over an affine neighborhood S' of s in S and applying Lemma 4, there is an affine open subscheme $U' \subset X \times_S S'$ which contains all generic points of $X \times_S S'$ and all generic points of the fibre X_s. The complement of U' in $X \times_S S'$ equals the support of finitely many prime divisors $D_1, \ldots, D_r$ of $X \times_S S'$. Removing from S' all closed points s' such that the support of some D_j is contained in $X_{s'}$, we may assume that U' is S'-dense in $X \times_S S'$. Proceeding this way with all closed points in S, and using a quasi-compactness argument, we obtain affine open subschemes $U^1, \ldots, U^n$ of X such that U^i is S^i-dense over an affine open part S^i of S and such that the S^i cover S. For simplicity, assume that S is irreducible with generic point ξ. Let D_ξ be an effective divisor on X_ξ with support

$$X_\xi - \bigcap_{i=1}^{n} U_\xi^i ,$$

let D be its schematic closure in X, and set $U := X - \operatorname{supp} D$. Then U is S-dense in X since all U_ξ^i are dense in X_ξ and since $\operatorname{supp} D$ cannot contain components of closed fibres of X. Furthermore, U is affine over S. Namely, $U \times_S S^i$ is contained in U^i; it differs from the affine scheme U^i by the support of a divisor. Therefore the inclusion $U \times_S S^i \hookrightarrow U^i$ is affine, as can be checked locally, and it follows that $U \times_S S^i$ must be affine itself; cf. [EGA II], 1.3.4. So we have constructed U as required in Proposition 2, thereby finishing the proof of Theorem 1. □

6.5 The Descent of Torsors

In this section we want to apply the descent techniques of 6.1 to torsors under group schemes. So far we have dealt only with the descent of schemes without considering a group structure or a structure of torsor on them; however, we will see that the methods of 6.1 apply immediately to the new situation. Namely, consider a faithfully flat and quasi-compact morphism of schemes $p: S' \longrightarrow S$ as well as an S'-group scheme G'. As in 6.1, set $S'' := S' \times_S S'$, and let $p_1, p_2 : S'' \longrightarrow S'$ be the projections. Recall that, in terms of schemes, a descent datum on G' with respect to p consists of an S''-isomorphism

$$\varphi : p_1^* G' \longrightarrow p_2^* G'$$

satisfying the cocycle condition. Using the canonical isomorphisms

$$p_i^*(G' \times_{S'} G') \cong p_i^* G' \times_{S''} p_i^* G' , \qquad i = 1, 2,$$

one obtains from φ a descent datum

$$\varphi \times \varphi : p_1^*(G' \times_{S'} G') \longrightarrow p_2^*(G' \times_{S'} G')$$

on $G' \times_{S'} G'$. Talking about descent data on group schemes, it is required that the descent datum φ on G' is *compatible with the group multiplication* $m: G' \times_{S'} G' \longrightarrow G'$; i.e., that the diagram

$$\begin{array}{ccc} p_1^*(G' \times_{S'} G') & \xrightarrow{\varphi \times \varphi} & p_2^*(G' \times_{S'} G') \\ \Big\downarrow {\scriptstyle p_1^*(m)} & & \Big\downarrow {\scriptstyle p_2^*(m)} \\ p_1^* G' & \xrightarrow{\varphi} & p_2^* G' \end{array}$$

is commutative. Viewing $p_i^* G'$ as the S''-group scheme obtained from G' by means of the base change $p_i : S'' \longrightarrow S'$, the condition simply says that the descent datum

$$\varphi : p_1^* G' \longrightarrow p_2^* G'$$

is an isomorphism of S''-*group* schemes. Then, if the descent is effective, i.e., if G' descends to an S-scheme G, Theorem 6.1/6 implies readily that the group structure descends from G' to G and, hence, that G is an S-group scheme.

The procedure is similar for torsors. Consider an S'-scheme X' which is a torsor under an S'-group scheme G'. Let φ be a descent datum on G' which is compatible with the group multiplication on G'. Then a descent datum ψ on X' is said to be compatible with the structure of X' as a torsor under G' if the action

$$G' \times_{S'} X' \longrightarrow X'$$

is compatible with the descent data φ and ψ. If φ and ψ are effective, G' descends to an S-group scheme G and X' to an S-scheme X which is a torsor under G.

In the following, we want to exploit the existence of ample invertible sheaves in order to treat the descent of torsors over discrete valuation rings. Since it is necessary to study the problems on generic fibres first, our considerations will include the more or less trivial case where the base consists of a field.

Theorem 1. *Let $R \longrightarrow R'$ be a faithfully flat extension of discrete valuation rings (resp. of fields). Let G' be an R'-group scheme which is smooth, separated, and of finite type over R', and let X' be an R'-torsor under G'. Furthermore, assume that there are descent data with respect to $R \longrightarrow R'$ on G' and X' such that these data are compatible with the group structure on G' and with the action of G' on X'. Then G' descends to an R-group scheme G, and X' descends to an R-torsor X under G. Furthermore, by the properties of descent, G and X are smooth, separated, and of finite type over R.*

Before we give the proof, let us discuss some applications of the theorem. First we go back to Section 5, where we have studied the problem of associating group schemes to birational group laws; cf. 5.1/5. In 5.2/3, which applies to strict birational group laws, we had worked out a solution for the case where the base consists of a strictly henselian local ring R which is noetherian and normal. Now, using descent, we can show that 5.2/3 remains true if we work over a discrete valuation ring or over a field, without assuming that the latter is strictly henselian. Thereby we will fill the gap which was left in the proof of 5.1/5; we refer to Section 6.6 for a more rigorous approach to the problem.

Corollary 2. *Let R be a discrete valuation ring or a field, and let m be a strict birational group law on an R-scheme U which is separated, smooth, faithfully flat, and of finite type over R. Then there exists an open immersion $U \hookrightarrow G$ with R-dense image into a smooth and separated S-group scheme G such that the group law on G restricts to m on U. The group scheme G is unique up to canonical isomorphism.*

Proof. Write R' for a strict henselization of R. Then, applying the base change $R \longrightarrow R'$ to our situation, we obtain a strict birational group law m' on the R'-scheme $U' = U \otimes_R R'$. It has a unique solution by 5.2/3; i.e., there is an open immersion $U' \hookrightarrow G'$ into an R'-group scheme G', just as we have claimed for U and m.

In order to prove the corollary, it is enough to extend the canonical descent datum on U' to a descent datum on G' which is compatible with the group structure on G'. Then Theorem 1 can be applied. As usual, set $R'' = R' \otimes_R R'$ and write p_1,

p_2 for the projections from Spec R'' to Spec R'. The canonical descent datum on U' consists of the canonical isomorphism

$$p_1^* U' \xrightarrow{\sim} p_2^* U' .$$

Working over the base R'', we see immediately from the uniqueness assertion in 5.1/3 that this isomorphism extends to an isomorphism of R''-group schemes

$$p_1^* G' \xrightarrow{\sim} p_2^* G' .$$

A similar argument shows that the isomorphism satisfies the cocycle condition; so we have a descent datum on G' as required. □

As a second application, we want to discuss the existence of Néron models for torsors in the local case. Since, over strictly henselian valuation rings, torsors under smooth group schemes are trivial, the problem is a question of descent.

Corollary 3. *Let $R \subset R' \subset R^{sh}$ be discrete valuation rings, where R^{sh} is a strict henselization of R, and let K, K' and K^{sh} denote the fields of fractions of R, R' and R^{sh}. Furthermore, let X_K be a K-torsor under a smooth K-group scheme G_K of finite type, and assume that, after the base change $K \longrightarrow K'$, there are Néron models G' of $G_{K'}$ and X' of $X_{K'}$ over R'. Then G' (resp. X') descends to a Néron model G of G_K (resp. X of X_K) over R. Furthermore, if the torsor X_K is unramified, i.e., if $X_K(K^{sh}) \neq \varnothing$, the structure of X_K as a torsor under G_K extends uniquely to a structure of X as a torsor under G.*

Postponing the proof for a moment, let us first explain why X might not be a torsor under G. The universal mapping property of Néron models implies that the action of G_K on X_K extends uniquely to an action of G on X giving rise to an isomorphism

$$G \times_R X \longrightarrow X \times_R X , \qquad (g, x) \longmapsto (gx, x) .$$

However, in general, X will not be a torsor under G, since the structural morphism $X \longrightarrow \operatorname{Spec} R$ might not be surjective; i.e., it can happen that the special fibre of X is empty. Due to 2.3/5, the latter is the case if and only if $X(R^{sh})$ is empty or, by the Néron mapping property, if and only if $X_K(K^{sh})$ is empty. The torsor X_K is called *ramified* if $X_K(K^{sh}) = \varnothing$, and *unramified* if $X_K(K^{sh}) \neq \varnothing$. Combining the assertion of 1.3/1 with the preceding corollary, we can say:

Corollary 4. *Let R, K, K^{sh} be as before, and let X_K be a K-torsor under a smooth K-group scheme G_K of finite type. Then the following conditions are equivalent:*

(a) *X_K admits a Néron model over R.*

(b) *$X_K(K^{sh})$ is bounded in X_K.*

(c) *X_K is ramified or $G_K(K^{sh})$ is bounded in G_K.*

Proof of Corollary 3. As far as the Néron model of X_K is concerned, the assertion is trivial if X' has empty special fibre and thus coincides with $X_{K'}$. So assume that the latter is not the case and, hence, that X' is a torsor under G'. We claim it is enough to verify that the canonical descent data on $G_{K'}$ and $X_{K'}$ extend to descent data on G' and X'. Namely, the extensions are unique since both G' and X' are flat and separated over R'. By the same reason, we obtain the compatibility of the descent data with the group structure of G' and the structure of X' as a torsor under G'. Then Theorem 1 is applicable, and it follows that the pair (G', X') descends to a pair (G, X) over R. That G and X satisfy the universal mapping property of Néron models is a consequence of 6.1/6 (a) and, again, of the fact that G' and X' are flat and separated over R'. So, as claimed, it is enough to construct extensions of the canonical descent data on $G_{K'}$ and $X_{K'}$. Next, observe that G' and X' are of finite type over R'. Since $R' \subset R^{sh}$, we see by a limit argument that G' and X' (as well as the group structure of G' and the structure of X' as a torsor under G') are already defined over an étale extension of R. So it is enough to consider the case where R' is étale over R.

Now write $R'' := R' \otimes_R R'$ and let $p_i : \operatorname{Spec} R'' \longrightarrow \operatorname{Spec} R'$, $i = 1, 2$, be the projections. Then, since the formation of Néron models is compatible with étale base change (cf. 1.2/2), we see that $p_i^*(X')$ is a Néron model of $p_i^*(X'_{K'})$ over $\operatorname{Spec} R''$. Thus, by the Néron mapping property, the canonical descent datum

$$\varphi_{K'} : p_1^*(X'_{K'}) \longrightarrow p_2^*(X'_{K'})$$

extends to an isomorphism

$$\varphi : p_1^*(X') \longrightarrow p_2^*(X')$$

which, in fact, constitutes a descent datum on X'. In the same way, the canonical descent datum on $G'_{K'}$ is extended to a descent datum on G. □

Remark 5. The assertion of Corollary 3 remains valid if, instead of a pair $R \subset R'$ where R' is contained in a strict henselization of R, one considers a pair of discrete valuation rings $R \subset R'$ such that a uniformizing element of R gives rise to a uniformizing element of R' and such that the residue extension of R'/R is trivial. For example, R' can be the maximal-adic completion of R (actually, it is only necessary to require that R' is of ramification index 1 over R; see 7.2/1). Namely, reviewing the proof of Corollary 3, the first part, which reduces the assertion to the problem of extending descent data from $G_{K'}$ to G' (resp. $X'_{K'}$ to X'), remains valid. That the required extensions of descent data exist is a consequence of Lemma 6.2/D.3.

It remains to give the *proof of Theorem* 1. For the applications in Corollaries 2 to 4 which have just been discussed, the theorem is not needed in its full generality. Namely, in the first case (Corollary 2), we know that

(a) *there exists an R'-dense open subscheme $U' \subset X'$, stable under the descent datum of X', such that the descent is effective on U',*

whereas in the second case (Corollaries 3 and 4) we know that

(b) *K', the field of fractions of R', is algebraic over K, the field of fractions of R.*

Both properties can simplify the proof substantially. In order to demonstrate this, we will first establish the theorem under the additional assumption (a), and then under (b). Finally, we will indicate how to reduce the general case to the situation (a). Also we want to mention that we have only to work out the descent for the torsor X', because G' can be handled in the same way by viewing it as a trivial torsor under itself.

As a first step we show that, independently of conditions (a) or (b), the descent we have to perform is always effective on generic fibres. So consider the extension $K \longrightarrow K'$ of the fields of fractions of $R \longrightarrow R'$. Since $X'_{K'}$ is of finite type over K', we may use a limit argument and thereby replace K' by a K-subalgebra C of finite type. Then the quotient $C/\mathfrak{m}$ by some maximal ideal $\mathfrak{m} \subset C$ is a finite extension of K. If $[C/\mathfrak{m} : K] = 1$, the morphism $\operatorname{Spec} C \longrightarrow \operatorname{Spec} K$ has a section, and the descent with respect to it is effective by 6.1/5. If $[C/\mathfrak{m} : K] > 1$, the same argument applies to $\operatorname{Spec}(C \otimes_K C/\mathfrak{m}) \longrightarrow \operatorname{Spec} C/\mathfrak{m}$ so that we may replace K' by $C/\mathfrak{m}$. Thereby we are reduced to the case where $[K' : K] < \infty$, and we may assume that K' is quasi-Galois, or since the descent is trivial for radicial extensions, that K' is Galois over K. Then the descent on $X'_{K'}$ is a Galois descent (see Example 6.2/B) and, in order to show it is effective, it is enough to know that finitely many given points of $X'_{K'}$ are always contained in an affine open subscheme of $X'_{K'}$. That the latter condition is fulfilled can be seen either from the quasi-projectivity of $X'_{K'}$ (use 6.4/1) or, in a more elementary way, by using standard translation arguments. So the descent is effective, and $X'_{K'}$ descends to a K-scheme X_K. This settles the assertion of Theorem 1 for the case where R and R' are fields.

Next, let us assume that condition (a) is satisfied. Then U' descends to an R-scheme U, where U_K is open in X_K. Applying Lemma 6.4/4 to U, we can find an R-dense affine open subscheme of U, and hence, by pulling it back to U', an R'-dense affine open subscheme of U' which is stable under the descent datum on X'. In other words, we can assume that U' is affine. We claim one can find an effective divisor D' on X' with support $X' - U'$ such that D' is stable under the descent datum on X'. Denoting the descent datum on X' by $\varphi : p_1^* X' \longrightarrow p_2^* X'$, the latter means that $p_1^* D'$ corresponds to $p_2^* D'$ under the isomorphism φ. In order to obtain such a divisor D', choose an effective divisor D_K on X_K with support $X_K - U_K$ (cf. [EGA IV$_4$], 21.12.7), and define D' as the schematic closure of the pull-back of D_K to $X'_{K'}$. By the properties of the schematic closure, the descent datum on X' extends to a descent datum on the pair $(X', \mathscr{L}')$ where $\mathscr{L}' := \mathscr{O}_X(D')$. Considering the action of the identity component of G' on X', we conclude from 6.4/2 and 6.4/3 that $\mathscr{L}'$ is ample. Hence, 6.1/7 shows that the descent is effective on X'. This settles the assertion of Theorem 1 if condition (a) is given.

Now let us assume that condition (b) is satisfied. We want to reduce to condition (a). Applying Lemma 6.4/4, there is an R'-dense affine open subscheme $\Omega' \subset X'$. In particular, $F'_{K'} := X'_{K'} - \Omega'_{K'}$ is nowhere dense in $X'_{K'}$ and, since K' is algebraic over K, its image F_K in X_K is nowhere dense. Set $U_K := X_K - F_K$. Then $U'_{K'} := U_K \otimes_K K'$ is a dense open subscheme of $\Omega'_{K'}$. Subtracting from X' the schematic closure of $X'_{K'} - U'_{K'}$ we arrive at an R'-dense open subscheme U' of X' whose generic fibre is $U'_{K'}$. Furthermore, by construction, U' is stable under the descent datum on X', and it is quasi-affine since $U' \subset \Omega'$. The latter inclusion is verified by using the fact that $X' - \Omega'$ is the support of a divisor and that, since Ω' is R'-dense in X', the schematic closure of $X'_{K'} - \Omega'_{K'}$ in X' coincides with $X' - \Omega'$. In particular, the descent is effective on U' by 6.1/6, and we have thus reduced assumption (b) to assumption (a).

In order to prove Theorem 1 in its general version, some preparations are necessary. Consider a smooth and separated scheme X of finite type over a discrete valuation ring R. Let K be the field of fractions of R, and let k be the residue field of R. Writing $A := \Gamma(X, \mathcal{O}_X)$, we have a canonical morphism

$$u : X \longrightarrow \operatorname{Spec} A$$

whose formation is compatible with flat base change. For each $f \in A$, we denote by A_f the localization of A by f and by

$$u_f : X_f \longrightarrow \operatorname{Spec} A_f$$

the morphism obtained from u by the base change $\operatorname{Spec} A_f \longrightarrow \operatorname{Spec} A$.

In this situation, u is of finite type since X is of finite type over R. Furthermore, $\operatorname{Spec} A$ is flat over R and normal since the same is true for X. Since the formation of global sections on X commutes with flat base change, there are canonical isomorphisms

$$A_K := A \otimes_R K \cong \Gamma(X_K, \mathcal{O}_X)$$

and, for $f \in A$,

$$A_f \cong \Gamma(X_f, \mathcal{O}_X) .$$

Moreover, we have a canonical injection

$$A_k := A \otimes_R k \hookrightarrow \Gamma(X_k, \mathcal{O}_{X_k}) .$$

So a global section $h \in A$ vanishes on the special fibre X_k if and only if $h \in \pi A$, where π is a uniformizing element of R.

Lemma 6. *Let $u : X \longrightarrow \operatorname{Spec} A$ be as above and assume that the generic fibre X_K is affine. Then $u_K : X_K \longrightarrow \operatorname{Spec} A_K$ is an isomorphism and, if $X_k \neq \emptyset$, there exists an element $f \in A$ such that $X_f \cap X_k \neq \emptyset$ and such that $u_f : X_f \longrightarrow \operatorname{Spec} A_f$ is an isomorphism.*

Proof. The first assertion is clear. Next, assume $X_k \neq \emptyset$. Using the separatedness of X, we can apply Lemma 6.4/4 and find an R-dense affine open subscheme $U \subset X$. Since $u : X \longrightarrow \operatorname{Spec} A$ is an isomorphism on generic fibres, there is an $f \in A_K$, we may assume $f \in A$, such that $(X_f)_K \subset U_K$. Furthermore, X_k is not empty, so we may assume $f \in A - \pi A$. Then consider the schematic closure of $X_K - (X_f)_K$ in X; it is contained in $X - X_f$. Similarly, since U is R-dense and affine in X, its complement $X - U$ is of pure codimension 1 by [EGA IV$_4$], 21.12.7, and we see that it equals the schematic closure of $X_K - U_K$ in X. So we obtain from $(X_f)_K \subset U_K$ the inclusions

$$X_K - (X_f)_K \supset X_K - U_K$$

and, hence,

$$X - X_f \supset X - U .$$

Therefore $X_f \subset U$ and, thus, $X_f = U_f$ is affine. Interpreting A_f as the ring of global sections on X_f, the morphism $u_f : X_f \longrightarrow \operatorname{Spec} A_f$ is an isomorphism. Consequently, since f does not vanish identically on X_k, the assertion of the lemma follows. □

It should be realized that, in the situation of Lemma 6, we cannot expect to find a global section $f \in A$ such that $u_f : X_f \longrightarrow \operatorname{Spec} A_f$ is an isomorphism and X_f is *R-dense* in X. For example, consider an irreducible conic $C \subset \mathbb{P}^2_R$ whose special fibre consists of two projective lines P_1 and P_2. Assume that C admits an R-valued point meeting P_2, but not P_1. Removing this point from C, we obtain an R-scheme X whose generic fibre is affine and whose special fibre consists of two components, one of them P_1. Since each global section of $\mathcal{O}_X$ must be constant on P_1, we see that any subscheme $X_f \subset X$, as in Lemma 6, must be disjoint from P_1. So X_f cannot be R-dense in this case.

Returning to the proof of Theorem 1, it is enough to construct an open subscheme $U' \subset X'$ as required in condition (a). In order to do this, we will forget about the special situation given in Theorem 1 and assume only that X' is a smooth and separated R'-scheme of finite type with a descent datum on it, which is effective on the generic fibre $X'_{K'}$. In particular, we may apply the above considerations to X' as a scheme over R' (and to suitable open subschemes of it). First we reduce to the case where the generic fibre of X' is affine; then Lemma 6 is applicable. Let $K \longrightarrow K'$ be the extension of fields of fractions corresponding to $R \longrightarrow R'$. We know already that the generic fibre $X'_{K'}$ descends to a K-scheme X_K. Choose an affine dense open subscheme $U_K \subset X_K$ and consider its pull-back $U'_{K'}$ to $X'_{K'}$. Then $X'_{K'} - U'_{K'}$ is thin in $X'_{K'}$, and its schematic closure is R'-thin in X'. If we remove it from X', we obtain an R'-dense open subscheme whose generic fibre is affine and which is stable under the descent datum on X'. We can replace X' by this subscheme and thereby assume that the generic fibre of X' is affine.

Now set $A' = \Gamma(X', \mathcal{O}_{X'})$ and consider the canonical morphism $u' : X' \longrightarrow \operatorname{Spec} A'$. Then the descent datum on X' yields a descent datum on $\operatorname{Spec} A'$ such that the morphism u' is compatible with these descent data. Let U' be the open subscheme of X' consisting of all points of X' where u is quasi-finite. We claim that

(i) the generic fibre of U' coincides with X'_K, and the special fibre of U' is non-empty,

(ii) U' is stable under the descent datum of X', and

(iii) U' is quasi-affine; in particular, the descent datum is effective on U'.

Namely, property (i) is a consequence of Lemma 6, whereas property (ii) follows from the fact that, for a morphism of finite type, quasi-finiteness at a certain point can be tested after surjective base change such as provided by the projections $\operatorname{Spec} R' \times_R \operatorname{Spec} R' \rightrightarrows \operatorname{Spec} R'$. In order to justify the latter claim, observe that quasi-finiteness can be tested on fibres. So it is enough to consider a field as base and a field extension as base change. In this situation, a dimension argument gives the desired assertion. Finally, property (iii) follows from Zariski's Main Theorem (in the version 2.3/2'); it implies that $u' : X' \longrightarrow \operatorname{Spec} A'$ restricts to an open immersion on U'. So U' is quasi-affine, and the descent is effective on U' by 6.1/4.

If U' is R'-dense in X', we have obtained an open subscheme of X' as required in condition (a). If U' is not R'-dense in X', remove from X' all components of the special fibre which meet U'. The resulting open subscheme of X', call it X'_1, is again stable under the descent datum. So, concluding as before, X'_1 contains an open subscheme U'_1 satisfying conditions (i) to (iii). Continuing this way, we can work up the finitely many components of X_k and thereby obtain finitely many open subschemes $U', U'_1, \ldots, U'_n \subset X'$ satisfying conditions (i) to (iii). Then the union of these subschemes is R'-dense in X' and, hence, gives rise to an open subscheme of X' as required in condition (a), thereby finishing the proof of Theorem 1. □

6.6 Applications to Birational Group Laws

In this section, we want to sharpen M. Artin's result on the construction of group laws from birational group laws, which is explained in [SGA 3_{II}], Exp. XVIII. Let S be a scheme, and consider an S-birational group law m on a smooth S-scheme X. It is shown in [SGA 3_{II}], Exp. XVIII, that, if m is strict in the sense of 5.2/1, there exists a solution $\bar{X}$ in the category of algebraic spaces such that $\bar{X}$ contains X as an S-dense open subspace; for the notion of algebraic spaces see Section 8.3. We will admit this result. However, if the base S is normal, it could also have been obtained by the construction technique of Section 5.3. The latter method yields even more, namely that $\bar{X}$ is a scheme for the étale topology of S. Using the descent techniques of Section 6.5, we want to show that $\bar{X}$ is already a scheme. So, we will mainly be concerned with the representability of a smooth group object in the category of algebraic spaces.

Theorem 1. *Let S be a scheme, and let m be an S-birational group law on a smooth and separated S-scheme X which is faithfully flat and of finite presentation over S. Then there exists a smooth and separated S-group scheme $\bar{X}$ of finite presentation* *with a group law $\bar{m}$, together with an S-dense open subscheme $X' \subset X$ and an open immersion $X' \hookrightarrow \bar{X}$ having S-dense image such that $\bar{m}$ restricts to m on X'.*

The group scheme $\bar{X}$ is unique up to canonical isomorphism. If the S-birational group law m is strict, the assertion is true with X' replaced by X.

Proof. Due to the uniqueness assertion 5.1/3, we may assume that S is affine and, using limit arguments, that S is noetherian. If the S-birational law is strict, it follows from the result of M. Artin that there exists a solution $\bar{X}$ of the strict law in the category of algebraic spaces containing X as an S-dense open subspace of $\bar{X}$. As we will see by the theorem below, the solution is represented by a scheme. Thereby, Theorem 1 will be proved for the case where the S-birational group law is strict. Now we want to treat the general case accepting the assertion of Theorem 1 for strict S-birational laws.

Let U be the largest open subscheme of S such that the S-birational group law has a solution over U; here and in the following, solutions are meant in the category of schemes. If $U \neq S$ choose the generic point s of an irreducible component of $S - U$. Since we consider only S-schemes of finite presentation, it suffices to verify that there exists a solution after the base change $\operatorname{Spec}(\mathcal{O}_{S,s}) \longrightarrow S$. So we may assume *that* S is a local scheme, and that s is the closed point of S; then $U = S - \{s\}$.

Assume first that, for each component X_s^i of X_s, there exists a section σ_i of X over S crossing the given component. Let $X(\sigma_i)$ be the union of all components of the fibres of X meeting the section σ_i; due to [EGA IV_3], 15.6.5, $X(\sigma_i)$ is an open subscheme of X. Denote by X_0 the union of the $X(\sigma_i)$; note that X_0 might not be S-dense in X. Then m induces an S-birational group law m_0 on X_0. Moreover, due to the construction, the components of the fibres of X are geometrically irreducible. Now one can proceed as in the proof of 5.2/2. The set Z (in the proof of 5.2/2) will provide an S-dense open subscheme X_0' of X_0 such that m_0 induces a strict law m_0' on X_0'. Namely, set

$$\Omega_1 = \bigcup_i \left(\bigcap_j p_1(Z \cap (X(\sigma_i) \times_S X(\sigma_j))) \right)$$

where $p_1 : X \times_S X \longrightarrow X$ is the first projection. Then Ω_1 is S-dense open in X_0, and $Z \cap (\Omega_1 \times_S X_0)$ is Ω_1-dense in $\Omega_1 \times_S X_0$. Defining Ω_2 in a similar way by using the second projection, the intersection $\Omega_1 \cap \Omega_2$ defines an S-dense open subscheme X_0' of X_0. As in 5.2/2, one shows that the restriction m_0' of m to X_0' is strict. As we have said above, there is a solution $\bar{X}_0'$ of the strict law m_0' which contains X_0' as an S-dense open subscheme. Since $\bar{X}_0' \times_S U$ is an open subscheme of the solution $\bar{X}_U$ of the restriction of m to U, one can glue $\bar{X}_0'$ and $\bar{X}_U$ along $\bar{X}_0' \times_S U$ in order to get a solution of m.

In the general case, one performs first an étale surjective extension $S^* \longrightarrow S$ of the base in order to get enough sections of X. So one obtains a solution $\bar{X}^*$ of the S^*-birational group law $m \times_S S^*$. Now consider the S^*-birational map

$$\iota : X \times_S S^* \dashrightarrow \bar{X}^* .$$

The canonical descent datum extends to a descent datum on $\bar{X}^*$ by the uniqueness of solutions; cf. 5.1/3. Furthermore, there exists a largest open subscheme X^* of

$X \times_S S^*$, where the map ι is defined and where ι is an open immersion; use the separatedness of $X \times_S S^*$ and of $\bar{X}^*$ as well as the birationality of ι. Since the domain of definition is compatible with flat base change (cf. 2.5/6), the formation of the largest open subscheme where ι is defined and where ι is an open immersion is compatible with flat base change. So X^* is stable under the descent datum and, hence, there exists an open subscheme X' of X which is S-dense in X such that $X' \times_S S^* = X^*$. Then it is easy to see that the S-birational law m on X restricts to a strict law on X'. □

In order to complete the proof of the preceding theorem, it remains to show the following result on the representability of algebraic spaces with group action.

Theorem 2. *Let S be a locally noetherian scheme and let G be a group object in the category of algebraic spaces over S. Assume that G is smooth over S and that G has connected fibres over S. Let X be a smooth algebraic space over S and let*

$$\sigma : G \times_S X \longrightarrow X$$

be a group action on X. Let Y be an open subspace of X. Then the image GY of $G \times_S Y$ in X is an open subspace of X. If GY equals X, the following assertions hold:

(a) *If Y is separated (resp. of finite type) over S, the same is true for X.*

(b) *If Y is a scheme, then X is a scheme.*

(c) *If S is affine and if Y is quasi-affine, any finite set of points of X is contained in an affine open subset of X.*

(d) *If S is normal and if Y is affine over S, any effective Weil divisor of X with support $X - Y$ is a Cartier divisor, and is S-ample. In particular, X is quasi-projective over S.*

Corollary 3. *Let S be a Dedekind scheme, and let G be a group object in the category of algebraic spaces over S. Assume that G is separated, smooth, and of finite type over S. Then G is a scheme.*

Proof of Corollary 3. Let Y be the open subspace of G consisting of all points which admit a scheme-like neighborhood. Due to Raynaud [6], Lemme 3.3.2, Y contains all the generic points of the fibres of G over S. Hence, Y is S-dense in G. So Theorem 2 yields that G is a scheme. □

Proof of Theorem 2. The group action σ is the composition of the maps

$$G \times_S X \xrightarrow{(p_1, \sigma)} G \times_S X \xrightarrow{p_2} X$$

where p_i is the projection onto the i-th factor, $i = 1, 2$. The first map is an isomorphism, and the second one is smooth, since G is smooth over S. Hence, the map σ is open, and the image GY is an open subspace of X.

(a) In order to prove the separatedness of X, we can use the valuative criterion. So, we may assume that S consists of a discrete valuation ring R with field of fractions K and residue field k. Then we have to show that any two R-valued points $x_1, x_2 \in X(R)$ which coincide on the generic fibre are equal. Let $\bar{x}_1, \bar{x}_2$ be the induced closed points. Since the sets

$$\bar{U}_i = \{\bar{g} \in G \times_S k, \bar{g}^{-1}\bar{x}_i \in Y \times_S k\}, \qquad i = 1, 2,$$

are open and non-empty, they are dense in $G \times_S k$. Due to the smoothness of G over S, there exist an étale surjective base extension $R \longrightarrow R'$ and a section $g \in G(R')$ inducing a point of $\bar{U}_1 \cap \bar{U}_2$. Thus $\bar{x}_i \in gY$ and, hence, $x_i \in gY$ for $i = 1, 2$. Since Y is separated over S, we see that $x_1 = x_2$.

In order to show that X is of finite type over S if Y is, it suffices to verify that X is quasi-compact if S is affine. Since the map

$$\sigma : G \times_S Y \longrightarrow X$$

is surjective, the assertion follows from the fact that G is quasi-compact, as can easily be deduced from Lemma 5.1/4.

(d) We may assume that S is affine. Due to assertion (a), X is of finite presentation and separated over S. Let D be an effective Weil divisor with support $X - Y$. Due to the theorem of Ramanujam-Samuel [EGA IV_4], 21.14.3, D is a relative Cartier divisor. Namely, as can be seen by an étale localization on X, this theorem carries over to the case of algebraic spaces. Next we want to show that $\mathscr{L} = \mathcal{O}_X(D)$ is S-ample. To do this, we need the fact that $\mathscr{L}^{\otimes n}$ satisfies the theorem of the square for large integers n if the generic fibres of X over S are geometrically irreducible, cf. Section 6.3. Namely, after étale localization of the base, X can be covered by open subspaces of type X_l where l varies over the global sections of $\mathscr{L}^{\otimes n}$. The X_l are affine as intersections of translates of Y; cf. the proof of 6.4/2 or Raynaud [4], Thm. V.3.10, p. 88. In order to verify that $\mathscr{L}^{\otimes n}$ satisfies the theorem of the square for large integers n, one proceeds as follows:

Similarly as in the proof of 6.3/2, one reduces to the case where S consists of a field. Then G is a scheme; cf. Section 8.3. We claim that X is a scheme, too. Let U be the set consisting of all points of X admitting a scheme-like neighborhood. Using finite Galois descent, one easily shows that U is invariant under G, since any finite set of points of U is contained in an affine open subscheme of U. In our case, due to the assumption $X = GY$, one has $U = X$. So, X is a scheme, and the assertion follows from Raynaud [4], Thm. IV 3.3 (d), p. 72.

Finally, since $Y \longrightarrow S$ is affine, the reduced subscheme with support $X - Y$ is a Weil divisor by [EGA IV_4], 21.12.7, and thus an S-ample Cartier divisor. Therefore $X \longrightarrow S$ is quasi-projective.

(c) First, let us show assertion (c) under the additional assumption that S is normal. Let $x_1, \ldots, x_n$ be finitely many points of X, and let $s_1, \ldots, s_n$ be their images in S. Since Y is quasi-affine, there exists an affine open subscheme Y^* of Y which gives rise to a dense open subscheme of the fibres $Y_{s_1}, \ldots, Y_{s_n}$. Then the points $x_1, \ldots, x_n$ are contained in the image X^* of $G \times_S Y^*$ under σ. We may replace X by X^*, and so we may assume that Y is affine. In this case, the assertion follows from assertion (d). Namely, X admits a relatively ample line bundle, since $X - Y$ with its reduced structure gives rise to a Weil divisor; cf. [EGA IV_4], 21.12.7. So, X is quasi-projective over S, and hence X satisfies assertion (c).

Now let us consider the general case. Using limit arguments, we may assume that S is of finite type over the ring of integers $\mathbb{Z}$. Let $\tilde{S}$ be the normalization of S, and set $\tilde{X} = X \times_S \tilde{S}$ and $\tilde{G} = G \times_S \tilde{S}$. Then $\tilde{X}$ is a scheme by what we have just proved, and any finite set of points of $\tilde{X}$ is contained in an affine open subscheme of $\tilde{X}$. Furthermore, $X' = X \times_S S'$ is a scheme after étale surjective base extension $S' \longrightarrow S$, since there are finitely many sections of G such that X can be covered by the translates of Y under these sections, as follows from 5.3/7; see also 6.4/2. In order to show the effectivity of the canonical descent datum on X' we make use of the following result which is contained in Raynaud [3], Cor. 3.8 and Thm. 4.2:

Let S be a locally noetherian scheme, let $S' \longrightarrow S$ be a faithfully flat quasi-compact morphism of schemes, and let $\tilde{S} \longrightarrow S$ be a finite surjective morphism of schemes. Let X be a sheaf for the fppf-topology of S (cf. Section 8.1). Assume that $X' = X \times_S S'$ is represented by an S'-scheme which is locally of finite presentation, and that $\tilde{X} = X \times_S \tilde{S}$ is represented by an $\tilde{S}$-scheme. Then

(i) *X is represented by an S-scheme of finite presentation if and only if, for each point $\tilde{x}$ of $\tilde{X}$, there exists an affine open subscheme of $\tilde{X}$ which contains all points of $\tilde{X}$ giving rise to the same point of X as $\tilde{x}$.*

(ii) *If $\tilde{X}$ satisfies the property that any finite set of points of $\tilde{X}$ is contained in an open affine subscheme, so does X.*

Thus we see that X is a scheme, and any finite set of points of X is contained in an affine open subscheme of X, since $\tilde{X}$ has this property.

Assertion (b) follows from (c). □

6.7 An Example of Non-Effective Descent

Let R be a discrete valuation ring with field of fractions K and residue field k. In the present section we will consider relative curves over R; i.e., flat R-schemes X whose fibres are of pure dimension 1. We assume that, in addition, X is normal and proper over R and that the generic fibre X_K is connected. Then X_K is regular (in fact, smooth over K if $\operatorname{char} K = 0$), and the set of singular points x of X (i.e., of those points where the local ring $\mathcal{O}_{X,x}$ is not regular) is a finite subset of the special fibre X_k; see [EGA IV_2], 5.8.6, and [EGA IV_2], 6.12.6. The example we want to present is based on the fact that, after replacing the base R by a henselization R^h, irreducible components of X_k can be contracted in X whereas, over a non-henselian ring R, such a procedure is not always possible.

To construct an R-curve with a non-effective descent datum on it, set $A = \mathbb{C}[\tau, \tau^{-1}]$, where τ is an indeterminate, and start out from a smooth and proper elliptic curve E over $S = \operatorname{Spec} A$ which has non-constant j-invariant. Alternatively, we can consider the ring $A = \mathbb{Q}[\tau, \tau^{-1}]$ and the elliptic curve with constant j-invariant $E \subset \mathbb{P}_S^2$ which is given by the equation

$$y^2 z = x^3 + \tau x z^2 .$$

Replacing A by the local ring $R = \mathcal{O}_{S,t}$ at a closed point $t \in S$ if $A = \mathbb{C}[\tau, \tau^{-1}]$ (resp. at a suitable closed point $t \in S$ corresponding to a maximal ideal $(\tau - t) \subset A$ with $t \in \mathbb{Q}^*$ if $A = \mathbb{Q}[\tau, \tau^{-1}]$), we will show in Proposition 5 that there exists a rational point $a_k \in E_k$ such that none of the multiples ra_k with $r > 0$ admits a lifting to an R-valued point of E. Blowing up a_k in E yields a proper curve X over R which is regular. Its special fibre X_k consists of two components, the strict transform $\tilde{E}_k$ of E_k and the inverse image of a_k which is a projective line P_k; both intersect transversally at a single point.

In this situation we will see in Lemma 6 that one cannot contract the component $\tilde{E}_k$ in X; i.e., there does not exist an R-morphism $u : X \longrightarrow Y$ of proper normal curves over R which is an isomorphism over $Y - \{y\}$ and which satisfies $\tilde{E}_k = u^{-1}(y)$. However, if we pass from R to a henselization R^h and consider the curve $X' = X \otimes_R R^h$ over R^h, the special fibre of X remains unchanged, and we will be able to conclude from Proposition 4 below that $\tilde{E}_k$ can be contracted in X'.

Let $u' : X' \longrightarrow Y'$ be such a contraction. There are canonical descent data on X' and on Y' with respect to $R \longrightarrow R^h$; namely on X', since it is obtained from X by means of the base change $R \longrightarrow R^h$, and on Y' since u' is an isomorphism on generic fibres and since each descent datum on the generic fibre of Y' extends uniquely to a descent datum on Y' by 6.2/D3. Furthermore, u' is compatible with these data. So if the descent datum on Y' were effective, $u' : X' \longrightarrow Y'$ would descend to an R-morphism $u : X \longrightarrow Y$, where Y is a proper normal curve by [EGA IV_2], 2.7.1 and 6.5.4. Since u' coincides with u on special fibres, the latter morphism would be a contraction of $\tilde{E}_k$ in X. However such a contraction cannot exist by Lemma 6 and, consequently, the descent datum on Y' cannot be effective.

Now, after we have given the description of the curve Y' and the non-effective descent datum on it, let us fill in the results mentioned above which are needed to make the example work. We begin with the explanation of contractions; see also M. Artin [1], [2]. So consider an arbitrary discrete valuation ring R and an R-curve X where, as we have said at the beginning of this section, X is assumed to be proper and normal and to have a connected generic fibre. Let $(X_i)_{i \in I}$ be the family of irreducible components of the special fibre X_k, providing them with the canonical reduced structure. For a strict subset $J \subset I$, a contraction of the components X_j, $j \in J$, in X consists of an R-morphism $u : X \longrightarrow Y$ of proper normal curves over R such that

(a) for each $j \in J$, the image $u(X_j)$ consists of a single point $y_j \in Y$, and
(b) u defines an isomorphism $X - \bigcup_{j \in J} X_j \xrightarrow{\sim} Y - \bigcup_{j \in J} \{y_j\}$.

Then u is automatically proper since X is proper over R and since Y is separated over R. Furthermore, using the Stein factorization [EGA III_1], 4.3.1, it is easily seen that u depends uniquely on the subset $J \subset I$ and that the fibres of u are connected. In order to give a criterion for the existence of contractions, we use the notion of effective relative Cartier divisors; cf. Section 8.2, in particular 8.2/6.

Theorem 1. *Let X be a proper normal R-curve with connected generic fibre X_K, let $(X_i)_{i \in I}$ be the family of irreducible components of the special fibre X_k, and consider a non-trivial effective relative Cartier divisor D on X. Let J be the set of all indices $j \in I$ such that* $\operatorname{supp}(D) \cap X_j = \varnothing$. *Then the canonical morphism*

$$u : X \longrightarrow Y := \operatorname{Proj}\left(\bigoplus_{m=0}^{\infty} \Gamma(X, \mathcal{O}_X(mD)) \right)$$

is a contraction of the components $X_j, j \in J$, and Y is a proper normal R-curve which is projective.

Before we give a proof, let us look at properties of Y which follow from its definition as a projective spectrum of a graded ring.

Lemma 2. *Let X be a proper scheme over a noetherian ring R and let $\mathscr{L}$ be an invertible sheaf on X such that, for some $n > 0$, the sheaf $\mathscr{L}^{\otimes n}$ is generated by its global sections. Then, for*

$$A = \bigoplus_{m=0}^{\infty} \Gamma(X, \mathscr{L}^{\otimes m}),$$

the scheme $Y = \mathrm{Proj}(A)$ is projective over R and the canonical morphism $u: X \longrightarrow Y$ has connected fibres. If, in addition, X is normal, Y is normal also.

Proof. Applying [EGA III_1], 3.3.1, we see that the ring A is of finite type over R. Thus $Y = \mathrm{Proj}(A)$ is projective over R; cf. [EGA II], 4.4.1.

For any section $l \in \Gamma(X, \mathscr{L}^{\otimes n})$, the morphism u gives rise to an isomorphism

$$A_{(l)} \xrightarrow{\sim} \Gamma(X_l, \mathscr{O}_X).$$

So $u_*(\mathscr{O}_X) = \mathscr{O}_Y$ and, since u is proper, it follows from [EGA III_1], 4.3.2, that the fibres of u are connected. Finally, if X is normal, the ring $\Gamma(X_l, \mathscr{O}_X)$ is seen to be integrally closed in its total ring of fractions. This implies that Y is normal. □

Now we come to the *proof of Theorem* 1. Set $\mathscr{L} := \mathscr{O}_X(D)$. We claim that $\mathscr{L}^{\otimes n}$ is generated by its global sections if n is large enough. Then Y will be projective and normal by the preceding lemma. In order to justify the claim, it is enough to find global sections generating $\mathscr{L}^{\otimes n}$ at the points of $\mathrm{supp}(D)$; the constant 1, as a global section of $\mathscr{O}_X$, will generate $\mathscr{L}^{\otimes n}$ elsewhere. So consider the exact sequence of $\mathscr{O}_X$-modules

$$0 \longrightarrow \mathscr{O}_X(-D) \longrightarrow \mathscr{O}_X \longrightarrow \mathscr{O}_D \longrightarrow 0.$$

Taking the tensor product with $\mathscr{L}^{\otimes n}$ yields the exact sequence

$$0 \longrightarrow \mathscr{L}^{\otimes n-1} \longrightarrow \mathscr{L}^{\otimes n} \longrightarrow \mathscr{O}_D \otimes \mathscr{L}^{\otimes n} \longrightarrow 0,$$

and we can use the following part of the associated cohomology sequence:

$$(*) \quad H^0(X, \mathscr{L}^{\otimes n}) \longrightarrow H^0(X, \mathscr{O}_D \otimes \mathscr{L}^{\otimes n}) \longrightarrow H^1(X, \mathscr{L}^{\otimes n-1}) \longrightarrow H^1(X, \mathscr{L}^{\otimes n}) \longrightarrow 0.$$

Note that $H^1(X, \mathscr{O}_D \otimes \mathscr{L}^{\otimes n}) = 0$ since D defines a closed subscheme of X which is affine; the latter is due to the fact that D is quasi-finite, proper and, hence, finite over R.

Next, consider the restriction D_K of D to the generic fibre X_K. Then D_K has a positive degree on X_K since D is effective and non-trivial, and we see that D_K is ample since X_K is irreducible. Therefore $H^1(X_K, \mathscr{L}^{\otimes n}) = 0$ for n big enough, and it follows that $H^1(X, \mathscr{L}^{\otimes n})$ is an R-torsion module of finite length since it is of finite type. The exact sequence $(*)$ implies that the length is decreasing for ascending n. Hence the length will become stationary and, for n big enough, the map

$$H^1(X, \mathscr{L}^{\otimes n-1}) \longrightarrow H^1(X, \mathscr{L}^{\otimes n})$$

is an isomorphism. But then

$$H^0(X, \mathscr{L}^{\otimes n}) \longrightarrow H^0(X, \mathscr{O}_D \otimes \mathscr{L}^{\otimes n})$$

is surjective. Thereby we see that $\mathscr{L}^{\otimes n}$ is generated by its global sections at the points of supp(D) and, hence, at all points of X, as claimed.

It remains to show that $u: X \longrightarrow Y$ is a contraction of the components $X_j, j \in J$. Fix such a component X_j. Then, since X_j is proper, each global section of $\mathscr{O}_X(nD)$ induces a constant function on X_j; i.e., an element of the finite extension $\Gamma(X_j, \mathscr{O}_{X_j})$ of k. Therefore the image $u(X_j)$ consists of a single point $y_j \in Y$. Next look at a component X_i with $i \in I - J$. Fix a point $x \in X_i \cap \operatorname{supp}(D)$ and, for some $n \in \mathbb{N}$ big enough, choose a global section l of $\mathscr{O}_X(nD)$ such that l generates $\mathscr{O}_X(nD)$ over a neighborhood U of x. Then $1/l$ may be viewed as a section in $\mathscr{O}_Y$ over Y_l or (by means of the pull-back under u) as a section in $\mathscr{O}_X$ over X_l. By its construction, $1/l$ vanishes on $U \cap \operatorname{supp}(D)$ and is non-zero on $U - \operatorname{supp}(D)$. Therefore the image $u(X_i)$ cannot consist of a single point so that u must be quasi-finite on X_i. Finally, using the facts that the fibres of $u: X \longrightarrow Y$ are connected and that Y is normal (see Lemma 2), one concludes with the help of Zariski's Main Theorem 2.3/2′ that u is a contraction of the components $X_j, j \in J$.

Corollary 3. *Let X be a proper normal R-curve with connected irreducible generic fibre X_K and let X_i, $i \in I$, be the irreducible components of the special fibre X_k. Let J be a strict subset of I. Then the following conditions are equivalent:*

(a) *There exists a contraction $X \longrightarrow Y$ of the components X_j, $j \in J$, where Y is projective over R.*

(b) *There exists a contraction $X \longrightarrow Y$ of the components X_j, $j \in J$, and there is a non-empty R-dense affine open subset $V \subset Y$ such that the images of the X_j as well as all singular points of Y are contained in V.*

(c) *There exists an effective relative Cartier divisor D on X with the property that* $\operatorname{supp}(D) \cap X_j = \varnothing$ *for all $j \in J$ and* $\operatorname{supp}(D) \cap X_i \neq \varnothing$ *for all $i \in I - J$.*

Proof. The implication (a) $\Longrightarrow$ (b) is clear since the set of singular points of Y is a finite subset of the special fibre Y_k and since Y is projective over R. To show the implication (b) $\Longrightarrow$ (c), choose an R-dense affine open subscheme $V \subset Y$ which contains the images of the components X_j, $j \in J$, as well as all singular points of Y. Then $Y - V$ gives rise to a relative Cartier divisor on Y whose inverse under $X \longrightarrow Y$ is a divisor on X as required in condition (c). Finally, the implication (c) $\Longrightarrow$ (a) follows from Theorem 1. □

Proposition 4. *In the situation of Corollary 3, assume that the valuation ring R is henselian. Then there exists an effective relative Cartier divisor D on X as required in condition (c) of Corollary 3. In particular, any strict subset of the set of irreducible components of X_k can be contracted in X.*

Proof. It is enough to construct an effective relative Cartier divisor D on X whose support meets only a single given component X_j of X_k. In order to do this, choose a closed point

$$x \in X_j - \bigcup_{i \neq j} X_i$$

which is regular on X; such a point exists since there are at most finitely many points where X is not regular. Using the fact that $\operatorname{prof} \mathcal{O}_{X,x} = 2$, one can find an affine open neighborhood $U = \operatorname{Spec} A$ of x such that there is a non-zero-divisor $\bar{f} \in A \otimes_R k$ which vanishes at x. Lifting $\bar{f}$ to $f \in A$, this element defines a closed subscheme $\Delta \subset U$ which we may interpret as an effective relative Cartier divisor on U. However, Δ might not be a closed subscheme of X; it can happen that its schematic closure $\bar{\Delta}$ cannot be interpreted as a relative Cartier divisor on X or that $\bar{\Delta}$ meets components C_i with $i \neq j$. So we cannot, in general expect, that Δ extends to a relative Cartier divisor on X satisfying the required properties.

But we know that $\Delta \longrightarrow \operatorname{Spec} R$ is quasi-finite. So, R being *henselian*, we can use 2.3/4 in order to obtain an open neighborhood $V \subset U$ of x such that $\Delta \cap V \longrightarrow \operatorname{Spec} R$ is finite. Then the immersion $\Delta \cap V \hookrightarrow X$ is finite, and its image is closed in X so that we may regard $\Delta \cap V$ as a relative Cartier divisor on X. The latter is of the required type. □

For the remainder of this section, we want to look at smooth and proper elliptic curves $E \subset \mathbb{P}_S^2$ (having a section) over a base scheme $S = \operatorname{Spec} A$ where $A = \mathbb{C}[\tau, \tau^{-1}]$ or $A = \mathbb{Q}[\tau, \tau^{-1}]$ and where τ is an indeterminate. So S is a Dedekind scheme; let K be its field of fractions. For $t \in \mathbb{C}^*$ (resp. $t \in \mathbb{Q}^*$), we will write t also for the closed point in S which corresponds to the ideal $(\tau - t) \subset A$. As usual, for closed points $t \in S$, the fibre of E over t is denoted by E_t.

Proposition 5. *Consider the following property of E at closed points $t \in S$:*

(P) *There exists a rational point $a_t \in E_t$ such that none of its multiples na_t, $n > 0$, (in the sense of the group law on E) lifts to an $\mathcal{O}_{S,t}$-valued point of E or, equivalently, of $E \otimes_A \mathcal{O}_{S,t}$.*

Then, if $A = \mathbb{C}[\tau, \tau^{-1}]$, and if E is a smooth and proper elliptic curve over $S = \operatorname{Spec} A$ with non-constant j-invariant, the property (P) *is true for all $t \in \mathbb{C}^*$. Furthermore, if $A = \mathbb{Q}[\tau, \tau^{-1}]$ and if $E \subset \mathbb{P}_S^2$ is given by the equation*

$$y^2 z = x^3 + \tau x z^2 ,$$

(P) *is true for some $t \in \mathbb{Q}^*$; for example, it holds for all primes* $p \equiv 5 (\operatorname{mod} 8)$, *where* $p < 1000$.

Proof. Let us start with the case $A = \mathbb{C}[\tau, \tau^{-1}]$. Fix a closed point $t \in S$ and set $R = \mathcal{O}_{S,t}$. Then, using the relative version of the Mordell-Weil theorem for function fields as contained in Lang and Néron [1], we see that the group $E(K)$ is finitely generated. By the valuative criteria of separatedness and of properness, the latter group is isomorphic to $E(R)$. Now let Γ be the image of $E(R)$ in $E_t(\mathbb{C})$ and let $\bar{\Gamma}$ be the subgroup of $E_t(\mathbb{C})$ consisting of all points b_t such that a multiple nb_t is contained in Γ. Then, since $E(R)$ is countable, the group $\bar{\Gamma}$ is countable. But $E_t(\mathbb{C})$ is not countable. So $E_t(\mathbb{C}) - \bar{\Gamma}$ contains a point a_t as required.

Next let us consider the case where $A = \mathbb{Q}[\tau, \tau^{-1}]$. We claim that $E(K)$ is finite. In order to justify this, we look for $t \in \mathbb{Q}^*$ at the specialization map

$$E(K) \xrightarrow{\sim} E(\mathcal{O}_{S,t}) \longrightarrow E_t(\mathbb{Q})$$

and use the following facts which we cite without proof:

(a) $E_t(\mathbb{Q})$ is finite for infinitely many $t \in \mathbb{Q}^*$; for example for all primes p with $p \equiv 7$ or $p \equiv 11 (\text{mod } 16)$; cf. Silverman [1], Chap. X, 6.2 and 6.2.1.

(b) The specialization map $E(K) \longrightarrow E_t(\mathbb{Q})$ is injective for allmost all $t \in \mathbb{Q}^*$; cf. Silverman [1], Appendix C, 20.3.

(c) There exist elements $t \in \mathbb{Q}^*$ such that $E_t(\mathbb{Q})$ is of rank ≥ 1, for example for all primes $p \equiv 5 (\text{mod } 8)$ less than 1000; cf. Silverman [1], Chap. X, 6.3.

It follows from (a) and from (b) that $E(K) \simeq E(\mathcal{O}_{S,t})$ is finite for all $t \in \mathbb{Q}^*$. Choosing t as in (c), one can find a rational point $a_t \in E_t(\mathbb{Q})$ which has infinite order. But then none of its multiples can admit a lifting to a point of $E(\mathcal{O}_{S,t})$. □

Now let E be a smooth and proper elliptic curve over a discrete valuation ring R such that the special fibre E_k contains a rational point a_k whose multiples na_k, $n > 0$, (in the sense of the group law on E) do not admit liftings to R-valued points of E. As we have just seen, examples of such curves do exist. By blowing up a_k in E, one obtains a proper curve X over R which is regular. Its special fibre X_k consists of the strict transform $\tilde{E}_k$ of E_k and of the inverse image of a_k which is a projective line P_k; both intersect transversally at a single point.

Lemma 6. *The strict transform $\tilde{E}_k$ of E_k under the blowing-up $X \longrightarrow E$ cannot be contracted in X. More precisely, there is no R-morphism $u: X \longrightarrow Y$ onto a proper normal R-curve Y which maps $\tilde{E}_k$ onto a point $y \in Y$ and which is an isomorphism over $Y - \{y\}$.*

Proof. Assume that such a contraction $u: X \longrightarrow Y$ exists. Then Y is regular at all its points except possibly for y, and the complement of any affine open neighborhood of y yields an effective relative Cartier divisor D on X, whose support meets P_k and is disjoint from $\tilde{E}_k$; cf. Corollary 3. Let D_k be the generic fibre of D and D' its schematic closure in E. Then D' is an effective relative Cartier divisor on E; let $d > 0$ be its degree. The support of D' is the projection of D on E; so the closed fibre D'_k is da_k. If e is the unit section of E, the invertible sheaf $\mathcal{L} = \mathcal{O}_E(D' - de)$ has degree 0 and, thus, corresponds to an element of $\mathrm{Pic}^0_{E/R}(R)$; cf. Section 9.2. Now, using the canonical isomorphism

$$E \longrightarrow \mathrm{Pic}^0_{E/R}, \qquad x \longmapsto \mathcal{O}_E(x - e),$$

it follows that $\mathcal{L}$ corresponds to a point $b \in E(R)$. Restricting ourselves to special fibres, we see that $b_k = da_k$. However, this contradicts the choice of $a_k \in E_k$. □

Chapter 7. Properties of Néron Models

Although the notion of a Néron model is functorial, it cannot be said that Néron models satisfy the properties, one would expect from a good functor. For example, Néron models do not, in general, commute with (ramified) based change; also, in the group scheme case, the behavior with respect to exact sequences can be very capricious. The situation stabilizes somewhat if one considers Néron models with semi-abelian reduction.

The purpose of the present chapter is to collect several properties of Néron models, and to give a number of examples which show that certain other, perhaps desirable, properties are in general not true. We prove a criterion for a smooth group scheme to be a Néron model and discuss the behavior of Néron models with respect to the formation of subgroups as well as with respect to base change and descent. Then we look at isogenies and Néron models with semi-abelian reduction. For example, we prove the criterion of Néron-Ogg-Shafarevich for good reduction. There is also a section dealing with various aspects of exactness properties. The chapter ends with a supplementary section where we explain the Weil restriction functor. If one works with respect to a finite and faithfully flat extension of Dedekind schemes $S' \longrightarrow S$, this functor respects Néron models. Furthermore, if K and K' are the rings of rational functions on S and S', the Weil restriction is used to describe the behavior of associated Néron models if one descends from a K'-group scheme $X_{K'}$ to a K-group scheme X_K.

7.1 A Criterion

Throughout this section we will denote by R a discrete valuation ring, by R^{sh} its strict henselization, and by K and K^{sh} the corresponding fields of fractions. Furthermore, k is the residue field of R, and k_s its separable algebraic closure. In the following we will consider R-group schemes G of finite type with a smooth generic fibre and with the property that each K^{sh}-valued point of G extends to an R^{sh}-valued point of G. We are interested in conditions under which G is a Néron model of its generic fibre G_K or, more generally, in the way of deriving a Néron model of G_K from G.

Theorem 1. *Let G be a smooth R-group scheme of finite type or a torsor under a smooth R-group scheme of finite type. Then the following conditions are equivalent:*

(i) *G is a Néron model of its generic fibre G_K.*

(ii) *G is separated and the canonical map $G(R^{sh}) \longrightarrow G(K^{sh})$ is surjective.*

(iii) *The canonical map* $G(R^{sh}) \longrightarrow G(K^{sh})$ *is bijective.*

Proof. It is enough to consider the case where G is a group scheme. Indeed, if G is a torsor we may assume by 6.5/3 that R is strictly henselian and, furthermore, that G is unramified. Then G admits a section over R and we can view G as a group scheme.

In the following, let us assume that G is a group scheme. The implications (i) $\Longrightarrow$ (ii) $\Longrightarrow$ (iii) are trivial, the second one by the valuative criterion of separatedness. Moreover, it is easy to see that condition (ii) implies condition (i). Namely, if G satisfies (ii), it is a weak Néron model of its generic fibre G_K. Hence the weak Néron property 3.5/3 and the extension theorem 4.4/1 show that G satisfies the definition of Néron models.

Turning to the remaining implication (iii) $\Longrightarrow$ (ii), we have to verify that (iii) implies the separatedness of G. Using Lemma 2 below, it is only to show that the unit section $\varepsilon : \operatorname{Spec} R \longrightarrow G$ is a closed immersion or, what amounts to the same, that $\operatorname{im} \varepsilon$ is closed in G. Restricting ε to generic fibres, we know that $\varepsilon_K : \operatorname{Spec} K \longrightarrow G_K$ is a closed immersion. Let F be the schematic image of ε_K in G. Then, pointwise, $\operatorname{im} \varepsilon$ and F coincide on G_K, and we have to show the same for the special fibre G_k of G. So consider a point $e_k \in F \cap G_k$. Working in an affine open neighborhood $U \subset G$ of e_k, let A be the ring of global sections on $F \cap U$. Then $R \subset A \subset K$ and, thus, $R = A$ since R is a discrete valuation ring. Hence the inclusion of $F \cap U$ into G gives rise to a point $e \in G(R)$ extending $\varepsilon_K \in G(K)$. However, condition (iii) implies $e = \varepsilon$. So F consists of only two points, namely, the points of $\operatorname{im} \varepsilon$, and it follows that $\operatorname{im} \varepsilon$ is closed in G. □

Lemma 2. *A group scheme G is separated over a base scheme S if and only if the unit section ε is a closed immersion.*

Proof. If G is separated, the diagonal morphism $\delta : G \longrightarrow G \times_S G$ is a closed immersion. Then the same is true for the unit section $\varepsilon : S \longrightarrow G = S \times_S G$, since ε is obtained from δ by means of the base change $\varepsilon : S \longrightarrow G$.

Conversely, viewing the diagonal in $G \times_S G$ as the inverse image of $\operatorname{im} \varepsilon$ with respect to the morphism

$$G \times_S G \longrightarrow G\,, \qquad (g,h) \longmapsto g \cdot h^{-1}\,,$$

it follows that G is separated if ε is a closed immersion. □

In order to demonstrate how Theorem 1 can be applied, let us give an example of an algebraic K-group which, although it is affine, admits a Néron model.

Example 3. Let R be a discrete valuation ring of equal characteristic $p > 0$, and let π be a uniformizing element of R. Consider the subgroup G of $\mathbb{G}_{a,R} \times_R \mathbb{G}_{a,R}$ which is given by the equation

$$x + x^p + \pi y^p = 0\,.$$

Then G is a smooth R-group scheme of finite type. Furthermore, looking at values of solutions of the above equation, one shows easily that the map $G(R^{sh}) \longrightarrow G(K^{sh})$ is surjective. Thus G is a Néron model of its generic fibre G_K. The group G_K is an example of a so-called K-wound unipotent group; i.e., of a connected unipotent algebraic K-group which does not contain $\mathbb{G}_{a,K}$ as a subgroup. Smooth commutative groups of this kind admit Néron models of finite type, at least in the case where R is excellent; cf. 10.2/1.

Next consider an R-group scheme G of finite type such that the generic fibre G_K is smooth. If the residue characteristic of R is zero, the special fibre G_k is smooth by Cartier's theorem, [SGA 3_{I}], Exp. VI_{B}, 1.6.1, so that, if G is flat, it will be smooth over R. However, since the latter result does not extend to the general case, we want to describe a procedure which, by means of the smoothening process, associates a smooth R-group scheme G' to G such that the canonical map $G'(R^{sh}) \longrightarrow G(R^{sh})$ is bijective. Let us call a morphism of R-group schemes $G' \longrightarrow G$, where G' is smooth and of finite type over R, a *group smoothening* of G if each R-morphism $Z \longrightarrow G$ from a smooth R-scheme Z admits a unique factorization through G'. Then, by the defining universal property, $G' \longrightarrow G$ is an isomorphism on generic fibres since G_K is smooth. In particular, if $G(R^{sh}) \longrightarrow G(K^{sh})$ is bijective, G' will be a Néron model of G_K by Theorem 1. Group smoothenings can be defined in the same way using a global Dedekind scheme as base. However, their existence can only be guaranteed in the local case; cf. Theorem 5 below.

Lemma 4. *Let G be an R-group scheme of finite type which has a smooth generic fibre. Denote by F_k the Zariski closure in G_k of the set of k_s-valued points in G_k which lift to R^{sh}-valued points of G. Then F_k, provided with its canonical reduced structure, is a closed subgroup scheme of G_k. Furthermore, let $u: Y \longrightarrow G$ be the dilatation of F_k in G. Using the notation δ for the defect of smoothness as in* 3.3, *we have*

$$\delta(a') \leq \max\{0, \delta(a) - 1\}$$

for each R^{sh}-valued point a of G and its lifting a' to Y.

Proof. Since the set of R^{sh}-valued points of G forms a group, it is clear that F_k is a subgroup scheme of G_k. In order to justify the second assertion, we use Lemma 3.4/1; it is only to show that $F_k \subset G_k$ is E-permissible, where $E = G(R^{sh})$. However this is clear. By construction, F_k is geometrically reduced and, hence, smooth over k, being a group scheme of finite type over a field. Furthermore, using 4.2/2, we see that the restriction of the sheaf of differentials $\Omega^1_{G/R}$ to G_k is free and, hence, that the restriction of $\Omega^1_{G/R}$ to F_k is free. Thus the two conditions characterizing E-permissibility are satisfied. □

It follows from 3.2/2(d) that the scheme Y of Lemma 4 is an R-group scheme again and that $u: Y \longrightarrow G$ is a group homomorphism. So a finite repetition of the construction leads to an R-group scheme G' which has generic fibre G_K and defect of smoothness 0, and thus is smooth at all its R^{sh}-valued points. In particular, G' is smooth at the unit section and therefore smooth everywhere since it is flat. We claim that the morphism $G' \longrightarrow G$ is a group smoothening of G. To justify this, consider an R-morphism $Z \longrightarrow G$ where Z is a smooth R-scheme. Writing k_s for the separable algebraic closure of k, the set of k_s-valued points of Z_k which lift to R^{sh}-valued points of Z is schematically dense in Z_k; cf. 2.3/5. Thus, we see that, in the situation of Lemma 4, the special fibre of Z is mapped into F_k. Then the desired factorization of $Z \longrightarrow G$ follows from 3.2/1(b), again. So we have derived the following facts on group smoothenings.

Theorem 5. *Let G be an R-group scheme of finite type with a smooth generic fibre G_K. Then there exists a group smoothening $G' \longrightarrow G$ of G. Due to its definition, G' is smooth and of finite type; it is characterized by the property that each R-morphism $Z \longrightarrow G$, where Z is smooth over R, factors uniquely through G'.*

Furthermore, if the map $G(R^{sh}) \longrightarrow G(K^{sh})$ is surjective and if G is separated, G' is a Néron model of G_K.

Proof. Only the assertion concerning the Néron model remains to be verified. If $G(R^{sh}) \longrightarrow G(K^{sh})$ is surjective and if G is separated, the same is true for $G'(R^{sh}) \longrightarrow G'(K^{sh})$ and G'. Thus G' is a Néron model of G_K by the criterion given in Theorem 1. □

As an application we want to examine how the Néron model G of a K-group scheme G_K behaves if we pass from G_K to a subgroup $H_K \subset G_K$.

Corollary 6. *Let S be a Dedekind scheme with ring of rational functions K. Furthermore, let G be an S-group scheme which is a Néron model of its scheme of generic fibres G_K, and let H_K be a smooth subgroup of G_K. Then H_K admits a Néron model H over S; more precisely, one can define H as a group smoothening of the schematic closure $\bar{H}$ of H_K in G. The schematic closure $\bar{H}$ itself is a Néron model of H_K if and only if it is smooth. In particular, the latter is the case if* $\operatorname{char} k(s) = 0$ *for all closed points $s \in S$.*

Proof. First, let us show that there exists a group smoothening of $\bar{H}$ over S. Since H_K is smooth, its schematic closure $\bar{H}$ is smooth over a dense open part S' of S. On the other hand, we know from Theorem 5 that, for each of the finitely many points $s \in S' - S$, the group scheme $\bar{H} \otimes_S \mathcal{O}_{S,s}$ admits a group smoothening. Then, similarly as explained in the proof of 1.4/1, we can glue $\bar{H} \otimes_S \mathcal{O}_{S,s}$ for $s \in S - S'$ to $\bar{H} \times_S S'$, thereby obtaining a global group smoothening H of $\bar{H}$ over S.

It remains to show that H is a Néron model of H_K. To do so, we may assume that S is local. Consider a smooth S-scheme Z and a K-morphism $Z_K \longrightarrow H_K$. Then, since $H_K \subset G_K$ and since G is a Néron model of G_K, this morphism extends uniquely to an S-morphism $Z \longrightarrow G$ which, by the definition of $\bar{H}$, must factor through $\bar{H}$. Furthermore, we conclude from Theorem 5 that $Z \longrightarrow \bar{H}$ extends uniquely to an R-morphism $Z \longrightarrow H$. The latter is unique as an extension of $Z_K \longrightarrow H_K$. So H is a Néron model of G_K and the remaining assertions are clear since $\bar{H}$ is flat over S. □

7.2 Base Change and Descent

One cannot expect that, for a faithfully flat extension of discrete valuation rings $R \subset R'$, the base change $\operatorname{Spec} R' \longrightarrow \operatorname{Spec} R$ transforms Néron models over R into Néron models over R'. In Example 7.1/3 of the preceding section we can see that, after adjoining a p-th root of the uniformizing element π of R to K, the boundedness of $G_K(K^{sh})$ and, hence, the existence of a Néron model of G_K is lost, since G_K becomes isomorphic to the additive group $\mathbb{G}_{a,K}$. On the other hand, it follows from 1.2/2 and 6.5/3 that Néron models behave well with respect to étale base change. The latter is true for a more general class of morphisms as we will see in this section (cf. 6.5/5 for a partial result of this type).

Consider a faithfully flat extension $R \subset R'$ of discrete valuation rings with fields of fractions K and K'. As usual we indicate strict henselizations by an exponent "sh" and we may assume that R^{sh} is a subring of R'^{sh}. Recall that R' is said to have ramification index 1 over R if a uniformizing element of R gives rise to a uniformizing element of R' and if the residue extension of R'/R is separable (cf. 3.6/1).

Theorem 1. *Let $R \subset R'$ and $K \subset K'$ be as above and consider a torsor X_K under a smooth K-group scheme G_K of finite type. Denote by $X_{K'}$ the torsor under $G_{K'}$ obtained by base change with K'.*

(i) *Assume that $X_{K'}$ admits a Néron model X' over R'. Then X_K admits a Néron model X over R, and there is a canonical R'-morphism $X \otimes_R R' \longrightarrow X'$, called morphism of base change.*

(ii) *Let R'/R be of ramification index 1. Then X_K admits a Néron model X over R if and only if $X_{K'}$ admits a Néron model X' over R'. If the latter is the case, the morphism of base change $X \otimes_R R' \longrightarrow X'$ is an isomorphism.*

Proof. If $X_{K'}$ admits a Néron model, $X_{K'}(K'^{sh})$ is bounded in $X_{K'}$. Using 1.1/5, we see that $X_K(K'^{sh})$ is bounded in X_K. But then $X_K(K^{sh})$ is bounded in X_K and a Néron model X of X_K exists by 6.5/4. Since $X \otimes_R R'$ is a smooth R'-model of $X_{K'}$, the identity on $X_{K'}$ extends to an R'-morphism $X \otimes_R R' \longrightarrow X'$ as required in assertion (i).

In the situation of assertion (ii) we have only to consider the case where X_K has a Néron model X. Furthermore, since Néron models are compatible with étale base change, we may assume that R and R' are strictly henselian. It has to be shown that $X \otimes_R R'$ is a Néron model of $X_{K'}$. To do this, it is enough to look at the case where the torsor $X_{K'}$ is unramified. So consider a K'-valued point of $X_{K'}$. Interpreting it as a point $a_K \in X_K(K')$ and working in an affine open neighborhood of its image in X_K, we can find an R-model $\tilde{X}$ of X_K of finite type such that a_K extends to a point $a \in \tilde{X}(R')$. Due to 3.6/4, we may assume that $\tilde{X}$ is smooth. But then, since X is a Néron model of X_K, we have a morphism $\tilde{X} \longrightarrow X$. Thus each $a_K \in X_K(K')$ extends to a point $a \in X(R')$ and, consequently, the canonical map $(X \otimes_R R')(R') \longrightarrow (X \otimes_R R')(K')$ is surjective. So $X \otimes_R R'$ is a Néron model of $X_{K'}$ by 7.1/1. □

It will be of interest in 10.1/3 that the argument for showing that $X \otimes_R R'$ is a Néron model of $X_{K'}$ can be changed slightly so that the use of 7.1/1 can be avoided. Namely, look at a discrete valuation ring R'' which is of ramification index 1 over R'. Then R'' has ramification index 1 also over R and, if K'' is the field of fractions of R'', the above given argument shows that the map $X(R'') \longrightarrow X(K'')$ is surjective. In particular, taking for R'' the local ring of a smooth R'-scheme Z' at a generic point of the special fibre Z'_k, we see that $X \otimes_R R'$ satisfies the weak Néron property. So if $X_{K'}$ is unramified, we may view $X \otimes_R R'$ as an R'-group scheme, which satisfies the Néron mapping property by the extension argument 4.4/1 for morphisms into group schemes. Thus $X \otimes_R R'$ is a Néron model of $X_{K'}$ in this case.

Corollary 2. *Over discrete valuation rings, the formation of Néron models (of torsors or group schemes) is compatible with extensions R'/R of ramification index* 1. *For example, R' can be the completion of R.*

Giving another application of Theorem 1, we show that the Néron mapping property can be strengthened.

Proposition 3. *Let X_K be a K-torsor under a smooth K-group scheme G_K of finite type, and assume that a Néron model X of X_K exists. Let A be an R-algebra of type $R\{t\}$ or $R[[t]]$ (strictly convergent or formal power series in a system of variables $t = (t_1, \ldots, t_n)$) where R is complete. Then each K-morphism*

$$u_K : \operatorname{Spec}(A \otimes_R K) \longrightarrow X_K$$

extends uniquely to an R-morphism $u : \operatorname{Spec} A \longrightarrow X$.

Proof. Let η be the generic point of the special fibre $\operatorname{Spec}(A \otimes_R k)$ of $\operatorname{Spec} A$. Then A_η is a discrete valuation ring which is of ramification index 1 over R. Writing F for the field of fractions of A_η, we see that u_K gives rise to an F-morphism $\operatorname{Spec} F \longrightarrow X_K \otimes_K F$. Applying Theorem 1, this morphism extends to an A_η-morphism $\operatorname{Spec} A_\eta \longrightarrow X \otimes_R A_\eta$ and, hence, to an R-rational map $u : \operatorname{Spec} A \dashrightarrow X$. In particular, the special fibre X_k is not empty and, thus, X_K cannot be a ramified torsor. We claim that u is a morphism. Then u extends u_K, and it is unique since X is separated.

If $X(R) \neq \varnothing$, we may view X as an R-group scheme, and one can conclude from Remark 4.4/3 that the R-rational map u is a morphism. In the general case, we choose a discrete valuation ring R' which is finite and étale over R and which satisfies the property that $X(R') \neq \varnothing$. The latter is possible since the torsor X_K is unramified. Set $A' = R'\{t\}$ or $A' = R'[[t]]$ depending on the type of power series we consider for A; note that R' is complete. Then it follows from the above special case that the composition of morphisms

$$\operatorname{Spec}(A' \otimes_R K) \xrightarrow{\text{pr}} \operatorname{Spec}(A \otimes_R K) \xrightarrow{u_K} X_K ,$$

where pr is the canonical projection, extends to an R-morphism $u' : \operatorname{Spec} A' \longrightarrow X$. In other words, the composition of the projection $\operatorname{Spec} A' \longrightarrow \operatorname{Spec} A$ with the R-rational map $u : \operatorname{Spec} A \dashrightarrow X$ is a morphism. But then, by 2.5/5, u is defined everywhere and, thus, is a morphism. □

Using the technique of Weil restriction to be explained in Section 7.6, one can describe in a precise way how, in the situation of Theorem 1 (i) and under the assumption that the extension of discrete valuation rings $R \subset R'$ is *finite*, a Néron model X of X_K can be constructed from a Néron model X' of $X_{K'}$, at least in the case of group schemes.

Proposition 4. *Let $S' \longrightarrow S$ be a flat and finite morphism of Dedekind schemes with rings of rational functions K and K'. Let G_K be a smooth K-group scheme of finite type and denote by $G_{K'}$ the K'-group scheme obtained from G_K by base change. Assume that the Néron model G' of $G_{K'}$ exists over S'. Then the Néron model G of G_K exists over S and can be constructed as a group smoothening of the schematic closure of G_K in the Weil restriction $\mathfrak{R}_{S'/S}(G')$.*

Proof. Using 7.6/6, we see that the Weil restriction $\mathfrak{R}_{S'/S}(G')$ exists as a scheme and that it is a Néron model of its scheme of generic fibres, i.e. of $\mathfrak{R}_{K'/K}(G'_{K'})$. Thus, considering the canonical closed immersion

$$\iota : G_K \longrightarrow \mathfrak{R}_{K'/K}(G_{K'}) ,$$

the assertion follows from 7.1/6. □

7.3 Isogenies

We want to investigate under what conditions an isogeny $G_K \longrightarrow G'_K$ between smooth and connected K-group schemes extends to an isogeny between associated Néron models. In order to attack this problem, we begin by recalling some well-known facts about homomorphisms between group schemes over a field k.

Lemma 1. *Let $f: G \longrightarrow G'$ be a homomorphism of group schemes which are smooth and of finite type over a field k. Assume that* $\dim G = \dim G'$. *Then the following conditions are equivalent:*

(a) *f is flat.*
(b) *$f(G^0) = G'^0$ where G^0 and G'^0 denote identity components of G and G'.*
(c) $\ker f$ *is finite.*
(d) *f is quasi-finite.*
(e) *f is finite.*

A commutative group scheme G which is smooth and of finite type over a field k is called *semi-abelian* if its identity component G^0 is an extension of an abelian variety by a (not necessarily deployed) affine torus. The latter fact can be checked over the algebraic closure $\bar{k}$ of k. Indeed, one knows from Chevalley's theorem 9.2/1 that $G^0_{\bar{k}}$ is uniquely an extension of an abelian variety by a connected affine group $H_{\bar{k}}$. Then $H_{\bar{k}}$ decomposes into the product of a torus part and a unipotent part, where the torus part is already defined over k; cf. [SGA $3_{\rm II}$], Exp. XIV, 1.1. So we see that G is semi-abelian if and only if the unipotent part of $H_{\bar{k}}$ is trivial. Over a general base scheme S, an S-group scheme G is called *semi-abelian* if it is smooth over S and if all its fibres are semi-abelian in the sense explained above.

Lemma 2. *Let G be a commutative S-group scheme which is smooth and of finite type over an arbitrary base scheme S. Let l be a positive integer.*

(a) *Suppose that G is semi-abelian. Then the l-multiplication $l_G : G \longrightarrow G$ is quasi-finite and flat.*

(b) *Suppose that* char $k(s)$ *does not divide l for all $s \in S$. Then the l-multiplication $l_G : G \longrightarrow G$ is étale.*

Proof. In order to verify the flatness of l_G in the situation (a) or (b), we can use the characterization of flatness in terms of fibres 2.4/2. So we may assume that S consists of a field k. Then, since l_G is surjective on abelian varieties and on tori, and in the situation (b), also on unipotent groups, it follows from the structure of commutative smooth and connected group schemes over k that $G^0 \subset \operatorname{im} l_G$. By Lemma 1 we see that l_G is quasi-finite and flat.

In the situation of assertion (b) we have just seen that l_G is flat. So we may use the criterion 2.4/8. Thus, just as before, we can assume that S consists of a field k. Then we can consider the Lie algebra $\operatorname{Lie}(G)$ and the endomorphism $\operatorname{Lie}(l_G) : \operatorname{Lie}(G) \longrightarrow \operatorname{Lie}(G)$ induced on it by l_G. Since $\operatorname{Lie}(l_G)$ is just the multiplication by l and since l is not divisible by char k, we see that it is bijective. So $l_G : G \longrightarrow G$ is étale by 2.2/10. □

For an S-group scheme G as in Lemma 2, we write ${}_lG$ for the kernel of the l-multiplication $l_G : G \longrightarrow G$. If char $k(s)$ does not divide l for all $s \in S$, we deduce from Lemma 2 that ${}_lG$, being the fibre of l_G over the unit section, is étale over S, whereas in the situation of Lemma 2 (a) we only know that ${}_lG$ is quasi-finite and flat over S.

In general, an S-group scheme H of finite type which is quasi-finite over S is not finite over S unless S consists of a field. However, if S is the spectrum of a *henselian* discrete valuation ring R and if H is quasi-finite and separated, one can consider its *finite part* H'. The latter is the open and closed subscheme of H consisting of the special fibre H_k and of all points of the generic fibre H_K which specialize into points of H_k. Namely, applying 2.3/4, one shows that H is the disjoint sum of two open and closed subschemes H' and H'', where H' is finite over S and where the special fibre of H'' is empty. The finite part H' of H is an open subgroup scheme of H.

Proposition 3. *Let R be a discrete valuation ring and let l be a positive integer such that the residue characteristic of R does not divide l. Then, for any smooth commutative R-group scheme G of finite type, the canonical map ${}_lG(R^{sh}) \longrightarrow {}_lG(k_s)$ is bijective, where R^{sh} is a strict henselization of R and where k_s is the residue field of R^{sh}.*

Proof. We may assume that R is strictly henselian. Since ${}_lG$ is étale over R by Lemma 2, its finite part is a disjoint union of copies of $S = \operatorname{Spec} R$; cf. 2.3/1. □

Definition 4. *Let* $f: G \longrightarrow G'$ *be a homomorphism of commutative group schemes of finite type over an arbitrary base scheme S. Then f is called an isogeny if, for each* $s \in S$, *the homomorphism* $f_s: G_s \longrightarrow G'_s$ *is an isogeny in the classical sense; i.e., if* f_s *is finite and surjective on identity components.*

Examples of isogenies are provided by l-multiplications on commutative group schemes G where l and G have to be chosen as required in Lemma 2 (a) or (b). In the situation of the definition, each f_s has a degree $\deg f_s$, which can be defined as the rank of the finite $k(s)$-group scheme $\ker f_s$. Recalling some facts on commutative finite group schemes H over a field k, we mention that H is étale if $\operatorname{char} k = 0$ (by Cartier's theorem) or, more generally, if $\operatorname{char} k$ does not divide the rank of H. If H is connected, its rank is a power of char k. Furthermore, the l-multiplication $l_H: H \longrightarrow H$ is the zero-homomorphism if l is a multiple of the rank of H.

We need a well-known result relating isogenies over fields to l-multiplications.

Lemma 5. *Let* $f: G \longrightarrow G'$ *be an isogeny between smooth and connected commutative group schemes of finite type over a field k. Assume either that* char k *does not divide* $\deg f$ *or that G is semi-abelian. Then there is an isogeny* $g: G' \longrightarrow G$ *such that* $g \circ f = l_G$ *where* $l = \deg f$.

Proof. Setting $l = \deg f$, we see that $\ker f \subset \ker l_G$. Then, f being flat and surjective, we have $G' = G/\ker f$ and, thus, homomorphisms

$$G \xrightarrow{f} G' \longrightarrow G/\ker l_G .$$

Since the l-multiplication $l_G: G \longrightarrow G$ is finite by Lemma 2, and since l_G factors through $G/\ker l_G$, the existence of g is clear. □

Now, working over a discrete valuation ring R and its field of fractions K, we can deal with the question of whether a homomorphism between R-group schemes is an isogeny as soon as it is an isogeny on generic fibres.

Proposition 6. *Let* G_K *and* G'_K *be smooth commutative and connected K-group schemes of finite type admitting Néron models G and G' over R. Consider an isogeny* $f_K: G_K \longrightarrow G'_K$ *and assume either that the residue characteristic of R does not divide* $\deg f_K$ *or that G is semi-abelian. Then* f_K *extends to an isogeny* $f: G \longrightarrow G'$, *and there is an isogeny* $g: G' \longrightarrow G$ *such that* $g \circ f = l_G$ *for* $l = \deg f_K$.

Proof. Using Lemma 5, there is an isogeny $g_K: G'_K \longrightarrow G_K$ satisfying $g_K \circ f_K = l_{G_K}$ for $l = \deg f_K$. Due to the Néron mapping property, f_K and g_K extend to homomorphisms $f: G \longrightarrow G'$ and $g: G' \longrightarrow G$ such that $g \circ f = l_G$. Then, by our assumptions on $l = \deg f_K$ or on G, we see from Lemma 2 that l_G is an isogeny, and it follows easily that f and g are isogenies. □

Corollary 7. *Let* $f_K: G_K \longrightarrow G'_K$ *be an isogeny of abelian varieties with Néron models G and G'. Then G is semi-abelian if and only if G' is semi-abelian.*

Proof. By the Néron mapping property, the isogeny f_K extends to a homomorphism $f: G \longrightarrow G'$. If G is semi-abelian, f is an isogeny by Proposition 6 and, consequently, G' is semi-abelian. Using an isogeny $g_K : G'_K \longrightarrow G_K$, one shows in the same way that G is semi-abelian if G' is semi-abelian. □

7.4 Semi-Abelian Reduction

Let G be a smooth group scheme of finite type over a Dedekind scheme S which, for simplicity, we will assume to be connected. We say that G has *abelian reduction* (resp. *semi-abelian reduction*) at a closed point $s \in S$ if the identity component G_s^0 is an abelian variety (resp. an extension of an abelian variety by an affine torus). In particular, if G is a Néron model of its generic fibre G_K, where K is the field of fractions of S, we will say that G_K has abelian (resp. semi-abelian) reduction at $s \in S$ if the corresponding fact is true for G. The latter amounts to the same as saying that the local Néron model $G \times_S \operatorname{Spec} \mathcal{O}_{S,s}$ of G_K at $s \in S$ has abelian (resp. semi-abelian) reduction.

If A_K is an abelian variety over K, then A_K is said to have *potential abelian reduction* (resp. *potential semi-abelian reduction*) at a closed point $s \in S$ if there is a finite Galois extension L of K such that A_L has abelian (resp. semi-abelian) reduction at all points over s. To be precise, we thereby mean that the Néron model A' of A_L over the normalization S' of S in L has abelian (resp. semi-abelian) reduction at all closed points $s' \in S'$ lying over s. Instead of abelian reduction, we will also talk about *good reduction*. Let us begin by mentioning the fundamental theorem on the potential semi-abelian reduction of abelian varieties.

Theorem 1. *Each abelian variety A_K over K has potential semi-abelian reduction at all closed points of S.*

The easiest way to obtain this result is via the potential semi-stable reduction of curves, as proved by Artin and Winters [1], a topic which is beyond the scope of the present book. So we will restrict ourselves to briefly indicating how the assertion of the theorem can be deduced from the corresponding results on curves.

Since abelian varieties have good reduction almost everywhere, see 1.4/3, the problem is a local one, and we may assume that S consists of a discrete valuation ring R. One starts with the case where A_K is the Jacobian $J_K = \operatorname{Pic}^0_{C_K/K}$ of a smooth and proper K-curve C_K. Then the theorem on the potential semi-stable reduction of curves asserts that, replacing K by a finite separable extension if necessary, we can extend C_K into a proper flat R-curve C whose geometric fibres have at most ordinary double points as singularities; cf. 9.2/7. For such a curve it is shown in 9.4/1 that the relative Jacobian $\operatorname{Pic}^0_{C/S}$ is a smooth and separated R-group scheme having semi-abelian reduction. Since $\operatorname{Pic}^0_{C/S}$ is an S-model of J_K, it follows from Proposition 3 below or from the more general discussion of the relationship between

Néron models and the relative Picard functor in 9.5/4 or 9.7/2 that $\mathrm{Pic}^0_{C/S}$ is the identity component of the Néron model of J_K. Thus J_K has semi-abelian reduction.

If A_K is a general abelian variety, one knows, see Serre [1], Chap. VII, § 2, n°13, that there is an exact sequence of abelian varieties

$$0 \longrightarrow A'_K \longrightarrow J_K \longrightarrow A_K \longrightarrow 0$$

where J_K is a product of Jacobians. Using the fact that J_K has potential semi-abelian reduction, it follows from the lemma below that A_K has potential semi-abelian reduction also. □

Lemma 2. *Let* $0 \longrightarrow A'_K \longrightarrow A_K \longrightarrow A''_K \longrightarrow 0$ *be an exact sequence of abelian varieties over K. Then* A_K *has semi-abelian (resp. abelian) reduction if and only if* A'_K *and* A''_K *have semi-abelian (resp. abelian) reduction.*

Proof. Due to Poincaré's complete reducibility theorem, see Mumford [3], Chap. IV, § 19, Thm. 1, there is an abelian subvariety $\tilde{A}''_K$ in A_K such that the canonical map $\tilde{A}''_K \times A'_K \longrightarrow A_K$ and, thus, also the composition $\tilde{A}''_K \longrightarrow A_K \longrightarrow A''_K$ are isogenies. So we see that A_K is isogenous to $A'_K \times A''_K$ and it follows from 7.3/7 that A_K has semi-abelian reduction if and only if the same is true for A'_K and A''_K. An application of 7.3/6 settles the case of abelian reduction. □

For the remainder of this section, let us assume that the base scheme S consists of a *discrete valuation ring* R with field of fractions K. We want to discuss properties of Néron models with abelian or semi-abelian reduction and to give criteria for the existence of Néron models with abelian or semi-abelian reduction over the given field K.

Proposition 3. *Let* A_K *be an abelian variety with Néron model A and let G be a smooth and separated R-group scheme which is an R-model of* A_K. *Assume that G has semi-abelian reduction. Then the canonical morphism* $G \longrightarrow A$ *is an open immersion; it is an isomorphism on identity components.*

Proof. We can assume that R is strictly henselian. Furthermore, it is enough to show that $G^0 \longrightarrow A^0$ is an isomorphism. So assume that $G = G^0$. Let l be a positive integer which is not divisible by the characteristic of the residue field k of R. Considering the kernels ${}_lG$ and ${}_lA$ of l-multiplications on G and A, we have a canonical commutative diagram

$$\begin{array}{ccc}
{}_lG(K) & \xrightarrow{\ \sim\ } & {}_lA(K) \\
\uparrow \cup & & \uparrow \wr \\
{}_lG(R) & \longrightarrow & {}_lA(R) \\
\downarrow \wr & & \downarrow \wr \\
{}_lG(k) & \longrightarrow & {}_lA(k)
\end{array}$$

where ${}_lG(R) \longrightarrow {}_lG(K)$ is injective since G is separated and where all other vertical maps are bijective; the upper one on the right-hand side because A is a Néron model of A_K and the lower ones by 7.3/3. So the middle horizontal map is injective, and the same is true for the lower horizontal one. Now, using the facts that G has semi-abelian reduction and that k is separably closed, it follows that the points in $G(k)$ which have finite order not divisible by char k are topologically dense in each connected subgroup of G_k. Therefore $G_k \longrightarrow A_k^0$ has a finite kernel. In particular, $G \longrightarrow A^0$ is quasi-finite and, thus, surjective by reasons of dimension. But then Zariski's Main Theorem 2.3/2′ shows that $G \longrightarrow A^0$ is an isomorphism. □

Corollary 4. *If an abelian variety A_K has semi-abelian reduction, then the formation of the identity component of the Néron model of A_K is compatible with faithfully flat extensions of discrete valuation rings R'/R.*

We have seen above that points of finite order play an important role when dealing with Néron models of abelian varieties. We want to use them in order to give a criterion for the existence of abelian or semi-abelian reductions over the given field K. As before, R will be a discrete valuation ring with field of fractions K and with residue field k. Let K_s be a separable algebraic closure of K and consider rings $R \subset R^h \subset R^{sh} \subset R_s \subset K_s$ where R^h is a henselization of R, where R^{sh} is a strict henselization of R, and where R_s is the localization of the integral closure of R in K_s at a maximal ideal lying over the maximal ideal of R^{sh}. As usual K^h and K^{sh} denote the fields of fractions of R^h and of R^{sh}. Then the inertia group of the maximal ideal of R_s coincides with the Galois group $\mathrm{Gal}(K_s/K^{sh})$; cf. 2.3/11. Fixing the above situation, we will call $I := \mathrm{Gal}(K_s/K^{sh})$ "the" inertia group of $\mathrm{Gal}(K_s/K)$.

Theorem 5. *Let A_K be an abelian variety over K with Néron model A over R, and let l be a prime different from char k. Then the following conditions are equivalent:*

(a) *A_K has abelian reduction; i.e., the identity component A_k^0 is an abelian variety over k.*

(b) *A is an abelian scheme over R.*

(c) *For each $\nu \geq 0$ the inertia group I of $\mathrm{Gal}(K_s/K)$ acts trivially on ${}_{l^\nu}A_K(K_s)$, the set of K_s-valued points of the kernel of the l^ν-multiplication $l^\nu_{A_K} : A_K \longrightarrow A_K$. In other words, the canonical map ${}_{l^\nu}A_K(K^{sh}) \longrightarrow {}_{l^\nu}A_K(K_s)$ is bijective.*

(d) *The Tate module $T_l(A_K) = \varprojlim {}_{l^\nu}A_K(K_s)$ is unramified over R; i.e., the inertia group I of $\mathrm{Gal}(K_s/K)$ operates trivially on $T_l(A_K)$.*

Proof. We begin by showing that conditions (a) and (b) are equivalent. If A_k^0 is an abelian variety, we can conclude from [EGA IV_3], 15.7.10, that A^0 is proper over R and, thus, is an abelian scheme over R. But then A^0 is a Néron model of its generic fibre by 1.2/8; thus, $A = A^0$. This verifies the implication (a) $\Longrightarrow$ (b); the converse is trivial.

The equivalence of (c) and (d) is clear. In order to verify the remaining implications, consider the canonical maps

$$(*) \qquad {}_{l^\nu}A(K_s) \hookleftarrow {}_{l^\nu}A(K^{sh}) \xleftarrow{\sim} {}_{l^\nu}A(R^{sh}) \xrightarrow{\sim} {}_{l^\nu}A(k_s)$$

where k_s is the residue field of R^{sh} and where the map on the right-hand side is bijective by 7.3/3. If A is an abelian scheme over R, the cardinality of both sets ${}_{l^\nu}A(K_s)$ and ${}_{l^\nu}A(k_s)$ is $l^{\nu \cdot 2n}$ where n is the dimension of A; cf. Mumford [3], p. 64. Therefore, all maps in (∗) are bijective and we see that (b) implies (c).

Conversely, assume that all maps in (∗) are bijective. Then the cardinality of ${}_{l^\nu}A(k_s)$ is $l^{\nu \cdot 2n}$ for each $\nu \geq 0$, and it follows from the structure of commutative group schemes of finite type (over an algebraically closed or perfect field k) that the identity component A_k^0 is an abelian variety. So we see that condition (c) implies condition (a). □

The equivalence of (a) and (d) in the above theorem is called the criterion of Néron-Ogg-Shafarevich for good reduction. To apply it, one may work over a strictly henselian base ring R. Then A_K has abelian reduction if and only if all l^ν-torsion points of A_K are rational over K. The criterion can be generalized to the semi-abelian reduction case; see [SGA 7_I], Exp. IX, 3.5. We include this generalization here without proof.

Theorem 6. *Let A_K be an abelian variety over K, and let l be a prime different from* char k. *Then the following conditions are equivalent*:

(a) *A_K has semi-abelian reduction over R.*

(b) *There is a submodule $T' \subset T := T_l(A_K(K_s))$ which is stable under the action of the inertia group I of* $\mathrm{Gal}(K_s/K)$ *such that I acts trivially on T' and on T/T'.*

7.5 Exactness Properties

In the following let S be a Dedekind scheme with ring of rational functions K. Except for the purposes of Proposition 1 below, we will only be concerned with the case where S consists of a discrete valuation ring R. Let G_K be a smooth K-group scheme of finite type, and let X_K be a torsor under G_K. Then the Néron model X of X_K, if it exists, may be viewed as a direct image $\iota_* X_K$ with respect to the canonical inclusion $\iota : \mathrm{Spec}\, K \longrightarrow S$. More precisely, X represents this direct image if one restricts to *smooth* schemes over S. This consideration suggests that the Néron model might behave reasonably well with respect to left exactness. However we will see that, except for quite special cases, there will be a defect of exactness, the defect of right exactness being much more serious than the one of left exactness. We will give some examples at the end of this section, after we have presented the general results. Let us begin with an assertion concerning the existence of Néron models.

Proposition 1. *Let S be a Dedekind scheme with ring of rational functions K and let*

$$(*) \qquad 0 \longrightarrow G_K' \longrightarrow G_K \longrightarrow G_K'' \longrightarrow 0$$

be an exact sequence of smooth K-group schemes of finite type (not necessarily commutative).

(a) *If G_K admits a Néron model over S, the same is true for G'_K, but not necessarily for G''_K.*

(b) *If G'_K and G''_K admit Néron models over S, the same is true for G_K.*

Proof. If G_K admits a Néron model, then G'_K admits a Néron model by 7.1/6. To justify the second part of assertion (a), we give an example showing that the existence of a Néron model for G_K does not imply the same for G''_K. Assume that S consists of a discrete valuation ring of equal characteristic $p > 0$ and, as in Example 7.1/3, let G_K be the subgroup of $\mathbb{G}_{a,K} \times_K \mathbb{G}_{a,K}$ given by the equation $x + x^p + \pi y^p = 0$, where π is a uniformizing element of R. Then G_K admits a Néron model over S and the projection of $\mathbb{G}_{a,K} \times_K \mathbb{G}_{a,K}$ onto its second factor gives rise to a smooth group epimorphism $G_K \longrightarrow \mathbb{G}_{a,K}$. Writing G'_K for its kernel, we have a short exact sequence

$$0 \longrightarrow G'_K \longrightarrow G_K \longrightarrow \mathbb{G}_{a,K} \longrightarrow 0$$

of smooth K-group schemes of finite type. The middle term admits a Néron model whereas the group $\mathbb{G}_{a,K}$ on right-hand side does not. The example is quite typical; the reason that a Néron model for G_K does not imply the existence of a Néron model for G''_K, comes mainly from the fact that the quotient of a K-wound unipotent group is not necessarily K-wound again.

Next, to prove assertion (b), assume that G'_K and G''_K admit Néron models G' and G'' over S, where S is an arbitrary Dedekind scheme again. First, if the given exact sequence (*) extends to an exact sequence of smooth S-group schemes of finite type

$$0 \longrightarrow G' \longrightarrow G \longrightarrow G'' \longrightarrow 0\,,$$

we claim that G is automatically a Néron model of G_K by the criterion given in 7.1/1. Namely, in order to verify this, we may assume that S consists of a strictly henselian discrete valuation ring R. Then it is enough to show that the canonical map $G(R) \longrightarrow G(K)$ is bijective. However, this follows easily from the commutative diagram

$$\begin{array}{ccccccccc} 0 & \longrightarrow & G'(R) & \longrightarrow & G(R) & \longrightarrow & G''(R) & \longrightarrow & 0 \\ & & \downarrow \wr & & \downarrow & & \downarrow \wr & & \\ 0 & \longrightarrow & G'(K) & \longrightarrow & G(K) & \longrightarrow & G''(K) & & \end{array}$$

by realizing that the first row is exact, due to the fact that the smoothness of $G \longrightarrow G''$ implies the surjectivity of $G(R) \longrightarrow G''(R)$; cf. 2.2/14.

In the general case we can apply a limit argument ([EGA IV_3], 8.8.2), and thereby extend (*) to an exact sequence of smooth group schemes of finite type over a dense open subscheme S' of S. Consequently, there is a Néron model of G_K over S'. Then, using 1.4/1, it is enough to construct the local Néron models of G_K at the finitely many remaining points of $S - S'$. So, in the proof of assertion (b), we are reduced to the case where S consists of a discrete valuation ring R. Since this problem does not seem to be accessible by elementary methods, we have to make use of a later criterion characterizing the existence of Néron models in terms of the structure of algebraic groups; cf. 10.2/1. It says that a smooth K-group scheme of finite type like G_K admits a Néron model if and only if, after the base change

$K \longrightarrow \hat{K}^{sh}$, the group G_K does not contain subgroups of type $\mathbb{G}_a$ or $\mathbb{G}_m$; here $\hat{K}^{sh}$ is the field of fractions of $\hat{R}^{sh}$, the strict henselization of the completion of R. Using this criterion, it is easily verified that G_K admits a Néron model over R if the same is true for G'_K and G''_K. □

Next, consider an exact sequence

$$0 \longrightarrow G'_K \longrightarrow G_K \longrightarrow G''_K \longrightarrow 0$$

and assume that the corresponding Néron models G', G, and G'' exist so that, due to the universal mapping property, there is an associated complex

$$0 \longrightarrow G' \longrightarrow G \longrightarrow G'' \longrightarrow 0 .$$

We want to examine under what conditions parts of the latter sequence are exact. To do this, it is enough to look at the local case. So, in the following, the base S will consist of a discrete valuation ring R with field of fractions K and with residue field k.

Proposition 2. *If* $\operatorname{char} k = 0$, *the closed immersion* $G'_K \longrightarrow G_K$ *gives rise to a closed immersion* $G' \longrightarrow G$ *of associated Néron models.*

Proof. Denote by H the schematic closure of G'_K in G. Then $G' \longrightarrow G$ factors through $H \subset G$ and we know from 7.1/6 that the induced morphism $G' \longrightarrow H$ is an isomorphism. □

Next, let us look at abelian varieties.

Proposition 3. *Consider an exact sequence of abelian varieties*

$$0 \longrightarrow A'_K \longrightarrow A_K \longrightarrow A''_K \longrightarrow 0$$

and the corresponding complex of Néron models

$$(\dagger) \qquad 0 \longrightarrow A' \longrightarrow A \longrightarrow A'' \longrightarrow 0 .$$

Let B_K *be an abelian subvariety of* A_K *such that* $A_K \longrightarrow A''_K$ *induces an isogeny* $u_K : B_K \longrightarrow A''_K$; *let* $n = \deg u_K$.

(a) *If* $\operatorname{char} k$ *does not divide* n, *then* $A' \longrightarrow A$ *is a closed immersion,* $A \longrightarrow A''$ *is smooth with kernel* A', *and the cokernel of* $A_k \longrightarrow A''_k$ *is killed by multiplication with* n. *If, in addition,* A *has abelian reduction,* $(\dagger)$ *is exact.*

(b) *If* A *has semi-abelian reduction, the sequence* $(\dagger)$ *is exact up to isogeny; i.e., it is isogenous to an exact sequence of commutative* S*-group schemes.*

Proof. The isogeny $u_K : B_K \longrightarrow A''_K$ gives rise to an isogeny $v_K : A'_K \times_K B_K \longrightarrow A_K$ of degree n. So there is an isogeny $w_K : A_K \longrightarrow A'_K \times_K B_K$ such that $w_K \circ v_K$ is multiplication by n. Let B be the Néron model of B_K. Then u_K, v_K, and w_K extend to R-morphisms $u : B \longrightarrow A''$, $v : A' \times_R B \longrightarrow A$, and $w : A \longrightarrow A' \times_R B$ such that $w \circ v$ is multiplication by n on $A' \times_R B$. Assuming the condition of (a), the multiplication by n is an étale isogeny on $A' \times_R B$, and u, v, and w are easily checked to be étale isogenies, too. Then $H := w^{-1}(A')$ is a smooth closed subgroup scheme of A which satisfies $H_K^0 = A'_K$. It follows that the schematic closure of A'_K in H or A is an open subgroup scheme of H and, thus, is smooth over R. So, by 7.1/6, it coincides with the Néron model A' of A'_K and we see that $A' \longrightarrow A$ is a closed immersion. The remaining assertions of (a) follow by using the étale isogeny u. One shows that $A \longrightarrow A''$ is flat, has kernel A' and, hence, is smooth. Furthermore, if A has abelian reduction, the same is true for A'' by 7.4/2 so that $A \longrightarrow A''$ is surjective.

Assertion (b) follows from the fact that $v : A' \times_R B \longrightarrow A$ and $u : B \longrightarrow A''$ are isogenies; use 7.3/6 and 7.3/7. □

Theorem 4. *Let* $0 \longrightarrow A'_K \longrightarrow A_K \longrightarrow A''_K \longrightarrow 0$ *be an exact sequence of abelian varieties and consider the associated sequence of Néron models* $0 \longrightarrow A' \longrightarrow A \longrightarrow A'' \longrightarrow 0$. *Assume that the following condition is satisfied*:

(∗) *R has mixed characteristic and the ramification index $e = v(p)$ satisfies $e < p - 1$, where p is the residue characteristic of R and where v is the valuation on R, which is normalized by the condition that v assumes the value 1 at uniformizing elements of R.*

Then the following assertions hold:

(i) *If A' has semi-abelian reduction, $A' \longrightarrow A$ is a closed immersion.*

(ii) *If A has semi-abelian reduction, the sequence $0 \longrightarrow A' \longrightarrow A \longrightarrow A''$ is exact.*

(iii) *If A has abelian reduction, the sequence $0 \longrightarrow A' \longrightarrow A \longrightarrow A'' \longrightarrow 0$ is exact and consists of abelian R-schemes.*

Proof. Let us first see how assertions (ii) and (iii) can be deduced from assertion (i). If A has semi-abelian or abelian reduction, the same is true for A' and A'' by 7.4/2. So $A' \longrightarrow A$ is a closed immersion by (i), and we can consider the quotient A/A'; it exists in the category of algebraic spaces, cf. 8.3/9. Furthermore, A/A' is smooth and separated and, thus, a scheme by 6.6/3. Now look at the canonical morphism $A/A' \longrightarrow A''$ which is an isomorphism on generic fibres. Since A has semi-abelian reduction, the same is true for A/A', and it follows from 7.4/3 that $A/A' \longrightarrow A''$ is an open immersion. So assertion (ii) is clear. Finally, if A has abelian reduction, the same is true for A/A'. So the latter is an abelian scheme by 7.4/5 and, thus, must coincide with the Néron model A'' of A''_K. Thereby we obtain assertion (iii).

It remains to verify assertion (i) under the assumption of condition (∗). As a key ingredient for the proof of this fact, we will need the following result on finite group schemes; cf. Raynaud [7], 3.3.6.

Lemma 5. *Let R be a discrete valuation ring satisfying condition (∗) of Theorem 4. Let $v : G' \longrightarrow G$ be a morphism of R-group schemes which are finite, flat, and commutative. Then, if $v_K : G'_K \longrightarrow G_K$ is an isomorphism, v is an isomorphism.*

The lemma implies a criterion for finite and flat R-group schemes to be étale. To state it in its simplest form, recall that a group scheme over a base scheme S is called constant if it is of the type H_S with an abstract group H.

Corollary 6. *Assume that R is as in condition (∗) of Theorem 4 and that, in addition, it is strictly henselian. Furthermore, consider a finite, flat, and commutative R-group scheme G whose generic fibre is constant. Then G is constant.*

Proof of Corollary 6. Let $G' \longrightarrow G$ be a group smoothening of G (see 7.1). Then G' coincides with its finite part and, thus, is finite over R since G is finite over R. Therefore $G' \longrightarrow G$ is an isomorphism by the lemma. Using the fact that G' is étale over R and that R is strictly henselian, G is constant. □

Now let us indicate how to obtain assertion (i) of Theorem 4 under the assumption of condition (∗). Since Néron models are preserved when R is replaced by its strict henselization or by its completion, we may assume that R is *strictly henselian* and *complete*.

We begin by showing that $u : A' \longrightarrow A$ is a monomorphism; i.e., that $N := \ker u$ is trivial. For this purpose it is enough to show that the special fibre N_k of N is trivial. If not, there is a prime l, not necessarily different from char k, such that ${}_lA'_k \cap N_k$ is non-trivial; as usual, ${}_lA'$ is the kernel of the l-multiplication on A'. Since A' has semi-abelian reduction, ${}_lA'$ is quasi-finite and flat over R; cf. 7.3/2. Now, R being henselian, we can consider the finite part G' of ${}_lA'$; see 7.3. It is enough to show that u is a monomorphism on G'. Let G be the schematic image of G' under u and consider the morphism $u' : G' \longrightarrow G$ given by u. Then u' is an isomorphism on generic fibres and thus, by the lemma, an isomorphism on G'. In particular, u' is a monomorphism, and it follows that u is a monomorphism.

If A' has abelian reduction, it is an abelian scheme by 7.4/5 and, thus, proper over R. So it follows that u is proper. But then, being a monomorphism, it must be a closed immersion. This ends the proof in the special case where A' has abelian reduction.

In the general case, some work remains to be done since there exist monomorphisms which are not immersions; cf. [SGA 3_{II}], Exp. VIII, 7 and Exp. XVI, 1. Let B be the schematic image of $u : A' \longrightarrow A$; it is a closed subgroup scheme of A which is flat over R. We will show that B or, what is enough, that B^0 is smooth. Then, due to the Néron mapping property, the morphism $A' \longrightarrow B$ admits an inverse and u is a closed immersion. In order to do so, we denote by an index n reductions modulo π^n, where π is a uniformizing element of R. Since u is a monomorphism, it is a closed immersion modulo π^n for all $n > 0$; cf. [SGA 3_{I}], Exp. VI_{B}, 1.4.2. So we can consider the exact sequence of R_n-schemes

$$0 \longrightarrow A_n'^0 \longrightarrow B_n^0 \longrightarrow Q_n \longrightarrow 0$$

where the quotient $Q_n = B_n^0/A_n'^0$ exists as an R-scheme by [SGA 3_{I}], Exp. VI_{A}, Thm. 3.2, and is flat by [SGA 3_{I}], Exp. VI_{B}, Thm. 9.2. Furthermore, Q_n is connected and, by reasons of dimension, finite over R_n. Taking inductive limits for n going to infinity, we obtain an exact sequence of formal group schemes over R

$$0 \longrightarrow \hat{A}' \longrightarrow \hat{B} \longrightarrow Q \longrightarrow 0$$

where Q is an R-scheme which is finite, flat, and connected. Let q be a power of p such that Q is annihilated by the q-multiplication on Q. Since $\hat{A}'$ is p-divisible, the above sequence restricts to an exact sequence

$$0 \longrightarrow {}_q\hat{A}' \longrightarrow {}_q\hat{B} \longrightarrow Q \longrightarrow 0$$

on the kernels of q-multiplications; the latter are finite flat R-group schemes by 7.3/2.

Furthermore, ${}_q\hat{A}'$ and ${}_q\hat{B}$ can be interpreted as the finite parts of the quasi-finite flat R-group schemes ${}_qA'^0$ and ${}_qB^0$.

Applying Grothendieck's orthogonality theorem [SGA 7_I], Exp. IX, Prop. 5.6, we see that the generic fibre of the quotient ${}_qA'/{}_q\hat{A}'$ is constant. Since A' and B coincide on generic fibres, it follows that the generic fibres of ${}_q\hat{B}/{}_q\hat{A}'$ and, thus, of Q are constant. But then Q is constant by Corollary 6 and, being connected, it must be trivial. So $\hat{A}'$ is isomorphic to $\hat{B}$ and, consequently, B^0 is smooth which remained to be shown. □

In the remainder of this section, we want to discuss the defect of exactness of Néron models by looking at some special examples.

Example 7. Let R be a complete discrete valuation ring with normalized valuation v. Let q be a non-zero element of R with $v(q) > 0$ and consider the Tate elliptic curves $E_K = \mathbb{G}_{m,K}/q^{\mathbb{Z}}$ and $E'_K = \mathbb{G}_{m,K}/(q^l)^{\mathbb{Z}}$ where l is a positive integer not divisible by char K. Since the l-multiplication on E_K factors through E'_K, it gives rise to an exact sequence

$$0 \longrightarrow G_K \longrightarrow E_K \longrightarrow E'_K \longrightarrow 0\,,$$

where G_K is a finite group scheme of order l, contained in the kernel of the l-multiplication on E_K; the latter is of order l^2. Let

$$0 \longrightarrow G \longrightarrow E \longrightarrow E' \longrightarrow 0$$

be the associated sequence of Néron models. We want to show that there can be a defect of exactness at G, at E, or at E', depending on l and on the residue characteristic of R.

Defect of exactness at G. Assume that R is of mixed characteristic, that $l = p =$ char k, and that all p-torsion points of E_K are rational over K. The latter condition implies that the ramification index e is at least $p - 1$; cf. Serre [4], Chap. IV, §4, Prop. 17. Then $G_K \simeq (\mathbb{Z}/p\mathbb{Z})_K$ and $G \simeq (\mathbb{Z}/p\mathbb{Z})_R$. Furthermore, the kernel of $E \longrightarrow E'$ is the group $\boldsymbol{\mu}_{p,R}$ of p-th roots of unity, and the morphism from G into the kernel of $E \longrightarrow E'$ coincides with a morphism $(\mathbb{Z}/p\mathbb{Z})_R \longrightarrow \boldsymbol{\mu}_{p,R}$ sending 1 to a primitive p-th root of unity of R. However, the latter is not a monomorphism since $p =$ char k. In particular, $G \longrightarrow E$ is not a monomorphism.

Defect of exactness at E. Keeping the situation we have developed above, we see that G cannot be mapped surjectively onto the kernel of $E \longrightarrow E'$ since the morphism $(\mathbb{Z}/p\mathbb{Z})_R \longrightarrow \boldsymbol{\mu}_{p,R}$ is not surjective.

Defect of exactness at E'. The group of connected components of the special fibre of E has order $v(q)$ whereas that of E' has order $l \cdot v(q)$. So, without restrictions on the residue characteristic of R, the morphism $E \longrightarrow E'$ cannot be surjective for arbitrary $l > 1$. □

Next we want to show that the assertion of Theorem 4 can be false if we do not require condition (∗) of this theorem.

Example 8 (Serre). We will construct a morphism $v: A' \longrightarrow A$ of abelian schemes over R which is not a monomorphism, but which has the property that $v_K: A'_K \longrightarrow A_K$ is a closed immersion. The valuation ring R is supposed to have mixed characteristic. So if $p = \operatorname{char} k$, we have to assume $e := v(p) \geq p - 1$ by Theorem 4. In the following we assume that R contains all p-th roots of unity so that e is a multiple of $p - 1$ by Serre [4], Chap. IV, §4, Prop. 17. Now, similarly as in Example 7, consider a morphism $u: (\mathbb{Z}/p\mathbb{Z})_R \longrightarrow \boldsymbol{\mu}_p$ sending 1 to a primitive p-th root of unity. Let E be an elliptic curve over R (i.e., an abelian scheme with elliptic curves as fibres) which contains $\boldsymbol{\mu}_p$ as a subscheme. Then u extends to a morphism $u: (\mathbb{Z}/p\mathbb{Z})_R \longrightarrow E$, which is a closed immersion on generic fibres, but which is not a monomorphism. Let E' be a second elliptic curve over R which contains $(\mathbb{Z}/p\mathbb{Z})_R$ as a subscheme (for example, a Serre-Tate-lifting of an elliptic curve over k containing $(\mathbb{Z}/p\mathbb{Z})_k$ as a subscheme). Then consider the co-cartesian diagram

$$\begin{array}{ccc} (\mathbb{Z}/p\mathbb{Z})_R & \xrightarrow{\ u\ } & E \\ \big\downarrow & & \big\downarrow \\ E' & \xrightarrow{\ u'\ } & F' \end{array}$$

where F' is the quotient of $E \times E'$ with respect to the action of $(\mathbb{Z}/p\mathbb{Z})_R$. Since the action is free, F' is an abelian scheme over R. Furthermore, u'_K is a closed immersion, but u' itself cannot be a monomorphism since u is not a monomorphism. □

Finally, we want to show that the condition on the semi-abelian reduction of A' in Theorem 4 cannot be cancelled.

Example 9. Consider discrete valuation rings $R \subset R'$ where $R = \mathbb{Z}_{(p)}$ and $R' = \mathbb{Z}_{(p)}[a]$ with a being a primitive p-th root of unity; p is a prime different from 2. Let $u': E' \longrightarrow F'$ be a morphism of abelian R'-schemes of the type constructed in Example 8; i.e., such that u' is not a monomorphism, but such that it is a closed immersion on generic fibres. Then apply the technique of Weil restriction of R' over R to u' (cf. Section 7.6) and consider the induced morphism $u^1: E^1 \longrightarrow F^1$. It follows from 7.6/6 that E^1 and F^1 are Néron models of their generic fibres, and from 7.6/2 that u^1 is a closed immersion on generic fibres. We claim that u^1 is not a monomorphism. Indeed, the image of the map $\operatorname{Lie}(u'): \operatorname{Lie}(E') \longrightarrow \operatorname{Lie}(F')$ cannot be locally a direct factor in $\operatorname{Lie}(F')$. The same is true for the Weil restriction of $\operatorname{Lie}(u')$, and the latter is canonically identified with $\operatorname{Lie}(u^1): \operatorname{Lie}(E^1) \longrightarrow \operatorname{Lie}(F^1)$. So $u^1: E^1 \longrightarrow F^1$ cannot be a closed immersion and, thus, not a monomorphism. Since $v(p) = 1 < p - 1$, where v is the normalized valuation on R, we see from Theorem 4 that E^1 cannot have semi-abelian reduction. □

7.6 Weil Restriction

The main purpose of this section is to discuss a criterion for the existence of Weil restrictions and to study the behavior of Néron models with respect to Weil restrictions.

Let $h: S' \longrightarrow S$ be a morphism of schemes. Then, for any S'-scheme X', the contravariant functor

$$\mathfrak{R}_{S'/S}(X') : (\mathrm{Sch}/S)^0 \longrightarrow (\mathrm{Sets}) , \qquad T \longmapsto \mathrm{Hom}_{S'}(T \times_S S', X') ,$$

is defined on the category (Sch/S) of S-schemes. If it is representable, the corresponding S-scheme, again denoted by $\mathfrak{R}_{S'/S}(X')$, is called the *Weil restriction* of X' with respect to h. Thus, the latter is characterized by a functorial isomorphism

$$\mathrm{Hom}_S(T, \mathfrak{R}_{S'/S}(X')) \xrightarrow{\sim} \mathrm{Hom}_{S'}(T \times_S S', X')$$

of functors in T where T varies over all S-schemes. There are several elementary properties of the functor $\mathfrak{R}_{S'/S}(X')$ and, hence, of Weil restrictions, which follow immediately from the definition. We will derive some of them once we have mentioned the adjunction formula in Lemma 1 below.

Imposing an appropriate condition on h such as being finite and locally free (which we mean as a synonym for being finite, flat, and of finite presentation), the existence of the Weil restriction of the affine n-space $\mathbb{A}^n_{S'}$ is trivial (cf. the beginning of the proof of Theorem 4). Then, in order to treat more general schemes, it is necessary to study the behavior of Weil restrictions with respect to open or closed immersions. In order not to worry about the representability of the functor $\mathfrak{R}_{S'/S}(X')$ too much, we will work entirely within the context of functors from schemes to sets. In particular, we will make no difference between an S-scheme X and its associated functor $\mathrm{Hom}_S(\cdot, X)$; in the same way we will proceed with S'-schemes.

It is convenient to define the functor $\mathfrak{R}_{S'/S}(X')$ not only for S'-schemes X', but, more generally, for arbitrary contravariant functors from the category (Sch/S') of S'-schemes to the category of sets. So consider a functor

$$F' : (\mathrm{Sch}/S')^0 \longrightarrow (\mathrm{Sets}) .$$

Then its direct image with respect to $h: S' \longrightarrow S$ consists of the functor

$$h_*F' : (\mathrm{Sch}/S)^0 \longrightarrow (\mathrm{Sets}) , \qquad T \longmapsto F'(T \times_S S') .$$

Using 4.1/1, we see easily that the functor

$$(\mathrm{Sch}/S) \longrightarrow (\mathrm{Sch}/S') , \qquad T \longmapsto T \times_S S' ,$$

plays the role of an adjoint of h_*; namely, the so-called adjunction formula is valid.

Lemma 1. *For any S-scheme T and any functor $F' : (\mathrm{Sch}/S')^0 \longrightarrow (\mathrm{Sets})$, there is a canonical bijection*

$$\mathrm{Hom}_S(T, h_*F') \xrightarrow{\sim} \mathrm{Hom}_{S'}(T \times_S S', F')$$

which is functorial in T and in F'.

As an application of the above formula, we want to derive some elementary properties of Weil restrictions. Let X' be an S'-scheme. Then the identity on $\mathfrak{R}_{S'/S}(X')$ gives rise to a functorial morphism

$$\mathfrak{R}_{S'/S}(X') \times_S S' \longrightarrow X'$$

if $\mathfrak{R}_{S'/S}(X')$ exists as an S-scheme. Likewise, if X is an S-scheme, the identity on $X \times_S S'$ defines a functorial morphism

$$X \longrightarrow \mathfrak{R}_{S'/S}(X \times_S S') .$$

On the other hand, each functorial morphism $F' \longrightarrow G'$ between contravariant functors from (Sch/S') to (Sets) induces a functorial morphism $h_* F' \longrightarrow h_* G'$. Furthermore, h_* commutes with fibred products, and it follows that $h_* F'$ is a group functor if the same is true for F'. In particular, the Weil restriction of a group scheme is, if it exists as a scheme, a group scheme again. Also it is easy to see that the notion of Weil restriction is compatible with base change; i.e., if $T \longrightarrow S$ is a morphism of base change, and if we write $T' := S' \times_S T$, then, for any S'-scheme X', there is a canonical isomorphism

$$\mathfrak{R}_{T'/T}(X' \times_S T') \simeq \mathfrak{R}_{S'/S}(X') \times_S T$$

of functors on (Sch/T).

In the following we need the terminology of relative representability of functors; cf. Grothendieck [1], Sect. 3. Let

$$F, G : (\mathrm{Sch}/S)^0 \longrightarrow (\mathrm{Sets})$$

be contravariant functors, and let $u : F \longrightarrow G$ be a functorial morphism. Then, for each functorial morphism $T \longrightarrow G$, where T is an arbitrary S-scheme, the fibred product $F_T = F \times_G T$ may be viewed as a functor from $(\mathrm{Sch}/T)^0$ to (Sets). One says that F is *relatively representable* over G via u if, for each $T \longrightarrow G$, the projection $F_T \longrightarrow T$ is a morphism in (Sch/S); i.e., if each F_T is representable by a T-scheme. Many notions on morphisms between schemes can easily be adapted to the context of relative representability. For example, u is called an open immersion, or a closed immersion, or a morphism of finite type, etc., if the corresponding property is true for each morphism of schemes $u_T : F_T \longrightarrow T$, obtained from $u : F \longrightarrow G$ by the "base change" $T \longrightarrow G$.

Proposition 2. *Let $u' : F' \longrightarrow G'$ be a morphism between functors from* $(\mathrm{Sch}/S')^0$ *to* (Sets).

(i) *Assume that u' is an open immersion and that $h : S' \longrightarrow S$ is proper. Then the associated morphism $h_*(u') : h_* F' \longrightarrow h_* G'$ is an open immersion.*

(ii) *Assume that u' is a closed immersion and that $h : S' \longrightarrow S$ is finite and locally free or, more generally, proper, flat, and of finite presentation. Then $h_*(u') : h_* F' \longrightarrow h_* G'$ is a closed immersion.*

Proof. Let us write $F = h_* F'$ and $G = h_* G'$, and let $T \longrightarrow G$ be a morphism, where T is an arbitrary S-scheme. Setting $T' := T \times_S S'$, we claim that $T \longrightarrow G$ factors canonically through $h_* T'$. Indeed, we have a canonical morphism $T \longrightarrow h_* T'$.

Furthermore, $T \longrightarrow G$ corresponds to a morphism $T' \longrightarrow G'$ and, hence, to a morphism $h_* T' \longrightarrow h_* G' = G$. That the composition with $T \longrightarrow h_* T'$ yields $T \longrightarrow G$ is easily verified with the help of 4.1/1. Consequently, we can view F_T as being obtained from $F_{h_* T'}$ by means of the base change $T \longrightarrow h_* T'$, a fact to be used below.

Furthermore, since h_* commutes with fibred products, there are isomorphisms

$$h_* F'_{T'} \simeq F \times_G h_* T' \simeq F_{h_* T'} ,$$

and we can look at the canonical commutative diagram

$$\begin{array}{ccc} F'_{T'} & \longrightarrow & T' \\ & & \downarrow \\ F_T & \longrightarrow & T \\ \downarrow & & \downarrow \\ F_{h_* T'} & \longrightarrow & h_* T' \end{array} .$$

In order to prove assertion (i), it has to be shown that the morphism in the middle row, which is obtained from the one in the lower row by the base change $T \longrightarrow h_* T'$, is an open immersion of schemes. We know already that the upper row is an open immersion of schemes; let U' be the image of $F'_{T'}$ in T', and set $V' := T' - U'$. Then V' is closed in T' and, since $T' \longrightarrow T$ is proper, its image V in T is closed again. Set $U := T - V$. Interpreting F_T as the fibred product of $F_{h_* T'}$ and T over $h_* T'$, we have

$$F_T = \mathrm{Hom}_{S'}(\cdot \times_S S', U') \times_{\mathrm{Hom}_{S'}(\cdot \times_S S', T')} \mathrm{Hom}_S(\cdot, T) .$$

Thus, if Z is an arbitrary S-scheme, $F_T(Z)$ consists of all S-morphisms $Z \longrightarrow T$ where $Z \times_S S' \longrightarrow T'$ factors through U'; i.e., of those S-morphisms $Z \longrightarrow T$ which factor through U. Hence F_T is represented by the open subscheme U of T and assertion (i) follows.

Next, let us verify assertion (ii) for the case where h is finite and locally free. Similarly as before, let V' be the closed subscheme of T' which is given by the closed immersion $F'_{T'} \longrightarrow T'$. Then we have to find a closed subscheme V of T such that, given any S-morphism $Z \longrightarrow T$, it factors through V if and only if $Z \times_S S' \longrightarrow T'$ factors through V'. The problem is local on S, T, and Z, so we may assume that all three schemes are affine, say with rings of global sections R, A, and C. Let $R \longrightarrow R'$ be the homomorphism between rings of global sections on S and S'. We may assume R' is a free R-module of rank n. Let $e_1, \ldots, e_n$ be a basis of R' over R; then these elements give rise to a basis of $A \otimes_R R'$ over R. Furthermore, let $\mathfrak{a}' \subset A \otimes_R R'$ be the ideal corresponding to V', and fix generators a'_i, $i \in I$, of $\mathfrak{a}'$. There are equations

$$a'_i = \sum_{j=1}^{n} c_{ij} e_j , \qquad i \in I ,$$

with coefficients $c_{ij} \in A$. These coefficients generate an ideal $\mathfrak{a} \subset A$, and we claim that the associated closed subscheme $V \subset T$ is as required. Namely, consider the homomorphism $\sigma : A \longrightarrow C$ which is associated to $Z \longrightarrow T$ as well as the homomorphism $\sigma' : A \otimes_R R' \longrightarrow C \otimes_R R'$ associated to $Z \times_S S' \longrightarrow T'$. Since

$$\ker \sigma' = (\ker \sigma) \otimes_R R' = \bigoplus_{i=1}^{n} (\ker \sigma) \cdot e_i ,$$

we see that $\mathfrak{a}' \subset \ker \sigma'$ if and only if $\mathfrak{a} \subset \ker \sigma$, i.e., that Z' is mapped into V' if and only if Z is mapped into V. So it follows that V represents the functor F_T.

If, more generally, h is proper, flat, and of finite presentation, one uses techniques from the construction of Hilbert schemes as in [FGA], n°221, Sect. 3, in order to show that there is a largest closed subscheme V of T such that an S-morphism $Z \longrightarrow T$ factors through V if and only if, after base change with $h : S' \longrightarrow S$, it factors through $V' \subset T'$. □

A functor $F : (\mathrm{Sch}/S)^0 \longrightarrow (\mathrm{Sets})$ is called a *sheaf with respect to the Zariski topology* (see 8.1) if, for each S-scheme T and for each covering $\{T_i\}$ of T, the sequence

$$\mathrm{Hom}_S(T, F) \longrightarrow \prod_i \mathrm{Hom}_S(T_i, F) \rightrightarrows \prod_{i,j} \mathrm{Hom}_S(T_i \cap T_j, F)$$

is exact. Of course, if F is a scheme, F is a sheaf in this sense.

Proposition 3. *If* $F' : (\mathrm{Sch}/S)^0 \longrightarrow (\mathrm{Sets})$ *is a sheaf with respect to the Zariski topology, then the same is true for* $F := h_* F'$.

Proof. Since, for any S-scheme T, we have

$$\mathrm{Hom}_S(T, F) = \mathrm{Hom}_{S'}(T \times_S S', F') ,$$

the assertion is obvious. □

We want to apply the above results to the case where F' consists of an S'-scheme X', and give a criterion of Grothendieck for the representability of $X := h_* X' = \mathfrak{R}_{S'/S}(X')$ by an S-scheme. Then, if X is representable, it defines the Weil restriction of X'.

Theorem 4. *Let* $h : S' \longrightarrow S$ *be a morphism of schemes which is finite and locally free, and let* X' *be an* S'*-scheme. Assume that, for each* $s \in S$ *and each finite set of points* $P \subset X' \otimes_S k(s)$, *there is an affine open subscheme* U' *of* X' *containing* P. *Then* $h_* X' = \mathfrak{R}_{S'/S}(X')$ *is representable by an* S*-scheme* X *and, thus, the Weil restriction of* X' *exists.*

Proof. We may assume that S and, hence, S' are affine, say with rings of global sections R and R' and that R' is a free R-module, say with generators $e_1, \ldots, e_n$. Let us first show that $h_* X'$ is representable if X' is affine. So assume X' is affine and view it as a closed subscheme of some scheme $\mathrm{Spec}\, R'[t]$, where t is a (finite or infinite) system of indeterminates. Applying Proposition 2, it is only necessary to consider the case where $X' = \mathrm{Spec}\, R'[t]$. Consider n copies of the system t and write $t_1, \ldots, t_n$ for these systems. Then, for any R-algebra A, there is a bijection

$$\mathrm{Hom}_{R'}(R'[t], A \otimes_R R') \longrightarrow \mathrm{Hom}_R(R[t_1, \ldots, t_n], A) ,$$

which is functorial in A. In order to define this map, consider an R'-homomorphism $\sigma' : R'[t] \longrightarrow A \otimes_R R'$. The latter is determined by the image $\sigma'(t)$ of t in $A \otimes_R R'$. Using the direct sum decomposition

$$A \otimes_R R' = \bigoplus_{i=1}^{n} (A \otimes_R R)e_i ,$$

we can write

$$\sigma'(t) = \sum_{i=1}^{n} \sigma(t_i) \otimes e_i$$

with systems $\sigma(t_1), \ldots, \sigma(t_n)$ of elements in A, and we can think of σ as of a homomorphism $\sigma : R[t_1, \ldots, t_n] \longrightarrow A$. Then it is easily seen that $\sigma' \longmapsto \sigma$ defines the desired bijection. Consequently, in this case the functor $h_* X'$ is representable by the S-scheme $\operatorname{Spec} R[t_1, \ldots, t_n]$, and it follows that the Weil restriction $\mathfrak{R}_{S'/S}(X')$ exists.

Next, let us consider the case where X' is not necessarily affine. Let $\{U_i'\}_{i \in I}$ be the system of all affine open subschemes of X'. Then, by what we have just seen, each $h_* U_i'$ is representable by an (affine) scheme U_i, and the open immersion $U_i' \hookrightarrow X'$ gives rise to a morphism $U_i \longrightarrow h_* X'$ which is an open immersion by Proposition 2. Viewing the U_i' as open subschemes of X', we have canonical gluing data for them, and these data give rise to gluing data for the U_i. So, gluing the U_i, we obtain an S-scheme Y. Since X' is a sheaf with respect to the Zariski topology, the same is true for $h_* X'$ (see Proposition 3) and there is a functorial morphism $Y \longrightarrow h_* X'$. The latter is an open immersion by Proposition 2.

In order to show that $Y \longrightarrow h_* X'$ is an equivalence of functors, it is enough to show that each functorial morphism $a : T \longrightarrow h_* X'$, where T is an arbitrary S-scheme, factors uniquely through Y or, what amounts to the same, that the latter is the case locally in a neighborhood of each point $z \in T$. Let (z_j) be the finite family of points in $T \times_S S'$ lying over z. Furthermore, let $a' : T \times_S S' \longrightarrow X'$ be the morphism corresponding to a, and set $x_j = a'(z_j)$. By our assumption, there is an affine open subscheme $U' \subset X'$ containing all points x_j. We know already that $h_* U'$ is representable by an S-scheme U and that the canonical morphism $U \longrightarrow h_* X'$ is an open immersion; the latter factors through Y by the definition of Y. Replacing T by a suitable open subscheme containing z, we may assume that $a' : T' \longrightarrow X'$ factors through U'. Then $a : T \longrightarrow h_* X'$ factors through U and, hence, through Y. The factorization is unique due to the fact that $Y \longrightarrow h_* X'$ is an open immersion. □

We want to mention some general properties of Weil restrictions, assuming that we are in the situation of Theorem 4.

Proposition 5. *Let $S' \longrightarrow S$ be a morphism of schemes which is finite and locally free, and let X' be an S'-scheme. Assume that the Weil restriction $X = \mathfrak{R}_{S'/S}(X')$ exists as an S-scheme, and consider the following properties for relative schemes*:

(a) *quasi-compact.*

(b) *separated,*

(c) *locally of finite type,*
(d) *locally of finite presentation,*
(e) *finite presentation,*
(f) *proper,*
(g) *flat,*
(h) *smooth.*

Then the above properties carry over from X' to X under the following additional assumptions:
property (a) *if S is locally noetherian or if $S' \longrightarrow S$ is étale,*
properties (b), (c), (d), (e), *and* (h) *without any further assumptions, and*
properties (f) *and* (g) *if $S' \longrightarrow S$ is étale.*

Proof. Let us begin with properties which carry over from X' to X without any additional assumptions, say with property (b). Since the Weil restriction of the diagonal morphism $X' \longrightarrow X' \times_{S'} X'$ yields the diagonal morphism $X \longrightarrow X \times_S X$ and since the Weil restriction respects closed immersions by Proposition 2, we see that X is separated if X' is separated.

Next, let us look at properties (c) and (d). That they carry over from X' to X follows from the construction of Weil restrictions in the affine case. Namely, if X' is a closed subscheme of the affine n-space $\mathbb{A}^n_{S'}$, and if $S' \longrightarrow S$ is a finite and free morphism of affine schemes, say of degree d, then it follows from Proposition 2 that X is a closed subscheme of $\mathfrak{R}_{S'/S}(\mathbb{A}^n_{S'}) \simeq \mathbb{A}^m_S$ where $m = nd$. So X is locally of finite type if the same is true for X'. Furthermore, the proof of Proposition 2 shows that the ideal defining X as a closed subscheme of $\mathbb{A}^m_S$ is finitely generated if the same is true for X' as a closed subscheme of $\mathbb{A}^n_{S'}$. So it follows that X is locally of finite presentation if the same is true for X'. The latter result can also be obtained by functorial arguments using the characterization [EGA IV_3], 8.14.2, of morphisms which are locally of finite presentation.

If X' satisfies property (e), we can view it as an S-scheme of finite presentation. Using a limit argument, we may assume that S is noetherian. Then X is locally of finite presentation, since property (d) carries over from X' to X, and quasi-compact over S since, as we will see below, also property (a) carries over from X' to X if the base S is noetherian. But then X is of finite presentation over S.

Finally, the characterization of smoothness in terms of the lifting property 2.2/6 shows by functorial reasons that X satisfies property (h) if X' does.

Now assume that $S' \longrightarrow S$ is étale and finite. In order to show that X satisfies properties (a), (f), or (g) if X' does, we may work locally on S, say in a neighborhood of a point $s \in S$. Furthermore, Weil restrictions commute with base change on S. So we may replace S by an étale neighborhood of s. But then, since locally up to étale base change étale morphisms are open immersions, see 2.3/8, we are reduced to the case where S' consists of a finite disjoint sum $\coprod S_i$ of copies S_i of S and where $S' \longrightarrow S$ is the canonical map. Then, in terms of fibred products over S,

$$\mathfrak{R}_{S'/S}(X') \simeq \prod_i \mathfrak{R}_{S_i/S}(X' \times_{S'} S_i) \simeq \prod_i X' \times_{S'} S_i ,$$

and it is trivial that X satisfies properties (a), (f), or (g) if X' does.

It remains to show that, under appropriate conditions, property (a) carries over from X' to X, a fact which is already known if $S' \longrightarrow S$ is étale. We claim that it is also true for radicial morphisms. To verify this, it is enough to prove that, for S' radicial over S, the Weil restriction $\mathfrak{R}_{S'/S}$ transforms any affine open covering (U_j') of X' into an affine open covering $(\mathfrak{R}_{S'/S}(U_j'))$ of X. Looking at fibres over S, we may assume that S is the spectrum of a field K. Then S' consists of a finite-dimensional local K-algebra K' whose residue field is purely inseparable over K. Now let (U_j') be an affine open covering of X'. To see that the sets $\mathfrak{R}_{K'/K}(U_j')$ really cover X, consider a geometric point $\operatorname{Spec} E \longrightarrow X$ where E is a field over K. Then the scheme $\operatorname{Spec}(E \otimes_K K')$ consists of a single point and the corresponding morphism $\operatorname{Spec}(E \otimes_K K') \longrightarrow X'$ must factor through a member of the open covering (U_j') of X'. Consequently, $\operatorname{Spec} E \longrightarrow X$ factors through a member of the family $(\mathfrak{R}_{K'/K}(U_j'))$ which justifies our claim.

Now assume that the base S is locally noetherian. In order to show that X satisfies property (a) if X' does, we may assume that S is noetherian. We will conclude by using a noetherian argument and a stratification of S. Let η be a generic point of S. Restricting ourselves to a neighborhood of η, we can assume that S is irreducible and, since quasi-compactness can be tested after killing nilpotent elements of structure sheaves, that S is reduced. Furthermore, we can assume that S and S' are affine, say $S = \operatorname{Spec} R$ and $S' = \operatorname{Spec} R'$. The fibre S'_η is the spectrum of the finite-dimensional K-algebra $K' = R' \otimes_R K$ where $K = k(\eta) = Q(R)$. Let L be the maximal étale K-subalgebra between K and K'. It is obtained as follows. Decompose K' into a finite direct product $\prod K_i'$ of local K-algebras K_i' and, for each i, choose a maximal separable extension field L_i between K and K_i'. Then the residue field of K_i' is purely inseparable over L_i and we have $L = \prod L_i$. Set $T := \operatorname{Spec}(R' \cap L)$ so that $S' \longrightarrow S$ factors through T. Over the generic point η, the finite morphism $T \longrightarrow S$ is étale. Thus, using the openness of the étale locus, we know that $T \longrightarrow S$ is étale over an open neighborhood of η. Restricting to this neighborhood, we may assume that $T \longrightarrow S$ is étale everywhere. Furthermore, for each $a \in K'$, there is an integer n such that a^n belongs to L. This property carries over to the fibres of $S' \longrightarrow T$ so that the latter morphism is radicial. Since $X = \mathfrak{R}_{T/S}(\mathfrak{R}_{S'/T}(X'))$, we see by what we have proved above for étale and for radicial morphisms that, working over a neighborhood of η, the scheme X is quasi-compact if X' is.

The argument just given shows that the original morphism $X \longrightarrow S$ is quasi-compact over a dense open subset of S if X' is quasi-compact over S'. Looking at the complement S_1 of this set and viewing it as a scheme with respect to the canonical reduced structure, we can perform the base change $S_1 \longrightarrow S$. It follows in the same way that $X \times_S S_1 \longrightarrow S_1$ is quasi-compact over a dense open subset of S_1. Continuing this way, the procedure will stop after finitely many steps due to the noetherian hypothesis. Thus, finally, it is seen that X is quasi-compact over S. □

We want to add, again in the situation of Theorem 4, that, for any S-scheme X, the canonical morphism $X \longrightarrow \mathfrak{R}_{S'/S}(X \times_S S')$ is a closed immersion, provided X and, thus, $\mathfrak{R}_{S'/S}(X \times_S S')$ are separated. This follows by means of descent from the fact that the composition of canonical morphisms

$$X \times_S S' \longrightarrow \mathfrak{R}_{S'/S}(X \times_S S') \times_S S' \longrightarrow X \times_S S'$$

is the identity on $X \times_S S'$.

Finally, let us state how Néron models behave with respect to Weil restrictions.

Proposition 6. *Let $S' \longrightarrow S$ be a finite and flat morphism of Dedekind schemes. Let* Spec K *and* Spec K' *denote the schemes of generic points of S and S'. Furthermore, consider a torsor X' (under a smooth S'-group scheme G') which is a Néron model of the scheme of generic fibres $X' \times_{S'}$* Spec K'. *Then the Weil restriction $X = \mathfrak{R}_{S'/S}(X')$ exists as an S-scheme and is a Néron model of the scheme of generic fibres $X \times_S$* Spec K.

Proof. Using the quasi-projectivity of torsors over Dedekind schemes (cf. 6.4/1), the existence of $X = \mathfrak{R}_{S'/S}(X')$ as an S-scheme follows from Theorem 4. Furthermore, it follows from Proposition 5 that X is separated, of finite type, and smooth. Finally, that X satisfies the Néron mapping property is a formal consequence of the definition of Weil restrictions, namely of the equation

$$\mathrm{Hom}_S(Z, X) = \mathrm{Hom}_{S'}(Z \times_S S', X') .$$

□

Chapter 8. The Picard Functor

Following Grothendieck's treatment [FGA], we introduce the relative Picard functor $\mathrm{Pic}_{X/S}$ and treat the notion of the rigidified relative Picard functor. The main purpose of this chapter is the presentation of various results on the representability of $\mathrm{Pic}_{X/S}$. We explain Grothendieck's theorem on the representability of $\mathrm{Pic}_{X/S}$ by a scheme and point out improvements due to Mumford [2] as well as those due to Altman and Kleiman [1]. In Section 8.3, we discuss the main steps of M. Artin's approach [5] to the representability of $\mathrm{Pic}_{X/S}$ by an algebraic space; for details, the reader is referred to his paper. At the end of the chapter, there is a collection of some results on smoothness as well as on finiteness properties of $\mathrm{Pic}_{X/S}$, as can be found in [SGA 6].

8.1 Basics on the Relative Picard Functor

For any scheme X, we denote by $\mathrm{Pic}(X) = H^1(X, \mathcal{O}_X^*)$ the group of isomorphism classes of invertible sheaves on X. It is called the absolute *Picard group* of X. Fixing a base scheme S and an S-scheme X, we can consider the contravariant functor

$$P_{X/S} : (\mathrm{Sch}/S)^0 \longrightarrow (\mathrm{Sets})\ , \qquad T \longmapsto \mathrm{Pic}(X \times_S T)\ ,$$

from the category (Sch/S) of S-schemes to the category of sets, which factors through the category of commutative groups. Using the procedure of sheafification, we want to associate a functor with $P_{X/S}$ which, under certain conditions, is representable; namely, the so-called *relative Picard functor*.

To begin with, let us discuss a necessary condition for a functor $F : (\mathrm{Sch}/S)^0 \longrightarrow (\mathrm{Sets})$ to be representable. Let $\mathfrak{M}$ be a class of morphisms in (Sch/S) which is stable under composition and under fibred products and which contains all isomorphisms. Then F is called a *sheaf with respect to* $\mathfrak{M}$ or an $\mathfrak{M}$*-sheaf* if, for any family of S-schemes $(T_i)_{i \in I}$, the canonical morphism

$$F(\coprod T_i) \longrightarrow \prod F(T_i)$$

is an isomorphism and if, for all morphisms $T' \longrightarrow T$ in $\mathfrak{M}$, the sequence

$$F(T) \longrightarrow F(T') \rightrightarrows F(T'')$$

is exact (where $T'' = T' \times_T T'$ and where the double arrows on the right are induced by the two projections from T'' onto T'). For example, we can consider the class $\mathfrak{M} = \mathfrak{M}_{\mathrm{Zar}}$ of all morphisms in (Sch/S) of type $\coprod T_i \longrightarrow T$, where the maps $T_i \longrightarrow T$

are open immersions and where $\{T_i\}_{i\in I}$ is an open covering of T. If F is a sheaf with respect to $\mathfrak{M}_{\mathrm{Zar}}$, it is said that F is a *sheaf with respect to the Zariski topology.* To give an equivalent condition, one can require that, for all open coverings $\{T_i\}_{i\in I}$ of T, the canonical sequence

$$F(T) \longrightarrow \prod_i F(T_i) \rightrightarrows \prod_{i,j} F(T_i \times_T T_j)$$

is exact.

There are further topologies of more general type; cf. [SGA 3_I], Exp. IV, 6.3.1. We mention the fpqc-topology, the fppf-topology, and the étale topology. If top is any of the abbreviations

fpqc (= faithfully flat and quasi-compact),

fppf (= faithfully flat and of finite presentation), or

ét (= étale surjective),

we write $\mathfrak{M}_{\mathrm{top}}$ for the class of all morphisms in (Sch/S) which are of type top and say that a functor $F : (\mathrm{Sch}/S)^0 \longrightarrow (\mathrm{Sets})$ is a sheaf with respect to the top-topology (or, simply, with respect to top), if it is a sheaf with respect to both $\mathfrak{M}_{\mathrm{Zar}}$ *and* $\mathfrak{M}_{\mathrm{top}}$.

Proposition 1. *Let F be a representable contravariant functor from* (Sch/S) *to* (Sets). *Then F is a sheaf with respect to* fpqc *and, hence, with respect to* fppf, ét, *and* Zar.

Proof. If F is represented by an S-scheme X, we have $F(T) = \mathrm{Hom}_S(T, X)$. Since morphisms to X can be defined locally, it follows for any open covering $\{T_i\}$ of T that the canonical sequence

$$\mathrm{Hom}_S(T, X) \longrightarrow \prod_i \mathrm{Hom}_S(T_i, X) \rightrightarrows \prod_{i,j} \mathrm{Hom}_S(T_i \times_T T_j, X)$$

is exact. So F is a sheaf with respect to the Zariski topology.

Furthermore, for any S-morphism $T' \longrightarrow T$ which is fpqc, the canonical sequence

$$\mathrm{Hom}_S(T, X) \longrightarrow \mathrm{Hom}_S(T', X) \rightrightarrows \mathrm{Hom}_S(T'', X)$$

is exact; namely, it is isomorphic to the sequence

$$\mathrm{Hom}_T(T, X_T) \longrightarrow \mathrm{Hom}_{T'}(T', X_{T'}) \rightrightarrows \mathrm{Hom}_{T''}(T'', X_{T''})$$

which, by descent theory, is exact, as shown in the proof of 6.1/6. Thus F is a sheaf with respect to fpqc. □

Returning to the functor

$$P_{X/S} : (\mathrm{Sch}/S)^0 \longrightarrow (\mathrm{Sets}) , \qquad T \longmapsto \mathrm{Pic}(X \times_S T) ,$$

it is easily seen that, in general, $P_{X/S}$ is not a sheaf, even with respect to the Zariski topology. As a consequence, we cannot expect its representability. Indeed, if $P_{X/S}$ were a sheaf with respect to the Zariski topology, a line bundle on $X \times_S T$ would be trivial as soon as it trivializes over (the pull-back of) an open covering of T.

However, this is not the case. So if we want to deal with a functor from which representability can be expected, we have to sheafify $P_{X/S}$; this can be done by using standard methods from sheaf theory.

In order to explain the procedure of sheafification, let us, again, consider a functor $F : (\mathrm{Sch}/S)^0 \longrightarrow (\mathrm{Sets})$ and a class $\mathfrak{M}$ of morphisms in (Sch/S) which is stable under composition and under fibred products and which contains all isomorphisms. To give a sheafification of F (within the context of sheaves with respect to $\mathfrak{M}$) means to construct a morphism $F \longrightarrow F^\dagger$ into a sheaf $F^\dagger$ such that each morphism from F into an arbitrary sheaf G (always with respect to $\mathfrak{M}$) admits a unique factorization through $F^\dagger$. The construction of $F^\dagger$ is straightforward. Let $T' \longrightarrow T$ be a morphism in $\mathfrak{M}$ and denote by $\bar{H}^0(T'/T, F)$ the subset of $F(T')$ consisting of all elements ξ which are characterized by the following property: if ξ_1 and ξ_2 are the "pull-backs" of ξ with respect to the two projections from $T'' = T' \times_T T'$ onto T', there is a morphism $\tilde{T} \longrightarrow T''$ in $\mathfrak{M}$ such that the images of ξ_1 and ξ_2 with respect to $F(T'') \longrightarrow F(\tilde{T})$ coincide in $F(\tilde{T})$. If T' varies over (Sch/S), the sets $\bar{H}^0(T'/T, F)$ form an inductive system. Provided $\mathfrak{M}$ is not "too big", the direct limit of this system exists, and we can set

$$F^\dagger(T) := \varinjlim \bar{H}^0(T'/T, F)\,.$$

It is verified without difficulties that $F^\dagger$ is a sheaf with respect to $\mathfrak{M}$ and that the canonical morphism $F \longrightarrow F^\dagger$ defines $F^\dagger$ as a sheafification of F.

The direct limits which have been used to define the sheaf $F^\dagger$ exist if we take for $\mathfrak{M}$ any of the classes $\mathfrak{M}_{\mathrm{Zar}}$, $\mathfrak{M}_{\mathrm{\acute{e}t}}$, or $\mathfrak{M}_{\mathrm{fppf}}$, whereas in the case $\mathfrak{M} = \mathfrak{M}_{\mathrm{fpqc}}$ some precautionary measures, like working in a fixed universe, are necessary. However, since the class $\mathfrak{M}_{\mathrm{fpqc}}$ is quite big, it may happen that sheafifications with respect to $\mathfrak{M}_{\mathrm{fpqc}}$ depend on the choice of the universe. It is for this reason that, when working with sheafifications, we will generally use the class $\mathfrak{M}_{\mathrm{fppf}}$ instead of $\mathfrak{M}_{\mathrm{fpqc}}$.

Now, in order to construct a sheafification of the functor

$$P_{X/S} : (\mathrm{Sch}/S)^0 \longrightarrow (\mathrm{Sets})\,, \qquad T \longmapsto \mathrm{Pic}(X \times_S T)\,,$$

say with respect to the fppf-topology, one first sheafifies $P_{X/S}$ with respect to $\mathfrak{M}_{\mathrm{fppf}}$. The resulting sheaf P_1 might not be a sheaf with respect to $\mathfrak{M}_{\mathrm{Zar}}$ since morphisms in $\mathfrak{M}_{\mathrm{Zar}}$ are not necessarily quasi-compact and, thus, not necessarily fppf. However, if T is affine, any morphism $\coprod T_i \longrightarrow T$ in $\mathfrak{M}_{\mathrm{Zar}}$ which corresponds to a finite open covering $\{T_i\}$ of T by basic open subschemes $T_i \subset T$ is fppf. Hence P_1 is already an fppf-sheaf on affine schemes. Therefore we can sheafify P_1 with respect to $\mathfrak{M}_{\mathrm{Zar}}$ without destroying sheaf properties with respect to $\mathfrak{M}_{\mathrm{fppf}}$ on affine schemes. It follows that the resulting functor is a sheaf with respect to the fppf-topology; it is the fppf-*sheaf associated to* $P_{X/S}$. Since $P_{X/S}$ is a group functor, the associated fppf-sheaf can be viewed as a group functor, too. In the same way, one can proceed with any other of the topologies introduced above.

Definition 2. *The* fppf-*sheaf associated to the functor*

$$P_{X/S} : (\mathrm{Sch}/S)^0 \longrightarrow (\mathrm{Sets})\,, \qquad T \longmapsto \mathrm{Pic}(X \times_S T)\,,$$

is called the relative Picard functor of X over S; it is denoted by $\operatorname{Pic}_{X/S}$. *For any S-scheme T, we call* $\operatorname{Pic}_{X/S}(T)$ *the relative Picard group of $X \times_S T$ over T.*

Using the structural morphism $f: X \longrightarrow S$ as well as the notion of higher direct images of f, we can define the relative Picard functor also by the formula

$$\operatorname{Pic}_{X/S}(T) = H^0(T, R^1 f_*(\mathbb{G}_m))$$

which has to be read with respect to the fppf-topology; note that $\mathbb{G}_m$ is the sheaf which associates to each scheme Z the group of units $\Gamma(Z, \mathcal{O}_Z^*)$. We will see below that the restriction to the fppf-topology in place of the fpqc-topology is not too serious since we are mainly interested in the case where $f: X \longrightarrow S$ is proper and fppf.

Sometimes it is useful to have an explicit description of elements of relative Picard groups. So consider an element $\xi \in \operatorname{Pic}_{X/S}(S)$ and assume for simplicity that S is affine or, more generally, quasi-compact. Otherwise one has to work locally with respect to an open affine covering of S. Then, in the quasi-compact case, ξ is represented by a line bundle $\xi' \in \operatorname{Pic}(X \times_S S')$ where S' is fppf over S. Furthermore, there must be an fppf-morphism $\tilde{S} \longrightarrow S'' = S' \times_S S'$ such that the pull-back of ξ' with respect to $\tilde{S} \longrightarrow S'' \longrightarrow S'$ is the same for both projections from S'' to S'. Conversely, each $\xi' \in \operatorname{Pic}(X \times_S S')$ satisfying the latter condition gives rise to an element $\xi \in \operatorname{Pic}_{X/S}(S)$. Two such elements $\xi_i' \in \operatorname{Pic}(X \times_S S_i')$, $i = 1, 2$, with S_i' fppf over S represent the same element $\xi \in \operatorname{Pic}_{X/S}(S)$ if and only if there exists an fppf-morphism $\tilde{S} \longrightarrow S_1' \times_S S_2'$ such that, on $\tilde{S}$, the pull-back of ξ_1' coincides with the pull-back of ξ_2'. Also it should be noted that, due to the sheaf property of $\operatorname{Pic}_{X/S}$, an element $\xi \in \operatorname{Pic}_{X/S}(S)$ is trivial if it is induced by the pull-back to X of a line bundle on S. The converse is not true, in general.

Proposition 3. *Assume that $f: X \longrightarrow S$ is proper and of finite presentation. Consider an element $\xi \in \operatorname{Pic}_{X/S}(S)$ which is induced by a line bundle $\mathcal{L}$ on X. Then ξ is trivial if and only if there is an open covering $\{S_i\}$ of S such that $\mathcal{L}$ trivializes over $X \times_S S_i$ for each i.*

Proof. The if-part of the assertion follows from the sheaf properties of $\operatorname{Pic}_{X/S}$. So it remains to justify the only-if-part. The direct image $f_*(\mathcal{O}_X)$ is a quasi-coherent $\mathcal{O}_S$-algebra. Assuming S to be affine and interpreting $f: X \longrightarrow S$ as a limit of morphisms of finite type between noetherian schemes, we can use the Stein factorization $X \xrightarrow{g} T \xrightarrow{h} S$ of f, where g satisfies $g_*(\mathcal{O}_X) = \mathcal{O}_T$ and where h, being a limit of finite morphisms, is integral. Furthermore, since the fibres of g are the connected components of the fibres of f, it follows that the fibres of h are set-theoretically finite. Now assume that $\mathcal{L}$ gives rise to the trivial element $\xi \in \operatorname{Pic}_{X/S}(S)$. We claim that the canonical homomorphism $g^*(g_*(\mathcal{L})) \longrightarrow \mathcal{L}$ is an isomorphism. Using descent, this fact can be tested after base change with an fppf-morphism. For example, we can assume that, after such a base change, $\mathcal{L}$ becomes trivial. Since the formation of $g_*(\mathcal{L})$ commutes with flat base change, the above isomorphism has only to be established for the trivial bundle $\mathcal{L}$. But then the claim follows from the fact that $g_*(\mathcal{O}_X) = \mathcal{O}_T$. So we see that $\mathcal{L}$ is the pull-back of the line bundle $g_*(\mathcal{L})$ on T. The latter is locally trivial over T. Since $h: T \longrightarrow S$ is integeral and, thus, a closed map, and since its fibres are set-theoretically finite, it follows that $g_*(\mathscr{L})$ is locally trivial also over S. Hence $\mathscr{L}$ is locally trivial over S. □

We assume in the following that $f: X \longrightarrow S$ is quasi-compact and quasi-separated. Then the Leray spectral sequence associated to f and $\mathbb{G}_m$ (see [SGA 4_{II}], Exp. V, § 3) gives the exact sequence

$$0 \longrightarrow H^1(S, f_*(\mathbb{G}_m)) \longrightarrow H^1(X, \mathbb{G}_m) \longrightarrow \mathrm{Pic}_{X/S}(S) \longrightarrow H^2(S, f_*(\mathbb{G}_m)) \longrightarrow H^2(X, \mathbb{G}_m)$$

where the cohomology groups are meant with respect to the fppf-topology. Since the descent with respect to fpqc-morphisms turns line bundles into line bundles, it follows that the group $H^1(X, \mathbb{G}_m)$ is the same for the fpqc-, the fppf-, the étale, and even for the Zariski topology. So we may use the Zariski topology and see $H^1(X, \mathbb{G}_m) = \mathrm{Pic}(X)$. Thus the obstruction of representing an element of $\mathrm{Pic}_{X/S}(S)$ by an element of $\mathrm{Pic}(X)$ is given by an element in $H^2(S, f_*(\mathbb{G}_m))$ which becomes zero in $H^2(X, \mathbb{G}_m)$. Just as in the case of $H^1(X, \mathbb{G}_m)$, one shows that $H^1(S, f_*(\mathbb{G}_m))$ is independent of the topologies mentioned above if $f_*(\mathcal{O}_X) = \mathcal{O}_S$ or, by means of the Stein factorization, if f is proper. In particular, we have $H^1(S, f_*(\mathbb{G}_m)) = \mathrm{Pic}(S)$ if $f_*(\mathcal{O}_X) = \mathcal{O}_S$.

In order to determine the cohomology group $H^2(X, \mathbb{G}_m)$, one can use the étale topology instead of the fppf-topology; cf. Grothendieck [3], pp. 171–183. The same is true for the cohomology group $H^2(S, f_*(\mathbb{G}_m))$ if $f_*(\mathcal{O}_X) = \mathcal{O}_S$ or, without this assumption, if f is proper. Namely, by means of the Stein factorization, it is possible to reduce to the case where $f_*(\mathcal{O}_X) = \mathcal{O}_S$. So, for example, if f is proper, the above exact sequence shows that the relative Picard functor $\mathrm{Pic}_{X/S}$ can be constructed by using the étale topology in place of the fppf-topology. In particular, the formula

$$\mathrm{Pic}_{X/S}(T) = H^0(T, R^1 f_*(\mathbb{G}_m))$$

remains valid if, on the right-hand side the fppf-topology is replaced by the étale topology.

The cohomology group $H^2(X, \mathbb{G}_m)$ is called the (cohomological) *Brauer group* of X. In particular, if we assume $f_*(\mathcal{O}_X) = \mathcal{O}_S$, the obstructions of representing elements in $\mathrm{Pic}_{X/S}(S)$ by line bundles on X are given by elements of the Brauer group $\mathrm{Br}(S)$ which become zero in the Brauer group $\mathrm{Br}(X)$. All obstructions of this type disappear if the map $H^2(S, \mathbb{G}_m) \longrightarrow H^2(X, \mathbb{G}_m)$ is injective; for example, if $f: X \longrightarrow S$ has a section or if the Brauer group $\mathrm{Br}(S)$ vanishes itself. For an affine scheme $S = \mathrm{Spec}\, R$, the group $Br(S)$ is zero in each of the following situations:

(a) R is a separably closed field.

(b) R is the field of fractions of a henselian discrete valuation ring with algebraically closed residue field; cf. Grothendieck [3], Thm. 1.1, or Milne [1], Chap. III, 2.22.

(c) R is a strictly henselian valuation ring; cf. Grothendieck [3], Prop. 2.1, or Milne [1], Chap. IV, 1.7 and 2.12.

The equation $f_*(\mathcal{O}_X) = \mathcal{O}_S$ is compatible with flat base change. We say that $f_*(\mathcal{O}_X) = \mathcal{O}_S$ holds *universally* if the equation is true after *any* base change over S. Using this terminology, we want to summarize the above considerations.

Proposition 4. *Let $f : X \longrightarrow S$ be quasi-compact and quasi-separated and assume that f satisfies $f_*(\mathcal{O}_X) = \mathcal{O}_S$ (resp. that $f_*(\mathcal{O}_X) = \mathcal{O}_S$ holds universally). Then, for each S-scheme T which is flat over S (resp. for each S-scheme T), the canonical sequence*

$$0 \longrightarrow \operatorname{Pic}(T) \longrightarrow \operatorname{Pic}(X \times_S T) \longrightarrow \operatorname{Pic}_{X/S}(T) \longrightarrow \operatorname{Br}(T) \longrightarrow \operatorname{Br}(X \times_S T)$$

is exact. If, in addition, f admits a section, the sequence

$$0 \longrightarrow \operatorname{Pic}(T) \longrightarrow \operatorname{Pic}(X \times_S T) \longrightarrow \operatorname{Pic}_{X/S}(T) \longrightarrow 0$$

is exact.

In particular, in the latter case, we can identify the relative Picard functor $\operatorname{Pic}_{X/S}$ in the usual way with the functor

$$(\mathrm{Sch}/S)^0 \longrightarrow (\mathrm{Sets}), \qquad T \longmapsto \operatorname{Pic}(X \times_S T)/\operatorname{Pic}(T).$$

If the existence of a global section is replaced by the condition that $f : X \longrightarrow S$ has local sections, one can still say that the formula

$$\operatorname{Pic}_{X/S}(T) = H^0(T, R^1 f_*(\mathbb{G}_m))$$

remains valid if one considers the Zariski topology on the right-hand side.

In order to see, in the above situation, that the relative Picard functor $\operatorname{Pic}_{X/S}$ is a sheaf even with respect to the fpqc-topology and in order to prepare the discussion of rigidificators, we want to look at the situation from another point of view. We assume that $f_*(\mathcal{O}_X) = \mathcal{O}_S$ holds universally and that f admits a section $\varepsilon : S \longrightarrow X$. For any line bundle $\mathcal{L}$ on X, let us call an isomorphism $\alpha : \mathcal{O}_S \xrightarrow{\sim} \varepsilon^*(\mathcal{L})$ a *rigidification* of $\mathcal{L}$. Furthermore, the pair $(\mathcal{L}, \alpha)$ will be referred to as a line bundle which is rigidified along the section ε. Then we can look at the functor $(P, \varepsilon) : (\mathrm{Sch}/S)^0 \longrightarrow (\mathrm{Sets})$ which associates to each S-scheme T the set $(P, \varepsilon)(T)$ of isomorphism classes of line bundles on $X_T = X \times_S T$ which are rigidified along the section $\varepsilon_T : T \longrightarrow X_T$. The functor (P, ε) has the advantage that it is automatically a sheaf with respect to the Zariski topology. Namely, using the fact that $f_*(\mathcal{O}_X) = \mathcal{O}_S$ is true universally, one shows easily that rigidified line bundles do not admit non-trivial automorphisms; hence the terminology of rigidifications is justified. Furthermore, it follows from descent theory that (P, ε) is a sheaf even with respect to the fpqc-topology. Namely, consider a sequence

$$(P, \varepsilon)(T) \longrightarrow (P, \varepsilon)(T') \rightrightarrows (P, \varepsilon)(T''),$$

where $T' \longrightarrow T$ is an fpqc-morphism and where $T'' = T' \times_T T'$. The map on the left-hand side is injective by 6.1/4. To show the exactness of the sequence, fix an element $(\mathcal{L}', \alpha') \in (P, \varepsilon)(T')$ whose images in $(P, \varepsilon)(T'')$ coincide. Then we have an isomorphism $p_1^* \mathcal{L}' \xrightarrow{\sim} p_2^* \mathcal{L}'$ between the two pull-backs of $\mathcal{L}'$ to T'' which is compatible with rigidifications. Hence this isomorphism is automatically a descent datum, and the descent is effective by 6.1/4. Thus the above sequence is exact, and (P, ε) is a sheaf with respect to fpqc. For each line bundle $\mathcal{L}$ on X, the bundle $\mathcal{L} \otimes f^*(\varepsilon^*(\mathcal{L}^{-1}))$ has a rigidification. Therefore we have

$$(P, \varepsilon)(T) = \operatorname{Pic}(X_T)/\operatorname{Pic}(T)$$

for all S-schemes T. Since (P, ε) is a sheaf with respect to the fpqc-topology and, thus, with respect to the fppf-topology, it is canonically isomorphic to the relative Picard functor $\mathrm{Pic}_{X/S}$. Thereby we see once more that the second assertion of Proposition 4 is true.

Using the above argument, it can easily be shown that the relative Picard functor $\mathrm{Pic}_{X/S}$ which has been defined within the framework of the fppf-topology is even a sheaf with respect to the fpqc-topology, provided $f: X \longrightarrow S$ is fppf and satisfies $f_*(\mathcal{O}_X) = \mathcal{O}_S$ universally. Namely, we may perform a base change with X over S and thereby assume that f has a section. Then, by considering rigidifications, it follows that $\mathrm{Pic}_{X/S}$ is a sheaf with respect to fpqc.

If the assumptions that the equation $f_*(\mathcal{O}_X) = \mathcal{O}_S$ holds universally and that there is a section $\varepsilon: S \longrightarrow X$ are not satisfied, it is sometimes useful to introduce a generalization of the notion of rigidifications so that, similarly as above, one can deal with rigidified line bundles.

Definition 5. *Let $f: X \longrightarrow S$ be proper, flat, and of finite presentation. Then a subscheme $Y \subset X$, which is finite, flat, and of finite presentation over S, is called a rigidificator of f or, more precisely, of the relative Picard functor* $\mathrm{Pic}_{X/S}$ *if*

$$(\mathrm{Sch}/S)^0 \longrightarrow (\mathrm{Sets}), \qquad T \longmapsto \Gamma(X_T, \mathcal{O}_{X_T}),$$

is a subfunctor of the functor

$$(\mathrm{Sch}/S)^0 \longrightarrow (\mathrm{Sets}), \qquad T \longmapsto \Gamma(Y_T, \mathcal{O}_{Y_T});$$

i.e., if the map $\Gamma(X_T, \mathcal{O}_{X_T}) \longrightarrow \Gamma(Y_T, \mathcal{O}_{Y_T})$, which is derived from the inclusion $Y_T \hookrightarrow X_T$, is injective for all S-schemes T.

If $f_*(\mathcal{O}_X) = \mathcal{O}_S$ holds universally, it is immediately clear that, for each section $\varepsilon: S \longrightarrow X$ of f, the closed subscheme $\varepsilon(S) \subset X$ is a rigidificator of f. Furthermore, let us mention without proof two non-trivial examples where rigidifications exist; cf. Raynaud [6], Prop. 2.2.3.

Proposition 6. *Let $f: X \longrightarrow S$ be as in Definition 5.*

(a) *If the fibres of f do not have embedded components, f admits a rigidificator locally over S with respect to the étale topology.*

(b) *If S is the spectrum of a discrete valuation ring, f has a rigidificator.*

Let Y be a rigidificator of $f: X \longrightarrow S$. Then an invertible sheaf on X which is rigidified along Y is defined as a pair $(\mathcal{L}, \alpha)$, where $\mathcal{L}$ is an invertible sheaf on X, and where α is an isomorphism $\mathcal{O}_Y \xrightarrow{\sim} \mathcal{L}_{|Y}$. Rigidified line bundles do not admit non-trivial automorphisms. Therefore the functor

$$(\mathrm{Pic}_{X/S}, Y): (\mathrm{Sch}/S)^0 \longrightarrow (\mathrm{Sets}),$$

which associates to an arbitrary S-scheme T the set of isomorphism classes of line bundles on X_T which are rigidified along Y_T, is a sheaf with respect to the Zariski topology and, by descent theory, even with respect to the fpqc-topology. Furthermore, $(\mathrm{Pic}_{X/S}, Y)$ is canonically a group functor.

In order to relate the functor $(\mathrm{Pic}_{X/S}, Y)$ to the relative Picard functor $\mathrm{Pic}_{X/S}$, it is necessary to look at rigidificators from another point of view. However, before we can do this, we have to discuss a basic result on the direct image of $\mathcal{O}_X$-modules which are locally of finite presentation; by the latter we mean (quasi-coherent) $\mathcal{O}_X$-modules which, locally, are isomorphic to the cokernel of a homomorphism of type $\mathcal{O}_X^m \longrightarrow \mathcal{O}_X^n$. Furthermore, we need the concept of cohomological flatness. Assume that $f: X \longrightarrow S$ is proper and of finite presentation, and consider an $\mathcal{O}_X$-module $\mathcal{F}$ of locally finite presentation, which is flat over S. Then $\mathcal{F}$ is said to be *cohomologically flat over S in dimension* 0 if the formation of the direct image $f_*(\mathcal{F})$ commutes with base change. If the condition is true for $\mathcal{F} = \mathcal{O}_X$, we say that f itself is cohomologically flat in dimension 0. The latter is the case if f is flat and if the geometric fibres of f are reduced; cf. [EGA III$_2$], 7.8.6.

Theorem 7. *Let $f: X \longrightarrow S$ be a proper morphism which is finitely presented. Furthermore, let $\mathcal{F}$ be an $\mathcal{O}_X$-module of locally finite presentation which is S-flat. Then there exists an $\mathcal{O}_S$-module $\mathcal{Q}$ of locally finite presentation, unique up to canonical isomorphism, such that there is an isomorphism of functors*

$$f_*(\mathcal{F} \otimes_{\mathcal{O}_S} \mathcal{M}) \xrightarrow{\sim} \mathcal{H}om_{\mathcal{O}_S}(\mathcal{Q}, \mathcal{M}),$$

which is functorial for all quasi-coherent $\mathcal{O}_S$-modules $\mathcal{M}$. In particular, there is an isomorphism of functors

$$\Gamma(X, \mathcal{F} \otimes_{\mathcal{O}_S} \mathcal{M}) \xrightarrow{\sim} \mathcal{H}om_{\mathcal{O}_S}(\mathcal{Q}, \mathcal{M}).$$

The $\mathcal{O}_S$-module $\mathcal{Q}$ is locally free if and only if $\mathcal{F}$ is cohomologically flat over S in dimension 0. In the latter case, $\mathcal{Q}$ and $f_*(\mathcal{F})$ are dual to each other and, in particular, $f_*(\mathcal{F})$ is locally free.

We will not repeat the *proof* of the theorem from [EGA III$_2$], 7.7.6. But to give some idea, we want to show how the assertions follow from the theorem on cohomology and base change as contained in Mumford [3], Chap. II, § 5. We may assume that S is affine, say $S = \operatorname{Spec} A$. Then the theorem on cohomology and base change says there is a finite complex

$$K^\cdot : 0 \longrightarrow K^0 \xrightarrow{\varphi} K^1 \longrightarrow K^2 \longrightarrow \ldots \longrightarrow K^n \longrightarrow 0$$

of finitely generated projective A-modules (we may assume of free A-modules, after restriction of S) as well as an isomorphism of functors

$$H^p(X, \mathcal{F} \otimes_A M) \simeq H^p(K^\cdot \otimes_A M), \qquad p \geq 0,$$

on the category of A-modules M. (Using Mumford's version of the base change, one has remove the noetherian hypothesis by a limit argument; furthermore, the above functors have to be considered on the category of all A-modules M and not just on the category of all A-algebras B.) Dualizing the map $\varphi: K^0 \longrightarrow K^1$ gives an exact sequence

$$0 \longleftarrow \operatorname{coker} \varphi^* \longleftarrow (K^0)^* \xleftarrow{\varphi^*} (K^1)^*,$$

and we claim there is a functorial isomorphism

$$(*) \qquad H^0(K^\cdot \otimes_A M) = \ker(\varphi \otimes M) \xrightarrow{\sim} \operatorname{Hom}_A(\operatorname{coker} \varphi^*, M)$$

of functors in M. Namely, applying the functor $\mathrm{Hom}_A(\cdot, M)$, which is left-exact, to the preceding exact sequence yields the exact sequence

$$0 \longrightarrow \mathrm{Hom}_A(\mathrm{coker}\,\varphi^*, M) \longrightarrow \mathrm{Hom}_A((K^0)^*, M) \longrightarrow \mathrm{Hom}_A((K^1)^*, M)\,.$$

Then we compare it with the exact sequence

$$0 \longrightarrow \ker(\varphi \otimes M) \longrightarrow K^0 \otimes_A M \xrightarrow{\varphi \otimes M} K^1 \otimes_A M\,.$$

The canonical homomorphisms $K^i \otimes_A M \longrightarrow \mathrm{Hom}_A((K^i)^*, M)$, $i = 1, 2$, are isomorphisms since K^0 and K^1 are free, and there is an isomorphism

$$H^0(K^{\cdot} \otimes_A M) \xrightarrow{\sim} \mathrm{Hom}_A(\mathrm{coker}\,\varphi^*, M)\,,$$

which is functorial in M. Hence the existence of the functorial isomorphism $(*)$ is proved. Writing $Q = \mathrm{coker}\,\varphi^*$ and using the theorem on cohomology and base change, the resulting functorial isomorphism

$$H^0(X, \mathscr{F} \otimes_A M) \xrightarrow{\sim} \mathrm{Hom}_A(Q, M)$$

implies the main assertion of our theorem. Since the tensor product is right-exact and since Hom is left-exact, the isomorphism $(*)$ shows that $\mathscr{F}$ is cohomologically flat over S in dimension 0 if and only if $Q = \mathrm{coker}\,\varphi^*$ is a projective, i.e., locally free A-module. If the latter is the case, $\ker \varphi$ is locally free since it is the dual of $\mathrm{coker}\,\varphi^*$.

□

If $f: X \longrightarrow S$ is proper, finitely presented, and flat, the assertion of the above theorem holds for the $\mathcal{O}_X$-module $\mathscr{F} = \mathcal{O}_X$. Restricting the resulting functorial isomorphism

$$f_*(\mathscr{F} \otimes_{\mathcal{O}_S} \mathscr{M}) \xrightarrow{\sim} \mathscr{H}om_{\mathcal{O}_S}(\mathscr{Q}, \mathscr{M})$$

to quasi-coherent $\mathcal{O}_S$-modules of type $\mathscr{M} = \mathcal{O}_T$ which are obtained from morphisms $T \longrightarrow S$, one ends up with functorial isomorphisms

$$\Gamma(X_T, \mathcal{O}_{X_T}) \xrightarrow{\sim} \mathrm{Hom}_{\mathcal{O}_S}(\mathscr{Q}, \mathcal{O}_T) \xrightarrow{\sim} \mathrm{Hom}_S(T, V)$$

where V denotes the S-scheme corresponding to the symmetric $\mathcal{O}_S$-algebra $\mathscr{S}ym_{\mathcal{O}_S}(\mathscr{Q})$ of $\mathscr{Q}$. Dropping the middle term, we get a functorial isomorphism between functors on the category of all S-schemes T. The scheme V is also referred to as the *total space of the module* $\mathscr{Q}$. We say that V is locally free if this is true for $\mathscr{Q}$ as an $\mathcal{O}_S$-module. The latter is equivalent to the fact that V is smooth over S. So we can state the following result.

Corollary 8. *Let $f: X \longrightarrow S$ be proper, finitely presented, and flat, and let $\mathscr{Q}$ be the $\mathcal{O}_S$-module associated to $f_*(\mathcal{O}_X)$ in the sense of Theorem 7. Then the functor*

$$(\mathrm{Sch}/S)^0 \longrightarrow (\mathrm{Sets})\,, \qquad T \longmapsto \Gamma(X_T, \mathcal{O}_{X_T})$$

is represented by the total space V of $\mathscr{Q}$. Furthermore, V is locally free if and only if f is cohomologically flat in dimension 0.

If, in addition to the above assumptions, f is finite, it is automatically cohomologically flat in dimension 0. In particular, the functor of global sections of a rigidificator is always represented by the total space of a module which is locally free. Using the assertion of the corollary, we can give a further characterization of rigidificators.

Proposition 9. *Let $f: X \longrightarrow S$ be proper, finitely presented, and flat, and consider a subscheme $Y \subset X$ which is finite, flat, and of finite presentation over S. Let V_X (resp. V_Y) be the S-scheme, which, as in Corollary 8, represents the functor of global sections on X (resp. Y). Then the following conditions are equivalent:*

(a) *Y is a rigidificator of f.*

(b) *The morphism $V_X \longrightarrow V_Y$, which is induced by the inclusion $Y \hookrightarrow X$, is a closed immersion.*

Proof. Let $\mathcal{Q}$ (resp. $\mathcal{Q}'$) denote the $\mathcal{O}_S$-module which is obtained by means of Theorem 7 from $f: X \longrightarrow S$ (resp. $Y \longrightarrow S$). Then, for all S-schemes T such that $\mathcal{O}_T$ is a quasi-coherent $\mathcal{O}_S$-module, the inclusion $Y \hookrightarrow X$ gives rise to a sequence

$$(*) \qquad 0 \longrightarrow \mathrm{Hom}_{\mathcal{O}_S}(\mathcal{Q}, \mathcal{O}_T) \longrightarrow \mathrm{Hom}_{\mathcal{O}_S}(\mathcal{Q}', \mathcal{O}_T)\,.$$

The latter is exact for all T if and only if Y is a rigidificator of f. Now the sequence $(*)$ corresponds to a sequence

$$(**) \qquad \mathcal{Q}' \longrightarrow \mathcal{Q} \longrightarrow 0$$

of $\mathcal{O}_X$-modules which is exact if and only if $(*)$ is exact for all T. On the other hand, $(**)$ yields a sequence between associated symmetric $\mathcal{O}_S$-algebras

$$(***) \qquad \mathcal{S}ym_{\mathcal{O}_S}(\mathcal{Q}') \longrightarrow \mathcal{S}ym_{\mathcal{O}_S}(\mathcal{Q}) \longrightarrow 0$$

which is exact if and only if it is exact in degree 1, i.e., if and only if $(**)$ is exact. This verifies the assertion of the proposition. □

As before, let $f: X \longrightarrow S$ be proper, finitely presented, and flat, and let V be the S-scheme representing the functor $T \longmapsto \Gamma(X_T, \mathcal{O}_{X_T})$ of global sections on X. Then V may be viewed as a functor to the category of rings and thus is a ring scheme. We claim:

Lemma 10. *The subfunctor of units $T \longmapsto \Gamma(X_T, \mathcal{O}_{X_T}^*)$ is represented by an open subscheme $V^* \subset V$. In particular, V^* is a group scheme.*

Proof. The assertion is clear if f is cohomologically flat in dimension 0. Namely, then V is locally free and we can use a norm argument. In the general case, one views V and V^* as functors and shows that the injection $V^* \hookrightarrow V$ is relatively representable by open immersions. In order to do this, consider an S-scheme T and a T-valued point $g: T \longrightarrow V$ as well as the associated cartesian diagram

$$\begin{array}{ccc} V^* \times_V T & \hookrightarrow & T \\ \big\downarrow & & \big\downarrow{\scriptstyle g} \\ V^* & \hookrightarrow & V\,. \end{array}$$

Then g corresponds to a global section in the structure sheaf of $X \times_S T$. Let U' be the maximal open subset of $X \times_S T$ where g is invertible. Since f is proper, the complement of U' projects onto a closed subset F of T. Therefore its complement $U := T - F$ is an open subscheme of T, and it is easily verified that $V^* \times_V T \longrightarrow T$ is represented by the open immersion $U \hookrightarrow T$. □

The canonical map $\mathcal{O}_S \longrightarrow f_*(\mathcal{O}_X)$ defines a morphism $\mathbb{G}_a \longrightarrow V$ which is a closed immersion as can be seen by using arguments as in the proof of Proposition 9. Restricting to the subschemes of units yields an immersion of group schemes $\mathbb{G}_m \longrightarrow V^*$ which is a closed immersion again. It is easily seen that $f_*(\mathcal{O}_X) = \mathcal{O}_S$ holds universally if and only if the map $\mathbb{G}_a \longrightarrow V$ or, equivalently, the map $\mathbb{G}_m \longrightarrow V^*$ is an isomorphism.

Finally, let Y be a rigidificator of $f : X \longrightarrow S$ and, as in Proposition 9, let V_X and V_Y denote the schemes representing the functors of global sections on X and on Y. Then the closed immersion $V_X \hookrightarrow V_Y$ gives rise to an immersion $V_X^* \hookrightarrow V_Y^*$, and there is a canonical map $V_Y^* \longrightarrow (\mathrm{Pic}_{X/S}, Y)$ to the Picard functor $(\mathrm{Pic}_{X/S}, Y)$ of line bundles which are rigidified along Y. Namely, fixing an S-scheme T, a global invertible section a on $Y \times_S T$ is mapped to the pair $(\mathcal{O}_{X_T}, \alpha)$ where the isomorphism $\alpha : \mathcal{O}_{X_T}|_{Y_T} \xrightarrow{\sim} \mathcal{O}_{X_T}|_{Y_T}$ is the multiplication by a. Adding the canonical map $(\mathrm{Pic}_{X/S}, Y) \longrightarrow \mathrm{Pic}_{X/S}$, one obtains the sequence

$$0 \longrightarrow V_X^* \hookrightarrow V_Y^* \longrightarrow (\mathrm{Pic}_{X/S}, Y) \longrightarrow \mathrm{Pic}_{X/S} \longrightarrow 0 .$$

Proposition 11. *The preceding sequence is exact in terms of sheaves with respect to the étale topology.*

The *proof* is straightforward; see Raynaud [6], 2.1.2 and 2.4.1. It is shown in the same article that $(\mathrm{Pic}_{X/S}, Y)$ is representable by an algebraic space; cf. our discussion of the representability of Picard functors in 8.3. Thus, even if $\mathrm{Pic}_{X/S}$ is not representable (by a scheme or by an algebraic space), but if there exists a rigidificator Y, there is a representable object which closely dominates the relative Picard functor.

8.2 Representability by a Scheme

There are two types of results concerning the representability of the relative Picard functor $\mathrm{Pic}_{X/S}$; namely, results on the representability by schemes and results on the representability by algebraic spaces. If one wants $\mathrm{Pic}_{X/S}$ to be a scheme, one has to ask strong conditions for the structural morphism $f : X \longrightarrow S$ whereas, if one allows to work more generally within the context of algebraic spaces, one can obtain the representability of $\mathrm{Pic}_{X/S}$ by an algebraic space under conditions which are not so restrictive and quite natural to ask.

In the present section, we will give an outline of Grothendieck's method for representing $\mathrm{Pic}_{X/S}$ by a scheme and, in the next section, we will roughly explain the idea of M. Artin's approach for representing $\mathrm{Pic}_{X/S}$ by an algebraic space. Let us start by stating the main results on the representability of $\mathrm{Pic}_{X/S}$ by a scheme.

Theorem 1 (Grothendieck [FGA], n°232, Thm. 3.1). *Let $f: X \longrightarrow S$ be projective and finitely presented. Assume that f is flat, and that the geometric fibres of f are reduced and irreducible. Then $\mathrm{Pic}_{X/S}$ is representable by a separated S-scheme which is locally of finite presentation over S.*

The proof of Theorem 1 consists mainly of methods from projective geometry. If one replaces the condition "projective" by "proper", these methods are not applicable for a general base S. Furthermore, the assumption on the fibres of f is an inevitable technical condition without which the proof cannot work. It is the very reason for getting representability by a scheme and for the fact that the representing S-scheme is separated.

To illustrate this point, let us look at an *example of Mumford.* He considered a projective flat family of geometrically reduced curves where $\mathrm{Pic}_{X/S}$ does not exist as a scheme. Namely let $S = \mathrm{Spec}\, \mathbb{R}[[t]]$, and let X be the S-subscheme of $\mathbb{P}_S^2$ given by the equation $X_1^2 + X_2^2 = tX_0^2$. One may view X as a conic which geometrically degenerates into two projective lines. The special fibre over the closed point of S is irreducible whereas, after the base change with $S' = \mathrm{Spec}\, \mathbb{C}[[t]]$, it decomposes into two lines which are conjugated under the Galois group $\mathbb{Z}/2\mathbb{Z}$ of S' over S. We claim that the Picard functor $\mathrm{Pic}_{X'/S'}$ is a scheme. Indeed, it is a disjoint union of subschemes representing the subfunctors $\mathrm{Pic}_{X'/S'}^d$, $d \in \mathbb{Z}$, of $\mathrm{Pic}_{X'/S'}$ which are given by line bundles of total degree d. Furthermore, each $\mathrm{Pic}_{X'/S'}^d$ is obtained by gluing copies of S' along the generic point; namely by gluing copies $S'_{a,b}$ with $a, b \in \mathbb{Z}$ and $a + b = d$ where the decompositions $d = a + b$ correspond to the possibilities of degenerations of a line bundle of degree d on the generic fibre into a line bundle with partial degrees a and b on the components of the special fibre. In particular, $\mathrm{Pic}_{X'/S'}$ is not separated and there are orbits of the Galois action on $\mathrm{Pic}_{X'/S'}$ which are not contained in an open affine subscheme. So, the descent datum given by the Galois action cannot be effective, and hence $\mathrm{Pic}_{X/S}$ is not representable by a scheme over S. A closer look at this example shows that the very reason for this is the fact that the irreducible components of the fibres of f are not geometrically irreducible. The same can be read from the following generalization of Grothendieck's result:

Theorem 2 (Mumford, unpublished). *Let $f: X \longrightarrow S$ be flat, projective, and finitely presented with geometrically reduced fibres. Assume that the irreducible components of the fibres of f are geometrically irreducible. Then $\mathrm{Pic}_{X/S}$ is representable by a (not necessarily separated) S-scheme which is locally of finite presentation over S.*

If the base scheme S is a field, one can prove the representability of $\mathrm{Pic}_{X/S}$ under weaker assumptions than those mentioned in Theorem 1. This was first done by Grothendieck for the projective case; cf. [FGA], n°232, Sect. 6. Later on Murre and Oort treated the proper case.

Theorem 3 (Murre [1] and Oort [1]). *Let X be a proper scheme over a field k. Then* $\mathrm{Pic}_{X/k}$ *is representable by a scheme which is locally of finite type over k.*

The theorem of Murre can also be deduced from the results on the representability of $\mathrm{Pic}_{X/S}$ by an algebraic space; cf. Section 8.3. Namely, a group object in the category of algebraic spaces over a field is representable by a scheme.

Finally, we want to introduce the notion of universal line bundles which is quite convenient to work with when $\mathrm{Pic}_{X/S}$ is representable. We assume that the structural morphism $f: X \longrightarrow S$ has a section ε and that $f_*\mathcal{O}_X = \mathcal{O}_S$ holds universally. In this case $\mathrm{Pic}_{X/S}$ is isomorphic to the functor

$$(P, \varepsilon): (\mathrm{Sch}/S)^0 \longrightarrow (\mathrm{Sets})$$

which associates to each S-scheme S' the set of isomorphism classes of line bundles on $X' = X \times_S S'$ which are rigidified along the induced section $\varepsilon' = \varepsilon \otimes \mathrm{id}_{S'}$; cf. Section 8.1. If $\mathrm{Pic}_{X/S}$ is a scheme, it also represents the functor (P, ε). So the identity on $\mathrm{Pic}_{X/S}$ gives rise to a line bundle $\mathscr{P}$ on $X \times_S \mathrm{Pic}_{X/S}$ which is canonically rigidified along the induced section. $\mathscr{P}$ is called the *universal line bundle* for $(X/S, \varepsilon)$. That this terminology is justified can be seen if we write down explicitly the condition of (P, ε) being representable:

Proposition 4. *Let $f: X \longrightarrow S$ be finitely presented and flat, and let ε be a section of f. Assume that $f_*\mathcal{O}_X = \mathcal{O}_S$ holds universally. If* $\mathrm{Pic}_{X/S}$ *is representable by a scheme, the universal line bundle $\mathscr{P}$ for $(X/S, \varepsilon)$ has the following property*:

For any S-scheme S', and for any line bundle $\mathscr{L}'$ on $X' = X \times_S S'$ which is rigidified along the induced section ε', there exists a unique morphism $g: S' \longrightarrow$ $\mathrm{Pic}_{X/S}$ *such that $\mathscr{L}'$, as a rigidified line bundle, is isomorphic to the pull-back of $\mathscr{P}$ under the morphism* $\mathrm{id}_X \times g$.

Note that $f_*\mathcal{O}_X = \mathcal{O}_S$ holds universally under the assumptions of Theorem 1; cf. [EGA III_2], 7.8.6.

Next we turn to the proof of Theorem 1. Since the relative Picard functor is a sheaf for the Zariski topology, its representability is a local problem on S. So we may assume that X is a closed subscheme of the projective space $\mathbb{P}^n_S$. In order to state what the proof yields in this special case, we have to introduce some further notions.

Following Altmann and Kleiman [1], a morphism of schemes $f: X \longrightarrow S$ is called *strongly projective* (resp. *strongly quasi-projective*) if it is finitely presented and if there exists a locally free sheaf $\mathscr{E}$ on S of constant finite rank such that X is S-isomorphic to a closed subscheme (resp. subscheme) of $\mathbb{P}(\mathscr{E})$. Let $\mathcal{O}_X(1)$ be the canonical (relatively) very ample line bundle on X. For any polynomial $\Phi \in \mathbb{Q}[t]$, one introduces the subfunctor $\mathrm{Pic}^{\Phi}_{X/S}$ of $\mathrm{Pic}_{X/S}$ which is induced by the line bundles with Hilbert polynomial Φ (with respect to $\mathcal{O}_X(1)$) on the fibres of X over S; cf. [EGA III_1], 2.5.3 for the definition of Hilbert polynomials. Then one can state the following stronger version of Theorem 1, which clearly suggests that Grothendieck's result deals with a problem inside the category of (quasi-) projective S-schemes.

Theorem 5. *Let $f: X \longrightarrow S$ be strongly projective, and let S be quasi-compact. Assume that f is flat, and that the geometric fibres of f are reduced and irreducible. Then, for every $\Phi \in \mathbb{Q}[t]$, the functor $\mathrm{Pic}^{\Phi}_{X/S}$ is representable by a strongly quasi-projective S-scheme. Furthermore, $\mathrm{Pic}_{X/S}$ is represented by the disjoint union of all $\mathrm{Pic}^{\Phi}_{X/S}$, where Φ ranges over $\mathbb{Q}[t]$.*

In the following we want to sketch the main steps of the *proof of Theorem* 5; in particular, we want to point out where the specific assumptions of the theorem are employed. The proof itself decomposes into three parts:

I) The notion of relative Cartier divisors gives rise to a functor

$$\mathrm{Div}_{X/S}: (\mathrm{Sch}/S)^0 \longrightarrow (\mathrm{Sets}),$$

which associates to an S-scheme S' the set of all relative Cartier divisors of the S'-scheme $X' := X \times_S S'$. There is a canonical morphism

$$\mathrm{Div}_{X/S} \longrightarrow \mathrm{Pic}_{X/S}$$

which is relatively representable. We will show a slightly weaker version of the latter statement which is enough for our purposes.

II) We will show that the functor $\mathrm{Div}_{X/S}$ is representable by an S-scheme. More precisely, we introduce Hilbert polynomials with respect to the fixed very ample line bundle $\mathcal{O}_X(1)$, and we look at the subfunctor $\mathrm{Div}^{\Phi}_{X/S}$ which consists of all relative Cartier divisors with Hilbert polynomial Φ. Then we will show that $\mathrm{Div}^{\Phi}_{X/S}$ is an open subfunctor of $\mathrm{Div}_{X/S}$ and that $\mathrm{Div}^{\Phi}_{X/S}$ is a strongly quasi-projective S-scheme. Furthermore, $\mathrm{Div}_{X/S}$ is the disjoint union of all schemes $\mathrm{Div}^{\Phi}_{X/S}$, where Φ ranges over $\mathbb{Q}[t]$. This part is the hardest of the whole proof, since the representability of the Hilbert functor is involved.

III) For suitable polynomials Φ, the functor $\mathrm{Pic}^{\Phi}_{X/S}$ is a quotient (as a sheaf for the fppf-topology) of an open subscheme of $\mathrm{Div}_{X/S}$ with respect to a proper smooth equivalence relation. We will show that such a quotient is representable by a scheme. Hence, $\mathrm{Pic}^{\Phi}_{X/S}$ is representable in such a special case. For general Φ, there exists an integer n_{Φ} such that the translate of $\mathrm{Pic}^{\Phi}_{X/S}$ by the element associated to $\mathcal{O}_X(n_{\Phi})$ is of the type as treated in the special case. So $\mathrm{Pic}^{\Phi}_{X/S}$ is representable again. More precisely, we will see that it is representable by a strongly quasi-projective S-scheme. Furthermore, $\mathrm{Pic}^{\Phi}_{X/S}$ is an open and closed subfunctor of $\mathrm{Pic}_{X/S}$, so $\mathrm{Pic}_{X/S}$ is represented by the disjoint union of all schemes $\mathrm{Pic}^{\Phi}_{X/S}$ where Φ ranges over $\mathbb{Q}[t]$.

Let us start with part I. An *effective Cartier divisor* on a scheme X is a closed subscheme D of X such that its defining sheaf of ideals $\mathcal{I}$ is an invertible $\mathcal{O}_X$-module; i.e., for each $x \in X$, the ideal $\mathcal{I}_x$ is generated by a regular element of $\mathcal{O}_X$. We denote by $\mathcal{O}_X(D)$ the associated line bundle

$$\mathcal{O}_X(D) = \mathcal{I}^{-1} = \mathcal{H}om_{\mathcal{O}_X}(\mathcal{I}, \mathcal{O}_X),$$

and by $s_D \in \Gamma(X, \mathcal{O}_X(D))$ the global section associated to the inclusion $\mathcal{I} \hookrightarrow \mathcal{O}_X$. We refer to s_D as the canonical section of $\mathcal{O}_X(D)$. It corresponds to the canonical inclusion $\mathcal{O}_X \hookrightarrow \mathcal{O}_X(D)$. Thus, an effective Cartier divisor gives rise to a pair $(\mathcal{L}, s)$ consisting of a line bundle $\mathcal{L}$ and a global section $s \in \Gamma(X, \mathcal{L})$ which induces a regular element s_x on each stalk $\mathscr{L}_x$, $x \in X$; i.e., the map $i_s : \mathcal{O}_x \longrightarrow \mathscr{L}_x$ sending the unit element 1_x of $\mathcal{O}_x$ to s_x is injective. Two pairs $(\mathscr{L}, s)$ and $(\mathscr{L}', s')$ are called equivalent if there exists an isomorphism $\varphi : \mathscr{L} \longrightarrow \mathscr{L}'$ such that $\varphi(s)$ and s' differ by a factor which is a global section of $\mathcal{O}_X^*$. Associating to a pair $(\mathscr{L}, s)$ the subscheme D of X which is defined by the sheaf of ideals $\mathscr{L}^{-1}$ viewed as a subsheaf of $\mathcal{O}_X$ via the morphism $i_s \otimes \mathscr{L}^{-1}$, we obtain a bijection between the set of all effective Cartier divisors on X and the set of all equivalence classes of pairs $(\mathscr{L}, s)$, where $\mathscr{L}$ is a line bundle on X, and where s is a global section of $\mathscr{L}$ inducing a regular element on each stalk of $\mathscr{L}$. We denote by $\Gamma(X, \mathscr{L})^*$ the subset of $\Gamma(X, \mathscr{L})$ consisting of all global sections of $\mathscr{L}$ which induce regular elements on each stalk $\mathscr{L}_x$, $x \in X$. Thus the set of effective Cartier divisors D on X inducing the same line bundle $\mathscr{L}$ corresponds bijectively to the set $\Gamma(X, \mathscr{L})^*/\Gamma(X, \mathcal{O}_X^*)$.

Now let $f : X \longrightarrow S$ be locally of finite presentation. An *effective relative Cartier divisor* on X over S is an effective Cartier divisor D on X which is flat over S. Further characterizations of effective relative Cartier divisors are given by the following lemma.

Lemma 6. *Let $\mathscr{I}$ be a quasi-coherent sheaf of ideals of $\mathcal{O}_X$ which is locally of finite presentation, and let D be the closed subscheme of X defined by $\mathscr{I}$. Let x be a point of D, and set $s = f(x)$. Then the following conditions are equivalent:*

(i) *$\mathscr{I}$ is invertible at x (i.e., $\mathscr{I}_x$ is generated by a regular element), and D is flat over S at x.*

(ii) *X and D are flat over S at x, and the restriction D_s of D onto the fibre X_s over s is an effective Cartier divisor on X_s at x.*

(iii) *X is flat over S at x, and $\mathscr{I}_x$ is generated by an element f_x which induces a regular element on X_s at x.*

Proof. To show the assertion (i) $\Longrightarrow$ (ii), let h be a local section of $\mathscr{I}$ which generates $\mathscr{I}_x$. Then h is a regular element of $\mathcal{O}_{X,x}$, and the multiplication by h gives rise to an exact sequence

$$0 \longrightarrow \mathcal{O}_{X,x} \longrightarrow \mathcal{O}_{X,x} \longrightarrow \mathcal{O}_{D,x} \longrightarrow 0 .$$

After tensoring with the residue field $k(s)$ of s over $\mathcal{O}_{S,s}$, we obtain the sequence

$$0 \longrightarrow \mathcal{O}_{X_s,x} \longrightarrow \mathcal{O}_{X_s,x} \longrightarrow \mathcal{O}_{D_s,x} \longrightarrow 0 .$$

Due to the flatness of D over S, this sequence is exact. Thus, h gives rise to a regular element of $\mathcal{O}_{X_s,x}$ and, hence, D_s is an effective Cartier divisor on X_s. In order to show that X is flat over S at x, we may use a limit argument ([EGA IV_3], 8.5.5 and 11.5.5.2) and thereby assume that S is locally noetherian. Looking at the long exact Tor-sequence, the flatness of D yields

$$h \cdot \mathrm{Tor}_n^{\mathcal{O}_{S,s}}(\mathcal{O}_{X,x}, k(s)) = \mathrm{Tor}_n^{\mathcal{O}_{S,s}}(\mathcal{O}_{X,x}, k(s))$$

for $n \geq 1$. Since S ist locally noetherian, and since X is locally of finite type over S, the modules $\mathrm{Tor}_n^{\mathcal{O}_{S,s}}(\mathcal{O}_{X,x}, k(s))$ are finitely generated over $O_{X,x}$. But then Nakayama's lemma implies $\mathrm{Tor}_n^{\mathcal{O}_{S,s}}(\mathcal{O}_{X,x}, k(s)) = 0$ for $n \geq 1$, because $x \in D$. Hence X is flat over S at x by Bourbaki [2], Chap. III, § 5, n°2, Thm. 1.

The assertion (ii) $\Longrightarrow$ (iii) follows from Nakayama's lemma, and the remaining implication (iii) $\Longrightarrow$ (i) is a consequence of [EGA IV_3], 11.3.7. □

It is clear from condition (ii) that the notion of effective relative Cartier divisors is stable under any base change $S' \longrightarrow S$. Thus, there is a functor

$$\mathrm{Div}_{X/S} : (\mathrm{Sch}/S)^0 \longrightarrow (\mathrm{Sets}) , \qquad S' \longmapsto \mathrm{Div}(X'/S')$$

where $\mathrm{Div}(X'/S')$ denotes the set of all effective relative Cartier divisors of $X' = X \times_S S'$ over S'. Associating to an effective relative Cartier divisor D the line bundle $\mathcal{O}_X(D)$, we obtain the canonical morphism

$$\mathrm{Div}_{X/S} \longrightarrow \mathrm{Pic}_{X/S} , \qquad D \longmapsto \mathcal{O}_X(D) .$$

As a first step towards the representability of $\mathrm{Pic}_{X/S}$, one proves that this morphism is relatively representable. Recall, this means that for each morphism $T \longrightarrow \mathrm{Pic}_{X/S}$ from an S-scheme T to $\mathrm{Pic}_{X/S}$, the morphism

$$\mathrm{Div}_{X/S} \times_{\mathrm{Pic}_{X/S}} T \longrightarrow T$$

obtained from $\mathrm{Div}_{X/S} \longrightarrow \mathrm{Pic}_{X/S}$ by the base change $T \longrightarrow \mathrm{Pic}_{X/S}$ is a morphism of schemes. However, we will show the latter only under the assumption that the map $T \longrightarrow \mathrm{Pic}_{X/S}$, as an element of $\mathrm{Pic}_{X/S}(T)$, is given by a line bundle on $X \times_S T$. This is enough for our application, because in part III we will apply it to the case where $T = \mathrm{Div}_{X/S}$ and where the map $T \longrightarrow \mathrm{Pic}_{X/S}$ is the canonical one. On the other hand, each map $T \longrightarrow \mathrm{Pic}_{X/S}$ corresponds to a line bundle on $X \times_S T$ if f has a section; cf. 8.1/4. So in this case we will really get the relative representability of $\mathrm{Div}_{X/S} \longrightarrow \mathrm{Pic}_{X/S}$.

Proposition 7. *Let $f : X \longrightarrow S$ be as in Theorem 5, and let T be an S-scheme. Let $\mathcal{L}$ be a line bundle on $X_T = X \times_S T$, and denote by $T \longrightarrow \mathrm{Pic}_{X/S}$ the morphism corresponding to $\mathcal{L}$. Then there exists an $\mathcal{O}_T$-module $\mathcal{F}$, which is locally of finite presentation, such that $\mathrm{Div}_{X/S} \times_{\mathrm{Pic}_{X/S}} T$ is represented by the projective T-scheme $\mathbb{P}(\mathcal{F})$.*

Furthermore, there is a canonical way to choose $\mathcal{F}$. If $\mathcal{L}$ is cohomologically flat in dimension zero, then $f_*(\mathcal{L})$ and $\mathcal{F}$ are locally free, and $\mathcal{F}$ is isomorphic to the dual of $f_*(\mathcal{L})$.

Proof. We may assume $T = S$. The fibred product $\mathrm{Div}_{X/S} \times_{\mathrm{Pic}_{X/S}} S$ is isomorphic to the functor $D_{\mathcal{L}} : (\mathrm{Sch}/S)^0 \longrightarrow (\mathrm{Sets})$ which associates to an S-scheme S' the set of all relative Cartier divisors D' on X'/S' such that $\mathcal{O}_{X'}(D')$ and $\mathcal{L}'$ give rise to the same element in $\mathrm{Pic}_{X/S}(S')$, where $\mathcal{L}'$ denotes the pull-back of $\mathcal{L}$ to X'. By Proposition 8.1/3 the latter condition is equivalent to the fact that $\mathcal{O}_{X'}(D')$ and $\mathcal{L}'$ are isomorphic locally over S'. Hence, as we have shown during our general discussion of Cartier divisors, there is a bijection

$$\Gamma(S', (f'_*\mathcal{L}')^*/f'_*(\mathcal{O}^*_{X'})) \longrightarrow D_{\mathcal{L}}(S')$$

where f' is obtained from f by the base change $S' \longrightarrow S$, and where $(f'_*\mathcal{L}')^*$ denotes the subsheaf of $(f'_*\mathcal{L}')$ consisting of all sections which induce regular elements on every fibre X_s of f. Thus, we have a bijection

$$\Gamma(S, (f_*\mathscr{L})^*/\mathcal{O}_S^*) \longrightarrow D_{\mathscr{L}}(S)$$

which is compatible with base change. Since f is proper and flat, there exists an $\mathcal{O}_S$-module $\mathscr{F}$ of locally finite presentation such that there is an isomorphism

$$(*) \qquad f_*\mathscr{L} \longrightarrow \mathscr{H}om_{\mathcal{O}_S}(\mathscr{F}, \mathcal{O}_S)$$

which is compatible with any base change $S' \longrightarrow S$; see Theorem 8.1/7. Furthermore, $\mathscr{F}$ is canonically determined by $\mathscr{L}$. Since the geometric fibres of f are reduced and irreducible, the local sections of $(f_*\mathscr{L})^*$ coincide with the local sections of $f_*\mathscr{L}$ which do not induce the zero section on any fibre X_s. Interpreting them as local homomorphisms $\mathscr{F} \longrightarrow \mathcal{O}_S$ via (*) and applying Nakayama's lemma, they correspond to those local homomorphisms $\mathscr{F} \longrightarrow \mathcal{O}_S$ which are surjective. Thus, the sections of $(f_*\mathscr{L})^*/\mathcal{O}_S^*$ correspond bijectively to the set of quasi-coherent quotients of $\mathscr{F}$ which are invertible, and hence to the sections of the projective bundle $\mathbb{P}(\mathscr{F})$; cf. [EGA II], 4.2.3. Since all maps under consideration are compatible with base change, $\mathscr{F}$ is as required. The last statement of the proposition has already been mentioned in 8.1/7. □

Thereby we have finished part I. Next, we discuss part II. The representability of $\mathrm{Div}_{X/S}$ will be derived from the representability of the Hilbert functor. The latter is defined as follows. For any S-scheme X denote by $\mathrm{Hilb}(X/S)$ the set of all closed subschemes D of X which are proper, finitely presented, and flat over S. Then

$$\mathrm{Hilb}_{X/S} : (\mathrm{Sch}/S)^0 \longrightarrow (\mathrm{Sets}), \qquad S' \longrightarrow \mathrm{Hilb}(X \times_S S'/S')$$

is a functor, the so-called *Hilbert functor of X over S*. We see from Lemma 6 that $\mathrm{Div}_{X/S}$ is an open subfunctor of $\mathrm{Hilb}_{X/S}$ if X is proper, finitely presented, and flat over S. Thus the representability of $\mathrm{Div}_{X/S}$ follows from the representability of $\mathrm{Hilb}_{X/S}$. We want to mention that, for the representability of $\mathrm{Hilb}_{X/S}$ by a scheme, it is essential that X is quasi-projective over S. Namely, there is an example by Hironaka of a proper and smooth manifold of dimension 3 over a field on which the group $\mathbb{Z}/2\mathbb{Z}$ acts freely. But the quotient with respect to this action does not exist in the category of schemes; cf. Hironaka [1]. One shows that, in this situation, the Hilbert functor cannot be represented by a scheme; namely, the equivalence relation given by the group action constitutes a closed subscheme R of $X \times_S X$ which is proper and flat with respect to the second projection. Thus R gives rise to an element $g \in \mathrm{Hilb}_{X/S}(X)$ and, if $\mathrm{Hilb}_{X/S}$ were representable as a scheme, the image of the morphism $X \longrightarrow \mathrm{Hilb}_{X/S}$ given by g would serve as a quotient of X under the group action.

For showing the representability of $\mathrm{Hilb}_{X/S}$, it is convenient to look at a more general situation. Given an $\mathcal{O}_X$-module $\mathscr{F}$ which is locally of finite presentation, one introduces the functor

$$\mathrm{Quot}_{(\mathscr{F}/X/S)} : (\mathrm{Sch}/S)^0 \longrightarrow (\mathrm{Sets})$$

which associates to an S-scheme S' the set of quotients $\mathscr{G}'$ of the pull-back $\mathscr{F}'$ of $\mathscr{F}$

to $X' = X \times_S S'$ where $\mathscr{G}'$ is required to be locally of finite presentation over $\mathscr{O}_{X'}$, to be flat over S', and to have proper support over S'. The key result on the representability of the functor $\mathrm{Quot}_{(\mathscr{F}/X/S)}$ is the following theorem of Grothendieck (cf. [FGA], n°221, Thm. 3.1); the strengthening from the projective to the strongly projective case is due to Altman and Kleiman [1], Thm. 2.6.

Theorem 8. *Let $f: X \longrightarrow S$ be strongly quasi-projective, and let $\mathscr{F}$ be an $\mathscr{O}_X$-module which is locally of finite presentation. Fix a (relatively) very ample line bundle $\mathscr{O}_X(1)$ associated to an embedding of X into a projective bundle over S. Assume that $\mathscr{F}$ is isomorphic to a quotient of an $\mathscr{O}_X$-module of the form $f^*\mathscr{B} \otimes \mathscr{O}_X(v)$ for some $v \in \mathbb{Z}$, where $\mathscr{B}$ is a locally free $\mathscr{O}_S$-module with a constant finite rank. Then* $\mathrm{Quot}_{(\mathscr{F}/X/S)}$ *is represented by a separated S-scheme which is a disjoint union of strongly quasi-projective S-schemes.*

If, in addition, f is proper, then $\mathrm{Quot}_{(\mathscr{F}/X/S)}$ *is a disjoint union of strongly projective S-schemes.*

Note that, for $\mathscr{F} = \mathscr{O}_X$, the functors $\mathrm{Quot}_{(\mathscr{F}/X/S)}$ and $\mathrm{Hilb}_{X/S}$ coincide. Furthermore, $\mathrm{Div}_{X/S}$ is a quasi-compact open subfunctor of $\mathrm{Hilb}_{X/S}$ if X is proper, finitely presented, and flat over S. Thus, if $\mathrm{Hilb}_{X/S}$ is represented by a disjoint union of strongly quasi-projective S-schemes, so is $\mathrm{Div}_{X/S}$.

When a very ample line bundle $\mathscr{O}_X(1)$ is fixed, $\mathrm{Quot}_{(\mathscr{F}/X/S)}$ can be covered in a canonical way by open subfunctors which will correspond to quasi-compact open subschemes of $\mathrm{Quot}_{(\mathscr{F}/X/S)}$ (resp. of $\mathrm{Hilb}_{X/S}$). Namely, for any $\mathscr{O}_X$-module $\mathscr{G}$ which is locally of finite presentation and has proper support, and for any point $s \in S$, one has the Hilbert polynomial $\chi(\mathscr{G}_s)(t)$; its value at any $n \in \mathbb{Z}$ is given by the Euler-Poincaré characteristic

$$\chi(\mathscr{G}_s(n)) = \sum_{i=0}^{\infty} (-1)^i \dim_{k(s)} H^i(X_s, \mathscr{G}_s(n))$$

of $\mathscr{G}(n)$ over the fibre X_s, where we have written $\mathscr{G}_s(n)$ for the restriction of $\mathscr{G} \otimes \mathscr{O}_X(n)$ to X_s. The Hilbert polynomial has rational coefficients; cf. [EGA III_1], 2.5.3. Furthermore, when $\mathscr{G}$ is flat over S, it is locally constant as a function of $s \in S$; cf. [EGA III_2], 7.9.11. So, for a polynomial $\Phi \in \mathbb{Q}[t]$, let $\mathrm{Quot}^{\Phi}_{(\mathscr{F}/X/S)}$ be the subfunctor of $\mathrm{Quot}_{(\mathscr{F}/X/S)}$ consisting of all quotients with a fixed Hilbert polynomial Φ. In the same way, one introduces the subfunctor $\mathrm{Hilb}^{\Phi}_{X/S}$ of $\mathrm{Hilb}_{X/S}$. It is clear that $\mathrm{Quot}^{\Phi}_{(\mathscr{F}/X/S)}$ constitutes an open and closed subfunctor of $\mathrm{Quot}_{(\mathscr{F}/X/S)}$ and that the subfunctors $\mathrm{Quot}^{\Phi}_{(\mathscr{F}/X/S)}$ cover $\mathrm{Quot}_{(\mathscr{F}/X/S)}$ if Φ ranges over $\mathbb{Q}[t]$. Thus, it suffices to prove the following theorem.

Theorem 8′. *Let X be S-isomorphic to a finitely presented subscheme of $\mathbb{P}(\mathscr{E})$ where $\mathscr{E}$ is a locally free $\mathscr{O}_S$-module of constant finite rank. Denote by $f: X \longrightarrow S$ the structural morphism and by $\mathscr{O}_X(1)$ the canonical (relatively) very ample line bundle on X. Let $\mathscr{F}$ be isomorphic to a quotient of $(f^*\mathscr{B}) \otimes \mathscr{O}_X(v)$ where $v \in \mathbb{Z}$ and where $\mathscr{B}$ is a locally free sheaf on S of constant finite rank, and assume that $\mathscr{F}$ is locally of finite presentation. Furthermore, fix a polynomial $\Phi \in \mathbb{Q}[t]$. Then, there exists an integer m_0 satisfying the following property:*

For each $m \geqq m_0$, the map

$$\mathrm{Quot}^{\Phi}_{(\mathscr{F}/X/S)} \longrightarrow \mathrm{Grass}_{\Phi(m)}(\mathscr{B} \otimes \mathscr{S}ym_{v+m}(\mathscr{E})) ,$$

which associates to an element $\mathscr{G}' \in \mathrm{Quot}^{\Phi}_{(\mathscr{F}/X/S)}(S')$ the direct image $f_*(\mathscr{G}'(m))$, constitutes a functor which is relatively representable by a quasi-compact immersion. In particular, $\mathrm{Quot}^{\Phi}_{(\mathscr{F}/X/S)}$ is representable by a strongly quasi-projective S-scheme.

If, in addition, X is a closed subscheme of $\mathbb{P}(\mathscr{E})$, the immersion of above is closed and $\mathrm{Quot}^{\Phi}_{(\mathscr{F}/X/S)}$ is strongly projective over S.

For a locally free $\mathcal{O}_S$-module $\mathscr{L}$ and a non-negative integer r, we denote by $\mathrm{Grass}_r(\mathscr{L})$ the contravariant functor from (Sch/S) to (Sets) which associates to an S-scheme S' the set of locally free quotients of $\mathscr{L} \otimes \mathcal{O}_{S'}$ of rank r. Then $\mathrm{Grass}_r(\mathscr{L})$ is representable by a closed subscheme of $\mathbb{P}(\mathscr{D})$, where $\mathscr{D}$ is the r-th exterior power of $\mathscr{L}$; cf. Grothendieck [2], § 2. Since we have not restricted ourselves to polynomials $\Phi \in \mathbb{Q}[t]$ which take values $\Phi(m)$ in the non-negative integers for large integers m, we define $\mathrm{Grass}_r(\mathscr{L})$ by the empty functor if $r \in \mathbb{Q} - \mathbb{N}$. Note that $\mathrm{Quot}^{\Phi}_{(\mathscr{F}/X/S)}$ is the empty functor if the polynomial Φ does not take values $\Phi(m)$ in the non-negative integers for large integers m.

For $\mathscr{F} = \mathcal{O}_X$, one has $\mathrm{Quot}^{\Phi}_{(\mathcal{O}_X/X/S)} = \mathrm{Hilb}^{\Phi}_{X/S}$. If X is proper and flat over S, we know that $\mathrm{Div}_{X/S}$ is an open subfunctor of $\mathrm{Hilb}_{X/S}$. So we denote by $\mathrm{Div}^{\Phi}_{X/S}$ the induced subfunctor of $\mathrm{Hilb}^{\Phi}_{X/S}$. Thus, Theorem 8′ implies the following corollary.

Corollary 9. *Let $f: X \longrightarrow S$ be strongly projective (resp. strongly quasi-projective), and let $\Phi \in \mathbb{Q}[t]$. Then $\mathrm{Hilb}^{\Phi}_{X/S}$ is representable by a strongly projective (resp. strongly quasi-projective) S-scheme.*

If, in addition, X is proper and flat over S, then $\mathrm{Div}^{\Phi}_{X/S}$ is representable by a strongly quasi-projective S-scheme.

Now let us give an *outline of the proof of Theorem* 8′. First one reduces to the case where X is the projective space $\mathbb{P}(\mathscr{E})$ associated to a locally free sheaf $\mathscr{E}$ of constant rank $e + 1$ on S, and where $\mathscr{F}$ is isomorphic to $f^*\mathscr{B}(v) := (f^*\mathscr{B}) \otimes \mathcal{O}_X(v)$ for some locally free sheef $\mathscr{B}$ on S which has constant rank b over S. Namely, $\mathrm{Quot}^{\Phi}_{(\mathscr{F}/X/S)}$ is a locally closed (resp. closed) subfunctor of $\mathrm{Quot}^{\Phi}_{(f^*\mathscr{B}(v)/\mathbb{P}(\mathscr{E})/S)}$ of finite presentation. In the latter case, there is a canonical isomorphism

$$\mathscr{B} \otimes \mathscr{S}ym_{v+m}(\mathscr{E}) \xrightarrow{\sim} f_*(\mathscr{F}(m))$$

for $m \in \mathbb{Z}$; cf. [EGA III_1], 2.1.15. We assume this situation from now on. Then a key point is the following observation of Mumford which simplifies the original proof of Grothendieck; cf. Mumford [2], Lecture 14.

Proposition 10. *There exists a constant m_0 depending on the integers e, b, v and on the coefficients of Φ, such that the following property is satisfied:*

Let S′ be an S-scheme, and let $\mathscr{G}' \in \mathrm{Quot}^{\Phi}_{(\mathscr{F}/X/S)}(S')$. Denote by $\mathscr{H}'$ the kernel of the canonical map $\mathscr{F}' \longrightarrow \mathscr{G}'$. Then, for all $m \geqq m_0$, the $\mathcal{O}_{X'}$-module $\mathscr{H}'(m)$ is generated by the local sections of $f'_*(\mathscr{H}'(m))$, and $R^i f'_*(\mathscr{H}'(m)$ vanishes for $i \geqq 1$. The same is true for $\mathscr{F}'(m)$ and $\mathscr{G}'(m)$.

A detailed proof of this proposition can be found in [SGA 6], Exp. XIII, § 1, for the case where S' defines a geometric point of S. The general case follows then by the theory of cohomology and base change; cf. Mumford [3], § 5.

Going back to the proof of Theorem 8′, keep the notation of Proposition 10. Then, for $m \geqq m_0$ and for each S-scheme S', the canonical map

$$f'_*(\mathcal{F}'(m)) \longrightarrow f'_*(\mathcal{G}'(m))$$

is surjective. Since $R^i f'_* \mathcal{G}'(m)$ vanishes for $m \geqq m_0$ and $i \geqq 1$, the direct image $f'_*(\mathcal{G}'(m)$ is a locally free $\mathcal{O}_{S'}$-module of rank $\Phi(m)$, due to [EGA III_2], 7.9.9. Thus, we get the canonical morphism

$$\mathrm{Quot}^{\Phi}_{(\mathcal{F}/X/S)} \longrightarrow \mathrm{Grass}_{\Phi(m)}(f_*(\mathcal{F}(m))$$

associating to a flat quotient $\mathcal{G}'$ of $\mathcal{F}'$ on X' the direct image $f_*(\mathcal{G}'(m))$. Moreover, one can reconstruct the subsheaf $\mathcal{H}'$ of $\mathcal{F}'$ from the canonical surjective map

$$f'_*(\mathcal{F}'(m)) \longrightarrow f'_*(\mathcal{G}'(m)) .$$

Thus, one can view $\mathrm{Quot}^{\Phi}_{(\mathcal{F}/X/S)}$ as a subfunctor of the Graßmannian functor $\mathrm{Grass}_{\Phi(m)}(\mathcal{B} \otimes \mathcal{S}ym_{\nu+m}(\mathcal{E}))$ which associates to an S-scheme S' the set of all locally free quotients of $f'_*(\mathcal{F}'(m))$ of rank $\Phi(m)$. It remains to see that the monomorphism

$$\mathrm{Quot}^{\Phi}_{(\mathcal{F}/X/S)} \longrightarrow \mathrm{Grass}_{\Phi(m)}(\mathcal{B} \otimes \mathcal{S}ym_{\nu+m}(\mathcal{E}))$$

is representable by a quasi-compact immersion. So denote by G the S-scheme $\mathrm{Grass}_{\Phi(m)}(\mathcal{B} \otimes \mathcal{S}ym_{\nu+m}(\mathcal{E}))$ and by $\mathcal{Q}$ the universal quotient of $\mathcal{B} \otimes \mathcal{S}ym_{\nu+m}(\mathcal{E})$. The latter is a quotient (as an $\mathcal{O}_G$-module) of the pull-back $(\mathcal{B} \otimes \mathcal{S}ym_{\nu+m}(\mathcal{E}))_G$ of $\mathcal{B} \otimes \mathcal{S}ym_{\nu+m}(\mathcal{E})$ to G, which is locally free of rank $\Phi(m)$. Let $\mathcal{F}_G$ be the pull-back of $\mathcal{F}$ on $X_G = X \times_S G$, and let $f_G : X_G \longrightarrow G$ be the map obtained from f by the base change $G \longrightarrow S$. By using the canonical isomorphism

$$(\mathcal{B} \otimes \mathcal{S}ym_{\nu+m}(\mathcal{E}))_G \longrightarrow (f_G)_*(\mathcal{F}_G(m)) ,$$

we obtain a canonical map

$$(f_G)_*(\mathcal{F}_G(m)) \longrightarrow \mathcal{Q} .$$

The kernel of this map generates a subsheaf $\mathcal{H}_G(m)$ of $\mathcal{F}_G(m)$. Denote by $\mathcal{H}_G$ the $\mathcal{O}_{X_G}$-module $\mathcal{H}_G(m) \otimes \mathcal{O}_{X_G}(-m)$ and by $\mathcal{G}_G$ the quotient $\mathcal{F}_G/\mathcal{H}_G$. By reducing to a noetherian base scheme S, one shows that there exists a (unique) subscheme Z of G such that a morphism $T \longrightarrow G$ factors through Z if and only if the pull-back $\mathcal{G}_T$ of $\mathcal{G}_G$ on $X \times_S T$ is flat over $T$and has Hilbert polynomial Φ on the fibres over T; cf. [FGA], n°221, Sect. 3. Furthermore, the inclusion $Z \hookrightarrow G$ is finitely presented. Hence, $\mathrm{Quot}^{\Phi}_{(\mathcal{F}/X/S)}$ is represented by Z which is strongly quasi-projective over S. Finally, Z is strongly projective because the valuative criterion is satisfied by [EGA IV_2], 2.8.1. □

Thereby we have finished part II. Finally we come to part III. We begin by recalling some definitions on equivalence relations in categories. Let C be a category such that direct products $X_1 \times X_2$ and fibred products $X_1 \times_Y X_2$ exist in C. A

C-equivalence relation on an object X of C is a representable subfunctor R of $X \times X$ such that, for each object T of C, the subset

$$R(T) \subset X(T) \times X(T)$$

is the graph of an equivalence relation on $X(T)$. Denote by $p_i : R \longrightarrow X$ the projection onto the i-th factor, $i = 1, 2$. A *categorical quotient of X with respect to the equivalence relation R* is a pair (Z, u) consisting of an object Z of C and a morphism $u : X \longrightarrow Z$ satisfying $up_1 = up_2$ such that, for any morphism $v : X \longrightarrow Y$ satisfying $vp_1 = vp_2$, there exists a unique morphism $\overline{v} : Z \longrightarrow Y$ such that $v = \overline{v}u$. If it exists, it is uniquely determined, and we will usually denote it by X/R. Furthermore, due to the definition of a fibred product, there is a canonical morphism

$$i : R \longrightarrow X \times_{X/R} X \ .$$

R is called an *effective equivalence relation on* X if the categorical quotient X/R exists and if the canonical morphism i is an isomorphism. In this case, X/R is called an *effective quotient*. Quite often, the canonical morphism i is not an isomorphism; this means that the equivalence relation given by the fibred product $X \times_{X/R} X$ over the categorical quotient X/R, is usually larger than the given relation R.

In the following, we consider the category of S-schemes. Then one can look at quotients also from the sheaf-theoretical point of view. Due to Proposition 8.1/1, any S-scheme X is a sheaf with respect to the fppf-topology (or the fpqc-topology). So, one can ask for the quotient of X with respect to R in the category of sheaves for the fppf-topology. Using the procedure of sheafification, one easily shows that such a quotient exists and that it is effective. Let us denote it by (X/R). Furthermore let us assume that the categorical quotient (in the category of S-schemes) X/R exists. So, viewing X and X/R as sheaves for the fppf-topology, one obtains canonical morphisms

$$X \longrightarrow (X/R) \longrightarrow X/R \ .$$

If (X/R) is represented by a scheme, (X/R) is the effective quotient of X with respect to R (for the category of S-schemes), and the canonical morphism $(X/R) \longrightarrow X/R$ is an isomorphism.

Example 11. *Let $f : X \longrightarrow Y$ be an* fppf-*morphism of S-schemes. Denote by $R(f)$ the subscheme $X \times_Y X$ of $X \times_S X$. Then $R(f)$ is an effective equivalence relation on X and (Y, f) is the effective quotient of X with respect to $R(f)$ in the category of S-schemes as well as in the category of sheaves for the* fppf-*topology.*

Proof. Since f is an fppf-morphism, Y is the quotient (in the category of sheaves for the fppf-topology) of X with respect to $R(f)$. Hence the assertion follows from what has been said before. □

For any property P applicable to morphisms, an equivalence relation R on an S-scheme X is said to satisfy the property P if P holds for the projections $p_i : R \longrightarrow X$.

We need the following general theorem on the existence of effective quotients with respect to proper flat equivalence relations.

Theorem 12. *Let $f: X \longrightarrow S$ be strongly quasi-projective, and let R be a proper flat equivalence relation on X which is finitely presented. Assume that the fibres of the projection $p_2: R \longrightarrow X$ have only a finite number of Hilbert polynomials with respect to an embedding of X into $\mathbb{P}(\mathscr{E})$, where $\mathscr{E}$ is a locally free $\mathcal{O}_S$-module of constant finite rank. Then R is effective, the quotient map $q: X \longrightarrow X/R$ is strongly projective and faithfully flat, and X/R is strongly quasi-projective over S.*

In particular, X/R is the effective quotient of X with respect to R in the category of sheaves for the fppf*-topology.*

The *proof* is easily done by using the existence of the Hilbert scheme; cf. Altman and Kleiman [1], § 2. Namely, set $H = \coprod \mathrm{Hilb}^{\Phi}_{X/S}$ where Φ ranges over the finitely many Hilbert polynomials of p_2; then H exists as a scheme and is strongly quasi-projective over S; cf. Corollary 9. Let D be the universal subscheme of $X \times_S H$. The projection $p: D \longrightarrow H$ is proper, flat, and finitely presented, and the equivalence relation R is a subscheme of $X \times_S X$ which is proper, flat, and finitely presented with respect to the second projection p_2. So, using the universal property of the Hilbert scheme, there is a unique morphism $g: X \longrightarrow H$ such that

$$R = (\mathrm{id}_X \times g)^* D .$$

Now the idea is to realize the quotient as the image of g.

For an S-scheme T and for points $x_1, x_2 \in X(T)$, write $x_1 \sim x_2$ whenever $(x_1, x_2) \in R(T)$. Then one shows

$$(*) \qquad x_1 \sim x_2 \Leftrightarrow gx_1 = gx_2 \Leftrightarrow (x_1, gx_2) \in D(T) .$$

Namely, set $R_i = (\mathrm{id}_X, x_i)^* R$ for $i = 1, 2$. Due to the definition of $\mathrm{Hilb}_{X/S}$, we have $gx_1 = gx_2$ if and only if for all T-schemes T', the set $R_1(T')$ coincides with $R_2(T')$ viewing both as subsets of $(X \times_S T)(T')$. Since R is an equivalence relation, the latter is equivalent to $(x_1, \mathrm{id}_T) \in R_2(T)$ and hence to $x_1 \sim x_2$. Thus, the first equivalence is clear. Due to the definition of g, the condition $(x_1, gx_2) \in D(T)$ is equivalent to $(x_1, x_2) \in R(T)$. Then the second equivalence is also clear.

Now, denote by Γ_g the graph of $g: X \longrightarrow H$. Since H is separated over S, the graph Γ_g is closed in $X \times_S H$. Furthermore, because Γ_g is isomorphic to X, it is of finite presentation over S. Since Γ_g is contained in D due to $(*)$, it is a closed subscheme of D. Moreover, Γ_g is of finite presentation over D, since D is of finite presentation over S. We want to show that Γ_g descends to a closed subscheme Z of H which is of finite presentation over H. So look at the projection $p: D \longrightarrow H$. Due to the definition of $\mathrm{Hilb}_{X/S}$, the map p is faithfully flat, proper, and finitely presented. Consider the canonical descent datum on D. In order to show Γ_g descends to a closed subscheme Z of H which is of finite presentation over H, it suffices to show that the closed subschemes $\Gamma_g \times_H D$ and $D \times_H \Gamma_g$ of $D \times_H D$ coincide. The latter is easily checked by looking at T-valued points and by using the equivalence $(*)$. The map $g: X \longrightarrow H$ factors through Z and, identifying X with Γ_g, the map $g: X \longrightarrow H$ is obtained from $p: D \longrightarrow H$ by the base change $Z \longrightarrow H$. Hence, $g: X \longrightarrow Z$ is faithfully flat, and strongly projective over Z, since D, being proper and strongly quasi-projective over H, is strongly projective over H. Because of (∗), we have a canonical isomorphism

$$R \longrightarrow X \times_Z X .$$

Then (Z, g) is an effective quotient of X with respect to R as explained in Example 11. Finally, $Z \longrightarrow S$ is strongly quasi-projective because Z is a closed subscheme of the strongly quasi-projective S-scheme H. □

Now we want to explain how the *proof of Theorem* 5 can be derived from the results we have discussed up to now. Let Φ be a polynomial with rational coefficients. Since the Hilbert polynomial of any $\mathcal{O}_X$-module, which is locally of finite presentation over X and flat over S, is locally constant, $\mathrm{Pic}^{\Phi}_{X/S}$ is an open and closed subfunctor of $\mathrm{Pic}_{X/S}$. Thus, it remains to show that $\mathrm{Pic}^{\Phi}_{X/S}$ is representable by a strongly quasi-projective S-scheme.

In order to do this, we need the notion of bounded families of coherent sheaves on the fibres of X over S. So, let S be a quasi-compact scheme and let X be an S-scheme of finite presentation. Let Λ be a family of isomorphism classes of coherent sheaves on the fibres of X over S; i.e., for each $s \in S$ and for each extension field K of $k(s)$, we are given a family of coherent sheaves $\mathcal{F}_K$ on X_K. Two sheaves $\mathcal{F}_K$ and $\mathcal{F}'_{K'}$ belong to the same class if there exist $k(s)$-embeddings of K and K' into a field L such that $\mathcal{F}_K \otimes_K L$ and $\mathcal{F}'_{K'} \otimes_{K'} L$ are isomorphic on X_L. The family Λ is called bounded if there exists an S-scheme T of finite presentation and a sheaf $\mathcal{F}$ on $X_T = X \times_S T$ which is locally of finite presentation such that Λ is contained in the family $(\mathcal{F}_{k(t)}; t \in T)$. There is the following proposition, cf. [SGA 6], Exp. XIII, Thm. 1.13.

Proposition 13. *Let S be quasi-compact, and let $X \longrightarrow S$ be strongly projective. Let Λ be a family of coherent sheaves on the fibres of X over S. Then the followig conditions are equivalent:*

(i) *Λ is bounded.*

(ii) *The set of Hilbert polynomials $\chi(\mathcal{F}_K)(t)$ is finite where $\mathcal{F}_K$ ranges over the elements of the family Λ, and there exist integers $n \in \mathbb{Z}$ and $N \in \mathbb{N}$ such that Λ is contained in the family of all classes of quotients of $\mathcal{O}_X(n)^N$.*

Furthermore we need the following result; cf. [SGA 6], Exp. XIII, Lemma 2.11.

Proposition 14. *Under the assumption of Theorem 5, a family Λ of line bundles $\mathcal{L}_K$ on the fibres of X over S is bounded if and only if the set of Hilbert polynomials $\chi(\mathcal{L}_K)(t)$ is finite.*

Now consider the morphism

$$\mathrm{Div}_{X/S} \longrightarrow \mathrm{Pic}_{X/S} .$$

Fix the polynomial Φ, and denote by $D(\Phi)$ the inverse image of $\mathrm{Pic}^{\Phi}_{X/S}$ in $\mathrm{Div}_{X/S}$. It is clear that $D(\Phi)$ is a disjoint union of connected components of $\mathrm{Div}_{X/S}$. Then it follows from Proposition 14 that there are only finitely many connected components of $\mathrm{Div}_{X/S}$ which are involved. Thus, due to Corollary 9, we see that $D(\Phi)$ is strongly quasi-projective over S.

Let us assume for a moment that the following condition on $\mathrm{Pic}^{\Phi}_{X/S}$ is satisfied: for any S-scheme S' and for any line bundle $\mathscr{L}'$ on $X' = X \times_S S'$ which induces an element of $\mathrm{Pic}^{\Phi}_{X/S}$, we have

$$R^i f'_*(\mathscr{L}'(n)) = 0 \quad \text{for} \quad i > 0 \quad \text{and} \quad n \geqq 0\,, \quad \text{and}$$
$$f'_*(\mathscr{L}'(n)) \neq 0 \quad \text{for } n \geqq 0\,.$$

Note that such line bundles are cohomologically flat in dimension zero. Furthermore, in this case, the map $D(\Phi) \longrightarrow \mathrm{Pic}^{\Phi}_{X/S}$ is an epimorphism (in terms of sheaves for the fppf-topology). Let $\mathscr{L}$ be the line bundle on $X \times_S D(\Phi)$ which corresponds to the universal (relative) Cartier divisor on $X \times_S D(\Phi)$. Then the map $D(\Phi) \longrightarrow \mathrm{Pic}^{\Phi}_{X/S}$ is induced by $\mathscr{L}$. If $f(\Phi)$ is the base change of f by $D(\Phi) \longrightarrow S$, the direct image of $\mathscr{L}$ under $f(\Phi)$ is locally free of rank $\Phi(0)$. Due to Proposition 7, the morphism

$$D(\Phi) \times_{\mathrm{Pic}^{\Phi}_{X/S}} D(\Phi) \longrightarrow D(\Phi)$$

is representable by the flat (even smooth) strongly projective morphism

$$\mathbb{P}(\mathscr{F}) \longrightarrow D(\Phi)\,,$$

where $\mathscr{F}$ is the dual of the direct image of $\mathscr{L}$ under $f(\Phi)$, since $\mathscr{L}$ is cohomologically flat in dimension zero. Now in order to show the representability of $\mathrm{Pic}^{\Phi}_{X/S}$, consider the following diagram

$$\begin{array}{ccc} D(\Phi) \times_{\mathrm{Pic}^{\Phi}_{X/S}} D(\Phi) & \longrightarrow & D(\Phi) \\ \downarrow & & \downarrow \\ D(\Phi) & \longrightarrow & \mathrm{Pic}^{\Phi}_{X/S}\,. \end{array}$$

It says that $\mathrm{Pic}^{\Phi}_{X/S}$ is isomorphic to the quotient (as sheaf for the fppf-topology) of $D(\Phi)$ by a proper and flat equivalence relation. Thus $\mathrm{Pic}^{\Phi}_{X/S}$ is representable by a strongly quasi-projective S-scheme; cf. Theorem 12.

Now it remains to remove the special assumption on $\mathrm{Pic}^{\Phi}_{X/S}$ which has been made above. If n is an integer, we denote by $\mathrm{Pic}^{\Phi}_{X/S} + n\xi$ the functor which associates to an S-scheme S' the subset

$$\{\mathscr{L}'(n)\,;\, \mathscr{L}' \in \mathrm{Pic}^{\Phi}_{X/S}(S')\}$$

of $\mathrm{Pic}_{X/S}(S')$. Note that this functor is of the form $\mathrm{Pic}^{\Psi}_{X/S}$ for a suitable polynomial $\Psi \in \mathbb{Q}[t]$. It suffices to show that there exists an integer n such that $\mathrm{Pic}^{\Phi}_{X/S} + n\xi$ fulfills the above assumptions. However, since $\mathrm{Pic}^{\Phi}_{X/S}$ is bounded due to Proposition 14, the latter follows from Propositions 13 and 10 by base change theory.

Thus we have finished part III, and thereby we conclude our discussion of Theorem 5.

8.3 Representability by an Algebraic Space

The most restrictive assumption in Grothendieck's theorem 8.2/1 on the representability of $\mathrm{Pic}_{X/S}$ is that the geometric fibres of $f : X \longrightarrow S$ have to be reduced and irreducible. As we have seen in the preceding section by looking at Mumford's example, even if X is projective and flat over S, there is an obstruction to $\mathrm{Pic}_{X/S}$ being a scheme, which is located in the fibres of f. However, in Mumford's example, there exists a surjective étale extension $S' \longrightarrow S$ such that the functor $\mathrm{Pic}_{X/S} \times_S S'$ is representable by a scheme over S'. Working within the category of algebraic spaces (the definition is given below), we can say that $\mathrm{Pic}_{X/S}$ is representable, since this category is stable under quotients by étale equivalence relations. This example suggests that, in comparison with Grothendieck's theorem, the assumptions on the S-scheme X can be weakened if one wants to represent $\mathrm{Pic}_{X/S}$ by an algebraic space.

Theorem 1 (M. Artin [5], Thm. 7.3). *Let $f : X \longrightarrow S$ be a morphism of algebraic spaces which is proper, flat, and finitely presented. Then, if f is cohomologically flat in dimension zero, the relative Picard functor* $\mathrm{Pic}_{X/S}$ *is represented by an algebraic space over S.*

A proper and flat morphism f is cohomologically flat in dimension zero if, for example, the geometric fibres of f are reduced; cf. [EGA III_2], 7.8.6. Furthermore, let us mention that there is a converse of Theorem 1 when the base S is reduced.

Remark 2. *Let $f : X \longrightarrow S$ be a morphism of schemes which is proper, flat, and finitely presented. Assume that S is reduced. Then* $\mathrm{Pic}_{X/S}$ *is an algebraic space if and only if f is cohomologically flat in dimension zero.*

Namely, in order to show the cohomological flatness of f when $\mathrm{Pic}_{X/S}$ is an algebraic space, one has only to verify that the dimension of $H^0(X_s, \mathcal{O}_{X_s})$ is locally constant on S; cf. [EGA III_2], 7.8.4. Then one can assume that S is a discrete valuation ring. Hence, the assertion follows from Raynaud [6], Prop. 5.2.

As we will see below, the method for the proof of Theorem 1 is completely different from the one used in the last section. It does not involve projective methods nor does it make use of the representability of the Hilbert functor or of the functor of relative Cartier divisors. Also we want to mention that the theorem does not contain 8.2/1. Only for the case where the base scheme S is a field, 8.2/1 and 8.2/3 are corollaries of Theorem 1, since a group object in the category of algebraic spaces over a field is represented by a scheme.

If, in the situation of Theorem 1, f is not cohomologically flat in dimension zero, the only option which is left is to work with rigidificators (cf. 8.1/5), and one can look for the representability of rigidified relative Picard functors; cf. Section 8.1.

Theorem 3 (Raynaud [1], Thm. 2.3.1). *Let $f : X \longrightarrow S$ be a proper, flat, and finitely presented morphism of algebraic spaces, and let Y be a rigidificator for* $\mathrm{Pic}_{X/S}$. *Then* *the rigidified Picard functor* $(\mathrm{Pic}_{X/S}, Y)$ *is representable by an algebraic space over* S, *and there exists a universal rigidified line bundle on* $(\mathrm{Pic}_{X/S}, Y)$.

The proofs of these theorems make use of a general principle for the construction of algebraic spaces which is due to M. Artin; cf. [5], Thm. 3.4. Namely, there is a criterion describing a necessary and sufficient condition for the representability of contravariant functors from (Sch/S) to (Sets) by algebraic spaces. It is for this criterion that the category of algebraic spaces yields a natural environment for questions on the representability of contravariant functors from (Sch/S) to (Sets). Within the category of algebraic spaces one can carry out many of the fundamental constructions, as contained in [FGA], under more general conditions, and one achieves results on the representability of certain functors under quite general assumptions.

Before we explain the criterion, let us briefly mention the basic definitions concerning algebraic spaces. As an introduction to the theory of algebraic spaces, we refer to M. Artin [3]. A detailed treatment can be found in Knutson [1].

In the following, let S be a scheme. Sometimes, for technical reasons, when we want to apply the approximation theorem 3.6/16, we have to assume that the base scheme S is locally of finite type over a field or over an excellent Dedekind ring.

Definition 4. *A (locally separated) algebraic space* X *over* S *is a functor*

$$X : (\mathrm{Sch}/S)^0 \longrightarrow (\mathrm{Sets})$$

with the following properties:

(i) X *is a sheaf with respect to the étale topology.*

(ii) *There exists a morphism* $\tau : U \longrightarrow X$ *of an* S*-scheme* U, *which is locally of finite presentation, to* X *such that* τ *is relatively representable by étale surjective morphisms of schemes.*

(iii) *The product* $U \times_X U$ *is represented by a subscheme of* $U \times_S U$ *such that the immersion* $U \times_X U \longrightarrow U \times_S U$ *is quasi-compact.*

Condition (ii) means that, for every S-scheme V and every morphism $V \longrightarrow X$, the product $U \times_X V$ is represented by a scheme and that the projection $U \times_X V \longrightarrow V$ is étale and surjective. Furthermore, it follows from (iii) that $U \times_X V \longrightarrow U \times_S V$ is a quasi-compact immersion. The algebraic space X is called *separated* over S if $U \times_X U$ is representable by a closed subscheme of $U \times_S U$.

Keeping the notations of Definition 4, the algebraic space X is the quotient of U by the equivalence relation $R = U \times_X U$ (in terms of sheaves with respect to the étale topology). Conversely, given an S-scheme U of locally finite presentation and a finitely presented subscheme R of $U \times_S U$ which defines an étale equivalence relation, one can show that the quotient of U by R (in terms of sheaves with respect to the étale topology) is an algebraic space. Thus we also could have defined algebraic spaces over S as quotients of S-schemes by étale equivalence relations.

A *morphism of algebraic spaces* over S is a morphism of functors. Viewing an algebraic space as a quotient of a scheme with respect to an étale equivalence relation, one can describe morphisms between algebraic spaces in terms of morphisms between schemes.

Proposition 5. *Let $f : X_1 \longrightarrow X_2$ be a morphism of algebraic spaces over S. Then, for each i, there exists a representation of X_i as a quotient of an S-scheme U_i by an étale equivalence relation (as above), and there is an S-morphism $g : U_1 \longrightarrow U_2$ such that one has the following commutative diagram*

$$\begin{array}{ccccc} (U_1 \times_{X_1} U_1) & \rightrightarrows & U_1 & \longrightarrow & X_1 \\ \downarrow{\scriptstyle g \times g} & & \downarrow{\scriptstyle g} & & \downarrow{\scriptstyle f} \\ (U_2 \times_{X_2} U_2) & \rightrightarrows & U_2 & \longrightarrow & X_2 . \end{array}$$

Furthermore, any morphism $g : U_1 \longrightarrow U_2$ inducing a commutative square as on the left-hand side gives rise to a morphism $f : X_1 \longrightarrow X_2$.

Associating to an S-scheme its functor of points, one gets a canonical map from the category of S-schemes to the category of algebraic spaces over S. This map gives rise to a fully faithful left exact embedding of categories. In the following, we will usually identify an S-scheme with its associated algebraic space over S.

Clearly, any property of S-schemes which is local for the étale topology, carries over to the context of algebraic spaces. One just requires that the property under consideration holds for the scheme U in Definition 4. This applies to the properties of being reduced, normal, regular, locally noetherian, etc.. Similarly, any property of morphisms of schemes which is local for the étale topology (on the source and on the target) carries over to the category of algebraic spaces. Thus, the properties of being flat, étale, locally of finite type, locally of finite presentation, etc. are defined. In particular, an algebraic space is provided with an étale topology in a natural way; a basis for this topology is given by the family of S-schemes U which are étale over X. The structure sheaves $\mathcal{O}_U$, where U is a scheme mapping étale to X, induce a sheaf (with respect to the étale topology) $\mathcal{O}_X$ on the algebraic space X. This sheaf is called the structure sheaf of X.

A morphism $Y \longrightarrow X$ of algebraic spaces over S is called an immersion (resp. open immersion, resp. closed immersion) if $Y \longrightarrow X$ is relatively representable by an immersion (resp. open immersion, resp. closed immersion). Thus, the notions of open and of closed subspaces of X are defined in the obvious way as equivalence classes of immersions. In particular, X carries a Zariski topology.

An algebraic space X over S is called quasi-compact if there exists a surjective étale morphism $U \longrightarrow X$ where U is a quasi-compact scheme. A morphism $X \longrightarrow Y$ of algebraic spaces is called quasi-compact if for any quasi-compact scheme V over Y, the fibre product $X \times_Y V$ is quasi-compact. Then we define a morphism $X \longrightarrow Y$ of algebraic spaces to be of finite type if it is quasi-compact and locally of finite type; and to be of finite presentation if it is quasi-compact, quasi-separated, and locally of finite presentation.

A morphism $X \longrightarrow Y$ of algebraic spaces is called proper if it is separated, of finite type, and universally closed. The latter has to be tested on the scheme level.

We mention that there is a valuative criterion for properness; cf. Deligne and Mumford [1], Thm. 4.19.

Now let us introduce the notion of points of an algebraic space.

Definition 6. *A point x of an algebraic space X over S is a morphism x*: $\operatorname{Spec} K \longrightarrow X$ *of algebraic spaces over S, where K is a field and where x is a categorical monomorphism. The field K is called the residue field of x, usually denoted by $k(x)$.*

Two points $x_i : \operatorname{Spec} K_i \longrightarrow X$, $i = 1, 2$, are called equivalent if there is an isomorphism $\sigma : \operatorname{Spec} K_1 \longrightarrow \operatorname{Spec} K_2$ such that $x_1 = x_2 \sigma$. We identify equivalent points. Since, in Definition 6, we have required x to be a monomorphism, it is easily seen that this notion of points is equivalent to the usual one when X is a scheme. Furthermore, if $U \longrightarrow X$ is a morphism where U is a scheme, then each point of U induces a point of X. So every non-empty algebraic space X over S has a point whose residue field is of finite type over S. One can even show that, for each point x of X, there exists an étale map $U \longrightarrow X$ from a scheme U and a point u of U mapping to x such that the induced extension of the residue fields $k(x) \longrightarrow k(u)$ is trivial. Such a pair (U, u) is called an étale neighborhood of (X, x) without residue field extension. By using Lemma 2.3/7, one easily sees that the family of all such étale neighborhoods is a directed inductive system. So we get the notion of a local ring at a point of an algebraic space.

Definition 7. *The local ring for the étale topology of an algebraic space X at a point x of X is defined by the inductive limit*

$$\mathcal{O}_{X,x} = \varinjlim \mathcal{O}_{U,u}$$

where the limit is taken over the family of all étale neighborhoods (U, u) of (X, x) without residue field extension.

As explained in Section 2.3, this ring is henselian. If x is a point of a scheme X, the henselization of the local ring of X at x (in terms of schemes with respect to the Zariski topology) serves as the local ring of X at x if X is viewed as an algebraic space.

Let us mention some conditions under which an algebraic space is already a scheme. So let us start with an S-scheme U and an étale equivalence relation R on U. If R is finite, then the quotient U/R (in terms of sheaves with respect to the étale topology) is represented by a scheme if and only if, for each point u of U, the set of points which, under R, are equivalent to u is contained in an affine open subscheme; cf. [FGA], n°212, Thm. 5.3. For example, if U is affine, then U/R is represented by the affine scheme defined by the kernel of the maps

$$\mathcal{O}_U(U) \rightrightarrows \mathcal{O}_R(R) .$$

In general, such a quotient is just an algebraic space and not necessarily a scheme, even if R is finite. But it can be shown that, for any algebraic space X over S, there exists a dense open subspace which is a scheme. If the base scheme S is a field, separated algebraic spaces over S of dimension 1 are schemes. Furthermore, group objects in the category of algebraic spaces over a field are schemes, as one easily shows by using the results of Section 6.6.

Next we want to describe M. Artin's criterion for a functor to be an algebraic space. We begin by reviewing some notions which are needed to state the general theorem. In the following, let S be a base scheme which is *locally of finite type over a field or over an excellent Dedekind ring*, and let

$$F : (\mathrm{Sch}/S)^0 \longrightarrow (\mathrm{Sets})$$

be a contravariant functor. If $T = \operatorname{Spec} B$ is an affine scheme over S, we will also write $F(B)$ instead of $F(T)$.

The functor F is said to be *locally of finite presentation* over S if, for every filtered inverse system of affine S-schemes $\{\operatorname{Spec} B_i\}$, the canonical morphism

$$\varinjlim F(B_i) \longrightarrow F(\varinjlim B_i)$$

is an isomorphism. Note that, if F is an S-scheme, then F is locally of finite presentation as a functor if and only if it is locally of finite presentation as a scheme over S; cf. [EGA IV_3], 8.14.2.

Furthermore, we need some definitions concerning deformations. Let s be a point in S whose residue field is of finite type over S, let k' be a finite extension of $k(s)$, and let ζ_0 be an element of $F(k')$. An *infinitesimal deformation* of ζ_0 is a pair (A, ξ) where A is an artinian local S-scheme with residue field k', and where ξ is an element of $F(A)$ inducing $\zeta_0 \in F(k')$ by functoriality. A *formal deformation of* ζ_0 is a pair $(\bar{A}, \{\xi_n\}_{n\in\mathbb{N}})$, where $\bar{A}$ is a complete noetherian local $\mathcal{O}_S$-algebra with residue field k', where the elements $\xi_n \in F(\bar{A}/\mathfrak{m}^{n+1})$ are compatible in the sense that ξ_n induces ξ_{n-1} by functoriality, and where ξ_0 coincides with ζ_0. Here $\mathfrak{m}$ is the maximal ideal of $\bar{A}$. If the sequence $\{\xi_n\}_{n\in\mathbb{N}}$ is induced by an element $\bar{\xi} \in F(\bar{A})$ via functoriality, then $(\bar{A}, \{\xi_n\}_{n\in\mathbb{N}})$ or $(\bar{A}, \bar{\xi})$ is called an *effective formal deformation* of ζ_0. A formal deformation $(\bar{A}, \{\xi_n\}_{n\in\mathbb{N}})$ of ζ_0 is said to be *versal* (resp. *universal*) if it has the following property:

Let (B', η') be an infinitesimal deformation of ζ_0 and, for an integer n, let the $(n+1)$-st power of the maximal ideal of B' be zero. Let B be a quotient of B', and denote by $\eta \in F(B)$ the element induced by η'. Then every map

$$(\bar{A}/\mathfrak{m}^{n+1}, \xi_n) \longrightarrow (B, \eta)$$

sending ξ_n to η can be factored (resp. uniquely factored) through (B', η') in the sense of morphisms of $\mathcal{O}_S$-algebras.

We mention that, in general, the canonical map

$$(*) \qquad F(\bar{A}) \longrightarrow \varprojlim F(\bar{A}/\mathfrak{m}^{n+1})$$

is not injective. Hence, if $(\bar{A}, \bar{\xi})$ is an effective formal deformation of ζ_0, the element $\bar{\xi} \in F(\bar{A})$ does not need to be uniquely determined by the sequence $\{\xi_n\}_{n\in\mathbb{N}}$ even if $(\bar{A}, \bar{\xi})$ is universal. Nevertheless, the ring $\bar{A}$ is uniquely determined (up to canonical isomorphism) if $(\bar{A}, \bar{\xi})$ is a universal deformation of ζ_0. But, for most of the functors we are interested in, the map $(*)$ is bijective for any noetherian complete local $\mathcal{O}_S$-algebra $\bar{A}$. For example, this is the case for the Hilbert functor $\mathrm{Hilb}_{X/S}$ or for the relative Picard functor $\mathrm{Pic}_{X/S}$ if X is proper over S, as one can show by using

Grothendieck's existence theorem on formal sheaves; cf. [EGA III_1], § 5. In particular, in these cases any formal deformation is effective.

Now let X be an algebraic space over S, and let x be a point of X which is of finite type over S. Denote by $k(x)$ the residue field of x and by ζ_0^x the inclusion of x into X. Let $\bar{A}^x$ be the completion of the local ring of X at x with respect to the maximal ideal, and let

$$\bar{\xi}^x : \operatorname{Spec} \bar{A}^x \longrightarrow X$$

be the canonical morphism. The pair $(\bar{A}^x, \bar{\xi}^x)$ will serve as an effective formal deformation of ζ_0^x which is universal. Thus, in order to show that a contravariant functor F from (Sch/S) to (Sets) is an algebraic space, one should first look for the existence of universal deformations at all points of F which are of finite type over S. Therefore, one introduces the following notion.

A contravariant functor $F : (\mathrm{Sch}/S)^0 \longrightarrow (\mathrm{Sets})$ is said to be *pro-representable* if the following data are given:

(a) an index set I,

(b) for each $x \in I$, an $\mathcal{O}_S$-field of finite type k^x and an element $\zeta_0^x \in F(k^x)$,

(c) for each $x \in I$, a formal deformation $(\bar{A}^x, \{\xi_n^x\}_{n \in \mathbb{N}})$ of $\zeta_0^x \in F(k^x)$,

satisfying the condition that, for each artinian local S-scheme T of finite type and for each $\eta \in F(T)$, there is a unique $x \in I$ and a unique map $T \longrightarrow \operatorname{Spec} \bar{A}^x$ sending $\{\xi_n^x\}$ to η.

Note that $(\bar{A}^x, \{\xi_n^x\}_{n \in \mathbb{N}})$ is a universal formal deformation of ζ_0^x. Furthermore, F is called *effectively pro-representable* if each sequence $\{\xi_n^x\}$ is induced by an element $\bar{\xi}^x \in F(\bar{A}^x)$. If F is effectively pro-representable, then the elements $x \in I$ are called the points of finite type of F. In the case where F is an algebraic space, the notion of points of finite type coincides with the one given in Definition 6; one associates to $x \in I$ the point of F given by the map $\zeta_0^x : \operatorname{Spec} k^x \longrightarrow F$. The universal deformations $(\bar{A}^x, \bar{\xi}^x)$ of ζ_0^x, $x \in I$, are called the *formal moduli* of F.

A morphism $\xi : X \longrightarrow F$ from an S-scheme X to the functor F is said to be *formally smooth* (resp. *formally étale*) *at a point* $x \in X$ if ξ fulfills the following lifting property: For every commutative diagram of morphisms

$$\begin{array}{ccc} X & \longleftarrow & Z_0 \\ \big\downarrow & & \big\downarrow \\ F & \longleftarrow & Z \end{array}$$

where Z is an artinian S-scheme, where Z_0 is a closed subscheme of Z defined by a nilpotent ideal, and where $Z_0 \longrightarrow X$ is a map sending Z_0 to x, there exists a factorization (resp. a unique factorization) $Z \longrightarrow X$ making the diagram commutative. One easily shows that, if $\xi : X \longrightarrow F$ is relatively representable by morphisms which are locally of finite presentation, ξ is formally étale at a point x of X if and only if, after any base change $Y \longrightarrow F$ by an S-scheme Y, the projection $X \times_F Y \longrightarrow Y$ is étale at every point of $X \times_F Y$ above x; use [EGA IV_4], 17.14.2.

Theorem 8 (M. Artin [5], Thm. 3.4). *Let S be a scheme which is locally of finite type over a field or over an excellent Dedekind ring. Let F be a functor from* $(\mathrm{Sch}/S)^0$ *to*

(Sets). *Then F is an algebraic space (resp. a separated algebraic space) over S if and only if the following conditions hold*:

[0] (sheaf axiom) *F is a sheaf for the étale topology.*

[1] (finiteness) *F is locally of finite presentation.*

[2] (pro-representability) *F is effectively pro-representable.*

[3] (relative representability) *Let T be an S-scheme of finite type, and let ξ, $\eta \in F(T)$. Then the condition $\xi = \eta$ is representable by a subscheme (resp. a closed subscheme) of $T \times_S T$.*

[4] (openness of versality) *Let X be an S-scheme of finite type, and let $\xi : X \longrightarrow F$ be a morphism. If ξ is formally étale at a point $x \in X$, then it is formally étale in a neighborhood of x.*

The necessity is not difficult to show and has already been discussed when introducing the above notions. For the sufficiency which is the more interesting part, one needs an approximation argument for algebraic structures over complete local rings; cf. M. Artin [5], Thm. 1.6. The rough idea for the proof of the sufficiency is the following.

One has to find a morphism $U \longrightarrow F$ from an S-scheme which is locally of finite presentation to F such that $U \longrightarrow F$ is relatively representable by étale surjective morphisms. We will first construct an étale neighborhood for each point of F which is of finite type over S. Consider such a point x of F, and let $(\bar{A}^x, \bar{\xi}^x)$ be the formal deformation pro-representing F at x. Then one constructs an algebraization of $(\bar{A}^x, \bar{\xi}^x)$; i.e., an S-scheme X of finite type, a closed point $x \in X$ with residue field $k(x) = k^x$, and an element $\xi \in F(X)$, such that the triple (X, x, ξ) gives rise to a versal formal deformation of $\bar{\zeta}_0^x$. More precisely, there is an isomorphism $\hat{\mathcal{O}}_{X,x} \cong \bar{A}^x$ such that ξ induces ξ_n^x in $F(\bar{A}^x/\mathfrak{m}^{n+1})$ for each $n \in \mathbb{N}$. The existence of such an algebraization follows easily from the approximation theorem 3.6/16 if the ring $\bar{A}^x$ of the formal modulus is isomorphic to a formal power series ring $\hat{\mathcal{O}}_{S,s}[[t_1, \ldots, t_n]]$, where $\hat{\mathcal{O}}_{S,s}$ is the completion of a local ring of S.—For example, this holds for the Picard functor of a relative curve.—In this case, $\bar{A}^x$ is the completion of an S-scheme X of finite type at a point x of finite type. Namely, write $\bar{A}^x$ as a union of $\mathcal{O}_S$-subalgebras B of finite type. Since F is assumed to be locally of finite presentation, the element $\bar{\xi}^x$ is represented by an element $\xi \in F(B)$ for some $\mathcal{O}_S$-subalgebra B of finite type. The inclusion $B \hookrightarrow \bar{A}^x$ yields a map $F(B) \longrightarrow F(\bar{A}^x)$ sending ξ to $\bar{\xi}^x$. Due to the approximation theorem, there is an étale neighborhood (X', x') of (X, x) without residue field extension such that there is a commutative diagram

$$\begin{array}{ccc} \operatorname{Spec} \bar{A}^x & \longleftarrow & \operatorname{Spec} \bar{A}^x/\mathfrak{m}^2\bar{A}^x \\ \big\downarrow & & \big\downarrow \\ \operatorname{Spec} B & \longleftarrow & X' \end{array}$$

sending the closed point of $\operatorname{Spec} \bar{A}^x/\mathfrak{m}^2\bar{A}^x$ to x'. The completion $\hat{\mathcal{O}}_{X',x'}$ is still isomorphic to the ring $\bar{A}^x$. Denote by $\xi' \in F(X')$ the image of ξ under the functorial map $F(B) \longrightarrow F(X')$. Due to the versality of $(\bar{A}, \bar{\xi}^x)$, there is an automorphism $\varphi : \bar{A}^x \longrightarrow \bar{A}^x$, which is the identity modulo $\mathfrak{m}^2\bar{A}^x$, and which sends ξ_n^x to ξ_n' for each

$n \in \mathbb{N}$ where ξ'_n is induced by ξ' via functoriality. Thus (X', x', ξ') is the required algebraization.

Now, let I be the set of points of F which are of finite type over S and, for $x \in I$, denote by (U^x, u^x, ξ^x) an algebraization of the formal modulus $(\bar{A}^x, \bar{\xi}^x)$. One easily shows that $\xi^x : U^x \longrightarrow F$ is formally étale at u^x. Due to condition [4], after shrinking U^x we may assume that ξ^x is étale at every point. Hence, since $U^x \longrightarrow F$ is relatively representable by condition [3], it is representable by étale maps. If we denote by U the disjoint union of the U^x, $x \in I$, the map

$$U = \coprod_{x \in I} U^x \longrightarrow F$$

is representable by étale surjective maps. Furthermore, due to condition [3], the equivalence relation $U \times_F U \longrightarrow U \times_S U$ is relatively representable by a subscheme (resp. by a closed subscheme) of $U \times_S U$. Thereby we see that F is an algebraic space as asserted in Theorem 8. □

Conditions [0] and [1] are natural, and they are satisfied quite often. For conditions [2] and [3], it is convenient to suppose that there is a deformation theory for the functor F so that one can rewrite the conditions in terms of deformation theory. Then it is often possible to decide whether a functor is pro-representable or relatively representable. Condition [4] is the one which is most difficult to verify, but it can also be interpreted by infinitesimal methods. We mention that there is a general theorem by M. Artin which relates the representability of a functor admitting a deformation theory to a list of conditions which can be checked in specific situations; for instance for the Hilbert functor or the relative Picard functor; cf. M. Artin [5], Thm. 5.4. Since many technical details are involved, we omit precise statements here.

To end our discussion, we want to indicate the procedure of proof for Theorem 1. Details can be found in M. Artin [5], Section 7; see also the appendix of M. Artin [7]. Since X is assumed to be of finite presentation over S, one can reduce to the case where the base scheme S is of finite type over the integers $\mathbb{Z}$. Then one applies the general criterion for a functor to be an algebraic space. The deformation theory for $\mathrm{Pic}_{X/S}$ is given by the exponential map. If $f: X \longrightarrow S$ is cohomologically flat in dimension zero, the deformation theory for $\mathrm{Pic}_{X/S}$ fulfills all conditions which are required in the list of the general statement. Thus $\mathrm{Pic}_{X/S}$ is pro-representable. Due to Grothendieck's existence theorem on formal sheaves, [EGA III_1], § 5, one obtains formal moduli for $\mathrm{Pic}_{X/S}$, i.e., $\mathrm{Pic}_{X/S}$ is effectively pro-representable. Then, due to M. Artin's approximation theorem, the formal moduli are algebraizable, and hence one gets local models for the space which will represent $\mathrm{Pic}_{X/S}$. Since these local models are unique up to étale morphism, they can be glued together to form an algebraic space over S.

Finally let us mention that the definition of algebraic spaces is not generalized by allowing flat equivalence relations of finite type in place of étale ones. This is due to the following fact; cf. M. Artin [7], Cor. 6.3.

If U is an S-scheme of finite type over a noetherian base scheme S, and if R is a flat equivalence relation of finite type on U, then the quotient U/R in terms of sheaves for the fppf-topology is represented by an algebraic space.

As a corollary, one obtains the following useful assertion.

Proposition 9. *Let H and G be group objects in the category of algebraic spaces over S and let $H \longrightarrow G$ be an immersion. Assume that H is flat over S. Then the quotient G/H in terms of sheaves for the fppf-topology is represented by an algebraic space.*

8.4 Properties

In this section we want to collect some results concerning the smoothness and certain finiteness properties of $\mathrm{Pic}_{X/S}$. Let us start with a theorem which is contained in [FGA], n°236, Thm. 2.10, for the case where $\mathrm{Pic}_{X/S}$ is a scheme; but it is immediately clear that it remains true if $\mathrm{Pic}_{X/S}$ is an algebraic space.

Theorem 1. *Let $f: X \longrightarrow S$ be a proper and flat morphism which is locally of finite presentation. Assume that f is cohomologically flat in dimension zero so that $\mathrm{Pic}_{X/S}$ is an algebraic space. Then the following assertions hold.*

(a) *There is a canonical isomorphism*

$$\mathrm{Lie}(\mathrm{Pic}_{X/S}) \xrightarrow{\sim} R^1 f_* \mathcal{O}_X$$

where $\mathrm{Lie}(\mathrm{Pic}_{X/S})$ is the Lie algebra of $\mathrm{Pic}_{X/S}$.

(b) *If S is the spectrum of a field K, then*

$$\dim_K \mathrm{Pic}_{X/K} \leq \dim_K H^1(X, \mathcal{O}_X)\,,$$

and equality holds if and only if $\mathrm{Pic}_{X/K}$ is smooth over K. In particular, the latter is the case if the characteristic of K is zero.

Proof. (a) Write $\mathcal{O}_S[\varepsilon]$ for the $\mathcal{O}_S$-algebra of the dual numbers over $\mathcal{O}_S$, and set $S[\varepsilon] = \mathrm{Spec}(\mathcal{O}_S[\varepsilon])$. Then one can interpret $\mathrm{Lie}(\mathrm{Pic}_{X/S})$ as the subfunctor of $\mathrm{Hom}_S(S[\varepsilon]), \mathrm{Pic}_{X/S})$ consisting of all morphisms which, modulo ε, reduce to the unit section of $\mathrm{Pic}_{X/S}$. Setting $X[\varepsilon] = X \times_S S[\varepsilon]$, one obtains the exact sequence

$$\begin{array}{c} 0 \longrightarrow \mathcal{O}_X \longrightarrow \mathcal{O}^*_{X[\varepsilon]} \longrightarrow \mathcal{O}^*_X \longrightarrow 0 \\ h \longmapsto 1 + h\cdot\varepsilon \end{array}$$

Since f is cohomologically flat in dimension zero, the canonical map $f_*\mathcal{O}_{X[\varepsilon]} \longrightarrow f_*\mathcal{O}_X$ is surjective. Therefore the sequence of sheaves with respect to the étale-topology

$$0 \longrightarrow R^1 f_* \mathcal{O}_X \longrightarrow R^1 f_* \mathcal{O}^*_{X[\varepsilon]} \longrightarrow R^1 f_* \mathcal{O}^*_X \longrightarrow R^2 f_* \mathcal{O}_X$$

is exact. Since Lie($\mathrm{Pic}_{X/S}$) corresponds to the kernel of the map $R^1 f_* \mathcal{O}^*_{X[\varepsilon]} \longrightarrow R^1 f_* \mathcal{O}^*_X$, it can be identified with $R^1 f_* \mathcal{O}_X$.

(b) follows from (a) and 2.2/15. □

Proposition 2. *Let $f: X \longrightarrow S$ be a proper and flat morphism which is locally of finite presentation. Let s be a point of S such that $H^2(X_s, \mathcal{O}_{X_s}) = 0$. Then there exists an open neighborhood U of s such that $\mathrm{Pic}_{X/S}|_U$ is formally smooth over U.*

In particular, in the case of a relative curve X over S, both $\mathrm{Pic}_{X/S}$ and $(\mathrm{Pic}_{X/S}, Y)$, where Y is a rigidificator for $\mathrm{Pic}_{X/S}$, are formally smooth over S.

Proof. Due to the semicontinuity theorem [EGA III_2], 7.7.5, there exists an open neighborhood U of s such that $H^2(X_s, \mathcal{O}_{X_s}) = 0$ for all $s \in U$. We may assume $U = S$. In order to prove that $\mathrm{Pic}_{X/S}$ is formally smooth over S, we have to establish the lifting property for $\mathrm{Pic}_{X/S}$. So consider an affine S-scheme Z and a subscheme Z_0 of Z which is defined by an ideal $\mathcal{N}$ of $\mathcal{O}_Z$ satisfying $\mathcal{N}^2 = 0$. Then we have to show that the map

$$R^1(f \times_S Z)_* \mathcal{O}^*_{X \times_S Z} \longrightarrow R^1(f \times_S Z_0)_* \mathcal{O}^*_{X \times_S Z_0}$$

is surjective. The cokernel of this map is a subsheaf of the $\mathcal{O}_Z$-module $R^2(f \times_S Z)_*(\mathcal{N} \otimes_{\mathcal{O}_S} \mathcal{O}_X)$. The latter vanishes, since $H^2(X_s, \mathcal{O}_{X_s}) = 0$ for all $s \in S$; use [EGA III_2], 7.7.10 and 7.7.5 (II). Thus we see that $\mathrm{Pic}_{X/S}$ satisfies the lifting property and, hence, is formally smooth over S.

In the case of a relative curve X over S, the assumption $H^2(X_s, \mathcal{O}_{X_s}) = 0$ is satisfied at all $s \in S$, so $\mathrm{Pic}_{X/S}$ is formally smooth over S. Furthermore, since there is no obstruction to lifting a rigidification, we see that $(\mathrm{Pic}_{X/S}, Y)$ is formally smooth over S, too. □

Now we will concentrate on finiteness assertions for $\mathrm{Pic}_{X/S}$. When proving Grothendieck's theorem 8.2/1, we had seen in 8.2/5 that $\mathrm{Pic}^{\Phi}_{X/S}$ is quasi-projective over S. But if we impose stronger conditions on the fibres of X, we can expect better results.

Theorem 3 ([FGA], n°236, Thm. 2.1). *Let $f: X \longrightarrow S$ be a proper (resp. projective) morphism which is locally of finite presentation. Assume that the geometric fibres of X are reduced and irreducible. Then $\mathrm{Pic}_{X/S}$ is a separated algebraic space (resp. separated scheme) over S.*

If, in addition, $f: X \longrightarrow S$ is smooth, then each closed subspace Z of $\mathrm{Pic}_{X/S}$ which is of finite type over S is proper (resp. projective) over S. In particular, if S consists of a field K, the identity component $\mathrm{Pic}^0_{X/K}$ of $\mathrm{Pic}_{X/K}$ is a proper scheme over K.

Proof. $\mathrm{Pic}_{X/S}$ is an algebraic space over S, due to 8.3/1. If X is projective over S, we know from 8.2/1 that $\mathrm{Pic}_{X/S}$ is a scheme over S and from 8.2/5 that each closed subspace Z which is of finite type over S is quasi-projective over S. The remaining assertions follow by using the valuative criteria for separatedness and properness.

Indeed, we may assume that S is the spectrum of a discrete valuation ring R, and that X admits a section over S. For showing the separatedness, we have to verify that a line bundle $\mathscr{L}$ on X which is trivial on the generic fibre is trivial. There exists a global section $f \in \Gamma(X, \mathscr{L})$ which generates $\mathscr{L}$ on the generic fibre. Since the local ring $\mathcal{O}_{X,\eta}$ of X at the generic point η of the special fibre is a discrete valuation ring such that the extension $R \longrightarrow \mathcal{O}_{X,\eta}$ is of ramification index 1, we may assume that f generates $\mathscr{L}$ at η. Then it is clear that f generates $\mathscr{L}$ on X and that $\mathscr{L}$ is trivial. Next assume that X is smooth over S. For the properness, we have to show that each line bundle on the generic fibre of X extends to a line bundle on X. Since the local rings of X are regular, the notions of Cartier divisor and Weil divisor coincide. Obviously, Weil divisors on the generic fibre of X extend to Weil divisors on X. So, each line bundle on the generic fibre extends to a line bundle on X.

If S consists of a field K, then $\mathrm{Pic}_{X/K}$ is a scheme by 8.2/3. Since any connected K-group scheme is of finite type as soon as it is locally of finite type, we see that $\mathrm{Pic}^0_{X/K}$ is of finite type and, thus, proper over K. □

Next we want to discuss finiteness assertions for $\mathrm{Pic}_{X/S}$ under more general assumptions. Since, in general, $\mathrm{Pic}_{X/S}$ will have infinitely many connected components, it cannot be of finite type over S. So the best one can expect is that there exists an open and closed subgroup $\mathrm{Pic}^\tau_{X/S}$ of $\mathrm{Pic}_{X/S}$ which is of finite type over S and which has the property that the quotient of $\mathrm{Pic}_{X/S}$ by $\mathrm{Pic}^\tau_{X/S}$ has geometric fibres which are finitely generated as abstract groups. We want to introduce the subgroup $\mathrm{Pic}^\tau_{X/S}$.

If S consists of a field, we know that the relative Picard functor $\mathrm{Pic}_{X/S}$ is a group scheme. Let $\mathrm{Pic}^0_{X/S}$ be its identity component. Then we set

$$\mathrm{Pic}^\tau_{X/S} = \bigcup_{n>0} n^{-1}(\mathrm{Pic}^0_{X/S})$$

where $n : \mathrm{Pic}_{X/S} \longrightarrow \mathrm{Pic}_{X/S}$ is the multiplication by n. Due to continuity, $\mathrm{Pic}^\tau_{X/S}$ is an open subscheme of $\mathrm{Pic}_{X/S}$.

For a general base S, we introduce $\mathrm{Pic}^0_{X/S}$ (resp. $\mathrm{Pic}^\tau_{X/S}$) as the subfunctor of $\mathrm{Pic}_{X/S}$ which consists of all elements whose restrictions to all fibres X_s, $s \in S$, belong to $\mathrm{Pic}^0_{X_s/k(s)}$ (resp. $\mathrm{Pic}^\tau_{X_s/k(s)}$). If $\mathrm{Pic}_{X/S}$ is an algebraic space (resp. a scheme), and if it is smooth over S along the unit section, then $\mathrm{Pic}^0_{X/S}$ is represented by an open subspace (resp. an open subscheme) of $\mathrm{Pic}_{X/S}$, cf. [EGA IV_3], 15.6.5.

Theorem 4 ([SGA 6], Exp. XIII, Thm. 4.7). *Let $f : X \longrightarrow S$ be a proper morphism which is locally of finite presentation, and let S be quasi-compact. Then*

(a) *The canonical inclusion $\mathrm{Pic}^\tau_{X/S} \hookrightarrow \mathrm{Pic}_{X/S}$ is relatively representable by an open and quasi-compact immersion.*

(b) *If $X \longrightarrow S$ is projective and if its geometric fibres are reduced and irreducible, the immersion $\mathrm{Pic}^\tau_{X/S} \hookrightarrow \mathrm{Pic}_{X/S}$ is open and closed.*

(c) *$\mathrm{Pic}^\tau_{X/S}$ is of finite type over S in the sense that the family of isomorphism classes of line bundles on the fibres of X which belong to $\mathrm{Pic}^\tau_{X/S}$ is bounded.*

The hardest part of the theorem is assertion (c). One can reduce it to the case where X is a closed subscheme of a projective space $\mathbb{P}^n_S$. In this case, one shows that all elements of $\mathrm{Pic}^{\tau}_{X/S}$ have the same Hilbert polynomial (with respect to the S-ample line bundle belonging to the embedding of X into $\mathbb{P}^n_S$), and then the assertion can be deduced from 8.2/5.

Next, we want to look at the special case where X is an abelian S-scheme, i.e., a smooth and proper S-group scheme with connected fibres.

Theorem 5. *Let A be a projective abelian S-scheme.*

(a) *Then $\mathrm{Pic}^{\tau}_{A/S}$ is a projective abelian S-scheme. It is denoted by A^* and is called the dual abelian scheme of A. In particular, A^* coincides with the identity component of $\mathrm{Pic}_{A/S}$.*

(b) *The Poincaré bundle on $A \times_S A^*$ gives rise to a canonical isomorphism $\iota : A \longrightarrow A^{**}$ where A^{**} is the dual abelian scheme of A^*.*

A proof of (a) can be found in Mumford [1], Corollary 6.8. For (b), since A and A^{**} are flat over S, it suffices to treat the case where S consists of an algebraically closed field. In this case, the assertion follows from Mumford [3], Section 13, p. 132.

In 1.2/8 we have seen that an abelian scheme over a Dedekind scheme is the Néron model of its generic fibre. Now, using the above theorem, one can show a much stronger mapping property for abelian schemes than the one required for Néron models.

Corollary 6. *Let A be an abelian S-scheme. Then any rational S-morphism $\varphi : T \dashrightarrow A$ from an S-scheme T to A is defined everywhere if T is regular.*

Proof. We may assume $T = S$. Then A is projective over S; cf. Murre [2], p. 16. Due to Theorem 5, we can identify A and A^{**}. So the map φ corresponds to a line bundle on $A^* \times_S S'$ where S' is a dense open subscheme of S. Since $S = T$ is regular and since $A^* \longrightarrow S$ is smooth, the scheme A^* is regular. So the line bundle extends to a line bundle on A^* and, thus, gives rise to an extension $S \longrightarrow A^{**}$ of $\varphi_{|S'}$. □

Now let us return to the general situation of a proper morphism $X \longrightarrow S$ of schemes. We want to discuss the group of connected components of $\mathrm{Pic}_{X/S}$ over a geometric point of S. Let s be a point of S and let $\bar{s}$ be a geometric point of S such that $k(\bar{s})$ is an algebraic closure of the residue field $k(s)$ at s. The group of connected components of $\mathrm{Pic}_{X_{\bar{s}}/k(\bar{s})}$ is called the *Néron-Severi group* of the geometric fibre $X_{\bar{s}} = X \times_S k(\bar{s})$ of X over s. It is denoted by $\mathrm{NS}_{X/S}(\bar{s})$ so that

$$\mathrm{NS}_{X/S}(\bar{s}) = \mathrm{Pic}_{X_{\bar{s}}/k(\bar{s})}(k(\bar{s})) / \mathrm{Pic}^0_{X_{\bar{s}}/k(\bar{s})}(k(\bar{s})) .$$

Theorem 7. ([SGA 6], Exp. XIII, Thm. 5.1). *Let $f : X \longrightarrow S$ be a proper morphism which is locally of finite presentation, and assume that S is quasi-compact. Then the Néron-Severi groups $\mathrm{NS}_{X/S}(\bar{s})$ of the geometric fibres of X are finitely generated. Their ranks as well as the orders of their torsion subgroups are bounded simultaneously.*

Remark 8. The Néron-Severi group is of arithmetical nature; i.e., the set of points where the Néron-Severi group is of a fixed type is not necessarily constructible.

For example, let $E \longrightarrow S$ be an elliptic curve with a non-constant j-invariant over an irreducible base S which is of finite type over a field. Then there are infinitely many geometric points $\bar{s}$ of S such that the geometric fibre $E_{\bar{s}}$ has complex multiplication, and there are infinitely many geometric points such that the geometric fibre $E_{\bar{s}}$ has no complex multiplication. Now consider the product $X = E \times_S E$. If $E_{\bar{s}}$ has no complex multiplication, the rank of the Néron-Severi group of $X_{\bar{s}}$ is 3. If $E_{\bar{s}}$ has complex multiplication, the rank of the Néron-Severi group of $X_{\bar{s}}$ is at least 4.

Chapter 9. Jacobians of Relative Curves

The chapter consists of two parts. In the first four sections we study the representability and structure of $\mathrm{Pic}_{X/S}$ for a relative curve X over a base S. Then, in the last three sections, we work over a base S consisting of a discrete valuation ring R with field of fractions K and, applying these results, we investigate the relationship between $\mathrm{Pic}_{X/S}$ and the Néron model of the Jacobian J_K of the generic fibre X_K.

The chapter begins with a discussion of the degree of divisors on relative curves. Then we give a detailed analysis of the Jacobian J_K of a proper curve X_K over a field, showing that the structure of J_K is closely related to geometric properties of X_K. The next two sections deal with the representability of Jacobians over a more general base. First, imposing strong conditions on the fibres of the curve and working over a strictly henselian base, we prove the representability by a scheme, using a method which was originally employed by Weil [2] and Rosenlicht [1]; see also Serre [1]. Then we explain results due to Deligne [1] and Raynaud [6], which are valid under far weaker conditions.

In the second half of the chapter, we follow Raynaud [6] and consider a proper and flat curve X over a discrete valuation ring R, assuming in most cases that X is regular at each of its points and that the generic fibre X_K is geometrically irreducible. Let P be the open subfunctor of $\mathrm{Pic}_{X/R}$ consisting of all line bundles of total degree 0 and let Q be the biggest separated quotient of P. We show that Q is a smooth R-group scheme whose generic fibre coincides with the Jacobian J_K of the generic fibre X_K. Thus if J is a Néron model of J_K, there is a canonical R-morphism $Q \longrightarrow J$. Without assuming the existence of J, we can prove under quite general conditions that, for example, if the residue field of R is perfect, then Q is already a Néron model of J_K. Thereby it is seen that the relative Picard functor leads to a second possibility of constructing Néron models. Also there are important situations where the identity component of $\mathrm{Pic}^0_{X/R}$ is already a separated scheme and where the canonical morphism $\mathrm{Pic}^0_{X/R} \longrightarrow J^0$ is an isomorphism. More precisely, we will see that the coincidence of $\mathrm{Pic}^0_{X/R}$ and J^0 is related to the fact that X has rational singularities.

In the above cases where Q is already a Néron model of J_K, it is possible to compute explicitly the group of components (of the special fibre) of this model, using the intersection form on X. In Section 9.6, we explain the general approach and carry out some computations in particular cases.

9.1 The Degree of Divisors

Let X be a proper curve over a field K. If x is a closed point of X and if f is a regular element of $\mathcal{O}_{X,x}$, we define the vanishing order of f at x by

$$\mathrm{ord}_x(f) := l_{\mathcal{O}_{X,x}}(\mathcal{O}_{X,x}/(f))$$

where $l_{\mathcal{O}_{X,x}}$ denotes the length of $\mathcal{O}_{X,x}$-modules. If, for example, x is a regular point of X, the local ring $\mathcal{O}_{X,x}$ is a discrete valuation ring and $\mathrm{ord}_x(f)$ corresponds to the order of f in $\mathcal{O}_{X,x}$ (with respect to the canonical valuation on $\mathcal{O}_{X,x}$). Since we have

$$\mathrm{ord}_x(f \cdot g) = \mathrm{ord}_x(f) + \mathrm{ord}_x(g)$$

for a product of regular elements $f, g \in \mathcal{O}_{X,x}$, we can define

$$\mathrm{ord}_x(f/g) = \mathrm{ord}_x(f) - \mathrm{ord}_x(g)$$

for any element f/g of the total ring of fractions of $\mathcal{O}_{X,x}$.

Now let D be a Cartier divisor on X. For a closed point $x \in X$, we set

$$\mathrm{ord}_x(D) = \mathrm{ord}_x(f_x/g_x)$$

where f_x/g_x is a local representation of D in a neighborhood of x. We can associate to D the Weil divisor

$$\sum_{x \in X} \mathrm{ord}_x(D) \cdot x \, .$$

The *degree of a Cartier divisor* D is defined by

$$\deg(D) = \sum_{x \in X} \mathrm{ord}_x(D) \cdot [k(x) : K] \, .$$

The degree function is additive, i.e.,

$$\deg(D_1 + D_2) = \deg(D_1) + \deg(D_2) \, .$$

If D is effective, we can write

$$\deg(D) = \dim_K H^0(X, \mathcal{O}_D)$$

where $\mathcal{O}_D$ denotes the structure sheaf of the subscheme associated to D. Thus we see that the degree of a Cartier divisor on X is not altered by a base change with a field extension K'/K.

Assuming for a moment that X is reduced, we can consider the normalization $\tilde{X} \longrightarrow X$ of X. Then one can pull back Cartier divisors D on X to Cartier divisors $\tilde{D}$ on $\tilde{X}$. We claim that

$$\deg(D) = \deg(\tilde{D}) \, .$$

Indeed, it suffices to justify the following assertion. Let $U = \mathrm{Spec}(A)$ be an affine open subscheme of X and let $\tilde{A}$ be the normalization of A. Then, for each regular element f of A, one has

$$\dim_K(A/(f)) = \dim_K(\tilde{A}/(f)) \, .$$

In order to prove this, look at the commutative diagram

$$\begin{array}{ccccccccc}
0 & \longrightarrow & A & \longrightarrow & \tilde{A} & \longrightarrow & C & \longrightarrow & 0 \\
 & & \downarrow{\scriptstyle f_A} & & \downarrow{\scriptstyle f_{\tilde{A}}} & & \downarrow{\scriptstyle f_C} & & \\
0 & \longrightarrow & A & \longrightarrow & \tilde{A} & \longrightarrow & C & \longrightarrow & 0
\end{array}$$

with exact rows, where the vertical maps are given by the multiplication with f. Since f is a regular element of both A and $\tilde{A}$, there is a long exact sequence

$$0 \longrightarrow \ker(f_C) \longrightarrow A/(f) \longrightarrow \tilde{A}/(f) \longrightarrow C/f \cdot C \longrightarrow 0 .$$

Using $\dim_K(C) < \infty$, it follows that $\dim_K(\ker(f_C)) = \dim_K(C/f \cdot C)$. Hence, the assertion is evident.

A Cartier divisor D on an arbitrary proper curve X is called *principal* if there exists a meromorphic function f on X such that $D = \operatorname{div}(f)$. For a principal divisor D, we have $\deg(D) = 0$. Two Cartier divisors D_1 and D_2 are said to be *linearly equivalent* if the difference $D_1 - D_2$ is principal. So we see that the degree of a Cartier divisor D is not altered if we replace D by a divisor which is linearly equivalent to D. Since each line bundle $\mathscr{L}$ on X corresponds to a Cartier divisor D which is unique up to linear equivalence, one can define the *degree of a line bundle* $\mathscr{L}$ by setting $\deg(\mathscr{L}) := \deg(D)$. The degree plays an important role in the Riemann-Roch formula.

Theorem 1. *Let X be a proper curve over a field K, and let $\mathscr{L}$ be a line bundle on X. Then the Euler-Poincaré characteristic*

$$\chi(\mathscr{L}) = \dim_K H^0(X, \mathscr{L}) - \dim_K H^1(X, \mathscr{L})$$

of $\mathscr{L}$ is related to the Euler-Poincaré characteristic of $\mathcal{O}_X$ by the formula

$$\chi(\mathscr{L}) = \deg(\mathscr{L}) + \chi(\mathcal{O}_X) .$$

Proof. One proceeds as in the case of a smooth curve by looking at an exact sequence

$$0 \longrightarrow \mathscr{L} \longrightarrow \mathscr{L} \otimes_{\mathcal{O}_X} \mathcal{O}_X(D) \longrightarrow \mathcal{O}_D \longrightarrow 0$$

where D is an effective Cartier divisor on X such that $\mathscr{L} \otimes_{\mathcal{O}_X} \mathcal{O}_X(D)$ is isomorphic to $\mathcal{O}_X(E)$ with an effective Cartier divisor E on X. Furthermore, one has the exact sequence

$$0 \longrightarrow \mathcal{O}_X \longrightarrow \mathcal{O}_X(E) \longrightarrow \mathcal{O}_E \longrightarrow 0 .$$

Calculating the Euler-Poincaré characteristic of both sequences, the assertion follows immediately from $\mathscr{L} \otimes_{\mathcal{O}_X} \mathcal{O}_X(D) \cong \mathcal{O}_X(E)$ and $\deg \mathscr{L} = \deg E - \deg D$. □

If $H^0(X, \mathcal{O}_X) = K$, for example, if X is geometrically reduced and connected, the Euler-Poincaré characteristic of $\mathcal{O}_X$ is given by $\chi(\mathcal{O}_X) = 1 - p_a$, where $p_a = \dim_K H^1(X, \mathcal{O}_X)$ is the arithmetic genus of the curve X.

If $X \longrightarrow S$ is a relative curve and if $\mathscr{L}$ is a line bundle on X, one can restrict $\mathscr{L}$ to the fibres of X over S. So, for each $s \in S$, we get a line bundle $\mathscr{L}_s$ on the fibre X_s, and the degree $\deg(\mathscr{L}_s)$ of $\mathscr{L}_s$ on the fibre X_s gives rise to a $\mathbb{Z}$-valued function on S.

Proposition 2. *Let $X \longrightarrow S$ be a flat proper S-curve of finite presentation and let $\mathscr{L}$ be a line bundle on X. For $s \in S$, denote by $\mathscr{L}_s$ the restriction of $\mathscr{L}$ to the curve X_s. Then the degree function*

$$\deg : S \longrightarrow \mathbb{Z} , \qquad s \longmapsto \deg(\mathscr{L}_s)$$

is locally constant on S.

Proof. The Euler-Poincaré characteristic of a flat family of coherent sheaves is locally constant on the base; cf. [EGA III_2], 7.9.4. Thus, using the Riemann-Roch formula, one sees that the degree function must be locally constant on S. □

Now let us return to the situation we started with. Let X be a proper curve over a field with (reduced) irreducible components $X_1, \ldots, X_r$. If $\mathscr{L}$ is a line bundle on X, we can restrict $\mathscr{L}$ to each component X_i, $i = 1, \ldots, r$, and we define the *partial degree* of $\mathscr{L}$ on X_i by

$$\deg_{X_i}(\mathscr{L}) = \deg(\mathscr{L}|_{X_i}) .$$

In order to explain the relationship between the total degree and the partial degrees, we need the notion of multiplicities of irreducible components.

Definition 3. *Let X be a scheme of finite type over a field K, let $\bar{K}$ be an algebraic closure of K, and set $\bar{X} = X \otimes_K \bar{K}$. Denote by $X_1, \ldots, X_r$ the (reduced) irreducible components of X and, for $i = 1, \ldots, r$, let $\eta_i \in X$ be the generic point corresponding to X_i. The multiplicity of X_i in X is the length of the artinian local ring $\mathcal{O}_{X,\eta_i}$. We denote it by d_i; so*

$$d_i = l(\mathcal{O}_{X,\eta_i}) .$$

The geometric multiplicity of X_i in X is the length of the artinian local ring $\mathcal{O}_{\bar{X},\bar{\eta}_i}$ where $\bar{\eta}_i$ is a point of $\bar{X}$ lying above η_i. We denote it by δ_i; so

$$\delta_i = l(\mathcal{O}_{\bar{X},\bar{\eta}_i})$$

If X is irreducible, we talk about the multiplicity (resp. the geometric multipicity) of X, thereby meaning the multiplicity (resp. the geometric multiplicity) of X in X. Furthermore, we denote by

$$e_i = l(\mathcal{O}_{\bar{X}_i,\bar{\eta}_i})$$

the geometric multiplicity of X_i.

Note that the definition is independent of the choice of $\bar{\eta}_i$, since all points of $\bar{X}$ above η_i are conjugated under the action of the Galois group of $\bar{K}$ over K. There are some elementary relations between the different notions of multiplicities which are easy to verify.

Lemma 4. *Keeping the notations of Definition* 3, *one has*

(a) $\delta_i = e_i \cdot d_i$ *for* $i = 1, \ldots, r$.

(b) $\delta_i = e_i$ *if and only if X is reduced at the point* η_i.

(c) $e_i = 1$ *if the characteristic of K is zero; otherwise it is a power of the characteristic of K.*

Using the notion of multiplicity of components, one can state a relationship between the (total) degree and the partial degrees of a line bundle.

Proposition 5. *Let X be a proper curve over a field K with (reduced) irreducible components $X_1, \ldots, X_r$. Denote by d_i the multiplicity of X_i in X, $i = 1, \ldots, r$. Then*

$$\deg(\mathscr{L}) = \sum_{i=1}^{r} d_i \cdot \deg(\mathscr{L}|_{X_i})$$

for each line bundle $\mathscr{L}$ on X.

Proof. It suffices to establish the formula for Cartier divisors D whose support does not contain any intersection point of the different components. Since both sides of the formula are additive for divisors, we have only to consider effective Cartier divisors. Then the assertion follows from the lemma below. □

Lemma 6. *Let A be a one-dimensional noetherian local ring and let $\mathfrak{p}_1, \ldots, \mathfrak{p}_r$ be the minimal prime ideals of A. Let M be a finitely generated A-module, and let a be an element of A which is not contained in any $\mathfrak{p}_i$. Denote by a_M the multiplication by a on M and define*

$$e_A(a, M) = l_A(\operatorname{coker}(a_M)) - l_A(\ker(a_M)) .$$

Then

$$e_A(a, M) = \sum_{i=1}^{r} l_{A_{\mathfrak{p}_i}}(M_{\mathfrak{p}_i}) \cdot e_A(a, A/\mathfrak{p}_i) .$$

Proof. Note that both sides are additive for exact sequences of A-modules. So we may assume $M = A/\mathfrak{p}$ for a prime ideal $\mathfrak{p}$ of A; cf. Bourbaki [2], Chap. IV, § 1, n°4, Thm. 1. If $\mathfrak{p}$ is maximal, both sides are zero. If $\mathfrak{p}$ is minimal, then $l_{A_\mathfrak{p}}(M_\mathfrak{p}) = 1$ and the localizations of M at the other minimal primes are zero. Thus, the formula is also clear in this case. □

The results about the degree of line bundles which are presented in the following will be used in Section 9.4 to establish the representability of $\operatorname{Pic}_{X/S}$ if X is a relative curve over a discrete valuation ring. Furthermore, they will be of interest in Section 9.5 where we will discuss the relationship between the Picard functor and Néron models of Jacobians.

Lemma 7. *Let K be a separably closed field. Let X be an irreducible K-scheme of finite type of dimension r and let δ be the geometric multiplicity of X. Then, for each closed point $x \in X$ and for each system of parameters $f = (f_1, \ldots, f_r)$ of the local ring $\mathcal{O}_{X,x}$, the following assertions hold:*

(a) $\dim_K \mathcal{O}_{X,x}/(f) \geq \delta$.

(b) *If f is a regular sequence, $\dim_K \mathcal{O}_{X,x}/(f)$ is a multiple of δ.*

(c) *If $\dim_K \mathcal{O}_{X,x}/(f) = \delta$, then f is a regular sequence.*

Furthermore, there exist x and f such that $\dim_K \mathcal{O}_{X,x}/(f) = \delta$.

Proof. After shrinking X, we may assume that f gives rise to a quasi-finite morphism

$$\varphi : X \longrightarrow Y := \mathbb{A}_K^r .$$

Denote by $\bar{K}$ the algebraic closure of K and by $\bar{\varphi}$ the morphism $\varphi \otimes_K \bar{K}$. Since K is assumed to be separably closed, there exists a unique point $\bar{x}$ of $\bar{X} = X \otimes_K \bar{K}$ above x. Consider now the henselization Y' of $\bar{Y} := \mathbb{A}_{\bar{K}}^r$ at the origin. Let X' be the local component of $\bar{X} \times_{\bar{Y}} Y'$ above $\bar{x}$. Then the map

$$\varphi' : X' \longrightarrow Y'$$

obtained from $\bar{\varphi}$ via base change is finite. Furthermore, φ' is flat if and only if f is a regular sequence; cf. [EGA 0_{IV}], 15.1.14 and 15.1.21. The local rings of X' at generic points are artinian of length δ and the generic points of X' lie above the generic point of Y'. Hence, the degree of X' over Y' is a non-zero multiple of δ. So, by Nakayama's lemma, the degree of the closed fibre of φ' is greater or equal to δ. Since the degree of the closed fibre is equal to $\dim_K \mathcal{O}_{X,x}/(f)$, we see that assertion (a) is true.

If f is a regular sequence, X' is flat over Y'. Then the degree of the special fibre of φ' is equal to the degree of X' over Y'. Thus, assertion (b) is clear.

If the degree of the special fibre is δ, it is equal to the degree of X' over Y'; then $\mathcal{O}_{Y'}(X')$ is free over $\mathcal{O}_{Y'}(Y')$ and, hence, flat. This shows that f is a regular sequence; so assertion (c) is true.

Next we want to show that the value δ can be attained. After replacing X by a dense open subset, we may assume that $\bar{X}_{\text{red}}$ is smooth over $\bar{K}$. So the module $\Omega^1_{\bar{X}_{\text{red}}/\bar{K}}$ is locally free. Furthermore, since $\Omega^1_{\bar{X}_{\text{red}}/\bar{K}}$ is a quotient of $\Omega^1_{\bar{X}/\bar{K}}$, we may assume that there exist elements $a_1, \ldots, a_r \in \Gamma(X, \mathcal{O}_X)$ such that the images of the differentials $da_1, \ldots, da_r$ in $\Omega^1_{\bar{X}_{\text{red}}/\bar{K}}$ give rise to a basis of this module. Consider now the morphism

$$a := (a_1, \ldots, a_r) : X \longrightarrow Y := \mathbb{A}^r_K$$

given by the functions $a_1, \ldots, a_r$. The restriction of the induced map $\bar{a} : \bar{X} \longrightarrow \bar{Y}$ to $\bar{X}_{\text{red}}$ is étale. After replacing X and Y by dense open subsets, we may assume that a is finite and flat. Let x be a point of X such that $a(x)$ is a rational point of Y. We may assume that $a(x)$ is the origin. Then $f := (a_1, \ldots, a_r)$ is as required. Namely, using notations as above, we have to show that the degree of the finite and flat morphism $\varphi' : X' \longrightarrow Y'$ is δ. Since the induced morphism

$$\varphi'_{\text{red}} : X'_{\text{red}} \xrightarrow{\sim} Y'$$

is an isomorphism, the degree of φ' coincides with the length of the local ring $\mathcal{O}_{X',\eta'}$ at the generic point η' of X', which is equal to δ. □

As a corollary of Lemma 7, we get a relation between the geometric multiplicity of a component X_i of X and the partial degree $\deg_{X_i}(\mathcal{L})$ of a line bundle $\mathcal{L}$ on X.

Corollary 8. *Let X be a proper curve over a field K and let $X_1, \ldots, X_r$ be its (reduced) irreducible components. Let $\mathcal{L}$ be a line bundle on X. Denote by e_i the geometric multiplicity of X_i, $i = 1, \ldots, r$. Then the partial degree $\deg_{X_i}(\mathcal{L})$ of $\mathcal{L}$ on X_i is a multiple of e_i for $i = 1, \ldots, r$.*

Proof. We may assume that $X = X_i$ is reduced and irreducible, and we may assume that $\mathcal{L} = \mathcal{O}_X(D)$ is associated to an effective Cartier divisor D on X which is concentrated at a single point x. Let f be a regular element of $\mathcal{O}_{X,x}$ which represents D at x, so we have

$$\dim_K \mathcal{O}_{X,x}/(f) = \deg(\mathcal{L}) = \deg_{X_i}(\mathcal{L}) .$$

Due to Lemma 7, if K is separably closed, the geometric multiplicity $\delta_i = e_i$ of $X = X_i$ divides $\dim_K \mathcal{O}_{X,x}/(f) = \deg(\mathcal{L})$. In the general case, consider a separable closure K' of K. The irreducible component $X = X_i$ decomposes into the irreducible components X'_{ij} of $X' = X \otimes_K K'$, but the geometric multiplicities e_{ij} of X'_{ij} coincide with e_i. Thus we see that e_i divides $\deg_{X'_{ij}}(\mathcal{L} \otimes_K K')$, for all j. Now it follows from Proposition 5 that e_i divides $\deg(\mathcal{L}) = \deg_{X_i}(\mathcal{L})$, since the degree function is compatible with extensions of the base field. □

If X is a scheme of finite presentation over a strictly henselian base S, Lemma 7 can be used to show the existence of subschemes of X which are finite and flat over S and which have small degrees over S.

Corollary 9. *Let S be a strictly henselian local scheme, let s be its closed point, and let X be a flat S-scheme which is locally of finite presentation. Let X_0 be an irreducible component of the special fibre X_s of X and let δ be the geometric multiplicity of X_0 in X_s. Then there exists an S-immersion $a : Z \longrightarrow X$, where Z is finite and flat over S of rank δ and where $a_s(Z_s)$ is a point of X_0 not lying on any other irreducible component of X_s.*

Proof. Let U be an open subscheme of X such that $U_s = U \times_S k(s)$ is non-empty and contained in X_0. Due to Lemma 7, there exist a closed point x of U_s and a regular system of parameters $\bar{f}$ of $\mathcal{O}_{U_s,x} = \mathcal{O}_{U,x} \otimes_{\mathcal{O}_{S,s}} k(s)$ such that

$$\dim_{k(s)} \mathcal{O}_{U_s,x}/(\bar{f}) = \delta .$$

After restricting U, one can lift $\bar{f}$ to a sequence f of elements of $\Gamma(U, \mathcal{O}_U)$. Then f is a regular sequence of $\mathcal{O}_{U,x}$; cf. [EGA 0_{IV}], 15.1.16. After restricting U, a local component Z of $V(f)$ which contains x is finite and flat over S, so Z fulfills the assertion; cf. [EGA 0_{IV}], 15.1.16. □

Corollary 10. *Let S be a strictly henselian local scheme with closed point s, and let X be a flat curve over S which is locally of finite presentation. Let X_0 be an irreducible component of the special fibre X_s with geometric multiplicity δ in X_s. Then there exists an effective Cartier divisor Z of degree δ on X such that Z meets X_0, but no other irreducible component of X_s. Furthermore, $\deg_{X_0}(Z) = e$ where e is the geometric multiplicity of X_0.*

Corollary 9 implies the following criterion for the representability of elements of $\mathrm{Pic}_{X/S}$ by line bundles.

Proposition 11. *Let $f: X \longrightarrow S$ be a quasi-separated morphism of finite presentation such that $f_* \mathcal{O}_X = \mathcal{O}_S$. Consider S-morphisms $Z_i \longrightarrow X$, $i = 1, \ldots, r$, where Z_i is finite and flat over S of degree n_i. Set $n = \gcd(n_1, \ldots, n_r)$. Then, for each flat S-scheme T and for each element $\xi \in \mathrm{Pic}_{X/S}(T)$, the multiple $n \cdot \xi$ is induced by a line bundle on $X_T = X \times_S T$.*

Proof. Since n is a linear combination of $n_1, \ldots, n_r$ with integer coefficients, it suffices to prove that each $n_i \cdot \xi$ is induced by a line bundle. Due to [EGA III_1], 1.4.15, and [EGA IV_1], 1.7.21, the assumption $f_* \mathcal{O}_X = \mathcal{O}_S$ remains true after flat base change. So we may assume $S = T$. The morphism $Z_i \longrightarrow X$ gives rise to a Z_i-section of $X \times_S Z_i$. So the pull-back of ξ in $\mathrm{Pic}_{X/S}(Z_i)$ is induced by a line bundle $\mathscr{L}$ on $X \times_S Z_i$; cf. 8.1/4. Then the norm of $\mathscr{L}$ with respect to the finite flat morphism $X \times_S Z_i \longrightarrow X$ gives rise to the element $n_i \cdot \xi$ in $\mathrm{Pic}_{X/S}(S)$; cf. [EGA IV_4], 21.5.6. □

As an application of Corollary 9 and Proposition 11, one obtains the following result.

Corollary 12. *Let S be a strictly henselian local scheme, let s be its closed point, and let $f: X \longrightarrow S$ be a flat morphism of finite presentation such that $f_* \mathcal{O}_X = \mathcal{O}_S$. Denote by δ the greatest common divisor of the geometric multiplicities in X_s of the irreducible components $X_1, \ldots, X_r$ of X_s. Then, for each flat S-scheme T, and each element ξ of* $\mathrm{Pic}_{X/S}(T)$, *the multiple $\delta \cdot \xi$ is induced by a line bundle on $X \times_S T$.*

9.2 The Structure of Jacobians

In the following let X be a proper curve over a field K. Then, due to 8.2/3 and 8.4/2, $\mathrm{Pic}^0_{X/K}$ is a smooth scheme; we will also refer to it as the *Jacobian* of X. In the present section, we want to discuss the structure of $\mathrm{Pic}^0_{X/K}$ as an algebraic group depending on data furnished by the given curve X. To start with, let us mention some general results on the structure of commutative algebraic groups.

Theorem 1 (Chevalley [1] or Rosenlicht [2]). *Let K be a field and let G be a smooth and connected algebraic K-group. Then there exists a smallest* (*not necessarily smooth*) *connected linear subgroup L of G such that the quotient G/L is an abelian variety.*

If K is perfect, L is smooth and its formation is compatible with extension of the base field.

Chevalley has treated the case where K is algebraically closed and has shown that there exists a smooth connected linear subgroup L of G such that the quotient G/L is an abelian variety. If the base field is perfect, the existence of such a subgroup follows by Galois descent from the case of algebraically closed fields. It is clear that such a group is the smallest connected linear subgroup of G with abelian cokernel, and that its formation is compatible with extension of the base field.

If the base field is not perfect, there exist a finite radicial extension K' of K and a connected smooth linear K'-subgroup H' of $G' = G \otimes_K K'$ such that the quotient G'/H' is an abelian variety. Let us first show that there exists a connected linear subgroup H of G such that $H \otimes_K K'$ contains H'. Let n be the exponent of the radicial extension K'/K. Then consider the n-fold Frobenius

$$F_n : G' \longrightarrow G'^{(p^n)} = G' \times_{K'} K'^{(1/p^n)}$$

(cf. [SGA 3_{I}], Exp. $\mathrm{VII}_{\mathrm{A}}$, 4.1); the second projection is induced by the inclusion $K' \longrightarrow K'^{(1/p^n)}$. Now let H'_n be the pull-back of the subgroup $H'^{(p^n)}$ of $G'^{(p^n)}$. If $\mathcal{I}'$ is the sheaf of ideals of $\mathcal{O}_{G'}$ associated to H', the sheaf of ideals associated to H'_n is generated by the p^n-th powers of the local sections of $\mathcal{I}'$. Since K'/K is of exponent n, we see that $\mathcal{I}'$ is generated by local sections of $\mathcal{O}_G$ and, hence, that H'_n is defined over K. Now it remains to show that there exists a smallest connected linear subgroup L of G having abelian cokernel. This follows immediately from the fact that an intersection of two linear subgroups of G is linear again and has abelian cokernel if each of them has abelian cokernel. □

For an arbitrary base field K, the connected linear subgroup L does not need to be compatible with field extensions. If the base field K is perfect and the group G is commutative, one has further information on the structure of the group L.

Theorem 2 ([SGA 3_{II}], Exp. XVII, Thm. 7.2.1). *Let K be a field and let G be a smooth and connected algebraic K-group of finite type. Assume that G is commutative and linear. Then G is canonically an extension of a unipotent algebraic group by a torus.*

If, in addition, the base field K is perfect, this extension splits canonically; i.e., G is isomorphic to a product of a unipotent group and a torus.

Now we come to the discussion of the structure of $\mathrm{Pic}^0_{X/K}$. We start with a result which is a direct consequence of 8.4/2 and 8.4/3.

Proposition 3. *Let X be a proper and smooth curve over a field K. Then the Jacobian $\mathrm{Pic}^0_{X/K}$ is an abelian variety.*

If the base field K is perfect, the curve X is smooth over K if and only if it is normal. The two notions are not equivalent over arbitrary fields, so it may happen that $\mathrm{Pic}^0_{X/K}$ is not proper although X is normal.

Proposition 4. *Let X be a proper curve over a field K. Assume that X is normal, geometrically reduced, and geometrically irreducible. Then $\mathrm{Pic}^0_{X/K}$ contains neither a subgroup of type $\mathbb{G}_a$ nor a subgroup of type $\mathbb{G}_m$.*

Proof. Since, for any separable field extension K'/K, the K'-curve $X \otimes_K K'$ is normal, we may assume that K is separably closed. Then there exists a rational point on X because X is geometrically reduced. So, for any K-scheme T, elements of $\mathrm{Pic}_{X/K}(T)$ can be represented by line bundles on $X \times_K T$; cf. 8.1/4. Now, let us assume that there is a subgroup G of $\mathrm{Pic}_{X/K}$ which is of type $\mathbb{G}_a$ or $\mathbb{G}_m$. The inclusion $G \hookrightarrow \mathrm{Pic}_{X/K}$ corresponds to a line bundle $\mathcal{L}$ on $X \times_K G$. Since X is normal, the line bundle $\mathcal{L}$ is isomorphic to the pull-back of a line bundle on X; cf. Bourbaki [2], Chap. VII, § 1, n°10, Prop. 17 and 18. Hence, the map $G \longrightarrow \mathrm{Pic}_{X/K}$ which is induced by $\mathcal{L}$ must be constant. So we get a contradiction and the assertion is proved. □

Now we turn to more general cases. Let us denote by X_{red} the largest reduced subscheme of X. By functoriality, we get a canonical map

$$\mathrm{Pic}^0_{X/K} \longrightarrow \mathrm{Pic}^0_{X_{\mathrm{red}}/K} .$$

So we can look at the kernel and at the image of this map. The algebraic group corresponding to the kernel can easily be described by the nilradical of $\mathcal{O}_X$.

Proposition 5. *Let X be a proper curve over a field K. Then the canonical map*

$$\mathrm{Pic}_{X/K} \longrightarrow \mathrm{Pic}_{X_{\mathrm{red}}/K}$$

is an epimorphism of sheaves for the étale topology. Its kernel is a smooth and connected unipotent group which is a successive extension of additive groups of type $\mathbb{G}_a$.

Proof. Let $X' \longrightarrow X$ be a closed subscheme which is defined by a sheaf of ideals $\mathcal{N}$ of $\mathcal{O}_X$ satisfying $\mathcal{N}^2 = 0$. It suffices to show that the canonical map

$$\mathrm{Pic}_{X/K} \longrightarrow \mathrm{Pic}_{X'/K}$$

is an epimorphism of sheaves for the étale topology and that its kernel is of the type described above. Let $f: X \longrightarrow \mathrm{Spec}\, K$ be the structural morphism. The exact sequence given by the exponential map

$$0 \longrightarrow \mathcal{N} \longrightarrow \mathcal{O}_X^* \longrightarrow \mathcal{O}_{X'}^* \longrightarrow 0$$

$$n \longmapsto 1 + n$$

gives rise to the exact sequence

$$R^1 f_* \mathcal{N} \longrightarrow R^1 f_* \mathcal{O}_X^* \longrightarrow R^1 f_* \mathcal{O}_{X'}^* \longrightarrow R^2 f_* \mathcal{N}$$

which has to be read as a sequence of sheaves for the étale topology. Because X is a curve, we have $R^2 f_* \mathcal{N} = 0$. Hence the canonical map

$$\mathrm{Pic}_{X/K} = R^1 f_* \mathcal{O}_X^* \longrightarrow \mathrm{Pic}_{X'/K} = R^1 f_* \mathcal{O}_{X'}^*$$

is an epimorphism. Since, for any K-scheme T, there is a canonical isomorphism

$$H^1(X, \mathcal{N}) \otimes_K \mathcal{O}_T(T) = R^1 f_* \mathcal{N}(T) ,$$

the group functor $R^1 f_* \mathcal{N}$ is represented by the vector group $H^1(X, \mathcal{N})$. Then it follows from the exact sequence above that the kernel of the map we are interested in is a quotient of the vector group $H^1(X, \mathcal{N})$. The latter is a successive extension of groups of type $\mathbb{G}_a$. So, as can easily be deduced from [SGA 3_{II}], Exp. XVII, Lemme 2.3, the kernel is as required. □

It remains to study $\mathrm{Pic}^0_{X/K}$ for reduced curves. Therefore, let us assume now that the curves under consideration are reduced. Before starting the discussion of the general case, we want to have a closer look at an example.

Definition 6. *Let S be any scheme, and let g be an integer. A semi-stable curve of genus g over S is a proper and flat morphism $f: X \longrightarrow S$ whose fibres $X_{\bar{s}}$ over geometric* *points $\bar{s}$ of S are reduced, connected, one-dimensional, and satisfy the following conditions*:

(i) *$X_{\bar{s}}$ has only ordinary double points as singularities,*

(ii) $\dim_{k(\bar{s})} H^1(X_{\bar{s}}, \mathcal{O}_{X_{\bar{s}}}) = g$.

A point x of a curve X over an algebraically closed field $\bar{K}$ is an ordinary double point if the completion $\hat{\mathcal{O}}_{X,x}$ of the local ring $\mathcal{O}_{X,x}$ of X at x is isomorphic to the quotient $\bar{K}[[\zeta, \xi]]/(\zeta\xi)$ of the formal power series ring $\bar{K}[[\zeta, \xi]]$ in two variables. For a curve X over a field K, one can formulate the condition of X being semi-stable, without performing the base extension by an algebraic closure $\bar{K}$ of K. Namely, a geometrically connected curve X over a field K is semi-stable if and only if for each non-smooth point of X there exists an étale neighborhood which is étale over the union of the coordinate axes in $\mathbb{A}^2_K$.

The interest in semi-stable curves comes from the semi-stable reduction theorem, see Deligne and Mumford [1] or Artin and Winters [1], which we want to mention without proof.

Theorem 7 (Semi-Stable Reduction Theorem). *Let R be a discrete valuation ring with fraction field K. Let X_K be a proper, smooth, and geometrically connected curve over K. Then there exist a finite separable field extension K' of K and a semi-stable curve X' over the integral closure R' of R in K' with generic fibre $X'_{K'} \cong X_K \otimes_K K'$. Furthermore, X' can be chosen to be regular.*

If X is a semi-stable curve over an algebraically closed field K, one can associate a graph $\Gamma = \Gamma(X)$ to it: the vertices of Γ are the irreducible components of X, say $X_1, \ldots, X_r$, and the edges are given by the singular points of X; namely, each singular point lying on X_i and on X_j defines an edge joining the vertices X_i and X_j. Note that $X_i = X_j$ is allowed.

Example 8. *Let X be a semi-stable curve over a field K. Then $\mathrm{Pic}^0_{X/K}$ is canonically an extension of an abelian variety by a torus T.*

More precisely, let $X_1, \ldots, X_r$ be the irreducible components of X, and let $\tilde{X}_i$ be the normalization of X_i, $i = 1, \ldots, r$. Then the canonical extension associated to $\mathrm{Pic}^0_{X/K}$ is given by the exact sequence

$$1 \longrightarrow T \hookrightarrow \mathrm{Pic}^0_{X/K} \xrightarrow{\pi^*} \prod_{i=1}^{r} \mathrm{Pic}^0_{\tilde{X}_i/K} \longrightarrow 1$$

where π^* is induced via functoriality by the morphisms $\pi_i : \tilde{X}_i \longrightarrow X$, $i = 1, \ldots, r$. The rank of the torus part T is equal to the rank of the cohomology group $H^1(\Gamma(X \otimes_K \bar{K}), \mathbb{Z})$.

Proof. Let $\pi : \tilde{X} \longrightarrow X$ be the normalization of X. The connected components of $\tilde{X}$ are the normalizations $\tilde{X}_i$ of the irreducible components X_i. They are proper and smooth over K, hence $\mathrm{Pic}^0_{\tilde{X}_i/K}$ is an abelian variety over K. Furthermore, the map π^* is compatible with field extensions. So we may assume that K is algebraically closed. Now look at the exact sequence

$$(*) \qquad 1 \longrightarrow \mathcal{O}^*_X \longrightarrow \pi_*\mathcal{O}^*_{\tilde{X}} \longrightarrow \pi_*\mathcal{O}^*_{\tilde{X}}/\mathcal{O}^*_X \longrightarrow 1\ .$$

The quotient $\mathcal{Q} = \pi_* \mathcal{O}_{\tilde{X}}^* / \mathcal{O}_X^*$ is concentrated at the singular points $x_1, \ldots, x_N$ of X. The associated long exact sequence

$$1 \longrightarrow H^0(X, \mathcal{O}_X^*) \longrightarrow H^0(X, \pi_* \mathcal{O}_{\tilde{X}}^*) \longrightarrow H^0(X, \mathcal{Q})$$
$$\longrightarrow H^1(X, \mathcal{O}_X^*) \longrightarrow H^1(X, \pi_* \mathcal{O}_{\tilde{X}}^*) \longrightarrow 1$$

can be written in the following way

$$(**) \qquad 1 \longrightarrow K^* \longrightarrow \prod_{i=1}^{r} K_i^* \longrightarrow \prod_{j=1}^{N} K_j^* \longrightarrow \mathrm{Pic}(X) \longrightarrow \mathrm{Pic}(\tilde{X}) \longrightarrow 1$$

where

$$K_i^* = H^0(\tilde{X}_i, \mathcal{O}_{\tilde{X}_i}^*) \cong K^* \text{ and } K_j^* = (K(\tilde{x}_{j1}) \times K(\tilde{x}_{j2})^*)/K^* \cong K^*$$

if $\tilde{x}_{j1}$ and $\tilde{x}_{j2}$ are the points of $\tilde{X}$ lying above the double point x_j. Using the long exact sequence of sheaves with respect to the étale topology which is associated to $(*)$, one sees that π^* is an epimorphism, since $R^1 f_* \mathcal{Q} = 0$ where $f: X \longrightarrow \mathrm{Spec}\, K$ is the structural morphism. Furthermore, the kernel of π^* is given by the quotient of the map $R^0 f_*(\pi_* \mathcal{O}_{\tilde{X}}^*) \longrightarrow R^0 f_*(\mathcal{Q})$. The latter is a quotient of a torus and, hence a torus. The assertion concerning the rank of the torus follows from the exact sequence $(**)$. □

Now let us return to the general situation of a reduced curve over a field K. As in the theorem of Chevalley, one can expect to describe the torus part and the unipotent part of $\mathrm{Pic}^0_{X/K}$ in geometric terms, at least if the base field is perfect. So, in the following, let K be a perfect field and let X be a proper curve over K which is reduced and geometrically connected. Denote by $\tilde{X} \longrightarrow X$ the normalization of X. We want to introduce an intermediate curve X' lying between X and $\tilde{X}$.

Since there is a dense open part of X which is smooth, there exist only finitely many non-smooth points of X. We will define X' by identifying all the points of $\tilde{X}$ lying above such a non-smooth point of X. In order to explain this procedure, we can work locally. So consider a non-smooth point x of X, and let $U = \mathrm{Spec}\, A$ be an affine open neighborhood of x such that x is the only non-smooth point of U. Let $\tilde{x}_1, \ldots, \tilde{x}_n$ be the points of $\tilde{X}$ lying above x, and let $\tilde{U} = \mathrm{Spec}\, \tilde{A}$ be the inverse image of U in $\tilde{X}$. Then we define the open affine subscheme $U' = \mathrm{Spec}\, A'$ of X' lying over U by taking for A' the amalgamated sum of the maps

$$\tilde{A} \longrightarrow \prod_{i=1}^{n} k(\tilde{x}_i) \qquad \text{and} \qquad k(x) \longrightarrow \prod_{i=1}^{n} k(\tilde{x}_i)\,.$$

So A' consists of all elements $f \in \tilde{A}$ which take the same value $r \in k(x)$ at all points $\tilde{x}_1, \ldots, \tilde{x}_n$. These local constructions fit together to build a proper curve X', and we get canonical morphisms

$$\tilde{X} \xrightarrow{f} X' \xrightarrow{g} X\,.$$

The map f maps the points $\tilde{x}_1, \ldots, \tilde{x}_n$ to a single point x' of X' with residue field $k(x)$. So g does not change the residue field. Let $\tilde{\mathfrak{m}}_i \subset \tilde{A}$ be the ideal of the point $\tilde{x}_i$, $i = 1, \ldots, n$. Then we obtain the exact sequences

$$\begin{array}{ccccccccc} 0 & \longrightarrow & \prod_{i=1}^{n} \mathfrak{m}_i & \longrightarrow & \prod_{i=1}^{n} \mathcal{O}_{\tilde{X},\tilde{x}_i} & \longrightarrow & \prod_{i=1}^{n} k(\tilde{x}_i) & \longrightarrow & 0 \\ & & \uparrow & & \uparrow & & \uparrow & & \\ 0 & \longrightarrow & \mathfrak{m}' & \longrightarrow & \mathcal{O}_{X',x'} & \longrightarrow & k(x') & \longrightarrow & 0 \end{array}$$

where $\mathfrak{m}'$ is the maximal ideal of $\mathcal{O}_{X',x'}$. The first vertical map is bijective, and the last one corresponds to the embedding of $k(x') = k(x)$ into the product of the residue fields $k(x_i)$, $i = 1, \ldots, n$. Due to the construction, it is clear that the map $X' \longrightarrow X$ is a universal homeomorphism. Moreover, X' is the largest curve between $\tilde{X}$ and X which is universally homeomorphic to X. One shows easily that the construction of X' is compatible with field extensions, since K is perfect. The singularities of X' are as mild as possible. Namely, after base extension by an algebraic closure $\bar{K}$ of K, the singularities of $X' \otimes_K \bar{K}$ are transversal crossings of a set of smooth branches (i.e., analytically isomorphic to the crossing of the coordinate axes in $\mathbb{A}^n$ for some n).

Proposition 9. *Let X be a proper reduced curve over a perfect field K. Let $g : X' \longrightarrow X$ be the largest curve between the normalization $\tilde{X}$ of X and X which is universally homeomorphic to X. Then the canonical map*

$$\psi : \mathrm{Pic}_{X/K} \longrightarrow \mathrm{Pic}_{X'/K}$$

is an epimorphism of sheaves for the étale topology. The kernel of ψ is a connected unipotent algebraic group which is trivial if and only if the canonical map $X' \longrightarrow X$ is an isomorphism.

Proof. Let $\mathscr{P} \subset \mathcal{O}_X$ (resp. $\mathscr{Q} \subset \mathcal{O}_{X'}$) be the sheaf of (reduced) ideals defining the non-smooth locus of X (resp. of X'). There exists an integer $e \in \mathbb{N}$ such that $g_*\mathscr{Q}^e \subset \mathscr{P}$. Consider the exact sequence

$$0 \longrightarrow \mathcal{O}_X^* \longrightarrow g_*\mathcal{O}_{X'}^* \longrightarrow (1 + g_*\mathscr{Q})/(1 + \mathscr{P}) \longrightarrow 0\,,$$

and set $\mathscr{C} := (1 + g_*\mathscr{Q})/(1 + \mathscr{P})$. It is a sheaf which is concentrated on the finitely many points of X which are not smooth; more precisely, its support consists of the points of X which are not ordinary multiple points. Let $f: X \longrightarrow \operatorname{Spec} K$ be the structural morphism. Since $R^1f_*\mathscr{C} = 0$ and $f_*\mathcal{O}_X^* = f_*g_*\mathcal{O}_{X'}^*$, the exact sequence of above gives rise to an exact sequence

$$1 \longrightarrow R^0f_*\mathscr{C} \longrightarrow R^1f_*\mathcal{O}_X^* \longrightarrow R^1f_*(g_*\mathcal{O}_{X'}^*) \longrightarrow 1$$

of sheaves for the étale topology. Thus, we see that

$$\mathrm{Pic}_{X/K} = R^1f_*\mathcal{O}_X^* \xrightarrow{\psi} \mathrm{Pic}_{X'/K} = R^1(f \circ g)_*\mathcal{O}_{X'}^* = R^1f_*(g_*\mathcal{O}_{X'}^*)$$

is an epimorphism. Due to Serre [1], Chap. V, n°15, Lemma 20, the group $R^0f_*\mathscr{C}$ and, hence, the kernel of ψ is represented by a unipotent group. For a further description of this group see Serre [1], Chap. V, n°16 and n°17. Moreover, the kernel of ψ is trivial if and only if the group $H^0(X, \mathscr{C})$ vanishes; i.e., if and only if $g_* \mathscr{Q} = \mathscr{P}$ or, equivalently, if and only if $X' \longrightarrow X$ is an isomorphism. $\square$

Proposition 10. *Let X be a proper reduced curve over a perfect field K, and let $\bar{K}$ be an algebraic closure of K. Let $X' \longrightarrow X$ be the largest curve between the normalization $\tilde{X}$ of X and X which is universally homeomorphic to X. Then the canonical map*

$$\varphi : \mathrm{Pic}_{X'/K} \longrightarrow \mathrm{Pic}_{\tilde{X}/K}$$

is an epimorphism of sheaves for the étale topology. The kernel of φ is a torus. The latter is trivial if and only if each irreducible component of $X \otimes_K \bar{K}$ is homeomorphic to its normalization and the configuration of the irreducible components of $X \otimes_K \bar{K}$ is tree-like; i.e., $H^1(X \otimes_K \bar{K}, \mathbb{Z}) = 0$.

Proof. The proof can be done similarly as in Example 8. We may assume $X = X'$. Let $\pi : \tilde{X} \longrightarrow X$ be the normalization of X. The connected components of $\tilde{X}$ are the normalizations $\tilde{X}_i$ of the irreducible components X_i. Let x_i, $i = 1, \ldots, N$, be the singular points of X, and let $\tilde{x}_{ij}$, $j = 1, \ldots, n_i$, be the points of $\tilde{X}$ lying above $\tilde{x}_i$. Consider the exact sequence

$$1 \longrightarrow \mathcal{O}_X^* \longrightarrow \pi_* \mathcal{O}_{\tilde{X}}^* \longrightarrow \pi_* \mathcal{O}_{\tilde{X}}^* / \mathcal{O}_X^* \longrightarrow 1 .$$

The quotient $\mathscr{Q} = \pi_* \mathcal{O}_{\tilde{X}}^* / \mathcal{O}_X^*$ is concentrated at the points x_i, $i = 1, \ldots, N$. The associated long exact sequence

$$\begin{aligned} 1 \longrightarrow H^0(X, \mathcal{O}_X^*) &\longrightarrow H^0(X, \pi_* \mathcal{O}_{\tilde{X}}^*) \longrightarrow H^0(X, \mathscr{Q}) \\ &\longrightarrow H^1(X, \mathcal{O}_X^*) \longrightarrow H^1(X, \pi_* \mathcal{O}_{\tilde{X}}^*) \longrightarrow 1 \end{aligned}$$

can be written in the following way

$$1 \longrightarrow \Gamma^* \longrightarrow \prod_{i=1}^{r} \Gamma_i^* \longrightarrow \prod_{i=1}^{N} \left(\prod_{j=1}^{n_i} K_{ij}^* \right) \Big/ K_i^* \longrightarrow \mathrm{Pic}(X) \longrightarrow \mathrm{Pic}(\tilde{X}) \longrightarrow 1$$

where $\Gamma^* = H^0(X, \mathcal{O}_X^*)$, $\Gamma_i^* = H^0(\tilde{X}_i, \mathcal{O}_{\tilde{X}}^*)$, $K_i^* = k(x_i)$, and $K_{ij}^* = k(\tilde{x}_{ij})$. As in Example 8, one shows that φ is an epimorphism for the étale topology and, moreover, that the kernel of φ is the quotient of the map $R^0 f_*(\pi_* \mathcal{O}_{\tilde{X}}^*) \longrightarrow R^0 f_*(\mathscr{Q})$ where $f : X \longrightarrow \mathrm{Spec}\, K$ is the structural morphism. The latter is a quotient of a torus and, hence, a torus.

It remains to show the last assertion. We may assume that K is algebraically closed. The kernel of φ is trivial if and only if the canonical map

$$\prod_{i=1}^{r} \Gamma_i^* \longrightarrow \prod_{i=1}^{N} \left(\prod_{j=1}^{n_i} K_{ij}^* \right) \Big/ K_i^*$$

is surjective. If the map is surjective, it is clear that, for any singular point x_i of X, the points $\tilde{x}_{ij}$, $j = 1, \ldots, n_i$, lie on pairwise different components of $\tilde{X}$. Hence, each irreducible component of X is homeomorphic to its normalization. Furthermore, the surjectivity implies $H^1(X, K^*) = 0$ which is equivalent to $H^1(X, \mathbb{Z}) = 0$. The converse implication follows by similar arguments. $\square$

Now we can deduce from Propositions 9 and 10 the structure of the linear part of $\mathrm{Pic}^0_{X/K}$.

Corollary 11. *Let X be a proper curve over a perfect field K and denote by $\tilde{X}$ the normalization of the largest reduced subscheme X_{red} of X. Then the canonical map*

$$\mathrm{Pic}_{X/K} \longrightarrow \mathrm{Pic}_{\tilde{X}/K}$$

is an epimorphism of sheaves for the étale topology. Its kernel consists of a smooth connected linear algebraic group L. The quotient of $\mathrm{Pic}^0_{X/K}$ by L is isomorphic to $\mathrm{Pic}^0_{\tilde{X}/K}$ which is an abelian variety.

Next we want to look at a reduced curve X over a perfect field K. As before, let X' denote the largest curve between X and its normalization $\tilde{X}$. Via functoriality, we get the following sequence of algebraic groups

$$\mathrm{Pic}_{X/K} \longrightarrow \mathrm{Pic}_{X'/K} \longrightarrow \mathrm{Pic}_{\tilde{X}/K} ,$$

where each map is an epimorphism of sheaves for the étale toplogy. Due to continuity, we obtain epimorphisms between the identity components

$$\mathrm{Pic}^0_{X/K} \longrightarrow \mathrm{Pic}^0_{X'/K} \longrightarrow \mathrm{Pic}^0_{\tilde{X}/K} .$$

Furthermore, if $\mathrm{Pic}^0_{X/K}$ does not contain a torus, $\mathrm{Pic}^0_{X'/K}$ does not either; for example, this can be deduced from Theorem 2. So, we obtain the following corollary.

Corollary 12. *Let X be a reduced proper curve over a perfect field K and let $\bar{K}$ be an algebraic closure of K.*

(a) *If $\mathrm{Pic}^0_{X/K}$ contains no unipotent connected subgroup, the singularities of $X \otimes_K \bar{K}$ are analytically isomorphic to the crossing of the coordinate axes in $\mathbb{A}^n$.*

(b) *If $\mathrm{Pic}^0_{X/K}$ contains no torus, each irreducible component of $X \otimes_K \bar{K}$ is homeomorphic to its normalization and the configuration of the irreducible components of $X \otimes_K \bar{K}$ is tree-like.*

(c) *If $\mathrm{Pic}^0_{X/K}$ is an abelian variety, the irreducible components of X are smooth and the configuration of the irreducible components of $X \otimes_K \bar{K}$ is tree-like.*

Finally we want to discuss the degree of line bundles belonging to $\mathrm{Pic}^0_{X/K}$. For example, if X is a connected proper and smooth curve over an algebraically closed field K, the elements of $\mathrm{Pic}^0_{X/K}(K)$ correspond to the line bundles of degree zero. Indeed, consider the universal line bundle $\mathscr{L}$ on $X \times_K \mathrm{Pic}_{X/K}$. Due to 9.1/2, the degree of the restriction $\mathscr{L}_\xi$ of $\mathscr{L}$ to the fibre over a point $\xi \in \mathrm{Pic}^0_{X/K}$ is zero. Conversely, a line bundle of degree zero is isomorphic to a line bundle $\mathcal{O}_X(D)$ where D is a Cartier divisor which can be written as

$$D = (x_1 - x_0) + \ldots + (x_n - x_0) ,$$

where $x_0, \ldots, x_n$ are closed points of X. Since X is connected, the image of the map

$$X \longrightarrow \mathrm{Pic}_{X/K} , \qquad x \longmapsto [\mathcal{O}_X(x - x_0)] ,$$

is contained in $\operatorname{Pic}^0_{X/K}$. Thus we see that each line bundle of degree zero gives rise to an element of $\operatorname{Pic}^0_{X/K}$. For arbitrary curves over fields, one has to look at the partial degrees on the irreducible components.

Corollary 13. *Let X be a proper curve over a field K and let $\bar{K}$ be an algebraic closure of K. Then $\operatorname{Pic}^0_{X/K}$ consists of all elements of $\operatorname{Pic}_{X/K}$ whose partial degree on each irreducible component of $X \otimes_K \bar{K}$ is zero.*

Proof. We may assume that K is algebraically closed. Let $X_1, \ldots, X_r$ be the (reduced) irreducible components of X. For $i = 1, \ldots, r$, let $\tilde{X}_i$ be the normalization of X_i. Then consider the canonical morphism

$$\operatorname{Pic}_{X/K} \longrightarrow \operatorname{Pic}_{\tilde{X}_i/K}$$

which is defined by functoriality. Due to continuity, the identity components are mapped into each other, so we have morphisms

$$\operatorname{Pic}^0_{X/K} \longrightarrow \operatorname{Pic}^0_{\tilde{X}_i/K} .$$

Since the degree of a Cartier divisor on X_i and the degree of its pull-back on $\tilde{X}_i$ coincide, we see that the partial degrees of elements of $\operatorname{Pic}^0_{X/K}(K)$ are zero. Due to Corollary 11, the canonical morphism

$$\operatorname{Pic}_{X/K} \longrightarrow \prod_{i=1}^{r} \operatorname{Pic}_{\tilde{X}_i/K}$$

is an epimorphism and its kernel is a connected subgroup of $\operatorname{Pic}_{X/K}$. So the kernel is contained in $\operatorname{Pic}^0_{X/K}$. Since the canonical map induces an epimorphism on the identity components, we see that line bundles on X whose partial degrees are zero belong to $\operatorname{Pic}^0_{X/K}$. □

Corollary 14. *Let X be a proper curve over an algebraically closed field K with r irreducible components $X_1, \ldots, X_r$. Then the Néron-Severi group of X is a free group of rank r.*

More precisely, the map given by the partial degrees

$$\operatorname{Pic}_{X/K}/\operatorname{Pic}^0_{X/K} \longrightarrow \mathbb{Z}^r , \qquad \mathscr{L} \longmapsto (\deg_{X_1}(\mathscr{L}), \ldots, \deg_{X_r}(\mathscr{L}))$$

is injective and has finite index.

9.3 Construction via Birational Group Laws

We want to explain how the proof of Grothendieck's theorem 8.2/1 can be modified in the case of relative curves in order to recover the Jacobian variety as constructed by Serre [1] and Weil [2]. We begin by repeating what Grothendieck's approach to the representability of $\operatorname{Pic}_{X/S}$ yields in the case of a relative curve X over a scheme S.

Theorem 1. *Let $X \longrightarrow S$ be a projective and flat curve which is locally of finite presentation. If the geometric fibres of X over S are reduced and irreducible,* $\mathrm{Pic}_{X/S}$ *is a smooth and separated S-scheme.*

More precisely, there is a decomposition

$$\mathrm{Pic}_{X/S} = \coprod_{n \in \mathbb{Z}} (\mathrm{Pic}_{X/S})^n$$

where $(\mathrm{Pic}_{X/S})^n$ *denotes the open and closed subscheme of* $\mathrm{Pic}_{X/S}$ *consisting of all line bundles of degree n; the scheme* $(\mathrm{Pic}_{X/S})^0$ *coincides with the identity component* $\mathrm{Pic}^0_{X/S}$ *of* $\mathrm{Pic}_{X/S}$. *Moreover,* $(\mathrm{Pic}_{X/S})^n$ *is quasi-projective over S and is a torsor under* $\mathrm{Pic}^0_{X/S}$ *for all $n \in \mathbb{Z}$.*

Proof. The representability of $\mathrm{Pic}_{X/S}$ is due to 8.2/1; see also 8.2/5. The smoothness follows from 8.4/2. Due to 9.1/2, the degree of line bundles belonging to a fixed connected component of $\mathrm{Pic}_{X/S}$ is constant, thus $\mathrm{Pic}_{X/S}$ breaks up into the disjoint union of the $(\mathrm{Pic}_{X/S})^n$, $n \in \mathbb{Z}$. In order to show that $(\mathrm{Pic}_{X/S})^n$ is a torsor under $\mathrm{Pic}^0_{X/S}$, it remains to show that $(\mathrm{Pic}_{X/S})^n$ and $\mathrm{Pic}^0_{X/S}$ become isomorphic after faithfully flat base extension. So we may assume that X has a section over S. Then it suffices to see that $(\mathrm{Pic}_{X/S})^0$ is isomorphic to $\mathrm{Pic}^0_{X/S}$. Since the geometric fibres of X over S are irreducible and reduced, the latter follows immediately from 9.2/13. □

Let us mention some conditions under which X is projective over S.

Remark 2. Let X be a proper flat curve over S which is locally of finite presentation and whose geometric fibres are reduced and irreducible curves of genus g. Assume that X is a relative complete intersection over S. Then the relative dualizing sheaf is a line bundle. If $g \geq 2$, it is S-ample and, hence, $X \longrightarrow S$ is projective. Likewise, if $g = 0$, the dual of the relative dualizing sheaf is S-ample and, hence, $X \longrightarrow S$ is projective; moreover it is smooth. If $g = 1$, it follows that $X \longrightarrow S$ is projective locally for the étale topology on S, since $X \longrightarrow S$ admits a section through the smooth locus after étale surjective base change, and since the line bundle of all meromorphic functions having only simple poles along the given section is relatively ample.

Now we turn to a more general situation where we can construct $\mathrm{Pic}^0_{X/S}$ via birational laws. In the following let $f: X \longrightarrow S$ be a quasi-projective morphism of schemes which is of finite presentation. We want to explain some basic facts on the relationship between the n-fold symmetric product $(X/S)^{(n)}$ and the Hilbert functor $\mathrm{Hilb}^n_{X/S}$, where $\mathrm{Hilb}^n_{X/S}$ is the Hilbert functor associated to the constant polynomial n. We can say that, for any S-scheme T, the set $\mathrm{Hilb}^n_{X/S}(T)$ consists of all subschemes D of $X \times_S T$ which are finite and locally free of rank n over T. The n-fold symmetric product $(X/S)^{(n)}$ is defined as the quotient of the n-fold product of X over S by the canonical action of the symmetric group. Let us start by discussing the representability of $(X/S)^{(n)}$.

For any commutative ring A and for any A-module M, define the symmetric n-fold tensor product of M by

$$\mathrm{TS}^n_A(M) := (M^{\otimes n})^{\mathfrak{S}_n} \subset M^{\otimes n}$$

where $M^{\otimes n}$ is the n-fold tensor product of M over A and where $\mathfrak{S}_n$ is the symmetric group acting on $M^{\otimes n}$ by permuting factors. If M is a free A-module, $\mathrm{TS}_A^n(M)$ is also free and there is a canonical way to choose a basis of $\mathrm{TS}_A^n(M)$ after fixing a basis of M. Thus, we see that $\mathrm{TS}_A^n(M)$ is compatible with any base change if M is a free A-module. Since any flat A-module is a limit of finitely generated free A-modules, $\mathrm{TS}_A^n(M)$ is a flat A-module and compatible with any base change if M is flat over A. If B is an A-algebra, $\mathrm{TS}_A^n(B)$ is a subalgebra of $B^{\otimes n}$. If X and S are affine, say $S = \operatorname{Spec} A$ and $X = \operatorname{Spec} B$, the symmetric product $(X/S)^{(n)}$ is represented by $\operatorname{Spec}(\mathrm{TS}_A^n(B))$. If X is quasi-projective over S, one can establish the representability of the symmetric product $(X/S)^{(n)}$ as an S-scheme by gluing such local pieces, since any finite set of points lying on a single fibre of X/S is contained in an open affine subscheme of X. Furthermore, as we have seen above, the symmetric product $(X/S)^{(n)}$ of a flat S-scheme X is flat over S and compatible with any base change.

A polynomial law f from an A-module M to an A-module N consists of the following data: for any commutative A-algebra A', there is a map

$$f_{A'}: M \otimes_A A' \longrightarrow N \otimes_A A'$$

such that, for any morphism $u: A' \longrightarrow A''$ of commutative A-algebras, the diagram

$$\begin{array}{ccc} M \otimes_A A' & \xrightarrow{f_{A'}} & N \otimes_A A' \\ \Big\downarrow{\scriptstyle M \otimes u} & & \Big\downarrow{\scriptstyle N \otimes u} \\ M \otimes_A A'' & \xrightarrow{f_{A''}} & N \otimes_A A'' \end{array}$$

is commutative. A polynomial law from M to N is called homogeneous of degree n if, in addition, for any $a' \in A'$ and for any $m' \in M \otimes_A A'$, the equation

$$f_{A'}(a' \cdot m') = (a')^n \cdot f_{A'}(m') .$$

holds. For example, the map

$$\gamma^n : M \longrightarrow \mathrm{TS}_A^n(M) , \qquad m \longmapsto m \otimes \cdots \otimes m \quad (n \text{ times})$$

gives rise to a homogeneous polynomial law of degree n. Furthermore, if M is a free A-module of finite rank, the map γ^n is universal; i.e., any homogeneous polynomial law f from M to N of degree n is induced by a unique A-linear map $\varphi: \mathrm{TS}_A^n(M) \longrightarrow N$. The latter means

$$f_{A'} = (\varphi \otimes A') \circ (\gamma^n \otimes A') ;$$

cf. [SGA 4_{III}], Exp. XVII, 5.5.2. Since a flat A-module is a limit of free A-modules, the map γ^n is universal if M is a flat A-module.

Let us fix $S = \operatorname{Spec} A$, $X = \operatorname{Spec} B$ and $f: X \longrightarrow S$. For any B-module L which is free of rank n over A, there is a canonical morphism

$$\det_L : \mathrm{TS}_A^n(B) \longrightarrow A$$

which is compatible with any base change $A \longrightarrow A'$. Indeed, viewing the multiplication on L by an element $b \in B$ as an A-linear map, the determinant yields a homogeneous polynomial law of degree n from B to A and, hence, a map of $\mathrm{TS}_A^n(B)$

to A. Furthermore, one can show that $\det_L$ is a morphism of A-algebras; cf. [SGA 4_{III}], Exp. XVII, 6.3.1.

If $f: X \longrightarrow S$ is affine and if $\mathscr{L}$ is an $\mathcal{O}_X$-module such that $f_*\mathscr{L}$ is locally free over S of rank n, one can construct a morphism

$$\sigma_{\mathscr{L}} : S \longrightarrow (X/S)^{(n)}$$

by gluing the local morphisms constructed above.

Now let $f: X \longrightarrow S$ be quasi-projective and consider an element $D \in \mathrm{Hilb}^n_{X/S}(T)$ for an S-scheme T, i.e., a subscheme D of $X \times_S T$ which is finite and locally free of rank n over T. Then $(f_T)_*\mathcal{O}_D$ is a locally free $\mathcal{O}_T$-module of rank n. So the above construction gives rise to a section

$$\sigma_{\mathcal{O}_D} : T \longrightarrow (D/T)^{(n)} \longrightarrow (X/S)^{(n)} .$$

Thus we get a canonical morphism

$$\sigma : \mathrm{Hilb}^n_{X/S} \longrightarrow (X/S)^{(n)} .$$

On the other hand, if $f: X \longrightarrow S$ is a separated smooth curve, each section s of f gives rise to a relative Cartier divisor $s(S)$ of X over S of degree 1. Namely, due to 2.2/7 the vanishing ideal of $\sigma(S)$ is locally principal. So we get a morphism

$$X^n \longrightarrow \mathrm{Hilb}^n_{X/S} , \qquad (s_1, \ldots, s_n) \longmapsto \sum s_i(S) ,$$

from the n-fold product of X over S to the Hilbert functor which is symmetric. Hence it factors through $(X/S)^{(n)}$. Note that, in this case, $\mathrm{Hilb}^n_{X/S}$ coincides with the subfunctor of $\mathrm{Div}^n_{X/S}$ consisting of all divisors with proper support. So it induces a morphism

$$\alpha : (X/S)^{(n)} \longrightarrow \mathrm{Hilb}^n_{X/S} .$$

Proposition 3 ([SGA 4_{III}], Exp. XVII, 6.3.9). *If $X \longrightarrow S$ is a smooth and quasi-projective morphism of relative dimension* 1, *then, for each $n \in \mathbb{N}$, the canonical morphisms*

$$\sigma : \mathrm{Hilb}^n_{X/S} \longrightarrow (X/S)^{(n)} \qquad \text{and} \qquad \alpha : (X/S)^{(n)} \longrightarrow \mathrm{Hilb}^n_{X/S}$$

are isomorphisms and inverse to each other.

Now let us consider the case where $f: X \longrightarrow S$ is a *faithfully flat projective curve of genus g whose geometric fibres are reduced and connected.* Denote by X' the smooth locus of X. Note that X' is S-dense in X and that, moreover, the canonical map

$$(X'/S)^{(g)} \longrightarrow (X/S)^{(g)}$$

is an open immersion with S-dense image, as one can easily verify by using the fact that $(X/S)^{(g)}$ commutes with any base change. Since X is proper over S, the functor $\mathrm{Hilb}^g_{X'/S}$ is an open subfunctor of $\mathrm{Hilb}^g_{X/S}$, and since X' is smooth over S, it is already an open subfunctor of $\mathrm{Div}^g_{X/S}$; cf. 8.2/6. Furthermore, since X is proper and flat over S, the functor $\mathrm{Div}^g_{X/S}$ is a subfunctor of $\mathrm{Hilb}^g_{X/S}$. Hence, we have a commutative diagram of canonical maps

$$\begin{array}{ccc}\mathrm{Hilb}^g_{X'/S} & \xrightarrow{\sim} & (X'/S)^{(g)} \\ \downarrow & & \downarrow \\ \mathrm{Div}^g_{X/S} & \longrightarrow & (X/S)^{(g)}\end{array} .$$

The S-scheme $(X'/S)^{(g)}$ is smooth. Indeed, by étale localization it is enough to treat the case $X' = \mathbb{A}^1_S$. But then the smoothness of $(X'/S)^{(g)}$ follows from the theorem on symmetric functions. Now, let $D \subset X \times_S (X'/S)^{(g)}$ be the effective relative Cartier divisor of degree g which is induced by the map $(X'/S)^{(g)} \longrightarrow \mathrm{Div}^g_{X/S}$. We will refer to D as the universal Cartier divisor of degree g. Let $W \subset (X'/S)^{(g)}$ be the subscheme of all points $w \in (X'/S)^{(g)}$ such that $H^1(X_w, \mathcal{O}_{X_w}(D_w))$ vanishes; so

$$W = \{w \in (X'/S)^{(g)}; H^1(X_w, \mathcal{O}_{X_w}(D_w)) = 0\} .$$

Then, due to the semicontinuity theorem [EGA III$_2$], 7.7.5, W is an open subscheme of $(X'/S)^{(g)}$, and the following lemma shows that $W \longrightarrow S$ is surjective.

Lemma 4. *Let X be a proper curve over a separably closed field K. Assume that X is geometrically reduced and connected. Then there exists an effective Cartier divisor D_0 of degree $g = \dim_K H^1(X, \mathcal{O}_X)$ on X whose support is contained in the smooth locus of X and which satisfies $H^0(X, \mathcal{O}_X(D_0)) = K$ and $H^1(X, \mathcal{O}_X(D_0)) = 0$.*

In particular, keeping the notations of above, the map $W \longrightarrow S$ is surjective.

Proof. The Riemann-Roch theorem implies $H^0(X, \mathcal{O}_X(D_0)) = K$ if $H^1(X, \mathcal{O}_X(D_0)) = 0$. So it suffices to show the existence of an effective Cartier divisor D_0 of degree g satisfying $H^1(X, \mathcal{O}_X(D_0)) = 0$. Let ω be a dualizing sheaf on X; cf. [FGA], n°149, Sect. 6, Thm. 3 bis. Then, for any Cartier divisor E of X, there is a canonical isomorphism

$$H^1(X, \mathcal{O}_X(E)) \xrightarrow{\sim} H^0(X, \omega(-E)) ,$$

where $\omega(-E)$ is the $\mathcal{O}_X$-module $\omega \otimes \mathcal{O}_X(-E)$. In particular, $\dim_K H^0(X, \omega) = g$. Proceeding by induction, we will show that there exist points $x_1, \ldots, x_g$ of the smooth locus of X such that

$$\dim_K H^0(X, \omega(-x_1 - \ldots - x_i)) = g - i , \qquad \text{for } i = 1, \ldots, g .$$

Since the $\mathcal{O}_X$-module ω has no embedded components, the support of a non-zero section of ω cannot consist of finitely many points. So one can choose a rational point x_{i+1} of the smooth locus of X such that there is an element of $H^0(X, \omega(-x_1 - \ldots - x_i))$ which does not vanish at x_{i+1}. Then,

$$D_0 = x_1 + \ldots + x_g$$

is an effective Cartier divisor as required. □

Due to [EGA III$_2$], 7.9.9, the direct image $(f_W)_* \mathcal{O}_{X \times_S W}(D)$ is locally free of rank 1, and the canonical morphism

$$((f_W)_* \mathcal{O}_{X \times_S W}(D))_w \otimes_{\mathcal{O}_{W,w}} k(w) \xrightarrow{\sim} H^0(X_w, \mathcal{O}_{X_w}(D_w))$$

is bijective; cf. Mumford [3], Sect. 5, Cor. 3.

The universal Cartier divisor D gives rise to a canonical map

$$\rho : W \longrightarrow \mathrm{Pic}^{(g)}_{X/S}$$

where $\mathrm{Pic}^{(g)}_{X/S}$ is the open subfunctor of $\mathrm{Pic}_{X/S}$ consisting of line bundles of (total) degree g; cf. Section. 9.1. Next we want to prove that ρ is an open immersion.

Lemma 5. *Keeping the notations of above, the canonical map*

$$\rho : W \longrightarrow \mathrm{Pic}^{(g)}_{X/S}$$

is an open immersion.

Proof. First of all let us show that ρ is a monomorphism. So, let L_1 and L_2 be elements of $W(T)$ for an S-scheme T giving rise to the same element in $\mathrm{Pic}^{(g)}_{X/S}(T)$. Let us denote by L_1 (resp. L_2) the associated divisors of $X \times_S T$, too. Due to 8.1/3, we may assume that the associated line bundles $\mathcal{O}_{X_T}(L_1)$ and $\mathcal{O}_{X_T}(L_2)$ are isomorphic. Since the direct images $(f_T)_*\mathcal{O}_{X_T}(L_i)$ are locally free of rank 1, it follows that L_1 and L_2 are equal and, hence, that ρ is a monomorphism. Now we prove that ρ is relatively representable by an open immersion. It has to be shown that, for any S-scheme T and for any morphism $\lambda : T \longrightarrow \mathrm{Pic}^{(g)}_{X/S}$, the induced morphism

$$\rho_T : W \times_{\mathrm{Pic}^{(g)}_{X/S}} T \longrightarrow T$$

is an open immersion. Since it suffices to check this after étale surjective base change, we may assume that the morphism λ is induced by a line bundle $\mathcal{L}$ on $X \times_S T$. The image of ρ_T is contained in the subset T' of T consisting of all points $t \in T$ satisfying $H^1(X_t, \mathcal{L}_t) = 0$. Since T' is open in T by [EGA III_2], 7.7.5, we may replace T by T'. In this case, $H^0(X_t, \mathcal{L}_t)$ is a $k(t)$-vector space of rank 1 for each $t \in T$. Moreover $(f_T)_*\mathcal{L}$ is locally free of rank 1 and a local generator of $(f_T)_*\mathcal{L}$ gives rise to a generator of $H^0(X_t, \mathcal{L}_t)$ on any fibre X_t. Therefore, a local generator of $(f_T)_*\mathcal{L}$ is uniquely determined up to a unit of the base. Hence, the local generators of $(f_T)_*\mathcal{L}$ give rise to a closed subscheme L of $X \times_S T$ whose defining ideal is locally generated by one element. Due to 8.2/6, there exists a largest open subscheme T'' of T such that the restriction of L to $X \times_S T''$ is an effective relative Cartier divisor. It is clear that ρ_T factors through T''. So we may replace T by T'' and we may assume that L is an effective relative Cartier divisor. Thus we can view λ as a section of $\mathrm{Div}^g_{X/S}$ and, hence, of $(X/S)^{(g)}$. Since W is an open subscheme of $(X/S)^{(g)}$, the map ρ_T can be represented by the open immersion of the inverse image $\lambda^{-1}(W)$ into T. □

Lemma 6. *Keeping the notations of above, there exist a surjective étale extension $S' \longrightarrow S$, an open subscheme W' of $W \times_S S'$ with geometrically connected fibres, and a section $\varepsilon' : S' \longrightarrow W'$ such that*

$$W' \longrightarrow \mathrm{Pic}^0_{X \times_S S'/S'}, \qquad w' \longmapsto \rho(w') - \rho \circ \varepsilon' \circ p'(w)$$

is an open immersion, where $p' : W' \longrightarrow S'$ is the structural morphism.

Proof. If there is a section $\varepsilon : S \longrightarrow W$, we can assume that the geometric fibres of W are connected after replacing W by an open subscheme; cf. [EGA IV_3], 15.6.5. Then we can transform the morphism

$$\rho : W \longrightarrow \mathrm{Pic}^{(g)}_{X/S}$$

by a translation into an open immersion

$$W \longrightarrow \mathrm{Pic}_{X/S}\,, \qquad w \longmapsto \rho(w) - \rho \circ \varepsilon \circ p(w)\,,$$

where $p: W \longrightarrow S$ is the structural morphism. Since the fibres of W over S are geometrically connected, the image of the above map is contained in $\mathrm{Pic}^0_{X/S}$. In the general case, one can perform a surjective étale extension $S' \longrightarrow S$ in order to get a section $S' \longrightarrow W$, because $W \longrightarrow S$ is smooth and surjective. Since the g-fold symmetric product $(X/S)^{(g)}$ commutes with the extension $S' \longrightarrow S$, one is reduced to the case discussed before. □

In the following, keep the notations of Lemma 6. Assume $S = S'$ and $W = W'$ and that there is a section $\varepsilon : S \longrightarrow W$. The group law of $\mathrm{Pic}_{X/S}$ induces an S-birational group law on W. We want to describe this S-birational group law on W in terms of divisors. So consider the projections

$$p_i : W \times_S W \longrightarrow W$$

for $i = 1, 2$, and let p be the structural morphism $p : W \longrightarrow S$. Since a morphism from an S-scheme T to W corresponds to an effective relative Cartier divisor of degree g on $X \times_S T$, namely, to the pull-back of the universal divisor D on $X \times_S W$, the projections p_1 and p_2 give rise to divisors D_1 and D_2 on $X \times_S W \times_S W$. Furthermore, let D_0 be the divisor on $X \times_S W \times_S W$ induced by ε. Then consider the locally free sheaf

$$\mathscr{L} = \mathscr{O}_{X \times_S W \times_S W}(D_1 - D_0 + D_2)\,.$$

on $X \times_S W \times_S W$. The pull-back of $\mathscr{L}$ via

$$(\mathrm{id}_W, \varepsilon \circ p) : W \longrightarrow W \times_S W$$

is isomorphic to $\mathscr{O}_{X \times_S W}(D)$. Since the fibres of W are geometrically irreducible, there is a p_1-dense open subscheme W_1 of $W \times_S W$ such that, for each point t of W_1, the restriction $\mathscr{L}_t$ of $\mathscr{L}$ to the fibre $X \times_S t$ satisfies $H^1(X_t, \mathscr{L}_t) = 0$. As before, we conclude that $(f_{W_1})_* \mathscr{L}$ is locally free of rank 1 over W_1 and that, for any $t \in W_1$, a generator of $H^0(X_t, \mathscr{L}_t)$ lifts to a local generator of $(f_{W_1})_* \mathscr{L}$ at t. A local generator of $(f_{W_1})_* \mathscr{L}$ is uniquely determined up to a unit of the base. Hence, the local generators of $(f_{W_1})_* \mathscr{L}$ give rise to a subscheme D_{12} of $X \times_S W_1$ whose defining ideal can locally be generated by one element. Since the pull-back of D_{12} by $(\mathrm{id}_W, \varepsilon \circ p)$ coincides with D which is an effective relative Cartier divisor, we see by Lemma 8.2/6 that there exists a p_1-dense open subscheme V_1 of W_1 such that $D_{12}|_{V_1}$ is an effective relative Cartier divisor of degree g. Since W is an open subfunctor of $\mathrm{Div}^g_{X/S}$, we see, after replacing V_1 by a smaller p_1-dense open subscheme of V_1, that $D_{12}|_{V_1}$ gives rise to a V_1-valued point of W. Proceeding similarly with the other projection, it is easy to show that the mapping

$$W \times_S W \dashrightarrow W\,, \qquad (D_1, D_2) \longmapsto D_{12}$$

gives rise to a strict S-birational group law; cf. 5.2/1.

In analogy to the classical case where the base S consists of a field, we call the S-group scheme associated to this S-birational group law the *Jacobian* of X over S if it exists. In the case where S consists of a field, it can easily be shown that the existence of the Jacobian implies the representability of $\mathrm{Pic}_{X/S}$; namely the latter is a disjoint sum of "translates" of the Jacobian. Furthermore, it is clear that the Jacobian coincides with $\mathrm{Pic}^0_{X/S}$. So, even for a general base, the Jacobian represents the subfunctor $\mathrm{Pic}^0_{X/S}$ as defined in Section 8.4. For example, if S is a local scheme which is normal and strictly henselian, the results of Section 5.3 can be used to show that the Jacobian is represented by a separated and smooth S-scheme. Summarizing our discussion, we can state the following result.

Theorem 7. *Let S be a normal strictly henselian local scheme and let $f\colon X \longrightarrow S$ be a flat projective morphism whose geometric fibres are reduced and connected curves. Then the Jacobian of X is a smooth and separated S-scheme. It coincides with* $\mathrm{Pic}^0_{X/S}$ *as defined in Section* 8.4.

If one admits Theorem 8.3/1, namely that $\mathrm{Pic}_{X/S}$ is an algebraic space over S, one can drop the assumption of S being normal in Theorem 7. Indeed, due to 8.4/2, $\mathrm{Pic}_{X/S}$ is smooth over S, since X is a relative curve. Hence, $\mathrm{Pic}^0_{X/S}$ is represented by an open subspace of $\mathrm{Pic}_{X/S}$. So in order to prove that $\mathrm{Pic}^0_{X/S}$ is a scheme, it suffices to show that $\mathrm{Pic}^0_{X/S}$ can be covered by the translates $\lambda W'$, where W' is the open subscheme of $\mathrm{Pic}^0_{X/S}$ constructed in Lemma 6, and where λ ranges over $W'(S)$. Since W' is smooth and faithfully flat over S, we have enough sections λ to cover $\mathrm{Pic}^0_{X/S}$ by translates $\lambda W'$; cf. 5.3/7. So every point of $\mathrm{Pic}^0_{X/S}$ has a scheme-like neighborhood. Hence $\mathrm{Pic}^0_{X/S}$ is a scheme.

If the geometric fibres of X over S are irreducible and reduced, and if there is a section $\sigma : S \longrightarrow X$ contained in the smooth locus of X, one can construct the whole Picard scheme $\mathrm{Pic}_{X/S}$ from $\mathrm{Pic}^0_{X/S}$ by translations. Namely,

$$\mathrm{Pic}_{X/S} = \coprod_{n \in \mathbb{Z}} (\mathrm{Pic}^0_{X/S} + n \cdot [\sigma(S)])\,,$$

where $[\sigma(S)]$ is the element of $\mathrm{Pic}_{X/S}$ associated to the Cartier divisor $\sigma(S)$; due to 2.2/7 the vanishing ideal of $\sigma(S)$ is an effective relative Cartier divisor of degree 1. It is not hard to show directly that the right-hand side represents the relative Picard functor in this case. So, for a normal and strictly henselian base, one obtains a different approach to the representability of $\mathrm{Pic}_{X/S}$ in the case of a flat projective curve X over S whose geometric fibres are reduced and irreducible.

In the case where the base S consists of a field, one has to perform a finite separable field extension $S' \longrightarrow S$ in order to get enough sections. Then the preceding construction yields the representability of $\mathrm{Pic}^0_{X/S} \times_S S'$ over the base S' and the representability over the given base is reduced to a problem of descent. If S consists of a field, this problem is not a serious one and can be overcome easily as was demonstrated by Serre and Weil. In Section 9.4, we will dicuss the representability of $\mathrm{Pic}^0_{X/S}$ by a separated S-scheme in the case of a more general base.

9.4 Construction via Algebraic Spaces

In the following, let $f: X \longrightarrow S$ be a proper and flat curve which is locally of finite presentation over the scheme S. So far we have discussed the case where the geometric fibres of X are reduced and connected. Now we want to study more general cases. Due to the general result 8.3/1, we know that $\mathrm{Pic}_{X/S}$ is an algebraic space if f is cohomologically flat in dimension zero. Recall that f is said to be cohomologically flat in dimension zero if, for every S-scheme S', the canonical morphism

$$(f_*\mathcal{O}_X) \otimes \mathcal{O}_{S'} \xrightarrow{\sim} f'_*\mathcal{O}_{X'}$$

is an isomorphism, where $X' = X \times_S S'$. For example, the condition is satisfied if the geometric fibres of X/S are reduced; cf. [EGA III_2], 7.8.6. The cohomological flatness of f is closely related to the condition that the arithmetic genus of the fibres of X is locally constant on S.

Indeed, if f is cohomologically flat in dimension zero, $f_*\mathcal{O}_X$ is a locally free $\mathcal{O}_S$-module by 8.1/7 and $\dim_{k(s)} H^0(X_s, \mathcal{O}_{X_s})$ is locally constant on S. Moreover, since the Euler-Poincaré characteristic of the fibres of X is locally constant on S by [EGA III_2], 7.9.4, the dimension $\dim_{k(s)} H^1(X_s, \mathcal{O}_{X_s})$ must be locally constant on S. Conversely, if the arithmetic genus of the fibres of X is locally constant on S, the same arguments as above show that $\dim_{k(s)} H^0(X_s, \mathcal{O}_{X_s})$ is locally constant on S. Then it follows from [EGA III_2], 7.8.4 that f is cohomologically flat in dimension zero at least if S is reduced.

If X is cohomologically flat over S in dimension zero, $\mathrm{Pic}_{X/S}$ is an algebraic space over S, but, in general, we cannot expect $\mathrm{Pic}_{X/S}$ to be a scheme, as Mumford's example shows; cf. Section 8.2. Since $\mathrm{Pic}_{X/S}$ is smooth over S by 8.4/2, $\mathrm{Pic}^0_{X/S}$ is represented by an open subspace of $\mathrm{Pic}_{X/S}$ which may be a scheme, even if $\mathrm{Pic}_{X/S}$ is not. The main task of this section will be to present conditions under which $\mathrm{Pic}^0_{X/S}$ is a scheme. We remind the reader that by Theorem 9.3/7 this is the case if the fibres of X are not too bad and if X admits many sections over S. Now let us first state the main results on the representability of $\mathrm{Pic}_{X/S}$ and of $\mathrm{Pic}^0_{X/S}$ in the case of relative curves, afterwards we will sketch their proofs.

Theorem 1 (Deligne [1], Prop. 4.3). *Let $X \longrightarrow S$ be a semi-stable curve which is locally of finite presentation. Then $\mathrm{Pic}_{X/S}$ is a smooth algebraic space over S. The identity component $\mathrm{Pic}^0_{X/S}$ is a smooth separated S-scheme and there is a canonical S-ample line bundle $\mathcal{L}(X/S)$ on $\mathrm{Pic}^0_{X/S}$. Furthermore, $\mathrm{Pic}^0_{X/S}$ is semi-abelian.*

If the base scheme S is the spectrum of a discrete valuation ring, one can prove the representability of $\mathrm{Pic}_{X/S}$ by an algebraic space and the representability of $\mathrm{Pic}^0_{X/S}$ by a separated S-scheme under far weaker assumptions on the fibres of X than in Theorem 1.

Theorem 2 (Raynaud [6], Thm. 8.2.1). *Let S be the spectrum of a discrete valuation ring. Let $f: X \longrightarrow S$ be a proper flat curve such that $f_*\mathcal{O}_X = \mathcal{O}_S$ and let X be normal. If the greatest common divisor of the geometric multiplicities of the irreducible* *components of X_s in X_s is 1 where s is the closed point of S, then*

(a) $\mathrm{Pic}_{X/S}$ *is an algebraic space over S,*

(b) $\mathrm{Pic}^0_{X/S}$ *is represented by a separated S-scheme.*

Corollary 3. *Let S be the spectrum of a discrete valuation ring. Let $f: X \longrightarrow S$ be a proper flat curve with connected generic fibre. Assume that X is regular and that there is a rational point on the generic fibre of X. Then* $\mathrm{Pic}_{X/S}$ *is an algebraic space over S and* $\mathrm{Pic}^0_{X/S}$ *is a separated S-scheme.*

Corollary 3 is easily deduced from Theorem 2. Indeed, due to the valuative criterion of properness, the given rational point on the generic fibre extends to a section σ of X over S. Due to 3.1/2, the image of σ is contained in the smooth locus of X. So there exists an irreducible component of the special fibre X_s of X having geometric multiplicity 1 in X_s. Therefore Theorem 2 applies and the assertion is clear. □

Now let us turn to the proofs. For the proof of Theorem 1, we need further information on $\mathrm{Pic}^0_{X/S}$ in the case of smooth relative curves.

Proposition 4. *Let $f: X \longrightarrow S$ be a proper smooth morphism of schemes whose geometric fibres are connected curves. Then* $\mathrm{Pic}^0_{X/S}$ *is an abelian S-scheme and there is a canonical S-ample rigidified line bundle $\mathscr{L}(X/S)$ on* $\mathrm{Pic}^0_{X/S}$.

The construction of $\mathscr{L}(X/S)$ is canonical in such a way that, for any base change $S' \longrightarrow S$, there is a canonical isomorphism of rigidified line bundles

$$\mathscr{L}(X/S) \otimes_{\mathcal{O}_S} \mathcal{O}_{S'} \xrightarrow{\sim} \mathscr{L}(X'/S')\,,$$

where X' denotes the S'-scheme $X \times_S S'$. One will use this fact to show the representability of $\mathrm{Pic}^0_{X/S}$ by an S-scheme in the more general case of semi-stable curves.

Proof of Proposition 4. In order to keep notations simple, let us write P instead of $\mathrm{Pic}^0_{X/S}$ in the following. Due to 6.1/7, it suffices to prove the assertion after étale surjective base change $S' \longrightarrow S$. So we may assume that $X \longrightarrow S$ is projective; cf. 9.3/2. Then $\mathrm{Pic}_{X/S}$ is a separated smooth S-scheme by 9.3/1 and the identity component P is quasi-projective over S. Since P is proper over S by 8.4/3, it is even projective over S. So it remains to explain the construction of the canonical S-ample sheaf $\mathscr{L}(X/S)$ on P.

It is enough to look at the universal case. So, since the base of the versal deformations of a smooth curve is smooth over $\mathbb{Z}$ (cf. Deligne and Mumford [1], Cor. 1.7), we may assume that S consists of a regular noetherian ring. Due to 8.2/1, the Picard functor $\mathrm{Pic}_{P/S}$ is a separated S-scheme and, due to 8.4/5, the identity component $\mathrm{Pic}^0_{P/S}$ is represented by an abelian S-scheme. Denote it by P^* and call it the dual of P. There is a universal line bundle $\mathscr{P}$ on $P \times_S P^*$, the Poincaré bundle, which is rigidified along the unit sections of P and P^* over S; cf. 8.2/4. For the construction of the canonical S-ample sheaf $\mathscr{L}(X/S)$ on P/S, we need the existence of the canonical isomorphism

$$\varphi : P \xrightarrow{\sim} P^*$$

which is given by the Θ-divisor. To define the Θ-divisor, assume first that $X \longrightarrow S$ has a section $\sigma : S \longrightarrow X$. Then one has a morphism

$$(X/S)^{(g-1)} \longrightarrow P = \mathrm{Pic}^0_{X/S}\,, \qquad D_T \longmapsto [D_T] - (g-1)[\sigma_T]\,,$$

where, for any S-scheme T and for any T-valued point D_T of $(X/S)^{(g-1)}$ (i.e., for any effective Cartier divisor on $X \times_S T$ of degree $g-1$), we denote by $[D_T]$ the element of $\mathrm{Pic}_{X/S}(T)$ corresponding to D_T and where σ_T denotes the relative Cartier divisor of $X \times_S T$ associated to the section $\sigma_T = \sigma \times_S T$. Let W^{g-1} be the schematic image of this morphism; note that it depends on the section σ. It is not hard to see that the induced map

$$(X/S)^{(g-1)} \longrightarrow W^{g-1}$$

is S-birational; cf. Lemma 9.3/5. Furthermore, W^{g-1} is an effective relative Cartier divisor on P, usually denoted by Θ_σ. If one replaces σ by a second section, Θ_σ has to be replaced by a translate. Now let us consider the morphism

$$\varphi_{\Theta_\sigma} : P \longrightarrow P^*\,, \qquad t \longmapsto \mathcal{O}_{P_T}(\tau_t^*(\Theta_\sigma)) \otimes (\mathcal{O}_{P_T}(\Theta_\sigma))^{-1}$$

where, for an S-scheme T, we denote by P_T the T-scheme $P \times_S T$ and where $\tau_t : P_T \longrightarrow P_T$ is the translation by the T-valued point $t \in P(T)$. This map is independent of the choice of σ; so we can drop the σ. If we do not have a section, we may perform an étale surjective base change in order to get a section and, hence, to obtain φ_Θ. Because φ_Θ is independent of the chosen section, it is already defined over the given base S by descent theory.

In order to check that the above map is an isomorphism, one can assume that the base scheme S consists of an algebraically closed field. In this case, the assertion is classical; cf. Weil [2], n°62, Cor. 2. Now we set

$$\mathcal{L}(\Theta) = m^*\mathcal{O}_P(\Theta) \otimes p_1^*(\mathcal{O}_P(\Theta))^{-1} \otimes p_2^*(\mathcal{O}_P(\Theta))^{-1}$$

where $m : P \times_S P \longrightarrow P$ is the group law of P and where $p_i : P \times_S P \longrightarrow P$ are the projections for $i = 1, 2$. Note that, a priori, this definition depends on the chosen section σ, but that in fact, due to the theorem of the square, $\mathcal{L}(\Theta)$ is independent of σ. Again, by descent theory, it is already defined over S. The morphism φ_Θ gives rise to an isomorphism

$$\mathrm{id}_P \times_S \varphi_\Theta : P \times_S P \xrightarrow{\sim} P \times_S P^*$$

such that there is an isomorphism of rigidified line bundles

$$\mathcal{L}(\Theta) \xrightarrow{\sim} (\mathrm{id}_P \times_S \varphi_\Theta)^*\mathcal{P}\,.$$

Consider now the pull-back of $\mathcal{P}$ by the map

$$(\mathrm{id}_P, \varphi_\Theta) : P \longrightarrow P \times_S P^*$$

and denote this line bundle on P by

$$\mathcal{L}(X/S) = (\mathrm{id}_P, \varphi_\Theta)^*\mathcal{P} = (\mathrm{id}_P, \mathrm{id}_P)^*\mathcal{L}(\Theta)$$

which is isomorphic to $\mathcal{O}_P(\Theta + (-1)^*\Theta)$. Then $\mathcal{L}(X/S)$ is rigidified along the unit section and one can show that $\mathcal{L}(X/S)$ is S-ample on P. □

For the *proof of Theorem* 1, we will use the canonical S-ample sheaf $\mathscr{L}(X/S)$ which was constructed in Proposition 4 for smooth curves X over S. Namely, due to Theorem 9.3/7 and the explanation following it, we know already that $\mathrm{Pic}^0_{X/S}$ is a scheme locally for the étale topology on S. Thus, we are concerned with a problem of descent. It suffices to verify the assertion concerning the canonical S-ample invertible sheaf $\mathscr{L}(X/S)$. Due to 6.1/7, it is enough to give the definition of $\mathscr{L}(X/S)$ after étale surjective base extension. Moreover, it suffices to look at the universal case. Since the base of the versal deformations of a fibre of X is smooth over $\mathbb{Z}$ (cf. Deligne and Mumford [1], Cor. 1.7), we may assume that S is regular. In this situation, we have to construct $\mathscr{L}(X/S)$. Denote by S_0 the open subscheme of S where X is smooth over S; note that S_0 is dense in S. Due to Proposition 4, there is a canonical line bundle $\mathscr{L}(X_0/S_0)$ on $\mathrm{Pic}^0_{X_0/S_0}$. Since S is regular, we can extend $\mathscr{L}(X_0/S_0)$ to a line bundle $\mathscr{L}(X/S)$ on $\mathrm{Pic}^0_{X/S}$ such that the pull-back of $\mathscr{L}(X/S)$ under the unit section is trivial on S. Since the geometric fibres of $\mathrm{Pic}^0_{X/S}$ are connected, the extension is unique. Then it follows from Raynaud [4], Thm. XI.1.13, page 170, that $\mathscr{L}(X/S)$ is S-ample, since the restriction of $\mathscr{L}(X/S)$ to S_0 is S_0-ample and since, for all points $s \in S$ of codimension 1, the restriction of $\mathrm{Pic}^0_{X/S}$ to $\mathrm{Spec}(\mathscr{O}_{S,s})$ is the identity component of the Néron model of its generic fibre; cf. 7.4/3 and 9.2/8. □

Finally we want to sketch the *proof of Theorem* 2. Denote the generic point of S by η and the closed point of S by s. Let P be the open subfunctor of $\mathrm{Pic}_{X/S}$ consisting of all line bundles of total degree zero.

Let $Y \hookrightarrow X$ be a rigidificator for $\mathrm{Pic}_{X/S}$; cf. 8.1/6. Then, due to 8.3/3, the functor $(\mathrm{Pic}_{X/S}, Y)$ is an algebraic space over S. Denote by (P, Y) the open subfunctor of $(\mathrm{Pic}_{X/S}, Y)$ consisting of all line bundles of total degree zero. Due to 8.4/2, (P, Y) is smooth over S. Let

$$r:(P, Y) \longrightarrow P$$

be the canonical morphism. There is a largest separated quotient Q of P (in the sense of sheaves for the fppf-topology), and one knows that Q is a smooth and separated S-group scheme; cf. 9.5/3. Let

$$q: P \longrightarrow Q$$

be the canonical morphism. It is clear that r and q are epimorphisms of sheaves with respect to the fppf-topology.

We want to show that q induces an isomorphism of P^0 to Q^0. Note that q_η is an isomorphism. First we want to see that $q \times_S S'$ admits a section over Q^0 where S' is a strict henselization of S. We may assume $S = S'$. Due to 9.1/12, there exists a universal line bundle $\mathscr{L}_\eta$ on $(X \times_S \mathrm{Pic}_{X/S})_\eta$. Let $(\mathscr{M}, \alpha)$ be the universal line bundle of (P, Y). Since $\mathscr{L}_\eta$ induces the universal line bundle of P_η, the line bundles $(\mathrm{id}_X \times q \circ r)^*\mathscr{L}_\eta$ and $\mathscr{M}_\eta$ define the same homomorphism to P_η. So, due to 8.1/4, there exists a line bundle $\mathscr{N}_\eta$ on $(P, Y)_\eta$ such that

$$(\mathrm{id}_X \times q \circ r)^*\mathscr{L}_\eta \cong \mathscr{M}_\eta \otimes f^*(\mathscr{N}_\eta)$$

Since (P, Y) is smooth over S and since S is regular, $\mathscr{N}_\eta$ extends to a line bundle $\mathscr{N}$ on (P, Y). After replacing $\mathscr{M}$ by $\mathscr{M} \otimes f^*\mathscr{N}$, we may assume that $\mathscr{M}$ extends $(\mathrm{id}_X \times q \circ r)^*\mathscr{L}_\eta$. By computing the associated divisor, one can show that, over the identity component $(P, Y)^0$, the line bundle $\mathcal{M}|_{X \times_S (P,Y)^0}$ descends to a line bundle $\mathcal{L}$ on $X \times_S Q^0$. Namely, as X is normal, $\mathcal{M}$ is determined by a Weil divisor D on $X \times (P, Y)^0$. Since $\mathcal{M}_\eta$ descends to $\mathcal{L}_\eta$, we may assume that D_η descends, too. So it suffices to look at "vertical" Weil divisors on $X \times (P, Y)^0$ with support contained in the special fibre. To treat the latter we remark that the sets of vertical Weil divisors (with support contained in special fibres) on X, on $X \times (P, Y)^0$, or, on $X \times Q^0$ are in one-to-one correspondence under the pull-back maps. Then $\mathcal{L}$ gives rise to a morphism $\lambda : Q^0 \longrightarrow P^0$. Since Q is separated and since $(q \circ \lambda)_\eta = \mathrm{id}_{Q^0_\eta}$, it follows that $q \circ \lambda = \mathrm{id}_{Q^0}$. Moreover, one shows easily that λ is a group homomorphism.

Next we claim that P is an algebraic space over S. Due to 8.3/1, it remains to see that f is cohomologically flat in dimension zero. By what we have said at the beginning of this section, it suffices to show that

$$\dim_{k(s)} H^1(X_s, \mathcal{O}_{X_s}) = \dim_{k(\eta)} H^1(X_\eta, \mathcal{O}_{X_\eta}) .$$

Due to 8.4/1, we know that $\dim_{k(s)} H^1(X_s, \mathcal{O}_{X_s})$ is equal to the dimension of $\mathrm{Pic}_{X_s/k(s)} = (\mathrm{Pic}_{X/S})_s$. Moreover we have $\dim P_\eta = \dim Q_\eta = \dim Q_s$. The latter holds, since Q is flat over S. So it remains to see that the canonical map $q_s : P_s \longrightarrow Q_s$ is locally quasi-finite or, that the kernel of $q_s|_{P_s^0}$ is finitely generated as an abstract group. Indeed, a group scheme of finite type over a field whose group of geometric points is finitely generated is finite; so the morphism $q_s|_{P_s^0}$ is quasi-finite, since P_s^0 is of finite type over $k(s)$. The kernel of $q_s|_{P_s^0}$ is smooth over $k(s)$ since, due to the existence of the section λ_s, it is a quotient of the smooth group P_s^0. So, assuming that S is strictly henselian, it remains to see that the set of $k(s)$-rational points of the kernel is finitely generated. Since the map $(P, Y)_s \longrightarrow P_s$ is smooth, the rational points of P_s are induced by rational points of $(P, Y)_s$. Since (P, Y) is smooth over S, the rational points of $(P, Y)_s$ are induced by S-valued points of (P, Y); in particular, by line bundles on X. Due to the existence of the section λ which is defined by a line bundle, we see that the $k(s)$-rational points of the kernel of $q_s|_{P_s^0}$ are induced by line bundles on X which are trivial on the generic fibre. Due to the assumption on X, such a line bundle $\mathcal{L}$ is associated to a Cartier divisor D having support on the special fibre only; hence $\mathcal{L} \cong \mathcal{O}_X(D)$. Thus we see that the kernel of the morphism $q_s|_{P_s^0}$ is finitely generated as an abstract group; namely, the group of Cartier divisors having support only on the special fibre is a subgroup of the free group generated by the irreducible components of the special fibre of X.

Now it is easy to complete the proof. In order to show that $q : P^0 \longrightarrow Q^0$ is an isomorphism, we may assume that S is strictly henselian. Recall that q is unramified and an isomorphism on generic fibres. Now look at the commutative diagram

$$\begin{array}{ccc} Q^0 & \xrightarrow{\ \lambda\ } & P^0 \\ & {}_{\mathrm{id}}\searrow \quad \swarrow_{q} & \\ & Q^0 & \end{array}$$

It follows from 2.2/9 that λ is étale. Then it is clear that λ and, hence, q are isomorphisms. □

Finally we want to mention that, in the case where X is regular, there is a direct proof of the cohomological flatness in Artin and Winters [1] which uses the intersection form.

9.5 Picard Functor and Néron Models of Jacobians

Let $S = \operatorname{Spec} R$ be a base scheme consisting of a discrete valuation ring R. As usual we denote by K the field of fractions of R and by k the residue field of R. In the following we will fix a proper and flat curve X over S; its generic fibre X_K is assumed to be *normal* as well as *geometrically irreducible*. Let $J_K = \operatorname{Pic}^0_{X_K/K}$ be the Jacobian of X_K. It is a smooth and connected K-group scheme of finite type and we can ask if there is a Néron model J of J_K. The purpose of the present section is to describe J, if it exists, in terms of the relative Picard functor $\operatorname{Pic}_{X/S}$. Thereby we will obtain a second method to construct Néron models, which is largely independent of the original method involving the smoothening process.

The key point of the whole construction is the fact that the relative Picard functor $\operatorname{Pic}_{X/S}$ satisfies a mapping property which is similar to the one enjoyed by Néron models. To explain this point, assume that X is regular and that X_K admits a section. Furthermore, consider a smooth S-scheme T and a K-morphism $u_K : T_K \longrightarrow \operatorname{Pic}_{X_K/K}$. Then, using 8.1/4, u_K corresponds to a line bundle ξ_K on $X_K \times_K T_K$, and the latter extends to a line bundle ξ on $X \times_S T$ since $X \times_S T$ is regular; see 2.3/9. Thus it follows that u_K extends to an S-morphism $u : T \longrightarrow \operatorname{Pic}_{X/S}$, where u is unique if $\operatorname{Pic}_{X/S}$ is separated. The same mapping property holds for $\operatorname{Pic}^0_{X/S}$ if the special fibre X_k is geometrically irreducible; use 9.1/2 and 9.2/13. So if, in addition, we know that $\operatorname{Pic}^0_{X/S}$ is a smooth and separated S-group scheme, for example if we are in the situation of Grothendieck's theorem 9.3/1, it follows that $\operatorname{Pic}^0_{X/S}$ is a Néron model of $J_K = \operatorname{Pic}^0_{X_K/K}$. In the latter case the assumption on X to have a section is not really necessary. Namely, if the special fibre of X is geometrically reduced (as is required in 9.3/1), then the smooth locus of X is faithfully flat over S by 2.2/16. Working over a strict henselization R^{sh} of R, it follows from 2.3/5 that $X \otimes_R R^{sh}$ admits a section. So, due to the fact that Néron models descend from R^{sh} to R by 6.5/3, we can state the following result.

Theorem 1. *Let X be a flat projective curve over S which is regular and which has geometrically reduced and irreducible fibres. Then $\operatorname{Pic}^0_{X/S}$ is a Néron model of its generic fibre; i.e., of the Jacobian J_K of X_K. In particular, the special fibre of the Néron model of J_K is connected.*

Before we construct Néron models of Jacobians J_K of a more general type, let us state the mapping property of the relative Picard functor $\operatorname{Pic}_{X/S}$ in the form we will need it later. The curve X is as mentioned at the beginning of this section.

Lemma 2. *Assume either that X_K admits a section or that K is the field of fractions of a henselian discrete valuation ring R with algebraically closed residue field k. Then each element of* $\mathrm{Pic}_{X/S}(K)$ *is represented by a line bundle on X_K. In particular, if X is regular, the canonical map* $\mathrm{Pic}_{X/S}(R) \longrightarrow \mathrm{Pic}_{X/S}(K)$ *is surjective.*

Proof. Let K' be the direct image of $\mathcal{O}_{X_K}$ with respect to the structural morphism $X_K \longrightarrow \operatorname{Spec} K$. Since X_K is geometrically irreducible, K' is a field and the extension K'/K is finite and purely inseparable. If X_K admits a section, K' coincides with K and the first assertion of the lemma follows from 8.1/4. On the other hand, if R is henselian and k is algebraically closed, there is a classical result of Lang saying that the cohomological Brauer group $\mathrm{Br}(K)$ vanishes (see Grothendieck [3], 1.1, or Milne [1], Chap. III, 2.22). In the same way we can show that $\mathrm{Br}(K')$ vanishes. Namely, K' can be viewed as the field of fractions of the integral closure R' of R in K' and R' is a discrete henselian valuation ring with algebraically closed residue field k; use 2.3/1′ or 2.3/4 (d) to show that R' is henselian. Thereby we see that there are no obstructions to representing elements of $\mathrm{Pic}_{X/S}(K)$ by line bundles on X_K; cf. 8.1/4.

If X is regular, each line bundle on X_K extends to a line bundle on X and the second assertion is clear also. □

If X is more general than in Theorem 1, but say, still regular, $\mathrm{Pic}^0_{X/S}$ might not be representable by a scheme or by an algebraic space. Moreover, even if $\mathrm{Pic}^0_{X/S}$ exists as a scheme and, thus, is a smooth scheme by 8.4/2 (for example, if X admits a section), the canonical map $\mathrm{Pic}^0_{X/S} \longrightarrow J$ to a possible Néron model J of J_K is not necessarily surjective. To remedy this, we replace $\mathrm{Pic}^0_{X/S}$ by the open and closed subsheaf $P \subset \mathrm{Pic}_{X/S}$ consisting of all line bundles of total degree 0 and pass to the biggest separated quotient Q of P. As we will see, the latter is a good candidate for a Néron model of J_K.

The subfunctor $P \subset \mathrm{Pic}_{X/S}$ may be viewed as the kernel of the degree morphism $\deg : \mathrm{Pic}_{X/S} \longrightarrow \mathbb{Z}$ and is formally smooth since the same is true for $\mathrm{Pic}_{X/S}$; cf. 8.4/2. Furthermore, the fibres of P over S are representable by smooth schemes (8.2/3 and 8.4/2) and, on the generic fibre, P coincides with $\mathrm{Pic}^0_{X/S}$ so that $P_K = J_K$.

In order to pass to the biggest separated quotient of P, we extend the notion of separatedness from S-schemes to contravariant functors $(\mathrm{Sch}/S)^0 \longrightarrow (\mathrm{Sets})$ by using the valuative criterion as a definition; thus a contravariant functor $F : (\mathrm{Sch}/S)^0 \longrightarrow (\mathrm{Sets})$ is called separated if, for any discrete valuation ring R' over R with field of fractions K', the canonical map $F(\operatorname{Spec} R') \longrightarrow F(\operatorname{Spec} K')$ is injective. If F is representable by a scheme or by an algebraic space and if the latter are locally of finite type over S (which, for algebraic spaces, is automatically the case by our definition), then the separatedness in terms of functors coincides with the usual notion of separatedness for schemes or algebraic spaces.

Now consider the quotient $Q = P/E$ (say, in the sense of fppf-sheaves) where E is the schematic closure in P of the unit section $S_K \longrightarrow \mathrm{Pic}_{X_K/K}$; then E is a subgroup functor of P. To define E if $\mathrm{Pic}_{X/S}$ is not necessarily representable by a scheme (or by an algebraic space), consider the sub-fppf-sheaf of $\mathrm{Pic}_{X/S}$ which is generated by all morphisms $Z \longrightarrow \mathrm{Pic}_{X/S}$ in $\mathrm{Pic}_{X/S}(Z)$ where Z is flat over S and where $Z_K \longrightarrow (\mathrm{Pic}_{X/S})_K = \mathrm{Pic}_{X_K/K}$ factors through the unit section of $\mathrm{Pic}_{X_K/K}$. Since the latter is a closed immersion, one recovers the usual notion of schematic closure if $\mathrm{Pic}_{X/S}$ exists as a scheme or as an algebraic space. Likewise, one can extend the notion of schematic closure in $\mathrm{Pic}_{X/S}$ to any closed subscheme of the generic fibre of $\mathrm{Pic}_{X/S}$. For example, we can view P as the schematic closure in $\mathrm{Pic}_{X/S}$ of the Jacobian $\mathrm{Pic}^0_{X_K/K} = J_K$.

Proposition 3. *As before, let X be a flat proper curve over S such that X_K is normal and geometrically irreducible. Then the quotient $Q = P/E$ is representable by a smooth and separated S-group scheme; it is the biggest separated quotient of P. Furthermore, the projection $P \longrightarrow Q$ is an isomorphism on generic fibres and, thus, the generic fibre of Q coincides with the Jacobian J_K of X_K.*

Proof. Instead of just dealing with the most general case, we will explain how to proceed depending on what is known about $\mathrm{Pic}_{X/S}$. That $P \longrightarrow Q$ is an isomorphism on generic fibres is due to the fact that, by the definition of E, the generic fibre E_K coincides with the generic fibre of the unit section $S \longrightarrow P$ since the generic fibre of P is separated. Furthermore, it is clear that Q is the biggest separated quotient of P if Q is representable by a separated scheme.

1st case: $\mathrm{Pic}_{X/S}$ *is a scheme.* In this situation P is a smooth group scheme whose identity component P^0 is separated by [SGA 3_{I}], Exp. VI_{B}, 5.5. So the intersection of E with P^0 is trivial and it follows that E is étale over S. More precisely, $E \longrightarrow S$ is a local isomorphism with respect to the Zariski topology. Then it is easily seen that the quotient $Q = P/E$ is representable by a smooth scheme and that the projection $P \longrightarrow Q$ is a local isomorphism with respect to the Zariski topology.

2nd case: $\mathrm{Pic}_{X/S}$ *is an algebraic space.* Since the unit section of P is locally closed, E is still étale over S, and it is clear that the quotient $Q = P/E$ exists as an algebraic S-group space which is smooth and separated. Furthermore, it follows from 6.6/3 that Q is an S-group scheme.

3rd case: $\mathrm{Pic}_{X/S}$ *is not necessarily representable by a scheme or by an algebraic space.* Then we can apply 8.1/6 and choose a rigidificator $Y \subset X$ of the structural morphism $f\colon X \longrightarrow S$. Associated to it is a sequence

$$0 \longrightarrow V_X^* \hookrightarrow V_Y^* \longrightarrow (\mathrm{Pic}_{X/S}, Y) \longrightarrow \mathrm{Pic}_{X/S} \longrightarrow 0$$

which is exact with respect to the étale topology; cf. 8.1/11. Considering only line bundles of total degree 0, this sequence restricts to a sequence

$$0 \longrightarrow V_X^* \hookrightarrow V_Y^* \longrightarrow (P, Y) \longrightarrow P \longrightarrow 0$$

which, again, is exact with respect to the étale topology. One knows from 8.3/3 and 8.4/2 that $(\mathrm{Pic}_{X/S}, Y)$ and, hence, (P, Y) is an algebraic space which is smooth over S.

Consider the exact sequence

$$V_Y^* \longrightarrow (P, Y) \xrightarrow{\ r\ } P \longrightarrow 0\,,$$

and let H be the schematic closure of the kernel of r_K. Then H is an algebraic subgroup space of (P, Y); it contains the kernel of r, as is easily seen by using the fact that V_Y^* is flat over S. Furthermore, the quotient $(P, Y)/H$ exists as an algebraic space by 8.3/9 since H is flat over S; it is separated due to the definition of H. We claim that $(P, Y)/H$ is canonically isomorphic to $Q = P/E$. To see this, we mention that, by continuity, r maps H into E. So r induces a morphism $\bar{r}: (P, Y)/H \longrightarrow P/E$. On the other hand, one concludes from $\ker(r) \subset H$ that the projection $(P, Y) \longrightarrow (P, Y)/H$ splits into morphisms

$$(P, Y) \xrightarrow{r} P \xrightarrow{q} (P, Y)/H .$$

Since $(P, Y)/H$ is separated and, thus, $E \subset \ker q$, we thereby obtain a morphism $\bar{q}: P/E \longrightarrow (P, Y)/H$ which is an inverse of $\bar{r}$. So Q is isomorphic to $(P, Y)/H$ and therefore is an algebraic group space. But then Q is a separated group scheme by 6.6/3, which is smooth by the analogue of [SGA 3_{I}], Exp. $\mathrm{VI_B}$, 9.2, for algebraic group spaces. □

In order to show that the smooth and separated S-group scheme Q of Proposition 3 is, in fact, a Néron model of J_K, we have to work under conditions like the ones given in Lemma 2 assuring that each K-valued point of Q extends to an R-valued point of Q (assuming R to be strictly henselian). Also we have to show that Q is of finite type over S.

Theorem 4. *Let X be a proper and flat curve over $S = \operatorname{Spec} R$ whose generic fibre is geometrically irreducible. Assume that, in addition, X is regular and either that the residue field k of R is perfect or that X admits an étale quasi-section. Then*:

(a) *If P denotes the open subfunctor of $\operatorname{Pic}_{X/S}$ given by line bundles of total degree 0 and if E is the schematic closure in P of the unit section $S_K \longrightarrow P_K$, then $Q = P/E$ is a Néron model of the Jacobian J_K of X_K.*

(b) *Let $X_1, \ldots, X_n$ be the irreducible components of the special fibre X_k and let δ_i be the geometric multiplicity of X_i in X_k; cf. 9.1/3. Assume that the greatest common divisor of the δ_i is 1. Then $\operatorname{Pic}^0_{X/S}$ is a separated scheme and, consequently, the projection $P \longrightarrow Q$ gives rise to an isomorphism $\operatorname{Pic}^0_{X/S} \xrightarrow{\sim} Q^0$. Thus, in this case, $\operatorname{Pic}^0_{X/S}$ coincides with the identity component of the Néron model of J_K.*

Remark 5. In the situation of the theorem, the assumption that X admits an étale quasi-section is automatically satisfied if the special fibre X_k is geometrically reduced or, more generally, if X_k contains an irreducible component which has geometric multiplicity 1 in X_k. Namely, then the smooth part of X must meet such a component and, passing to a strict henselization of S, we have a section by 2.3/5. On the other hand, if X admits an étale quasi-section over S, say a true section after we have replaced S by an étale extension, then, X being regular, this section factors through the smooth locus of X; see 3.1/2. In particular, there are irreducible components which have geometric multiplicity 1 in X_k so that the condition in Theorem 4 (b) is automatically satisfied.

Now let us start with the *proof of Theorem* 4. The main part will be to show that Q is of finite type over S. We will use the remainder of the present section to establish this fact; see Lemmata 7 and 11 below. But let us first explain how to obtain assertions (a) and (b) if we know that Q is of finite type.

The formation of the schematic closure E is compatible with flat extensions of valuation rings. Likewise, the regularity of X remains invariant under étale base change by 2.3/9. Thus, in order to show that Q is a Néron model of J_K, we may assume that R is strictly henselian.

It is already known from Proposition 3 that Q is a smooth and separated S-group scheme with generic fibre J_K. Furthermore, it follows from Lemma 2 and 9.1/2 that the canonical map $P(R) \longrightarrow P(K)$ is surjective. So we see that the canonical map $Q(R) \longrightarrow Q(K)$ is surjective and, hence, bijective since Q is separated. Thus, if Q is of finite type, it is a Néron model of J_K by the criterion 7.1/1. This verifies assertion (a). Using the representability result 9.4/2 for $\mathrm{Pic}^0_{X/S}$, assertion (b) is a consequence of assertion (a).

It remains to show that the quotient $Q = P/E$ is of finite type over S. We will present two methods to obtain this result. The first one is based on the existence theorem for Néron models 10.2/1 and uses the fact that the Néron-Severi group of the special fibre of $\mathrm{Pic}_{X/S}$ is finitely generated. But it works only under the additional assumption that the generic fibre X_K is geometrically reduced (which is the case if X admits an étale quasi-section; see 3.1/2). Relying on the existence of a Néron model J of J_K, there is a canonical morphism $Q \longrightarrow J$ and it is to show that the latter is an isomorphism. The second method is independent of the theory of Néron models and uses the intersection form which is associated to the irreducible components of the special fibre X_k. It works in the general situation of Theorem 4 and, as we will see in Section 9.6, provides a means of computing the group of connected components (of the special fibre) of the Néron model J of J_K.

Q is of finite type, a first proof via the existence of a Néron model J of J_K. We start by translating the existence theorem for Néron models 10.2/1 to our situation, a result which we will prove in Chapter 10 and which is independent of Chapter 9.

Proposition 6. *Let X_K be a proper curve over K which is geometrically reduced and irreducible. Let J_K be its Jacobian. Then J_K admits a Néron model J of finite type over S if any of the following conditions is satisfied:*

(a) *X_K is smooth,*

(b) *$X_K \times_K \hat{K}$ is normal, where $\hat{K}$ is the completion of K,*

(c) *X_K is normal and R is excellent.*

Proof. If X_K is smooth, J_K is an abelian variety by 9.2/3. So J_K has a Néron model J of finite type.

If only condition (b) is known, J_K is not necessarily an abelian variety. However, condition (b) is compatible with separable extensions of the field $\hat{K}$. So, for any separable field extension L over $\hat{K}$, we know from 9.2/4 that J_L does not contain subgroups of type $\mathbb{G}_a$ or $\mathbb{G}_m$. Therefore we can conclude from 10.2/1 that J_K has a Néron model J of finite type.

Finally, condition (c) implies condition (b) since $\hat{K}$ is separable over K in this case. □

Let us apply Proposition 6 in order to show that, in the situation of Theorem 4 and under the additional assumption of X_K being geometrically reduced, the

Jacobian J_K of X_K admits a Néron model of finite type. Since X is proper over S, all closed points of X belong to the special fibre X_k. Therefore, if $\hat{R}$ is the completion of R, the local rings at closed points of $X_{\hat{R}}$ may be viewed as completions of local rings of X and, thus, the hypothesis on the regularity of X remains unchanged if we replace R by its completion $\hat{R}$. So, in particular, $X_{\hat{R}}$ is regular and, thus, J_K admits a Néron model J of finite type by Proposition 6. Now it is quite easy to prove that Q is of finite type.

Lemma 7. *In the situation of Theorem 4, assume that X_K is geometrically reduced. Then $Q = P/E$ is of finite type over S.*

Proof. As we have just seen, J_K admits a Néron model J. Since the formation of Q and of J is compatible with étale base change, we may assume that the base ring R is strictly henselian. Furthermore, recall that Q is a smooth and separated S-group scheme such that the canonical map $Q(R) \longrightarrow Q(K)$ is bijective. It is enough to show that the canonical morphism $v : Q \longrightarrow J$ restricts to an isomorphism $Q^0 \xrightarrow{\sim} J^0$. Namely, using the bijectivity of $Q(R) \longrightarrow J(R)$, this implies that the groups $Q(R)/Q^0(R)$ and $J(R)/J^0(R)$, which by 2.3/5 can be interpreted as the groups of connected components of the special fibres of Q and J, coincide and thus are finite. Consequently, Q will be of finite type.

So let us show that v induces an isomorphism $Q^0 \longrightarrow J^0$. The group of connected components $Q(R)/Q^0(R) = Q(k)/Q^0(k)$ may be viewed as a quotient of a subgroup of the Néron-Severi group of the special fibre of $\mathrm{Pic}_{X/S}$ and, thus, is finitely generated (in the sense of abstract groups); see 9.2/14. Since the map $v : Q \longrightarrow J$ is surjective on R-valued points and, hence, on k-valued points, it follows that the quotient $J_k^0/v(Q_k^0)$ is a connected smooth algebraic group over k whose group of k-valued points is finitely generated. However, then $J_k^0/v(Q_k^0)$ must be of dimension zero and, thus, is trivial as is easily seen by considering the multiplication with an integer n not divisible by char k. Therefore $Q^0 \longrightarrow J^0$ is surjective and quasi-finite. But then, being an isomorphism on generic fibres, it must be an isomorphism by Zariski's Main Theorem 2.3/2′ so that the desired assertion on Q follows. □

Q is of finite type, a second proof via the intersection form associated to the special fibre X_k. This approach requires a detailed analysis of divisors on X which have support on the special fibre X_k only.

Lemma 8. *Let X be a proper flat curve over $S = \operatorname{Spec} R$ such that X is normal and such that X_K is geometrically irreducible. Assume that R is a strictly henselian discrete valuation ring. Let D be the group of Cartier divisors on X which have support on the special fibre X_k, let D_0 be the subgroup of all divisors in D which are principal, and let E be as in Theorem 4. Then the canonical map $D/D_0 \longrightarrow E(R)$ is bijective.*

Proof. The injectivity of the map follows from 8.1/3. To show the surjectivity, we consider the Stein factorization

$$X \xrightarrow{g} Y \xrightarrow{h} S$$

of the structural morphism $f: X \longrightarrow S$, where $g_*(\mathcal{O}_X) = \mathcal{O}_Y$ and where $h: Y \longrightarrow S$ is finite. Then Y is the spectrum of a normal ring R' which is finite over R. Since X_K is geometrically irreducible and since X is normal, it follows that $K' = R' \otimes_R K$ is a finite purely inseparable field extension of K and that R' is the integral closure of R in K'. So, similarly as in the proof of Lemma 2, it is seen that R' is a strictly henselian discrete valuation ring and that each $a \in E(R)$ is represented by a line bundle $\mathcal{L}$ on X.

Now fix a point $a \in E(R)$ and a representing line bundle $\mathcal{L}$ on X. Since the restriction of $\mathcal{L}$ to the generic fibre X_K is trivial, $\mathcal{L}$ is of the form $\mathcal{O}_X(\Delta)$ where Δ is a Cartier divisor on X having support on the special fibre of X. Thus a is represented by $\Delta \in D$. □

Let $(X_i)_{i \in I}$ be the family of reduced irreducible components of the special fibre X_k. As in 9.1/3, we write d_i for the multiplicity of X_i in X_k and e_i for the geometric multiplicity of X_i. Then e_i is a power of the characteristic of k and $\delta_i = d_i e_i$ is the geometric multiplicity of X_i in X_k; cf. 9.1/4.

For any line bundle $\mathcal{L}$ on X, one can consider its degree $\deg_i(\mathcal{L})$ on the component X_i; it is a multiple of the geometric multiplicity e_i of X_i; cf. 9.1/8. In particular, we can consider the map

$$\rho: \operatorname{Pic}(X) \longrightarrow \mathbb{Z}^I, \qquad \mathcal{L} \longmapsto (e_i^{-1} \cdot \deg_i(\mathcal{L}))_{i \in I}$$

which, composed with the canonical map $D \longrightarrow \operatorname{Pic}(X)$ yields a map $\alpha: D \longrightarrow \mathbb{Z}^I$, where D is as in Lemma 8.

Lemma 9. *Let R, X, D, D_0, and E be as in Lemma 8. Then there is a canonical complex*

$$0 \longrightarrow D_0 \hookrightarrow D \xrightarrow{\alpha} \mathbb{Z}^I \xrightarrow{\beta} \mathbb{Z} \longrightarrow 0$$

where β is given by $\beta(a_1, \ldots, a_r) := \sum a_i \delta_i$. The latter gives rise to a surjection

$$\sigma: \ker \beta / \operatorname{im} \alpha \longrightarrow Q(S)/Q^0(S)$$

which is bijective if $P \longrightarrow Q = P/E$ induces a surjection

$$\operatorname{Pic}^0_{X/S}(S) \longrightarrow Q^0(S)$$

between S-valued points of the identity components of $\operatorname{Pic}_{X/S}$ *and Q. Furthermore, if* $\operatorname{im} \alpha$ *has rank* $\operatorname{card}(I) - 1$, *then* $\ker \beta / \operatorname{im} \alpha$ *and, thus, also $Q(S)/Q^0(S)$ is finite.*

Proof. To begin with, recall that divisors in D have total degree 0 and that therefore $\beta \circ \alpha = 0$ by 9.1/4 and 9.1/5. So the sequence in question is a complex. Furthermore, the map $\rho: \operatorname{Pic}(X) \longrightarrow \mathbb{Z}^I$ is surjective by 9.1/10. Since R is strictly henselian and since $\operatorname{Pic}_{X/S}$ can be defined by using the étale topology in place of the fppf-topology, we can interpret $\operatorname{Pic}(X)$ as $\operatorname{Pic}_{X/S}(S)$. So $P(S)$ is mapped surjectively onto $\ker \beta$ and, due to 9.2/13, we have the exact sequence

$$0 \longrightarrow \operatorname{Pic}^0_{X/S}(S) \longrightarrow P(S) \longrightarrow \ker \beta \longrightarrow 0 .$$

Using Lemma 8 we can interpret $\operatorname{im}\alpha$ as the image of $E(S)$ under the map ρ: $\operatorname{Pic}(X) \longrightarrow \mathbb{Z}^I$. Therefore we have a canonical isomorphism

$$P(S)/(\operatorname{Pic}^0_{X/S}(S) + E(S)) \xrightarrow{\sim} \ker\beta/\operatorname{im}\alpha\ .$$

Taking the above isomorphism as an identification, we define σ as the canonical map

$$(*) \qquad P(S)/(\operatorname{Pic}^0_{X/S}(S) + E(S)) \longrightarrow Q(S)/Q^0(S)\ .$$

To show that it is surjective, it is enough to show that the canonical map

$$(**) \qquad P(S)/P^0(S) \longrightarrow Q(S)/Q^0(S)$$

is surjective. We will prove the latter fact by relating (**) to the canonical map

$$(***) \qquad P_k(k)/P_k^0(k) \longrightarrow Q_k(k)/Q_k^0(k)\ .$$

The map (***) is surjective. Namely, k is separably closed, and P_k is smooth, as follows from the formal smoothness of P. Thus, (***) may be interpreted as mapping connected components of P_k to connected components of Q_k. So it is surjective, due to the surjectivity of $P_k \longrightarrow Q_k$.

Since we know already from Proposition 3 that Q is a smooth group scheme and since the base S is strictly henselian, it follows from 2.3/5 that the restriction map

$$Q(S)/Q^0(S) \longrightarrow Q_k(k)/Q_k^0(k)$$

is bijective. The same is true for

$$P(S)/P^0(S) \longrightarrow P_k(k)/P_k^0(k)$$

if P is a scheme or an algebraic space which is locally of finite type over S. Namely, then the formal smoothness of P says that P is, in fact, smooth. So (**) will be surjective in this case.

In the general case, we must work with a rigidificator Y and consider the associated exact sequence

$$0 \longrightarrow V_X^* \hookrightarrow V_Y^* \longrightarrow (P, Y) \longrightarrow P \longrightarrow 0$$

of 8.1/11. It is enough to show that

$$P(S)/P^0(S) \longrightarrow P_k(k)/P_k^0(k)$$

is surjective, or, that the composition

$$(P, Y)(S) \longrightarrow (P, Y)_k(k) \longrightarrow P_k(k)$$

is surjective. The first map $(P, Y)(S) \longrightarrow (P, Y)_k(k)$ is surjective by 2.3/5 since (P, Y) is smooth (8.4/2). Furthermore, $(P, Y)_k$ is an extension of the smooth group scheme P_k by the quotient $(V_Y^*)_k/(V_X^*)_k$. The latter is smooth since V_Y^* is smooth; cf. [SGA 3_{I}], Exp. VI_{B}, 9.2. Thus, by the same reference, we see that the morphism $(P, Y)_k \longrightarrow P_k$ is smooth and it follows, again from 2.3/5, that $(P, Y)_k(k) \longrightarrow P_k(k)$ is surjective. This shows that the map (**) is surjective.

The injectivity of σ under the assumption that $\operatorname{Pic}^0_{X/S}(S) \longrightarrow Q^0(S)$ is surjective is easily derived from the exact sequence

$$0 \longrightarrow E(S) \longrightarrow P(S) \longrightarrow Q(S) .$$

Finally, the submodule $\ker \beta \subset \mathbb{Z}^I$ has rank $\operatorname{card}(I) - 1$. If the same is true for $\operatorname{im} \alpha$, it follows that $\ker \beta / \operatorname{im} \alpha$ and, thus, also $Q(S)/Q^0(S)$ is finite. □

Let us assume now that X is regular. Under this assumption we can give an explicit description of the $\mathbb{Z}$-submodule $\operatorname{im} \alpha \subset \mathbb{Z}^I$ considered in the preceding lemma. To do so we introduce the intersection matrix $((X_i \cdot X_j))_{i,j \in I}$ where the intersection number $(X_i \cdot X_j)$ is defined as the degree on X_j of the line bundle which is associated to X_i as a Cartier divisor on X. Thereby we obtain a symmetric bilinear intersection pairing $D \times D \longrightarrow \mathbb{Z}$ on the group $D \simeq \mathbb{Z}^I$ of divisors on X which have support on the special fibre X_k; see also [SGA 7_{II}], Exp. X, 1.6. The map α is closely related to the intersection pairing; namely, $\alpha : D \simeq \mathbb{Z}^I \longrightarrow \mathbb{Z}^I$, as a $\mathbb{Z}$-linear map, is described by the matrix $(e_i^{-1}(X_i \cdot X_j))_{i,j \in I}$ which is called the *modified intersection matrix*.

Lemma 10. *Let R, X, and D be as in Lemmata 8 and 9 and assume that, in addition, X is regular. Let d_i be the multiplicity of X_i in X_k, i.e., the multiplicity of X_i in the divisor (π) =* "special fibre of X", *and let d be the greatest common divisor of the d_i, $i \in I$. Then, for any divisor $\sum n_i X_i \in D$, we have*

$$\left(\sum n_i X_i\right)^2 = -\sum_{i<j} \frac{1}{d_i d_j} (n_i d_j - n_j d_i)^2 (X_i \cdot X_j) .$$

Therefore the intersection form $D \times D \longrightarrow \mathbb{Z}$ is negative semi-definite and its kernel is generated by the divisor $\Delta = \sum d_i d^{-1} X_i \in D$. Furthermore, the $\mathbb{Z}$-module $\operatorname{im} \alpha$ of Lemma 9 is isomorphic to $D/\mathbb{Z}\Delta$ and thus has rank $\operatorname{card}(I) - 1$.

Proof. Tensoring with $\mathbb{Q}$, we can extend the bilinear pairing $D \times D \longrightarrow \mathbb{Z}$ to a bilinear pairing $D \otimes \mathbb{Q} \times D \otimes \mathbb{Q} \longrightarrow \mathbb{Q}$. Therefore we may work with rational coefficients. Set $Y_i = d_i X_i$ and $m_i = n_i d_i^{-1}$. Since $(\pi) = \sum d_j X_j = \sum Y_j$ and since $(Y_i \cdot (\pi)) = (X_i \cdot (\pi)) = 0$ for all i, we can write

$$\begin{aligned}
\left(\sum_i n_i X_i\right)^2 &= \left(\sum_i m_i Y_i\right)^2 = \sum_i m_i \left(Y_i \cdot \sum_j m_j Y_j\right) \\
&= \sum_i m_i \left(Y_i \cdot \sum_j m_j Y_j - m_i \sum_j Y_j\right) \\
&= \sum_i m_i \left(Y_i \cdot \sum_{j \neq i} (m_j - m_i) Y_j\right) \\
&= \sum_{i \neq j} m_i (m_j - m_i)(Y_i \cdot Y_j) \\
&= \sum_{i<j} (m_i - m_j)(m_j - m_i)(Y_i \cdot Y_j) \\
&= -\sum_{i<j} (m_i - m_j)^2 (Y_i \cdot Y_j) \\
&= -\sum_{i<j} \frac{1}{d_i d_j} (n_i d_j - n_j d_i)^2 (X_i \cdot X_j)
\end{aligned}$$

All assertions of the lemma follow easily from this computation since the special fibre of X is connected. The latter is due to the fact that X is proper over S and that the generic fibre of X is connected. □

Now it is easy to complete the proof of Theorem 4 and to show that the group scheme Q is of finite type over R.

Lemma 11. *Assume that X is a flat proper curve over R which is regular and which has geometrically irreducible generic fibre X_K. Then the smooth and separated S-group scheme $Q = P/E$ is of finite type.*

Proof. We may assume that R is strictly henselian. Then it follows from Lemmata 9 and 10 that $\ker \beta/\operatorname{im} \alpha$ and thus $Q(S)/Q^0(S)$ are finite. The latter implies that Q is of finite type since it is locally of finite type; cf. [SGA 3_I], Exp. VI_B, 3.6. □

Remark 12. In the assertion of Theorem 4, we may replace the condition that X be regular by the condition that all local rings of $X \times_R \operatorname{Spec}(R^{sh})$ are factorial (R^{sh} being a strict henselization of R); only this is needed for the proof of Lemma 2. In particular, it is enough to require the strict henselizations of all local rings of X to be factorial.

Remark 13. The above approach to the proof of Theorem 4 via the relative Picard functor and via the intersection form provides a second method of constructing Néron models, which is fairly independent of the one presented in earlier chapters. However, if one starts with a proper and smooth curve X_K over K, say under the assumption that R is excellent and that its residue field k is perfect, then in order to apply Theorem 4 to the Jacobian J_K of X_K, one first has to construct a proper R-model X of X_K which is regular; i.e., one has to use the process of desingularization for curves over R; see Abhyankar [1] or Lipman [1]. Alternatively, for a smooth curve X_K, one can apply the semi-stable reduction theorem and thereby construct a semi-abelian Néron model of J_K, after replacing R by its integral closure in a finite extension of K. Then the technique of Weil restriction leads to a Néron model of J_K over R; cf. 7.2/4. Proceeding either way, one constructs Néron models for Jacobians of smooth curves and eventually for general abelian varieties. But it should be kept in mind that the original construction of Néron models which we have given in Chapters 3 and 4 is more elementary in the sense that it uses just the smoothening process and not the theory of Picard functors as well as the existence of desingularizations or semi-stable reductions.

9.6 The Group of Connected Components of a Néron Model

In the following we assume that the base scheme $S = \operatorname{Spec} R$ consists of a *strictly henselian discrete valuation ring* R. Then, if J is an R-group scheme which is a Néron model of its generic fibre J_K, we can talk about the group $J(R)/J^0(R)$ of connected components of J or, more precisely, of the special fibre of J. The purpose of the present section is to give explicit computations for this group in the situation of Theorem 9.5/4, where we deal with Néron models J of Jacobians and where J can be described in terms of the relative Picard functor of a proper and flat S-curve X. As a key ingredient, we will use Lemma 9.5/9 of the previous section.

The notations will be as in 9.5/4. So X is a flat proper curve over S which is regular and whose generic fibre is geometrically irreducible. Furthermore, let $(X_i)_{i \in I}$ be the family of reduced irreducible components of the special fibre X_k, and let d_i (resp. e_i, resp. $\delta_i = d_i e_i$) be the multiplicity of X_i in X_k (resp. the geometric multiplicity of X_i, resp. the geometric multiplicity of X_i in X_k); cf. 9.1/3. Usually we will set $I = \{1, \ldots, r\}$. Also recall that the intersection number $(X_i \cdot X_j)$ between irreducible components of X_k has been defined as the degree on X_i of the line bundle given by X_j as a Cartier divisor on X; it is divisible by the multiplicity e_i.

Theorem 1. *Let S be the spectrum of a strictly henselian discrete valuation ring R and, as in 9.5/4, let X be a flat proper curve over S which is regular and whose generic fibre is geometrically irreducible. Furthermore, assume either that the residue field k of R is perfect (and, thus, algebraically closed) or that X admits an étale quasi-section (and, thus, a true section).*

Let J_K be the Jacobian of X_K, and let $(X_i)_{i \in I}$ be the family of (reduced) irreducible components of X_k. Then, considering the maps

$$D \simeq \mathbb{Z}^I \xrightarrow{\alpha} \mathbb{Z}^I \xrightarrow{\beta} \mathbb{Z}$$

of 9.5/9, where α is given by the modified intersection matrix $(e_i^{-1}(X_i \cdot X_j))_{i,j \in I}$ and where $\beta(a_1, \ldots, a_r) = \sum a_i \delta_i$, the group of connected components $J(R)/J^0(R)$ of the Néron model J of J_K is canonically isomorphic to the quotient $\ker \beta / \operatorname{im} \alpha$.

Proof. It follows from 9.5/4 that the Néron model J of J_K exists and coincides with the quotient $Q = P/E$, where P is the kernel of the degree morphism $\deg : \operatorname{Pic}_{X/S} \longrightarrow \mathbb{Z}$ and where E is the schematic closure of the generic fibre of the unit section $S \longrightarrow \operatorname{Pic}_{X/S}$. Furthermore, Lemma 9.5/9 provides a canonical surjection

$$\ker \beta / \operatorname{im} \alpha \longrightarrow Q(S)/Q^0(S) = J(S)/J^0(S)$$

which we have to show is bijective. As stated in 9.5/9, the bijectivity will follow if the canonical map

$$(*) \qquad \operatorname{Pic}^0_{X/S}(S) \longrightarrow Q^0(S)$$

is surjective. So let us prove the latter fact.

The easiest case is the one where X admits a section or, more generally (see 9.5/5), where the gcd of the geometric multiplicities δ_i of the components X_i in X_k is 1. Then it follows from 9.5/4 (b) that $\operatorname{Pic}^0_{X/S}$ is a separated scheme and that the canonical morphism $\operatorname{Pic}^0_{X/S} \longrightarrow Q^0$ is an isomorphism. So the bijectivity of $(*)$ is trivial in this case.

It remains to treat the case where the residue field k is algebraically closed. To do this, we may assume that, in addition to our assumptions, the base ring R is complete. Namely, the assumptions of the theorem are not changed if R is replaced by its completion; for the regularity of X this has been explained after 9.5/6. Furthermore, note that the special fibre X_k remains the same if R is replaced by its completion and that the formation of Q is compatible with such a base change since it commutes with flat extensions of discrete valuation rings.

The canonical morphism $P \longrightarrow Q$ is an isomorphism on generic fibres. Furthermore, the map $P(S) \longrightarrow P(K)$ is surjective by 9.5/2 and $Q(S) \longrightarrow Q(K)$ is bijective since Q is a Néron model of its generic fibre. So the canonical map

$$P(S) \longrightarrow Q(S)$$

is seen to be surjective. In order to derive the surjectivity of (*) from this fact, we will use the Greenberg functor; see Greenberg [1]. Having no information on the representability of P at hand, it is necessary to work within the context of rigidificators.

Therefore, choose a rigidificator $Y \subset X$, and let (P, Y) be the open and closed subfunctor of the Picard functor of rigidified line bundles $(\mathrm{Pic}_{X/S}, Y)$ which equals the kernel of the degree morphism. We claim that the canonical map $(P, Y)(S) \longrightarrow P(S)$ is surjective. Namely, each element of $P(S)$ is given by a line bundle $\mathscr{L}$ on X and the pull-back of $\mathscr{L}$ to Y is trivial. The latter is true because Y is finite over S and because S is a local scheme. Hence, the composite map $(P, Y)(S) \longrightarrow Q(S)$ is surjective. For our purposes, it is enough to show that it restricts to a surjection $(P, Y)^0(S) \longrightarrow Q^0(S)$. Then, a fortiori, $P^0(S) \longrightarrow Q^0(S)$ will be surjective. Therefore, using the fact that (P, Y) is a smooth algebraic space (see 8.3/3 and 8.4/2) and that $(P, Y)(S)/(P, Y)^0(S)$ can be viewed as a quotient of a subgroup of the Néron-Severi group of the special fibre of X and, thus, is of finite type by 9.2/14, we have reduced the problem to showing the following assertion:

Lemma 2. *Let R be a complete discrete valuation ring with algebraically closed residue field k. Let $G \longrightarrow H$ be an R-morphism of smooth commutative algebraic R-group spaces with the property that $G(R)/G^0(R)$ is finitely generated (in the sense of abstract groups). Then, if $G(R) \longrightarrow H(R)$ is surjective, the same is true for $G^0(R) \longrightarrow H^0(R)$.*

By means of the Greenberg functor, we will be able to reduce the assertion to the corresponding one where R is replaced by the algebraically closed field k and where we consider a k-morphism $G \longrightarrow H$ of smooth commutative k-group schemes of finite type such that $G(k)/G^0(k)$ is finitely generated. Then, if $G(k) \longrightarrow H(k)$ is surjective, it is easy to see that the map $G^0(k) \longrightarrow H^0(k)$ is surjective. Namely, proceeding indirectly, assume that $G^0(k) \longrightarrow H^0(k)$ is not surjective. Then $G^0 \longrightarrow H^0$ cannot be an epimorphism since we are working over an algebraically closed field k. So the image of G^0 in H^0 is a closed subgroup M such that H^0/M is of positive dimension. Its group of k-valued points may be viewed as a quotient of a subgroup of $G(k)/G^0(k)$ and thus, by our assumption on $G(k)/G^0(k)$, is finitely generated. However, then H^0/M cannot have positive dimension as is easily seen by considering the multiplication on H^0/M by an integer which is not divisible by char k. Hence we have derived a contradiction and it follows that $G^0(k) \longrightarrow H^0(k)$ is surjective as claimed.

Next let us recall some basic facts on the Greenberg functor from Greenberg [1]; see also Serre [3], § 1. Let π be a uniformizing element of R and set $R_n := R/(\pi^n)$. Then the Greenberg functor Gr_n of level n associates to each R_n-scheme Y_n of locally finite type a k-scheme $\mathfrak{Y}_n = \mathrm{Gr}_n(Y_n)$ of locally finite type in such a way that, functorially in Y_n, we have $Y_n(R_n) = \mathfrak{Y}_n(k)$. For example, in the equal characteristic case, R_n may be viewed as a finite-dimensional k-algebra and the Greenberg functor Gr_n associated to R_n is just the Weil restriction functor (see 7.6) with respect to the morphism $\mathrm{Spec}\, R_n \longrightarrow \mathrm{Spec}\, k$. Weil restrictions are always representable by schemes in this case, due to the fact that R_n is an artinian local ring with residue field k.

In the unequal characteristic case, R_n cannot be viewed as a k-algebra and the notion of Weil restriction has to be generalized. Then, k being perfect, R is canonically an algebra of module-finite type over the ring of Witt vectors $W(k)$ and $W(k)$ is a complete discrete valuation ring of mixed characteristic, just as R is; see Bourbaki [2], Chap. 9, §§ 1 and 2, in particular, § 1, n°7, Prop. 8, and § 2, n°5, Thm. 3. So, in terms of $W(k)$-modules, R_n is a direct sum of rings of Witt vectors of finite length over k. Using the definition of Witt vectors, we can identify the set of R_n with a product k^m in such a way that the ring structure of R_n corresponds to a ring structure on k^m which is given by polynomial maps. Thereby it is immediately clear that we may interpret R_n as the set of k-valued points of a ring scheme $\mathcal{R}_n$ over k where, as a k-scheme, $\mathcal{R}_n$ is isomorphic to $\mathbb{A}_k^m$.

Similarly as in the case of Weil restrictions, one defines $\mathrm{Gr}_n(Y_n)$ for any R_n-scheme Y_n on a functorial level before one tries to prove its representability by a k-scheme. Namely, consider the functor h^* which associates to any k-scheme T the locally ringed space $h^*(T)$ consisting of T as a topological space and of $\mathcal{H}om_k(T, \mathcal{R}_n)$ as structure sheaf. Then

$$h^*(\mathrm{Spec}\, A) = \mathrm{Spec}(R_n \otimes_{W(k)} W(A))$$

for any k-algebra A. In particular, taking $A = k$, we see that $h^*(T)$ is a locally ringed space over $\mathrm{Spec}\, R_n$. It is shown in Greenberg [1] that, for R_n-schemes Y_n of locally finite type, the contravariant functor

$$\mathrm{Gr}_n(Y_n)\colon (\mathrm{Sch}/k) \longrightarrow (\mathrm{Sets})\,, \qquad T \longmapsto \mathrm{Hom}_{R_n}(h^*(T), Y_n)$$

is representable by a k-scheme $\mathfrak{Y}_n$ which, again, is locally of finite type. So $\mathfrak{Y}_n = \mathrm{Gr}_n(Y_n)$ is characterized by the equation

$$\mathrm{Hom}_k(T, \mathfrak{Y}_n) = \mathrm{Hom}_{R_n}(h^*(T), Y_n)$$

and, in particular, setting $T := \mathrm{Spec}\, k$, we obtain $\mathfrak{Y}_n(k) = Y_n(R_n)$, the property of the Greenberg functor Gr_n we have mentioned at the beginning.

The canonical projection $R_{n+1} \longrightarrow R_n$ gives rise to a functorial transition morphism $\mathrm{Gr}_{n+1} \longrightarrow \mathrm{Gr}_n$. Furthermore, the Greenberg functor Gr_n respects closed immersions, open immersions, and fibred products. In fact, by establishing the first two of these compatibility properties, the representability of $\mathfrak{Y}_n = \mathrm{Gr}_n(Y_n)$ is reduced to the trivial case where $Y_n = \mathbb{A}_{R_n}^m$ and where $\mathfrak{Y}_n = (\mathcal{R}_n)^m$. Furthermore, it can be shown that the Greenberg functor respects smooth and étale morphisms. So this functor extends in a natural way from schemes to algebraic spaces. Working with group objects in the sense of algebraic spaces, we see that $\mathfrak{Y}_n$ will be an algebraic group space and, thus, by 8.3, a group scheme over k if Y is an algebraic group space over R_n. Moreover, for smooth group objects, the Greenberg functor respects identity components.

After this digression, let us turn to the *proof of Lemma* 2. Let $R_n = R/(\pi^n)$ be as above. Applying the base change $R \longrightarrow R_n$ and then the Greenberg functor of level n, we can associate to $G \longrightarrow H$ a morphism of k-group schemes of locally finite type $\mathfrak{G}_n \longrightarrow \mathfrak{H}_n$ such that the maps

$$G(R_n) \longrightarrow H(R_n)\,, \qquad \mathfrak{G}_n(k) \longrightarrow \mathfrak{H}_n(k)$$

can be identified. Since $G(R) \longrightarrow H(R)$ is surjective by our assumption and since $H(R) \longrightarrow H(R_n)$ is surjective by the lifting property 2.2/6 characterizing smoothness, we see that $G(R_n) \longrightarrow H(R_n)$ and, thus, $\mathfrak{G}_n(k) \longrightarrow \mathfrak{H}_n(k)$ is surjective. Furthermore, it follows that $\mathfrak{G}_n(k)/\mathfrak{G}_n^0(k)$, as a quotient of $G(R)/G^0(R)$, is finitely generated. Thus, as we have explained before, $\mathfrak{G}_n^0(k) \longrightarrow \mathfrak{H}_n^0(k)$ and therefore also $G^0(R_n) \longrightarrow H^0(R_n)$ must be surjective.

The map $G^0(R) \longrightarrow H^0(R)$ can be interpreted as the projective limit of the surjective maps $\mathfrak{G}_n^0(k) \longrightarrow \mathfrak{H}_n^0(k)$, $n \in \mathbb{N}$. In order to show the surjectivity of

$$\varprojlim \mathfrak{G}_n^0(k) \longrightarrow \varprojlim \mathfrak{H}_n^0(k)\,,$$

it is enough to show that the system $(\mathfrak{N}_n)$, where $\mathfrak{N}_n$ is the kernel of the morphism $\mathfrak{G}_n^0 \longrightarrow \mathfrak{H}_n^0$, satisfies the Mittag-Leffler condition. However, this is clear since each $\mathfrak{G}_n^0$ is a k-scheme of finite type and, thus, satisfies the noetherian chain condition. So we have finished the proof of Lemma 2 and thereby also the proof of Theorem 1. □

The assertion of Theorem 1 reduces the computation of the group of connected components $J(R)/J^0(R)$ to a problem of linear algebra. In the remainder of the present section, we want to give some formulas for the order of $J(R)/J^0(R)$ as well as determine this group explicitly in some special cases. Let us start with some easy consequences of Theorem 1.

Corollary 3. *Assume that the conditions of Theorem 1 are satisfied. Set $I = \{1,\ldots,r\}$ and let $n_1,\ldots,n_{r-1}$, 0 be the elementary divisors of the modified intersection matrix $A = (e_i^{-1}(X_i \cdot X_j))_{i,j\in I}$. Then the group of connected components $J(R)/J^0(R)$ of the Néron model J of J_K is isomorphic to $\mathbb{Z}/n_1\mathbb{Z} \oplus \ldots \oplus \mathbb{Z}/n_{r-1}\mathbb{Z}$. Its order is the greatest common divisor of all $(r-1)\times(r-1)$-minors of A.*

Proof. Since the image of $\beta : \mathbb{Z}^r \longrightarrow \mathbb{Z}$ has no torsion and, thus, is free of rank 1, it follows that $\ker\beta$ is a direct factor in $\mathbb{Z}^r$, free of rank $r-1$. We know from 9.5/10 that the submodule $\operatorname{im}\alpha \subset \ker\beta$ is of rank $r-1$ also and, thus, can be described by non-zero elementary divisors $n_1,\ldots,n_{r-1}$. But then $n_1,\ldots,n_{r-1}$, 0 are the elementary divisors of $\operatorname{im}\alpha$ viewed as a submodule of $\mathbb{Z}^r$ and the assertions of the corollary are clear. □

If, in the above situation, all geometric multiplicities e_i are trivial, i.e., if $e_i = 1$ for all i, then the modified intersection matrix A coincides with the intersection matrix $((X_i \cdot X_j))_{i,j\in I}$. Considering the associated intersection pairing on the group $D \simeq \mathbb{Z}^I$ of all Cartier divisors on X which have support on the special fibre X_k, we know from 9.5/10 that the pairing is negative semi-definite and has a kernel $\Delta \cdot \mathbb{Z}$ of rank 1, where $\Delta = \sum d_i d^{-1} X_i$, as a divisor in D; the element d is the gcd of the multiplicities d_i. Dividing out the kernel, we get a quadratic form on $D/\Delta\mathbb{Z} \simeq \mathbb{Z}^I/\ker\alpha$ whose discriminant yields the order of the group of connected components $J(R)/J^0(R)$.

Corollary 4 (Lorenzini [1], 2.1.2). *Assume that the conditions of Theorem 1 are satisfied and that, in addition, all geometric multplicities e_i, $i \in I$, are equal to 1. Let $I = \{1,\dots,r\}$. Then, for all indices $i, j \in I$, the absolute value of*

$$a_{ij}^* \,(\gcd(d_1,\dots,d_r))^2 d_i^{-1} d_j^{-1} ,$$

where a_{ij}^* is the $(r-1)\times(r-1)$-minor of index (i,j) of the intersection matrix $A = ((X_i \cdot X_j))$, is independent of i and j. It equals the order of the group of connected components $J(R)/J^0(R)$.

The *proof* is by establishing a lemma from linear algebra (see Lemma 5 below) which allows to compute the gcd of the $(r-1)\times(r-1)$-minors of the intersection matrix A. To apply it, set $d_i' := d_i d^{-1}$. Then the assertion of Corollary 4 follows from Corollary 3. For the purposes of the lemma, we will use an exponent "t" to denote transposition of matrices.

Lemma 5. *Let $A = (a_{ij}) \in \mathbb{Z}^{r\times r}$ define a semi-definite quadratic form of rank $r-1$. Let its kernel be generated over $\mathbb{Z}$ by the vector $d' = (d_r',\dots,d_r')^t \in \mathbb{Z}^r$ and let $A^* = (a_{ij}^*)$ be the adjoint matrix of A. Then there exists a positive integer v such that*

$$A^* = \pm v \cdot d' \cdot d'^t .$$

Furthermore, v is the gcd *of all $(r-1)\times(r-1)$-minors of A.*

Proof. Since $\gcd(d_1',\dots,d_r') = 1$, the assertion on the greatest common divisor of the $(r-1)\times(r-1)$-minors of A follows from the formula for A^*. So it is enough to establish this formula. To do this, note that the kernel of A as a semi-definite quadratic form on $\mathbb{Z}^r$ coincides with the kernel of A as a $\mathbb{Z}$-linear map $\mathbb{Z}^r \longrightarrow \mathbb{Z}^r$. Then, using the equation

$$A \cdot A^* = \det(A) \cdot \text{unit matrix} = 0 ,$$

we see that all columns of A^* belong to the kernel of A. So there is a vector $c = (c_1,\dots,c_r)^t \in \mathbb{Z}^r$ satisfying $A^* = d \cdot c^t$. Since A^* is symmetric, we have $c \cdot d'^t = d' \cdot c^t$ and, thus, $A \cdot c \cdot d'^t = 0$. This implies $A \cdot c = 0$ since $d' \neq 0$ so that c belongs to the kernel of A. Hence there is an element $v \in \mathbb{Z}$ satisfying $c = v \cdot d'$. Replacing v by its absolute value if it is negative, we have $A^* = \pm v \cdot d' \cdot d'^t$ as required. □

If one wants to prove more specific assertions on the group of connected components $J(R)/J^0(R)$, it is important to have information on the configuration of the components X_i of the special fibre X_k. The latter can be described using graphs. There are several possibilities to associate a graph to X_k depending on how multiple intersections of components as well as multiplicities of intersection points are treated. We will deal with two cases, the one where the graph of X_k, in the weakest possible sense, is a tree and the one where X is a semi-stable curve. As a general assumption, we require that we are in the situation of Theorem 1 and that, in addition, the multiplicities e_i, $i \in I$, are equal to 1. For example, the latter is the case if k is algebraically closed. The index set I will always be the set $\{1,\ldots,r\}$.

The case where the graph of X_k is a tree (cf. Lorenzini [1]). The graph Γ we want to associate to X_k is constructed in the following way: the vertices of Γ are the components X_i of X_k, and a vertex X_i is joined to a vertex X_j different from X_i if the intersection number $(X_i \cdot X_j)$ is non-zero. In particular, the precise number of intersection points in $X_i \cap X_j$ is not reflected in the graph Γ. We define the multiplicity s_i of X_i, as a vertex of Γ, as the number of edges joining X_i; so

$$s_i = \operatorname{card}\{j \in I; i \neq j \quad \text{and} \quad (X_i \cdot X_j) \neq 0\} .$$

Furthermore, we need the multiplicity d_i of X_i in the special fibre X_k (which coincides with the geometric multiplicity δ_i of X_i in X_k since $e_i = 1$), the number $d = \gcd(d_1,\ldots,d_r)$, and the quotients $d_i' = d_i d^{-1}$ which are relatively prime.

Proposition 6. *In the situation of Theorem 1, assume that the graph Γ is a tree and that the geometric multiplicities e_i are equal to 1. Then, writing $a_{ij} = (X_i \cdot X_j)$, the group of connected components $J(R)/J^0(R)$ has order*

$$\prod_{a_{ij} \neq 0, i<j} a_{ij} \cdot \prod_{i=1}^{r} (d_i')^{s_i - 2} .$$

Furthermore, if all d_i' are equal to 1, we have

$$J(R)/J^0(R) \simeq \prod_{a_{ij} \neq 0, i<j} \mathbb{Z}/a_{ij}\mathbb{Z} .$$

The assertion will be reduced to Corollary 3 by means of the following result:

Lemma 7. *Let $A = (a_{ij}) \in \mathbb{Z}^{r \times r}$ be a symmetric matrix, which is negative semi-definite of rank $r-1$, and let the vector $(d_1',\ldots,d_r')^t \in \mathbb{Z}^r$ with positive entries d_i' generate the kernel of A. Furthermore, let Γ be the graph associated to A in the manner we have described for intersection matrices above. Then, if Γ is a tree, the greatest common divisor of all $(r-1) \times (r-1)$-minors of A is given by the product*

$$\prod_{a_{ij} \neq 0, i<j} a_{ij} \cdot \prod_{i=1}^{r} (d_i')^{s_i - 2} .$$

Furthermore, if $d_i' = 1$ for all i, the elements a_{ij} occurring in the first factor constitute the non-zero elementary divisors of A.

Proof. Let us first assume $d_i' = 1$ for all i. Then, since the vector $(d_1',\ldots,d_r')$ belongs to the kernel of the intersection matrix $A = (a_{ij})$, it follows that the sum of all columns of A is zero. The same is true for the sum of all rows of A since A is symmetric. Consider a terminal edge C of Γ; i.e., an edge with attached vertices, say

X_1 and X_2, such that $s_1 = 1$ and $s_2 = 2$. Then the intersection matrix A has the following form where $a_{ij} = a_{ji}$ and where empty space indicates zeros:

$$\begin{bmatrix} a_{11} & a_{12} & & & & & \\ a_{21} & a_{22} & * & \cdot & \cdot & \cdot & * \\ & * & * & \cdot & \cdot & \cdot & * \\ & \cdot & \cdot & \cdot & \cdot & \cdot & \cdot \\ & \cdot & \cdot & \cdot & \cdot & \cdot & \cdot \\ & * & * & \cdot & \cdot & \cdot & * \end{bmatrix}$$

Now add the first column to the second column and, likewise, the first row to the second row. Using the fact that the sum of the columns or rows in A vanishes, we have $a_{11} = -a_{12} = -a_{21}$. Thus, we see that this operation kills the entries a_{12} and a_{21} so that the resulting matrix is of the form

$$\begin{bmatrix} -a_{12} & & & & & & \\ & a'_{22} & * & \cdot & \cdot & \cdot & * \\ & * & * & \cdot & \cdot & \cdot & * \\ & \cdot & \cdot & \cdot & \cdot & \cdot & \cdot \\ & \cdot & \cdot & \cdot & \cdot & \cdot & \cdot \\ & * & * & \cdot & \cdot & \cdot & * \end{bmatrix}$$

where $a'_{22} = a_{22} + a_{21}$. Let Γ' be the graph obtained from Γ by removing the terminal edge C we are considering as well as the vertex X_1. Then Γ' is a tree again and it can be viewed as a graph which corresponds to the lower bloc, call it A', of the above matrix, where A' has again the property that the sum of its columns or rows vanishes. Thus we can proceed with A' and Γ' in the same way as we have done before with A and Γ. Since Γ is a tree, the procedure of removing terminal edges and vertices stops after finitely many steps with a graph which is reduced to a single vertex and with an associated (1×1)-matrix which is zero. At the same time we have converted A by means of elementary column and row operations into a diagonal matrix; the diagonal elements, except for the last entry which is zero, consist of all elements $-a_{ij}$, $i < j$, such that X_i is joined to X_j by an edge of Γ. This verifies the assertion of the proposition in the case where all d'_i are equal to 1.

In order to verify the remaining assertion on the greatest common divisor of all $(r-1) \times (r-1)$-minors of A in the general case, we consider the matrix $B = (a_{ij} d'_i d'_j)$. It is negative semi-definite of rank $r - 1$ again and has the property that the sum of its columns or rows is zero. So, using the graph Γ, we can determine its elementary divisors as before. In particular, the gcd of all $(r-1) \times (r-1)$-minors of B equals the product

$$\mu := \prod_{a_{ij} \neq 0, i<j} a_{ij} \cdot \prod_{i=1}^{r} (d'_i)^{s_i} .$$

Let ν be the gcd of all $(r-1)\times(r-1)$-minors of A. Writing A_{11} and B_{11} for the matrices obtained from A and B by removing the first column and the first row, we see from Lemma 5 that

$$\det A_{11} = \pm\nu(d_1')^2\,, \qquad \det B_{11} = \pm\mu\,.$$

Thus

$$\mu = \pm\det B_{11} = \pm(d_2'\ldots d_r')^2 \det A_{11} = \pm(d_1'\ldots d_r')^2\nu\,,$$

and the desired assertion follows from the above equation for μ. □

Remark 8. The graph Γ associated to the special fibre X_k of a curve X as above is a tree if the Néron model J of the Jacobian J_K of X_K has potential abelian reduction or, more generally, if the special fibre J_k does not contain a non-trivial torus. Namely, using the notation of 9.5/4, we have $J_k = P_k/E_k$, where E_k^0 is a unipotent group by Raynaud [6], 6.3/8. So if J_k does not contain a non-trivial torus, the same is true for P_k and, thus, for $\mathrm{Pic}_{X_k/k}$. Then the configuration of the components X_i of X_k is "tree-like" by 9.2/12. However, it should be noted that the graph Γ as we have defined it can be a tree also in some cases where the configuration of the components of X_k is not "tree-like". For example, X_k can be a semi-stable curve consisting of two components which intersect each other in several points. In this case, it follows from 9.2/10 again that J_k contains a non-trivial torus.

We want to apply Proposition 6 in order to show that the order of the group of connected components $J(R)/J^0(R)$ is bounded if J_K has potential good reduction. See Lorenzini [1] for more precise bounds and McCallum [1] for a generalization to abelian varieties.

Theorem 9. *Let R be a strictly henselian discrete valuation ring with algebraically closed residue field k and with field of fractions K. Furthermore, let X_K be a proper smooth curve over K, which is geometrically connected, has a Jacobian J_K with potential good reduction, and admits a regular minimal model X over R.*

Then, for each integer $g > 0$, there exists a bound $M(g)$ such that, for each choice of R, K, and k, and for each curve X_K of genus g as above, the order of the group of connected components $J(R)/J^0(R)$ of the Néron model J of J_K is bounded by $M(g)$.

Proof. We will use the methods of Artin and Winters [1]; the notation is as before. The connected components of X_k are denoted by X_i, and d_i is the multiplicity of X_i in X_k. Furthermore, let d be the gcd of the d_i and set $d_i' = d_i d^{-1}$. Let X_k' be the scheme given by $\sum d_i' X_i$, the latter being viewed as a Cartier divisor on X. Then

$$(*) \qquad H^0(X_k', \mathcal{O}_{X_k'}) = k$$

by Artin and Winters [1], Lemma 2.6, since the gcd of the d_i' is 1.

We want to compute the arithmetic genus of X_k'. Let $\mathfrak{K}$ be a relative canonical divisor on X. Then we can compute the Euler-Poincaré characteristic of $\mathcal{O}_{X_k'}$ as

$$-\chi(\mathcal{O}_{X_k'}) = (X_k'\cdot(X_k' + \mathfrak{K}))/2 = (X_k'\cdot\mathfrak{K})/2 = (X_k\cdot\mathfrak{K})/2d = (g-1)/d\,;$$

the last equality is due to the fact that the degree of $\mathfrak{K}$ is the same on the generic and on the special fibre of X. So, using the equality (*), the arithmetic genus g' of X_k' is given by

$$g' = 1 - \chi(\mathcal{O}_{X_k'}) = 1 + (X_k' \cdot \mathfrak{K})/2 = 1 + (g-1)/d .$$

In particular, g' coincides with the abstract genus introduced by Artin and Winters [1], 1.3, and we have $g' \leq g$. If $H^0(X_k, \mathcal{O}_{X_k}) \neq k$, which may be the case if $d > 1$, and if we compute the arithmetic genus of X_k, it can happen that the latter is greater than g. This is the reason why one has to introduce the curve X_k'.

Now, in order to determine the order of the group of connected components $J(R)/J^0(R)$, one applies Corollary 3 and determines the greatest common divisor of all $(r-1) \times (r-1)$-minors of the intersection matrix $((X_i \cdot X_j))$; let us denote it by ν. The intersection matrix is the same for X_k and for X_k'. Thus, also the graph Γ is the same for both curves, and it follows from our explanations given in Remark 8 that Γ is a tree since J_K has potential good reduction. We want to show that the integer ν remains invariant if we contract an exceptional curve C of the second kind in the sense of Artin and Winters [1], 1.4, in X_k. Such a curve C corresponds to the middle edge of a chain

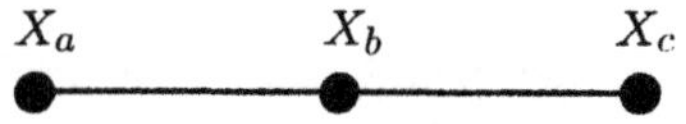

in Γ such that $d_a' = d_b' = d_c'$ and $(X_a \cdot X_b) = (X_b \cdot X_c) = 1$ and such that $s_b = 2$; i.e., there is no ramification at the vertex X_b. Contracting X_b modifies Γ to the extent that we have to replace the above chain by

where now $d_{a'}' = d_a'$, $d_{c'}' = d_c'$, and $(X_{a'} \cdot X_{c'}) = 1$, all other intersection numbers remaining untouched. It follows from the formula in Lemma 7 that the integer ν remains unchanged under such a contraction process. In a similar way one shows that contractions of exceptional curves of the first kind, as considered in Artin and Winters [1], Lemma 1.18, cannot cause ν to increase.

We now use the fact proved in Artin and Winters [1], Thm. 1.6, that, up to contraction of exceptional curves of the first and second kind, there are only finitely many possible types of graphs and intersection matrices for a given genus g' and, thus, for the finitely many genera $g' < g$. So there are only finitely many possible values for the integer ν and, hence, for the order of the group of connected components $J(R)/J^0(R)$. □

The case of semi-stable curves. In the following we will assume that all geometric multiplicities $\delta_i = d_i e_i$ are equal to 1. So, in addition to $e_i = 1$, we have $d_i = 1$ for all $i \in I$. We do not require from the beginning that the special fibre X_k of the curve X is semi-stable; we will restrict ourselves to this case later. The graph we want to consider here is the so-called intersection graph Γ of X_k. Its vertices are given by the irreducible components X_i of the special fibre X_k as before, whereas, different from the graph used above, its edges correspond to the intersection points of such components; i.e., X_i and X_j, $i \neq j$, are joined by as many edges as there are irreducible components in the intersection $X_i \cap X_j$.

We want to compute the group $J(R)/J^0(R)$ of connected components of the Néron model of the Jacobian J_K of X_K by describing the group $\ker\beta/\operatorname{im}\alpha$ of Theorem 1 in terms of the graph Γ. To do this, choose an orientation on Γ and consider the (augmented) simplicial homology complex

$$0 \longrightarrow C_1(\Gamma, \mathbb{Z}) \xrightarrow{\partial_1} C_0(\Gamma, \mathbb{Z}) \xrightarrow{\partial_0} \mathbb{Z}$$

of Γ with coefficients in $\mathbb{Z}$. Then $\operatorname{im}\partial_1 = \ker\partial_0$ since Γ is connected. Identifying $C_0(\Gamma, \mathbb{Z})$ with $\mathbb{Z}^I$, the map ∂_0 coincides with $\beta : \mathbb{Z}^I \longrightarrow \mathbb{Z}$. Thus, if M is any $\mathbb{Z}$-submodule of $C_1(\Gamma, \mathbb{Z})$ lifting $\operatorname{im}\alpha$, i.e., whose image under ∂_1 coincides with $\operatorname{im}\alpha \subset \mathbb{Z}^I \simeq C_0(\Gamma, \mathbb{Z})$, we see that

$$J(R)/J^0(R) \simeq \ker\beta/\operatorname{im}\alpha \simeq C_1(\Gamma, \mathbb{Z})/(M + H_1(\Gamma, \mathbb{Z})),$$

where the first cohomology group $H_1(\Gamma, \mathbb{Z})$ is the kernel of the map ∂_1.

A canonical lifting M of $\operatorname{im}\alpha$ can be obtained by choosing canonical liftings ζ_i of the generators $\xi_i = ((X_i \cdot X_j))_{j \in I}$, $i \in I$, of $\operatorname{im}\alpha$. Namely, define ζ_i as a sum $\sum_\rho c_{i\rho}\eta_{i\rho}$ where the $c_{i\rho}$ are integers which will be specified below and where the $\eta_{i\rho}$ vary over all edges joining the vertex X_i with a second vertex X_j. Up to its sign, the multiplicity $c_{i\rho}$ is the local intersection number of X_i and X_j at the irreducible component x of $X_i \cap X_j$ which corresponds to $\eta_{i\rho}$. The sign of $c_{i\rho}$ is "+" or "−" depending on the orientation of $\eta_{i\rho}$. We use "+" if $\eta_{i\rho}$ originates at X_i and ends at X_j and "−" otherwise. Then, since X_k, as a Cartier divisor on X, is principal, we have $\sum_{j \in I}(X_i \cdot X_j) = 0$ for all $i \in I$ and we see that $M := \sum_{i \in I}\zeta_i\mathbb{Z}$ is a lifting of $\operatorname{im}\alpha$ so that

$$(*) \qquad J(R)/J^0(R) \simeq C_1(\Gamma, \mathbb{Z})/(M + H_1(\Gamma, \mathbb{Z})).$$

We want to give an explicit example.

Proposition 10. *Let X be a proper and flat curve over S, which is regular and has a geometrically irreducible generic fibre X_K as well as a geometrically reduced special fibre X_k. Assume that X_k consists of the irreducible components $X_1, \ldots, X_r$ and that the local intersection numbers of the X_i are 0 or 1 (the latter is the case if different components intersect at ordinary double points). Furthermore, assume that the intersection graph Γ is of the type*

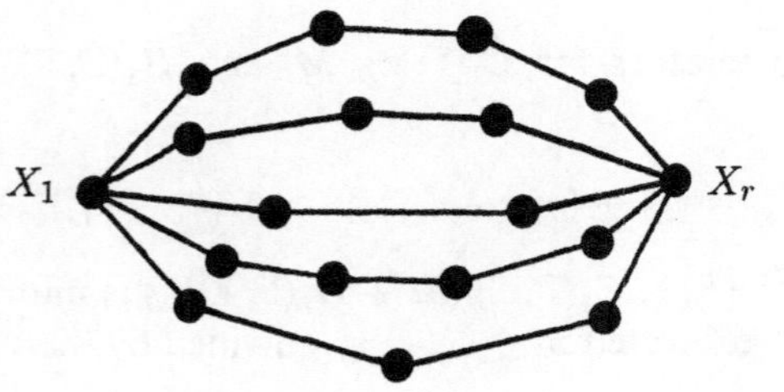

I. e., Γ consists of l arcs of edges starting at X_1 and ending at X_r. For each $\lambda = 1, \ldots, l$, let the λ-th arc consist of the edges $\eta_{\lambda 1}, \ldots, \eta_{\lambda m_\lambda}$, where m_λ is its length. Then the group $J(R)/J^0(R)$ has order

$$\sigma_{l-1}(m_1,\ldots,m_l) := \sum_{\lambda=1}^{l} \prod_{\mu\neq\lambda} m_\mu .$$

More precisely, $J(R)/J^0(R)$ is trivial if $l = 1$. For $l \geq 2$ it is isomorphic to the group

$$(\mathbb{Z}/g_1\mathbb{Z}) \oplus (\mathbb{Z}/g_2 g_1^{-1}) \oplus \ldots \oplus (\mathbb{Z}/g_{l-2} g_{l-3}^{-1}\mathbb{Z}) \oplus (\mathbb{Z}/\sigma_{l-1}(m_1,\ldots,m_l) g_{l-2}^{-1}\mathbb{Z})$$

where g_i is the greatest common divisor of all summands occurring in the i-th elementary symmetric polynomial

$$\sigma_i(m_1,\ldots,m_l)\,, \qquad i = 1,\ldots,l-2\,.$$

Proof. We use the formula (∗). A basis of $C_1(\Gamma,\mathbb{Z})$ is given by the elements

$$\begin{array}{lll} \eta_{11}\,, & \ldots, & \eta_{l1} \\ \eta_{12}-\eta_{11}\,, & \ldots, & \eta_{l2}-\eta_{l1} \\ \vdots & & \\ \eta_{1m_1}-\eta_{1m_1-1}\,, & \ldots, & \eta_{lm_l}-\eta_{lm_l-1} \end{array}$$

Next we write down generators for the canonical lifting M of $\operatorname{im}\alpha$:

$$\begin{array}{lll} \sum_{\lambda=1}^{l} \eta_{\lambda 1}\,, & & \\ \eta_{12}-\eta_{11}\,, & \ldots, & \eta_{l2}-\eta_{l1} \\ \vdots & & \\ \eta_{1m_1}-\eta_{1m_1-1}\,, & \ldots, & \eta_{lm_l}-\eta_{lm_l-1} \\ -\sum_{\lambda=1}^{l} \eta_{\lambda m_\lambda}\,, & & \end{array}$$

and for $H_1(\Gamma,\mathbb{Z})$:

$$\sum_{j=1}^{m_\lambda} \eta_{\lambda j} - \sum_{j=1}^{m_1} \eta_{1j}\,; \qquad \lambda = 2,\ldots.l\,.$$

Using the above generators for $C_1(\Gamma,\mathbb{Z})$, M, and $H_1(\Gamma,\mathbb{Z})$, as well as the fact that

$$\eta_{\lambda j} = \eta_{\lambda 1} + (\eta_{\lambda 2}-\eta_{\lambda 1}) + \ldots + (\eta_{\lambda j}-\eta_{\lambda j-1})\,,$$

if follows that $J(R)/J^0(R) \simeq C_1(\Gamma,\mathbb{Z})/(M + H_1(\Gamma,\mathbb{Z}))$ is isomorphic to the quotient of the free $\mathbb{Z}$-module generated by $\eta_{11},\ldots,\eta_{l1}$, divided by the submodule generated by the relations

$$\sum_{\lambda=1}^{l} \eta_{\lambda 1}\,, \qquad m_\lambda\eta_{\lambda 1} - m_1\eta_{11}\,; \qquad \lambda = 2,\ldots,l\,.$$

The relations are described by the matrix

$$A = \begin{bmatrix} 1 & -m_1 & -m_1 & \cdots & -m_1 \\ 1 & m_2 & 0 & \cdots & 0 \\ 1 & 0 & m_3 & \cdots & 0 \\ \cdot & \cdot & \cdot & \cdots & \cdot \\ \vdots & \vdots & \vdots & & \vdots \\ \cdot & \cdot & \cdot & \cdots & \cdot \\ 1 & 0 & 0 & \cdots & m_l \end{bmatrix}.$$

Computing the determinant of A by developing it via the first column, we get

$$\det A = \sigma_{l-1}(m_1, \ldots, m_l) .$$

Thus, by the theory of elementary divisors, this is already the group order of $J(R)/J^0(R)$. To determine the elementary divisors of A explicitly, we use the criterion involving the gcd of minors; cf. Bourbaki [1], Chap. 7, §4, n°5, Prop. 4.

The gcd of all coefficients of A is 1; so this is the first elementary divisor. For $1 < \lambda < l$, the gcd of all $(\lambda \times \lambda)$-minors is the gcd of all products occurring as summands in the $(\lambda - 1)$-st elementary symmetric polynomial $\sigma_{\lambda-1}(m_1, \ldots, m_l)$; hence it is $g_{\lambda-1}$. Therefore the elementary divisors of A are

$$1, g_1, g_2 g_1^{-1}, \ldots, g_{l-2} g_{l-3}^{-1}, \sigma_{l-1}(m_1, \ldots, m_l) g_{l-2}^{-1}$$

and, consequently, $J(R)/J^0(R)$ is as claimed. □

Corollary 11. *Let X be a flat proper curve over S. Assume that the generic fibre X_K is smooth and that the special fibre X_k is geometrically reduced and consists of two irreducible components X_1 and X_2 which intersect transversally at l rational points $x_1, \ldots, x_l$. Thus, for each $\lambda = 1, \ldots, l$, the curve X is, up to étale localization at x_λ, described by an equation of type $uv = \pi^{m_\lambda}$. If X has no other singularities, then, just as in the situation of Proposition 9, the group of connected components of the Néron model J of the Jacobian J_K of X_K is isomorphic to the group*

$$(\mathbb{Z}/g_1\mathbb{Z}) \oplus (\mathbb{Z}/g_2 g_1^{-1}\mathbb{Z}) \oplus \ldots \oplus (\mathbb{Z}/g_{l-2} g_{l-3}^{-1}\mathbb{Z}) \oplus (\mathbb{Z}/\sigma_{l-1}(m_1, \ldots, m_l) g_{l-2}^{-1}\mathbb{Z})$$

where g_i is the greatest common divisor of all summands occurring in the i-th elementary symmetric polynomial

$$\sigma_i(m_1, \ldots, m_l) , \qquad i = 1, \ldots, l-2 .$$

The assertion is a direct consequence of the preceding proposition since the minimal desingularization of X is of the type considered in Proposition 10. Curves of this type occur within the context of modular curves; see the appendix by Mazur and Rapoport to the article Mazur [1].

Remark 12. If in the situation of Proposition 10 the graph Γ of the special fibre of X is of type

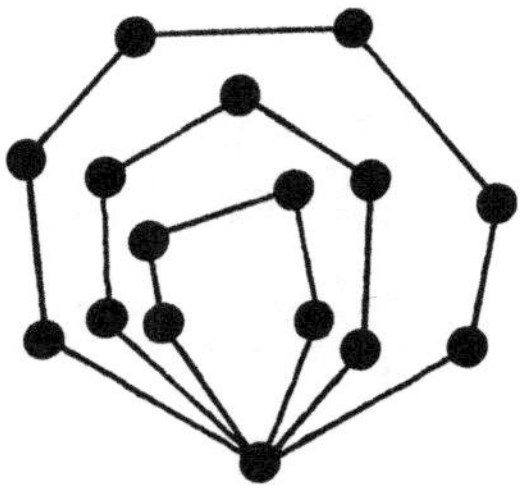

i.e., consists of l loops of length $m_1, \dots, m_l$ starting at X_1 each, the group of connected components of J can be computed as exercised in the proof of Proposition 10. One shows

$$J(R)/J^0(R) = \mathbb{Z}/m_1\mathbb{Z} \oplus \dots \oplus \mathbb{Z}/m_l\mathbb{Z} .$$

Thereby one obtains an analogue of Corollary 11 for curves X whose special fibre is irreducible and has at most ordinary double points as singularities.

9.7 Rational Singularities

Let $S = \operatorname{Spec} R$ be a base scheme consisting of a discrete valuation ring R. As usual, K is the field of fractions and k is the residue field of R. Starting with a proper and flat S-curve X which is normal and has geometrically irreducible generic fibre, we want to relate the fact that a Néron model J of the Jacobian J_K of X_K exists and that the canonical morphism $\operatorname{Pic}^0_{X/S} \longrightarrow J^0$ is an isomorphism to the fact that X has singularities of a certain type, namely *rational singularities*. To explain the latter terminology, assume that X admits a desingularization $f: X' \longrightarrow X$ (which, by Abhyankar [1] or Lipman [1] exists at least in the case where R is excellent). There are only finitely many points where X is not regular. X is said to have rational singularities if $R^1f_*(\mathcal{O}_{X'}) = 0$. It can be shown that the latter condition is independent of the chosen desingularization.

Theorem 1. *Let X be a flat proper curve over S which is normal and which has geometrically irreducible generic fibre X_K. Let $X_1, \dots, X_n$ be the irreducible components of the special fibre X_k. Assume that X admits a desingularization $f: X' \longrightarrow X$ and, furthermore, that the following conditions are satisfied*:

(i) *The residue field k of R is perfect or X admits an étale quasi-section.*

(ii) *The greatest common divisor of the geometric multiplicities δ_i of X_i in X_k (cf. 9.1/3) is 1.*

Then, by (i), *the Jacobian J_K of X_K admits a Néron model J of finite type and, by* (ii), *the identity component $\operatorname{Pic}^0_{X/S}$ of the relative Picard functor is a scheme. Further-*

more, the canonical morphism $\mathrm{Pic}^0_{X/S} \longrightarrow J^0$ *is an isomorphism if and only if* X *has rational singularities.*

Proof. It is easily seen that conditions (i) and (ii) carry over from X to X'. For example, if X admits an étale quasi-section over S, the same is true for X' by the valuative criterion of properness since $f: X' \longrightarrow X$ is proper. Thus it follows from condition (i) and from 9.5/4 that J_K, which is also the Jacobian of X'_K, has a Néron model J of finite type. Furthermore, the canonical morphism $\bar{v}: P'/E' \longrightarrow J$ is an isomorphism where P' is the subfunctor of $\mathrm{Pic}_{X'/S}$ given by line bundles of total degree 0 and where E' is the schematic closure of the generic fibre of the unit section of $\mathrm{Pic}_{X'/S}$.

On the other hand, using 9.4/2, condition (ii) implies that $\mathrm{Pic}^0_{X/S}$ and $\mathrm{Pic}^0_{X'/S}$ are represented by separated schemes. So we get canonical maps between S-group schemes

$$\mathrm{Pic}^0_{X/S} \longrightarrow \mathrm{Pic}^0_{X'/S} \xrightarrow{\sim} J^0,$$

the latter map being an isomorphism by 9.5/4. So $\mathrm{Pic}^0_{X/S} \longrightarrow J^0$ is an isomorphism if and only if $\mathrm{Pic}^0_{X/S} \longrightarrow \mathrm{Pic}^0_{X'/S}$ is an isomorphism and the latter is the case if and only if $\mathrm{Lie}(\mathrm{Pic}^0_{X/S}) \longrightarrow \mathrm{Lie}(\mathrm{Pic}^0_{X'/S})$ is an isomorphism. Writing $R[\varepsilon]$ for the algebra of dual numbers over R, we can interpret $\mathrm{Lie}(\mathrm{Pic}^0_{X/S})$ as the subfunctor of $\mathrm{Hom}_S(\mathrm{Spec}\, R[\varepsilon], \mathrm{Pic}^0_{X/S})$ consisting of all morphisms which modulo ε reduce to the unit section of $\mathrm{Pic}^0_{X/S}$. Then, as we have seen in the proof of 8.4/1, it follows that $\mathrm{Lie}(\mathrm{Pic}^0_{X/S})$ can be identified with the cohomology group $H^1(X, \mathcal{O}_X)$. Proceeding in the same way with X', we see that $\mathrm{Lie}(\mathrm{Pic}^0_{X/S}) \longrightarrow \mathrm{Lie}(\mathrm{Pic}^0_{X'/S})$ is an isomorphism if and only if the canonical map $H^1(X, \mathcal{O}_X) \longrightarrow H^1(X', \mathcal{O}_{X'})$ is an isomorphism.

Now let us look at the Leray sequence associated to $f: X' \longrightarrow X$. It starts as follows:

$$0 \longrightarrow H^1(X, \mathcal{O}_X) \longrightarrow H^1(X', \mathcal{O}_{X'}) \longrightarrow H^0(X, R^1 f_*(\mathcal{O}_X)) \longrightarrow H^2(X, \mathcal{O}_X)$$

Since X is a curve, we have, in fact, a short exact sequence

$$0 \longrightarrow H^1(X, \mathcal{O}_X) \longrightarrow H^1(X', \mathcal{O}_{X'}) \longrightarrow H^0(X, R^1 f_*(\mathcal{O}_X)) \longrightarrow 0.$$

So $H^1(X, \mathcal{O}_X) \longrightarrow H^1(X', \mathcal{O}_{X'})$ is an isomorphism if and only if $H^0(X, R^1 f_*(\mathcal{O}_X)) = 0$. Since $R^1 f_*(\mathcal{O}_X)$ is concentrated at a finite number of closed points of X, the latter is equivalent to $R^1 f_*(\mathcal{O}_X) = 0$; i.e., to the fact that X has rational singularities. This establishes the desired equivalence. □

For semi-stable curves over S (cf. 9.2/6), assumptions (i) and (ii) of Theorem 1 are automatically satisfied. So, using 9.2/8, we see:

Corollary 2. *Let* X *be a semi-stable curve over* S *which is proper, flat, and normal, and which has a geometrically irreducible generic fibre* X_K. *Then the Jacobian* J_K *of* X_K *has a Néron model* J *and the canonical morphism* $\mathrm{Pic}^0_{X/S} \longrightarrow J^0$ *is an isomorphism. In particular,* J *has semi-abelian reduction.*

In the situation of the theorem we can say that $\mathrm{Pic}^0_{X/S}$ is independent of the choice of the S-model X of X_K as long as we limit ourselves to proper, normal, and flat S-curves which have rational singularities. Namely, then $\mathrm{Pic}^0_{X/S}$ coincides with the identity component of the Néron model J of the Jacobian J_K of X_K.

We want to give an application to the modular curve $X_0(N)$. To recall the description of this curve, let N be a positive integer and write U_N for the open subscheme of $\mathrm{Spec}\,\mathbb{Z}$ where N is invertible. Then $X_0(N)|_{U_N}$ is a proper and smooth curve over U_N; it is the compactified coarse moduli space associated to the stack of couples (E, C) of the following type: E is an elliptic curve over some U_N-scheme S and C is a subgroup scheme of E which is finite, étale, and cyclic of order N. For $N = 1$ one obtains the projective line P over $\mathbb{Z}$, to be interpreted as the compactification of the affine line where the j-invariant of elliptic curves serves as a parameter.

Writing $X_0(N)$ for the normalization of P in $X_0(N)|_{U_N}$, the curve $X_0(N)$ is proper over $\mathbb{Z}$ and extends the curve we had already over U_N. For example, if p is a prime strictly dividing N, the curve $X_0(N)$ has semi-stable reduction at p. More precisely, the fibre of $X_0(N)$ over p consists of two smooth components which intersect transversally at the supersingular points; cf. Deligne and Rapoport [1], Chap. VI, Thm. 6.9, or the appendix by Mazur and Rapoport to Mazur [1], Thm. 1.1.

If p^2 divides N, the geometry of fibres is more complicated and certain components have non-trivial multiplicities. In this case one can use the modular interpretation à la Drinfeld which yields information on $X_0(N)$, particularly at bad places. Namely, $X_0(N)$ is the coarse moduli space associated to a certain modular stack which is relatively representable and regular over $\mathbb{Z}$; cf. Katz and Mazur [1], 5.1.1. Then, if x is a closed point of $X_0(N)$, the henselization at x is a quotient of a regular local ring by a finite group whose order divides 12. From this one deduces by means of a norm argument that the singularities of the fibres of $X_0(N)$ over any prime $p > 3$ are rational. Furthermore, over each prime p, there are irreducible components which have geometric multiplicity 1 in the fibre over p; cf. Katz and Mazur [1], 13.4.7. So, using 9.4/2, and Theorem 1, as well as a globalization argument of the type provided in 1.2/4, we obtain:

Proposition 3. *The modular curve $X_0(N)$ is cohomologically flat over $\mathbb{Z}$ and $\mathrm{Pic}^0_{X_0(N)/\mathbb{Z}}$ is representable by a group scheme. Furthermore, outside $p = 2$ and 3, it is the identity component of the Néron model of the Jacobian of $X_0(N) \otimes_{\mathbb{Z}} \mathbb{Q}$.*

Chapter 10. Néron Models of Not Necessarily Proper Algebraic Groups

For this last chapter we introduce a new type of Néron models, so-called Néron lft-models. To define them, we modify the definition of Néron models by dropping the condition that they are of finite type. Then, due to the smoothness, Néron lft-models are locally of finite type. This is the reason why we use the abbreviation "lft". For example, tori do admit Néron lft-models whereas, for non-zero split tori, Néron models (in the original sense) do not exist.

We begin by collecting basic properties of Néron lft-models and by explaining some examples. Then, for the local case, we prove a necessary and sufficient condition for a smooth algebraic K-group G_K to admit a Néron model (resp. a Néron lft-model). In the special case where the valuation ring is strictly henselian and excellent, it states that G_K admits a Néron model (resp. a Néron lft-model) if and only if G_K does not contain a subgroup of type $\mathbb{G}_a$ or $\mathbb{G}_m$ (resp. of type $\mathbb{G}_a$). In the last section, we attempt to globalize our results for excellent Dedekind schemes. An example of Oesterlé shows that one cannot expect a local-global-principle for the existence of Néron models. However, in the case of Néron lft-models, we feel that such a principle is true and formulate it as a conjecture: G_K admits a Néron lft-model if G_K does not contain a subgroup of type $\mathbb{G}_a$. Finally, admitting the existence of desingularizations, we are able to show that the existence of Néron models (in the original sense) is related to the fact that G_K does not contain a non-trivial unirational subvariety.

10.1 Generalities

If R is a discrete valuation ring with field of fractions K, the set of K-valued points of the multiplicative group $\mathbb{G}_{m,K}$ is not bounded in $\mathbb{G}_{m,K}$. Thus $\mathbb{G}_{m,K}$ does not have a Néron model of finite type over R. We will see, however, that there exists a unique R-model of $\mathbb{G}_{m,K}$ which is a smooth R-group scheme and satisfies the Néron mapping property, but which is not of finite type. This is one of the reasons why we want to generalize the notion of Néron models.

Definition 1. *Let S be a Dedekind scheme with ring of rational functions K. Let X_K be a smooth K-scheme. A smooth and separated S-model X is called a Néron* lft-*model of X_K if X satisfies the Néron mapping property*; *cf.* 1.2/1.

Since we do not require X to be of finite type over S, such models are just locally of finite type (lft) over S. As in the case of Néron models, it follows from the Néron mapping property that Néron lft-models are unique and that their formation is compatible with localization and étale base change. In particular, the analogue of 1.2/4 remains valid: an S-scheme X which is locally of finite type is a Néron lft-model of X_K over S if and only if $X \otimes_S \mathcal{O}_{S,s}$ is a Néron lft-model of X_K over Spec $\mathcal{O}_{S,s}$ for each closed point $s \in S$. The Néron lft-model X of a group scheme X_K is a group scheme again. In this case the identity component X^0 is of finite type. Namely, locally on S, there exists an S-dense open affine subscheme U of X^0 and the map $U \times_S U \longrightarrow X^0$ induced by the group law is surjective. Furthermore, it follows from 6.4/1 that any finite set of points of a fibre of X is contained in an affine open subscheme of X.

In the following we want to generalize certain results on Néron models to the case of Néron lft-models. Let us start with the criterion 7.1/1.

Proposition 2. *Let R be a discrete valuation ring and let G be a smooth and separated R-group scheme. Then the following conditions are equivalent:*

(a) *G is a Néron* lft-*model of its generic fibre.*

(b) *Let $R \longrightarrow R'$ be a local extension of discrete valuation rings where R' is essentially smooth over R. Then, if K' is the field of fractions of R', the canonical map $G(R') \longrightarrow G(K')$ is surjective. (Recall that R' is said to be essentially smooth over R if it is the local ring of a smooth R-scheme).*

Proof. The implication (a) $\Longrightarrow$ (b) is a consequence of the Néron mapping property. For the implication (b) $\Longrightarrow$ (a), consider a smooth R-scheme Z and a K-morphism $Z_K \longrightarrow G_K$ of the generic fibres. Due to the assumption, this map extends to an R-rational map $Z \dashrightarrow G$ and, hence, to an R-morphism $Z \longrightarrow G$ by Weil's extension theorem 4.4/1. Thus we see that G satisfies the Néron mapping property. □

Note that, in Proposition 2, it is not sufficient to ask the extension property for étale integral points, as it is in 7.1/1 in the case of Néron models. Next we want to formulate 7.2/1 (ii) for Néron lft-models; the second proof we have given in Section 7.2 carries over without changes.

Proposition 3. *Let R be a discrete valuation ring and let $R \longrightarrow R'$ be an extension of ramification index 1 with fields of fractions K and K'. Assume that G_K is a smooth K-group scheme. If G is a Néron* lft-*model of G_K over R, then $G \otimes_R R'$ is a Néron* lft-*model of $G_K \otimes_K K'$ over R'.*

Moreover, there is an analogue of 7.2/4.

Proposition 4. *Let $S' \longrightarrow S$ be a finite flat extension of Dedekind schemes with rings of rational functions K' and K. Let G_K be a smooth K-group scheme and denote by $G_{K'}$ the K'-group scheme obtained by base change. Let H_K be a closed subgroup of G_K which is smooth. Assume that $G_{K'}$ admits a Néron* lft-*model G' over S'. Then the Néron* lft-*model of H_K over S exists and can be constructed as a group smoothening of the schematic closure of H_K in the Weil restriction $\mathfrak{R}_{S'/S}(G')$.*

Proof. Since any finite set of points of G' is contained in an affine open subscheme of G', the Weil restriction $\mathfrak{R}_{S'/S}(G')$ is represented by an S-scheme which is separated and smooth; cf. 7.6/4 and 7.6/5. By functoriality it is clear that $\mathfrak{R}_{S'/S}(G')$ is the Néron lft-model of $\mathfrak{R}_{K'/K}(G'_{K'})$ over S; cf. 7.6/6. There is a canonical closed immersion

$$\iota : H_K \longrightarrow \mathfrak{R}_{K'/K}(G'_{K'}) .$$

Denote by $\bar{H}$ the schematic closure of H_K in $\mathfrak{R}_{S'/S}(G')$. Then $\bar{H}$ is flat over S. Similarly as exercised in Section 7.1 by applying the smoothening process to the closed fibres of $\bar{H}$, we get a morphism $H \longrightarrow \bar{H}$ from a smooth R-group scheme H to $\bar{H}$ by successively blowing up subgroup schemes in the closed fibres. Indeed, $\bar{H} \cap \mathfrak{R}_{S'/S}(G')^0$ is of finite type over S, since the identity component $\mathfrak{R}_{S'/S}(G')^0$ of $\mathfrak{R}_{S'/S}(G')$ is of finite type over S. So $\bar{H} \cap \mathfrak{R}_{S'/S}(G')^0$ has at most finitely many non-smooth fibres over S. Using translations, one sees that the same is true for $\bar{H}$ and, furthermore, that the non-smooth locus of $\bar{H}$ is invariant under translations. Then it is clear that the process of group smoothenings will work as in the finite type case, since it suffices to control the defect of smoothness over $\bar{H} \cap \mathfrak{R}_{S'/S}(G')^0$. As in 7.1/6, one verifies that H is the Néron lft-model of H_K over R. □

Example 5. *Let S be a Dedekind scheme with ring of rational functions K. The multiplicative group $\mathbb{G}_{m,K}$ over K admits a Néron* lft*-model G over S. Its identity component is isomorphic to $\mathbb{G}_{m,S}$.*

Proof. In order to give a precise description of G, one proceeds as follows.. Let s be a closed point of S and let π_s be a generator of the ideal corresponding to the closed point $s \in S$ over an open neighborhood $U(s)$ of s. So, for each $\nu \in \mathbb{Z}$, we can view π_s^ν as a $(U(s) - \{s\})$-valued point of $\mathbb{G}_{m,S}$. Then, let $\pi_s^\nu \cdot \mathbb{G}_{m,S}$ be a copy of $\mathbb{G}_{m,S} \times_S U(s)$, viewed as the translate of $\mathbb{G}_{m,S}$ by π_s^ν in the Néron lft-model we want to construct. The translations by the sections π_s^ν, $\nu \in \mathbb{Z}$, define gluing data between $\mathbb{G}_{m,S}$ and the $\pi_s^\nu \cdot \mathbb{G}_{m,S}$ over $U(s) - \{s\}$ in a canonical way. So we can define

$$G = \bigcup_{s \in |S|} \bigcup_{\nu \in \mathbb{Z}} (\pi_s^\nu \cdot \mathbb{G}_{m,S})$$

as the result of the gluing of $\mathbb{G}_{m,S}$ with the copies $(\pi_s^\nu \cdot \mathbb{G}_{m,S})$ where $|S|$ is the set of closed points of S.

In order to show that G is a Néron lft-model of $\mathbb{G}_{m,K}$ over S, note first that G is a smooth and separated S-group scheme with generic fibre $\mathbb{G}_{m,K}$. So we have only to verify the Néron mapping property for G. Since the construction of G is compatible with localization of S, we may assume that S consists of a discrete valuation ring R; cf. the analogue of 1.2/4. Due to Proposition 2, it suffices to show for any extension $R \longrightarrow R'$ of ramification index 1 that each K'-valued point extends to an R'-valued point of G. Since the construction of G is compatible with such ring extensions, we may assume $R = R'$. But then it is clear that the canonical map $G(R) \longrightarrow G(K)$ is bijective, so that we are done. □

The example we have just given can be generalized to tori over K.

Proposition 6. *Let S be a Dedekind scheme with ring of rational functions K. Any torus T_K over K admits a Néron* lft-*model over S.*

Proof. We may assume that S is affine and that it consists of a Dedekind ring R. If the torus is split, the assertion follows from the above example. In the general case, there exists a finite separable field extension K'/K such that $T_{K'} = T_K \otimes_K K'$ is split. If R' is the integral closure of R in K', then $T_{K'}$ admits a Néron lft-model over R'. Now the assertion follows from Proposition 4. □

Also we can handle the case of extensions of certain algebraic K-groups by tori. For technical reasons we will restrict ourselves to split tori, although this restriction is unnecessary as can be seen by using 10.2/2.

Proposition 7. *Let S be a Dedekind scheme with ring of rational functions K. Let G_K be a smooth connected algebraic K-group which is an extension of a smooth algebraic K-group H_K by a split torus T_K. Assume that* $\mathrm{Hom}(H_K, \mathbb{G}_{m,K}) = 0$; *for example, the latter is the case if H_K is an extension of an abelian veriety by a unipotent group. Then, if H_K admits a Néron* lft-*model over S, the same is true for G_K.*

Proof. Since T_K is a split torus, say of rank r, the extension G_K of H_K by T_K is given by primitive line bundles $\mathscr{L}_1, \ldots, \mathscr{L}_r$ on H_K; cf. Serre [1], Chap. VII, n°15, Thm. 5. Although Serre considers only the case where H_K is an abelian variety, the result extends to our situation, since each homomorphism of H_K to $\mathbb{G}_{m,K}$ is constant. A line bundle $\mathscr{L}$ on a group scheme G is called primitive if there is an isomorphism

$$m^*\mathscr{L} \cong p_1^*\mathscr{L} \otimes p_2^*\mathscr{L}$$

where m is the group law of G and where $p_i: G \times G \longrightarrow G$ are the projections, $i = 1, 2$. Since the local rings of the Néron model H of H_K are factorial, the line bundles $\mathscr{L}_\rho$, $\rho = 1, \ldots, r$, extend to primitive line bundles on the identity component H^0 of H. Thus, they give rise to an extension

$$1 \longrightarrow T^0 \longrightarrow G^0 \longrightarrow H^0 \longrightarrow 1$$

whose generic fibre is the extension we started with. Then G^0 will be the identity component of the Néron lft-model G of G_K whereas G itself has to be constructed by gluing "translates" of G^0.

In order to do this, let us start with the construction of the local Néron lft-model at a closed point s of S. Let R_s^{sh} be a strict henselization of the local ring R_s and let K_s^{sh} be its field of fractions. Then set

$$\Lambda_s = G(K_s^{sh})/G^0(R_s^{sh}) \supset I_s = T(K_s^{sh})/T^0(R_s^{sh}),$$

where I_s is isomorphic to $\mathbb{Z}^r$. Due to Hilbert's Theorem 90, the quotient Λ_s/I_s is canonically isomorphic to the group $H(K_s^{sh})/H^0(R_s^{sh})$. In the case where Λ_s can be represented by a set $\{\lambda_s\}$ of K-valued points of G_K, we can, similarly as in Example 5, define a smooth and separated R-group scheme

$$G(s) = \bigcup_{\lambda_s \in \Lambda_s} (\lambda_s \cdot G^0)$$

as the result of a gluing where the gluing data are concentrated on the generic fibre and are given by the translations with the sections λ_s. Then each K-valued point of G_K extends to an R-valued point of G. Since this construction is compatible with any extension $R \longrightarrow R'$ of ramification index 1, each K'-valued point of G_K extends to an R'-valued point of G where K' is the ring of fractions of R'. Then, using Proposition 2, one shows that $G(s)$ satisfies the Néron mapping property. Hence, it is the Néron lft-model of G_K over R_s. If the sections $\{\lambda_s\}$ are not defined over R_s, one shows by means of descent that the group $G(s)$ which can be defined over a strict henselization R_s^{sh} of R_s is already defined over the given ring R_s and, hence, is a Néron lft-model of G_K over R_s. In the global case, the Néron lft-model G of G_K is given by gluing the local models $G(s)$, $s \in |S|$, where $|S|$ is the set of all closed points of S; hence

$$G = \bigcup_{s \in |S|} G(s) .$$

In order to explain the gluing procedure, consider a "component" $G(s)'$ of $G(s)$; thereby we mean an open subscheme consisting of G_K and of a connected component of $G(s)$. Then $G(s)'$ is of finite type over R_s and, hence, it extends over an open neighborhood $U(s)$ of s. Since G_K is connected, we may assume that $G(s)'$ coincides with G^0 over $U(s) - \{s\}$. So this way we obtain gluing data between G^0 and each component $G(s)'$ of $G(s)$ and, hence, between G^0 and $G(s)$. It is clear that these data give rise to gluing data for the family $(G(s); s \in |S|)$. In particular, the pull-back of G to the local scheme $\operatorname{Spec} \mathcal{O}_{S,s}$ is isomorphic to $G(s)$. Thus, it is clear that G satisfies the Néron mapping property and, hence, is a Néron lft-model of G_K over S. □

Unipotent K-groups may contain a subgroup of type $\mathbb{G}_a$. So they do not necessarily admit Néron lft-models as we will see by the following proposition. But we mention that, if K is not perfect, there are smooth connected unipotent groups, so-called K-wound unipotent groups, which do not contain the additive group $\mathbb{G}_{a,K}$. In Section 10.2 we will discuss the existence of Néron models for such groups.

Proposition 8. *Let S be a Dedekind scheme with ring of rational functions K. If G_K admits a Néron* lft-*model, then G_K does not contain a subgroup of type $\mathbb{G}_a$.*

Proof. Since Néron lft-models are compatible with localizations and étale extensions of the base scheme, we may assume that S consists of a strictly henselian discrete valuation ring R with uniformizing parameter π. Proceeding indirectly, we may assume by Proposition 4 that $G_K = \mathbb{G}_{a,K}$ and that G_K admits a Néron lft-model G. Let us fix a coordinate function ξ_0 for G_K, say $G_K = \operatorname{Spec} K[\xi_0]$. Then set $G^n = \operatorname{Spec} R[\xi_n]$ for $n \in \mathbb{N}$, where the ξ_n are indeterminates, and consider the morphisms

$$G^n = \operatorname{Spec} R[\xi_n] \longrightarrow G^{n+1} = \operatorname{Spec} R[\xi_{n+1}]$$

induced by sending ξ_{n+1} to $\pi \cdot \xi_n$. These morphisms induce the zero map on the special fibres. We regard each G^n as a smooth R-model of G_K via the isomorphism

$$G^n \otimes_R K \longrightarrow G_K$$

induced by the map $K[\xi_0] \longrightarrow K[\xi_n]$ sending ξ_0 to $\pi^{-n}\xi_n$. Thus, we get commutative diagrams

$$\begin{array}{ccc} G_K^n & \longrightarrow & G_K^{n+1} \\ & \searrow \quad \swarrow & \\ & G_K & \end{array}$$

Due to the Néron mapping property, these diagrams extend to communtative diagrams

$$\begin{array}{ccc} G^n & \longrightarrow & G^{n+1} \\ & \searrow \quad \swarrow & \\ & G & \end{array}$$

The morphisms induce the zero map on special fibres. So we see that each S-valued point of G specializes into the zero section, since such a point can be regarded as an S-valued point of some G^n. Hence, we arrive at a contradiction. □

Next we will discuss a criterion relating the existence of global Néron lft-models to the existence of local Néron lft-models.

Proposition 9. *Let S be a Dedekind scheme with ring of rational functions K. Let G_K be a smooth connected algebraic K-group. Assume that, for each closed point s of S, the local Néron* lft-*model of G_K over $\mathcal{O}_{S,s}$ exists. Then the following conditions are equivalent*:

(a) *G_K admits a global Néron* lft-*model over S.*

(b) *There exists a dense open subscheme U of S and, over U, a smooth group scheme with connected fibres which coincides with the identity component of the local Néron* lft-*model G_K for each closed point s of U.*

(c) *There exists a coherent (locally free) $\mathcal{O}_S$-module $\mathcal{L}$ which, over each local ring of S, coincides with the Lie algebra of the local Néron* lft-*model of G_K.*

Proof. The implication (a) $\Longrightarrow$ (c) is trivial. To show the implication (c) $\Longrightarrow$ (b), let $G(s)^0$, for any closed point s of S, be the identity component of the local Néron lft-model of G_K over $\mathcal{O}_{S,s}$. Since $G(s)^0$ is quasi-compact, there exist an open neighborhood $U(s)$ of s and a smooth $U(s)$-group scheme $G^0_{U(s)}$ with connected fibres such that $G^0_{U(s)}$ induces $G(s)^0$ over the local ring $\mathcal{O}_{S,s}$. Furthermore, due to the assumption (c), we may assume that the Lie algebra of $G^0_{U(s)}$ coincides with the Lie algebra of the local Néron lft-model at each point t of $U(s)$. Then, for each $t \in U(s)$, the canonical map

$$G^0_{U(s)} \times_{U(s)} \operatorname{Spec} \mathcal{O}_{S,t} \longrightarrow G(t)^0$$

is étale and, hence, an isomorphism, since it is an isomorphism on generic fibres. So condition (b) is clear.

For the implication (b) $\Longrightarrow$ (a) we will first construct the identity component of the Néron lft-model. So let G^0_U be the U-group scheme given by condition (b). If s

is a closed point of S not contained in U, the identity component $G(s)^0$ of the Néron lft-model of G_K over $\mathcal{O}_{S,s}$ is of finite type over $\mathcal{O}_{S,s}$ and, hence, extends to a smooth group scheme $G^0_{U(s)}$ with connected fibres over an open neighborhood $U(s)$ of s. Since G^0_U and $G^0_{U(s)}$ coincide on the generic fibre, they coincide over an open neighborhood of s in $U \cap U(s)$. So we get gluing data and, hence, a smooth S-group scheme G^0 with connected fibres which coincides with the identity components of the local Néron lft-models at closed points of S. Now, a Néron lft-model G of G_K is obtained by gluing the local Néron lft-models $G(s)$, $s \in |S|$, where $|S|$ is the set of all closed points of S; i.e.,

$$G = \bigcup_{s \in |S|} G(s) .$$

The procedure is the same as in Proposition 7. Also the Néron mapping property is verified as exercised in the proof of Proposition 7. □

Since a smooth group scheme with connected fibres over a Dedekind scheme is quasi-compact, the proof of the implication (b) ⟹ (a) of the above proposition shows the following fact:

Corollary 10. *Let S be a Dedekind scheme with ring of rational functions K. Let G_K be a smooth connected algebraic K-group. Assume that there exists a global Néron* lft-*model of G_K over S. Then G_K admits a Néron model over S if and only if the groups of connected components of the local Néron* lft-*models are finite and, for almost all closed points of S, are trivial.*

Finally, we want to give an example showing that the existence of local Néron models does not imply the existence of a global Néron model.

Example 11 (Oesterlé [1]). Let R be an excellent Dedekind ring with field of fractions K of positive characteristic p, let K'/K be a radicial field extension of order p^n, and let R' be the integral closure of R in K'. Let G_K be the Weil restriction of the multiplicative group $\mathbb{G}_{m,K'}$ with respect to K'/K. Consider the quotient $U_K = G_K/\mathbb{G}_{m,K}$ where $\mathbb{G}_{m,K}$ is viewed as a subgroup of G_K via the canonical closed immersion

$$\mathbb{G}_{m,K} \longrightarrow G_K = \mathfrak{R}_{K'/K}(\mathbb{G}_{m,K'}) .$$

For each closed point s of $\operatorname{Spec} R$, we will see that the local Néron model exists and that its group of connected components is a cyclic group of order e_s where e_s is the index of ramification of the extension R'_s/R_s. Moreover, U_K admits a global Néron lft-model over R which, in general, will not be of finite type over R if R has infinitely many maximal ideals.

As a typical case, one may take for R the ring of an affine normal curve over a perfect field. In this case, the ramification index at each closed point coincides with the degree of the radicial extension $[K':K]$. In particular, U_K does not admit a global Néron model if the extension K'/K is not trivial.

So let us justify the fact on U_K we have claimed above. Due to Hilbert's Theorem 90, we have

$$U_K(K) = (K')^*/K^* .$$

If R is a discrete valuation ring and $R \longrightarrow R'$ is of ramification index e, the group $U_K(K)$ can be written in the form

$$(K')^*/K^* = (R')^*/R^* \times (\mathbb{Z}/e\mathbb{Z}) .$$

Similarly as for the generic fibre, we have a canonical map

$$\mathbb{G}_{m,R} \longrightarrow G_R := \mathfrak{R}_{R'/R}(\mathbb{G}_{m,R'}) ,$$

which is a closed immersion. Thus, we can define the quotient

$$U^0 = G_R/\mathbb{G}_{m,R} .$$

which is a smooth separated algebraic space; cf. 8.3/9. Due to 6.6/3, it even is a smooth R-group scheme. Moreover, we have

$$U^0(R) = (R')^*/R^* .$$

For each closed point s of Spec(R), the local Néron model $U(s)$ is obtained by gluing $U^0 \otimes_R R_s$ with e_s copies of it along the generic fibre where the gluing data are given via the translation on the generic fibre by representatives of $U(K)/U^0(R_s)$. Then, as in Example 5, it is easy to see that U_s satisfies the Néron mapping property. By Proposition 9, we see that there exists a global Néron lft-model of U_K over R.

One can show that the global Néron lft-model of U_K is isomorphic to the quotient of the Weil restriction of the Néron model of $\mathbb{G}_{m,K'}$ by the Néron model of $\mathbb{G}_{m,K}$. □

10.2 The Local Case

In the following, let R be a discrete valuation ring with field of fractions K and let G_K be a *smooth commutative algebraic K-group*. So, in particular, G_K is of finite type over K. We want to discuss criteria for the existence of a Néron model (resp. of a Néron lft-model) of G_K over R depending on its structure as algebraic group. To fix the notations, let R^{sh} be the strict henselization of R with field of fractions K^{sh}, let $\hat{R}^{sh}$ be the strict henselization of the completion $\hat{R}$ of R, and let $\hat{K}^{sh}$ be the field of fractions of $\hat{R}^{sh}$. Since certain parts of our considerations will require an excellent base ring, recall that the strict henselization of an excellent discrete valuation ring is excellent again by 3.6/2. So $\hat{R}^{sh}$ is excellent. In particular, the extension $\hat{K}^{sh}/\hat{K}$ is separable. Furthermore, if R is excellent, R^{sh} is excellent and the extension $\hat{K}^{sh}/K$ is separable.

We will first concentrate on Néron models. We know already that G_K admits a Néron model if and only if the set of its K^{sh}-valued points is bounded in G_K. Now we want to formulate a necessary and sufficient condition for the existence of a

Néron model for G_K in terms of the group structure of G_K. Let us begin with some definitions. If X is a separated K-scheme of finite type, a *compactification* of X is an open immersion $X \hookrightarrow \bar{X}$ of X into a proper K-scheme $\bar{X}$ such that X is schematically dense in $\bar{X}$. The subscheme $\bar{X} - X$ will be referred to as the infinity of the compactification. Due to Nagata [1], [2], compactifications always exist. If, in addition, X and $\bar{X}$ are regular, we will call $\bar{X}$ a *regular compactification* of X. For a regular K-scheme X, there exists a regular compactification if the characteristic of K is zero or if the dimension of X is ≤ 2; cf. Hironaka [2] and Abhyankar [1].

Theorem 1. *Let R be a discrete valuation ring with field of fractions K, and let G_K be a smooth commutative algebraic K-group. Then the following conditions are equivalent:*

(a) *G_K has a Néron model over R.*
(b) *$G_K \otimes_K \hat{K}^{sh}$ contains no subgroup of type $\mathbb{G}_a$ or $\mathbb{G}_m$.*
(c) *$G_K \otimes_K \hat{K}^{sh}$ admits a compactification without a rational point at infinity.*
(d) *$G_K(\hat{K}^{sh})$ is bounded in G_K.*
(e) *$G_K(K^{sh})$ is bounded in G_K.*

If, in addition, R is excellent, the above conditions are equivalent to

(b′) *$G_K \otimes_K K^{sh}$ contains no subgroup of type $\mathbb{G}_a$ or $\mathbb{G}_m$.*
(c′) *$G_K \otimes_K K^{sh}$ admits a compactification without a rational point at infinity.*

For example, a K-wound commutative unipotent algebraic K-group admits a Néron model over R if R is excellent. Namely, such a group does not contain subgroups of type $\mathbb{G}_a$ or $\mathbb{G}_m$ and this property remains true after any separable field extension; cf. Tits [1], Chap. IV, Prop. 4.1.4.

If G_K is the Jacobian J_K of a normal proper curve X_K over K assumed to be geometrically reduced and irreducible, then, due to 9.2/4, there is no subgroup of type $\mathbb{G}_a$ or $\mathbb{G}_m$ in $J_K \otimes_K L$, for any separable field extension L of K. So, if K is the field of fractions of an excellent discrete valuation ring R, our theorem implies that J_K admits a Néron model over R; cf. 9.5/6. Furthermore, there is a natural compactification of J_K without a rational point at infinity; cf. Example 9.

Before starting with the proof of Theorem 1, we want to deduce a criterion for the existence of Néron lft-models.

Theorem 2. *Let R be a discrete valuation ring with field of fractions K and let G_K be a smooth commutative algebraic K-group. Then the following conditions are equivalent:*

(a) *G_K admits a Néron* lft-*model over R.*
(b) *$G_K \otimes_K \hat{K}^{sh}$ contains no subgroup of type $\mathbb{G}_a$.*

If, in addition, R is excellent, these conditions are equivalent to

(b′) *G_K contains no subgroup of type $\mathbb{G}_a$.*

Let us first deduce Theorem 2 from Theorem 1. The implications (a) $\Longrightarrow$ (b) and (a) $\Longrightarrow$ (b′) follow from 10.1/3 and 10.1/8. Next let us show the implication (b′) $\Longrightarrow$ (a) under the assumption that R is excellent. Let T_K be the maximal torus of G_K; cf. [SGA 3_{II}], Exp. XIV, Thm. 1.1. Then we have an exact sequence of algebraic

K-groups

$$1 \longrightarrow T_K \longrightarrow G_K \longrightarrow H_K \longrightarrow 1 ,$$

where H_K is an extension of an abelian variety by a linear group and where the latter is an extension of a unipotent group U_K by a finite multiplicative group; cf. 9.2/1 and [SGA 3_{II}], Exp. XVII, Thm. 7.2.1. Due to [SGA 3_{II}], Exp. XVII, Thm. 6.1.1(A)(ii), the K-groups H_K and, hence, U_K do not contain a subgroup of type $\mathbb{G}_a$, since the same is true for G_K. Then it follows from Tits [1], Chap. IV, Prop. 4.1.4, that $U_K \otimes_K K'$ and, hence by [SGA 3_{II}], Exp. XVII, Lemme 2.3, that $H_K \otimes_K K'$ does not contain a subgroup of type $\mathbb{G}_a$ for any separable field extension K' of K. However, there exists a finite separable field extension K' of K such that $T_K \otimes_K K'$ is split. So, if R' is the integral closure of R in K', the K'-group $H_K \otimes_K K'$ admits a Néron model over R' by Theorem 1, since R' is excellent. Hence, $G_K \otimes_K K'$ admits a Néron lft-model over R' by 10.1/7. Then it follows from 10.1/4 that G_K admits a Néron lft-model over R. For the proof of (b) $\Longrightarrow$ (a), we may assume $R = \hat{R}^{sh}$ by 10.1/4. In particular, R is excellent now and, hence, the assertion follows from the implication (b′) $\Longrightarrow$ (a) which has just been proved. □

Now we come to the *proof of Theorem* 1. Some parts of it have already been proved:

(a) $\Longrightarrow$ (b) Néron models are compatible with base change of ramification index 1; cf. 7.2/2. Hence $G_K \otimes_K \hat{K}^{sh}$ admits a Néron model of finite type over $\hat{R}^{sh}$. So the set of $\hat{K}^{sh}$-valued points of G_K is bounded in G_K and, hence, $G_K \otimes_K \hat{K}^{sh}$ cannot contain a subgroup isomorphic to $\mathbb{G}_a$ or $\mathbb{G}_m$.

(b) $\Longrightarrow$ (b′) is trivial.

(c) $\Longrightarrow$ (d) follows from 1.1/10, since $\hat{R}^{sh}$ is excellent.

(c′) $\Longrightarrow$ (e) follows from 1.1/10, since R^{sh} is excellent.

(d) $\Longrightarrow$ (e) is trivial.

(e) $\Longrightarrow$ (a); cf. Theorem 1.3/1.

The remainder of this section is devoted to the proof of the implications

(b) $\Longrightarrow$ (c) and (b′) $\Longrightarrow$ (c′).

Let us first explain the meaning of conditions (c) and (c′)

Proposition 3. *Let X be a smooth and separated K-scheme of finite type. Consider the following conditions:*

(a) *There exists a compactification $\bar{X}$ of X such that there is no rational point in $\bar{X} - X$.*

(b) *For any affine smooth curve C over K with a rational point s, each K-morphism $C - \{s\} \longrightarrow X$ extends to a K-morphism $C \longrightarrow X$.*

(c) *The canonical map $X(K[[\xi]]) \longrightarrow X(K((\xi)))$ is bijective, where ξ is an indeterminate and where $K((\xi))$ is the field of fractions of $K[[\xi]]$.*

Then one has the following implications: (a) $\Longrightarrow$ (b) $\Longleftrightarrow$ (c). *If, in addition, X admits a regular compactification $\bar{X}'$, conditions* (a), (b), (c) *are equivalent and, moreover, they are equivalent to*

(d) *$(\bar{X}' - X)(K)$ is empty.*

Proof. (a) $\Longrightarrow$ (b) is trivial, since such a morphism $C - \{s\} \longrightarrow X$ extends to a morphism $C \longrightarrow \bar{X}$ and since the image of s gives rise to a rational point of $\bar{X}$.

(b) $\Longrightarrow$ (c). Let R be the localization of $K[\xi]$ at the origin and let $a \in X(K((\xi)))$. If $\bar{X}$ is a compactification of X, one can view a as a $K[[\xi]]$-valued point of $\bar{X}$. Since R is excellent, it follows from 3.6/9 that there exists a local étale extension R' of R with residue field K and an R'-valued point a' of $\bar{X}$ inducing the given point a on the closed fibre. Furthermore, we may assume that the generic fibre of a' is contained in X. Rewriting the situation in terms of curves, it means that there are an étale map $\varphi : C \longrightarrow \mathbb{A}^1_K$ of an affine curve to the affine line, a rational point s of C lying above the origin, and a morphism $\alpha : C \longrightarrow \bar{X}$ such that the local ring of C at s is isomorphic to R' and such that α induces the R'-valued point a'. Due to (b), the image of α is contained in X. Thus, we see that a is a $K[[\xi]]$-valued point of X and the implication (b) $\Longrightarrow$ (c) is clear.

(c) $\Longrightarrow$ (b). The completion of the local ring of C at s is isomorphic to a formal power series ring $K[[\xi]]$. Hence the assertion follows as in 2.5/5.

(b) $\Longrightarrow$ (d). Let x be a rational point of $\bar{X}' - X$. By taking hyperplane sections, one can construct an irreducible subvariety C of $\bar{X}'$ of dimension one such that C is not contained in $\bar{X}' - X$, such that the point x lies on C, and such that C is smooth at x. We may assume that C is smooth over K. Hence, the inclusion $C \longrightarrow \bar{X}$ yields a contradiction to (b).

(d) $\Longrightarrow$ (a) is evident. □

In order to complete the proof of Theorem 1, it suffices to show that a commutative algebraic K-group G which contains no subgroup of type $\mathbb{G}_a$ or $\mathbb{G}_m$ admits a G-equivariant compactification $\bar{G}$ without a rational point at infinity. A compactification $\bar{G}$ is called *G-equivariant* if G acts on $\bar{G}$ and if the action is compatible with the group law on G. Let us start with some technical definitions.

Definition 4. *Let G be an algebraic K-group which acts on a K-scheme X of finite type. A subscheme Z of X is called a K-orbit under the action of G if there exist a finite field extension K' of K and a K'-valued point x' of $Z \otimes_K K'$ such that $Z \otimes_K K'$ is the orbit of x' under $G \otimes_K K'$.*

Definition 5 (Mumford [1], Chap. 1.3). *Let G be an algebraic K-group with an action σ on a K-scheme X. Let $\pi : L \longrightarrow X$ be a line bundle on X. A G-linearization is a bundle action λ of G on L which is compatible with the G-action on X; i.e., the diagram*

$$\begin{array}{ccc} G \times_K L & \xrightarrow{\ \lambda\ } & L \\ {\scriptstyle \mathrm{id}_G \times \pi}\downarrow & & \downarrow{\scriptstyle \pi} \\ G \times_K X & \xrightarrow{\ \sigma\ } & X \end{array}$$

is commutative.

For example, look at the canonical action of GL_{n+1} on $\mathbb{P}^n$ and at the canonical ample line bundle $\mathcal{O}_{\mathbb{P}^n}(1)$. There is a canonical GL_{n+1}-linearization on $\mathcal{O}_{\mathbb{P}^n}(1)$, but the action of the projective linear group PGL_n cannot be lifted to a PGL_n-linearization of $\mathcal{O}_{\mathbb{P}^n}(1)$.

Now consider a scheme T and a flat T-group scheme G of finite presentation which acts on a T-scheme X of finite presentation. Let P be a torsor under G over T. Then G acts freely on $X \times_T P$ by setting

$$g \circ (x, p) = (g \circ x, g \circ p) .$$

Denote by $(X \times_T P)/G$ the quotient (in terms of sheaves for the fppf-topology) of $X \times_T P$ with respect to the G-action. The quotient commutes with any base change $T' \longrightarrow T$. If $P \longrightarrow T$ admits a section, there is an isomorphism $(X \times_T P)/G \longrightarrow X$. So, $(X \times_T P)/G$ becomes isomorphic to X and, hence, is representable after a base change with an fppf-morphism, since $P \longrightarrow T$ is of this type. If L is a line bundle on X with a G-linearization, then $M = (L \times_T P)/G$ gives rise to a line bundle on $(X \times_T P)/G$ provided that $(X \times_T P)/G$ is a scheme. Due to 6.1/7, we have the following lemma.

Lemma 6. *If L is T-ample, then $(X \times_T P)/G$ is a T-scheme and $M = (L \times_T P)/G$ is T-ample.*

Now let T be the affine scheme of a field K and let G be a smooth K-group scheme. If, in addition, X is projective, the quotient $(X \times_K P)/G$ is always a scheme. Namely, after a finite Galois extension K'/K, there exists a K'-valued point of P. So, the quotient is representable after the extension K'/K. Since finite Galois descent is effective for quasi-projective schemes, we see that $(X \times_K P)/G$ is represented by a quasi-projective K-scheme.

The proof of the implications (b) $\Longrightarrow$ (c) and (b') $\Longrightarrow$ (c') in Theorem 1 will be provided by Theorem 7 below. Namely, if G is not connected, then $(\bar{G}^0 \times G)/G^0$ yields a compactification of G as required, where $\bar{G}^0$ is a compactification of the identity component G^0 as in condition (d) below.

Theorem 7. *Let K be a field and let G be a connected (not necessarily smooth) commutative algebraic K-group. Then the following conditions are equivalent:*

(a) *G contains no subgroup of type $\mathbb{G}_a$ or $\mathbb{G}_m$.*

(b) *G admits a compactification $\bar{G}$ without a rational point at infinity.*

(c) *G admits a G-equivariant projective compactification $\bar{G}$ such that, for each K-torsor P under G, there is no rational point in $(\bar{G} \times_K P)/G - (G \times_K P)/G$.*

(d) *G admits a G-equivariant projective compactification $\bar{G}$ such that there is no K-orbit of $\bar{G}$ under G contained in $\bar{G} - G$.*

If, in addition, G is linear, these conditions are equivalent to

(d') *G admits a G-equivariant compactification $\bar{G}$ together with a G-linearized ample line bundle such that there is no K-orbit of $\bar{G}$ under G contained in $\bar{G} - G$.*

Remark 8. (i) For a smooth K-wound unipotent algebraic group, the existence of an equivariant projective compactification without rational points at infinity has also been established by Tits (unpublished).

(ii) Presumably, the commutativity of G in Theorem 7 is not necessary. In particular, one can expect that a smooth algebraic K-group which does not contain a subgroup of type $\mathbb{G}_a$ or $\mathbb{G}_m$ admits an equivariant projective compactification without rational points at infinity. The latter is mainly a question of linear groups. It can be answered positively if G is semi-simple; cf. Borel and Tits [1].

Before starting the proof of Theorem 7, let us have a look at Jacobians where, in certain cases, canonical compactifications exist; cf. Altman and Kleiman [1] and [2].

Example 9 (Altman and Kleiman [1], Thm. 8.5). Let X be a proper curve over a field K, assumed to be geometrically reduced and irreducible, and let $J = \mathrm{Pic}^0_{X/K}$ be its Jacobian. Let $\bar{J}$ be the fppf-sheaf induced by the functor which associates to a K-scheme S the set of isomorphism classes of modules on $X \times_K S$ which are locally of finite presentation and S-flat, and which induce torsion-free modules of rank 1 and degree 0 on the fibres of $X \times_K S$ over S. Then $\bar{J}$ is a projective K-scheme containing J as an open subscheme. If, in addition, X is normal, there is no rational point contained in $\bar{J} - J$.

Indeed, we may assume that K is separably closed, so X has a rational point. Then a rational point of $\bar{J}$ represents a torsion-free rank-1 module of degree 0 on X. Since X is a normal curve, such a module is invertible and, hence, represents a point of J. Moreover, since J is smooth, any K-orbit of $\bar{J}$ under J is smooth, too. So, by the same argument as above, it is clear that there is no K-orbit of $\bar{J}$ contained in $\bar{J} - J$.

Let X be locally planar (i.e., the sheaf of differentials is locally generated by at most two elements); for example, this is the case, if X is normal and if K admits a p-basis of length at most 1. Then J is schematically dense in $\bar{J}$ and, hence, $\bar{J}$ is a compactification of J in our sense; cf. Rego [1]. The canonical action of J on itself by left translation extends to an action of J on $\bar{J}$ and, hence, $\bar{J}$ is a J-equivariant compactification of J. In the general case, the schematic closure of J in $\bar{J}$ is an equivariant compactification in our sense.

Now let us prepare the proof of Theorem 7. The implications

$$(d') \Longrightarrow (d) \Longrightarrow (c) \Longrightarrow (b) \Longrightarrow (a)$$

are quite easy whereas the proof of (a) $\Longrightarrow$ (d′) (resp. of (a) $\Longrightarrow$ (d)) will be explained in the remainder of this section. If G is smooth over a perfect field K, it is an extension of an abelian variety by a smooth connected linear group L which is a product of a torus and a unipotent group, cf. 9.2/1 and 9.2/2. Furthermore, the unipotent part is a successive extension of groups of type $\mathbb{G}_a$; cf. [SGA 3_{II}], Exp. XVII, Cor. 4.1.3. Thus, condition (a) implies that the unipotent part of L is trivial and, hence, that G is an extension of an abelian variety by a torus in this case. So, when we are given a smooth K-group G, the later considerations concerning unipotent groups are only of interest in the case where the base field K is not perfect.

Due to the structure of commutative algebraic groups, we will reduce the general situation by "dévissage" to the following special cases:

—K-wound unipotent (not necessarily smooth) algebraic K-groups; i.e., connected unipotent K-groups which do not contain subgroups of type $\mathbb{G}_a$.

—anisotropic tori; i.e., tori which do not contain subgroups of type $\mathbb{G}_m$.

We will begin by discussing the K-wound unipotent case. If the group under consideration is smooth and killed by multiplication with p, one has a rather explicit description of it.

Proposition 10 (Tits [1], Chap. III, Section 3). *Let K be a field of characteristic $p > 0$ with infinitely many elements. Let G be a smooth connected commutative algebraic K-group of dimension $n - 1$ such that $p \cdot G = 0$. Then G is K-isomorphic to a closed subgroup of $\mathbb{G}_a^n$ defined by a p-polynomial*

$$F(T_1, \ldots, T_n) = \sum_{i=1}^{n} \sum_{j=0}^{m_i} c_{ij} \cdot T_i^{p^j} \in K[T_1, \ldots, T_n] .$$

If, in addition, G contains no subgroup of type $\mathbb{G}_a$, one can choose $F(T_1, \ldots, T_n)$ in such a way that the polynomials

$$\sum_{j=0}^{m_i} c_{ij} \cdot T_i^{p^j} \in K[T_i]$$

are non-zero, $i = 1, \ldots, n$, and that the principal part

$$f(T_1, \ldots, T_n) = \sum_{i=1}^{n} c_{im_i} \cdot T_i^{p^{m_i}}$$

of $F(T_1, \ldots, T_n)$ has no non-trival rational zero in $\mathbb{A}_K^n$.

Using the specific situation of Proposition 10, it is easy to find an equivariant compactification for smooth unipotent commutative groups which are K-wound and are killed by multiplication with p.

Proposition 11. *Let K be a field of characteristic $p > 0$. Let G be a smooth connected commutative algebraic K-group which is killed by multiplication with p. If G is K-wound, then G admits a G-equivariant compactification $\bar{G}$ together with a G-linearized ample line bundle such that there is no K-orbit of $\bar{G}$ under G in $\bar{G} - G$.*

Proof. We may assume that K has infinitely many elements; otherwise G is trivial. Keep the notations of the last proposition and assume that the exponents occurring in the principle part of the p-polymonial satisfy

$$m_1 \leq m_2 \leq \ldots \leq m_n .$$

Let P be the quasi-homogeneous space over K with coordinates

$$Y_i , \qquad i = 0, 1, \ldots, n ,$$

having weights

$$w_i = p^{m_n - m_i} , \qquad i = 0, \ldots, n ,$$

where we have set $m_0 = m_n$. The open subspace U_0 of P where Y_0 is not zero can be viewed as the group $\mathbb{G}_a^n$ with coordinates

$$T_i = Y_i / Y_0^{w_i} , \qquad i = 1, \ldots, n .$$

The action of U_0 on itself extends to an action on P by setting

$$U_0 \times_K P \longrightarrow P , \qquad ((t_i), (y_0, y_i)) \longmapsto (y_0, y_i + t_i \cdot y_0^{w_i}) .$$

We regard G as a closed subscheme of U_0 given by a p-polynomial $F(T_1, \ldots, T_n)$. Now, let $X_0, \ldots, X_n$ be the coordinates of the projective space $\mathbb{P}_K^n$ and let

$$u : P \longrightarrow \mathbb{P}_K^n$$

be the morphism sending X_i to $(Y_i)^{p^{m_i}}$. Denote by V_0 the open subscheme of $\mathbb{P}_K^n$ where X_0 does not vanish. We can view V_0 as the group $\mathbb{G}_a^n$ with coordinates

$$S_i = X_i/X_0 , \qquad i = 1, \ldots, n .$$

The morphism u induces a morphism

$$u_0 : U_0 \longrightarrow V_0$$

of algebraic K-groups and the morphism u is equivariant. In terms of coordinates of rational points the equivariance means the commutativity of the following diagram

$$\begin{array}{ccccc} U_0 \times_K P & \longrightarrow & P & ((t_i), (y_0, y_i)) \longmapsto (y_0, y_i + t_i \cdot (y_0)^{w_i}) \\ \Big\downarrow{\scriptstyle u_0 \times u} & & \Big\downarrow{\scriptstyle u} & \\ V_0 \times \mathbb{P}_K^n & \longrightarrow & \mathbb{P}_K^n & ((s_i), (x_0, x_i)) \longmapsto (x_0, x_i + s_i \cdot x_0) \end{array}$$

where $s_i = t_i^{p^{m_i}}$ for $i = 1, \ldots, n$ and where $x_i = (y_i)^{p^{m_i}}$ for $i = 0, \ldots, n$. The canonical sheaf $\mathcal{O}_{\mathbb{P}_K^n}(1)$ has a V_0-linearization. Hence, $u^*(\mathcal{O}_{\mathbb{P}_K^n}(1))$ is an ample invertible sheaf on P which has a U_0-linearization.

The schematic closure $\bar{G}$ of G in P is given by the polynomial

$$(Y_0)^{p^{m_n}} \cdot F(Y_1/Y_0^{w_1}, \ldots, Y_n/Y_0^{w_n})$$

which can be viewed as a weighted homogeneous polynomial in the variables $Y_0, \ldots, Y_n$. Due to the choice of the weights, the principal part $f(Y_1, \ldots, Y_n)$ of $F(Y_1, \ldots, Y_n)$ is a weighted homogeneous polynomial and describes the set of the points at infinity of the compactification $\bar{G}$. So, we have

$$\bar{G} - G = \{y \in P, f(y) = 0\} .$$

Due to Proposition 10, there is no rational point in $\bar{G} - G$. Moreover, G acts trivially on $\bar{G} - G$. So $\bar{G}$ cannot contain a K-orbit under G at infinity. □

In order to generalize Proposition 11 to smooth unipotent commutative K-wound groups which are not necessarily killed by multiplication with p, we will need the following lemma.

Lemma 12. *Let G be a connected unipotent commutative algebraic K-group. Assume that G is smooth and K-wound. Then there exists a filtration*

$$0 = G_0 \subset G_1 \subset G_2 \subset \ldots \subset G_n = G$$

such that the successive quotients have the same properties as G, and, in addition, are killed by multiplication with p.

Proof. Let n be the smallest integer such that G is annihilated by p^n. We will proceed by induction on n. Let N (resp. I) be the kernel (resp. the image) of the p-multiplication on G. Then I is a smooth connected subgroup of G and, hence, K-wound. The group N is not necessarily smooth. So, consider the largest smooth subgroup M of N. Then M is K-wound as a subgroup of G and, since M is the largest smooth subgroup of N, the quotient N/M is K-wound, too. Since the image of the multiplication by p^{n-1} is contained in N and is smooth, the quotient G/M is killed by multiplication with p^{n-1}. Moreover, G/M is K-wound, since it is an extension of I by N/M both of which are K-wound. Then we can set $G_1 = M$ and the induction hypothesis is applicable to G/M. □

Proceeding by dévissage, we are now able to prove Theorem 7 for unipotent groups which are smooth. But when treating general commutative groups, we will also be concerned with unipotent groups which occur as unipotent radicals. Such unipotent groups do not need to be smooth. Therefore, we need the following lemma.

Lemma 13. *Let G be a connected unipotent commutative algebraic K-group which is not necessarily smooth.*

(a) *There exists an immersion of G into a connected unipotent commutative algebraic K-group G' which is smooth.*

(b) *If G is K-wound, one can choose G' to be K-wound, too.*

Proof. (a) We will first show that G can be embedded into a smooth unipotent commutative group. Denote by F_n the kernel of the n-fold Frobenius morphism on G. Due to [SGA 3_{I}], Exp. $\mathrm{VII_A}$, Prop. 8.3, there exists an integer $n \in \mathbb{N}$ such that the quotient G/F_n is smooth. Thus, it suffices to show the assertion for the group F_n. So we may assume that G is a finite connected unipotent group. Hence, it is a successive extension of groups of type α_p; cf. [SGA 3_{II}], Exp. XVII, Prop. 4.2.1. Consider now the Cartier dual G^* of G, which is a successive extension of groups of type α_p also. Hence, the algebra $A = \Gamma(G^*, \mathcal{O}_{G^*})$ is local. The algebraic group U representing the group functor

$$(\mathrm{Sch}/K)^0 \longrightarrow (\mathrm{Groups})\,, \qquad T \longmapsto \Gamma(T \times_K G^*, \mathcal{O}^*_{T \times_K G^*})$$

is smooth. Interpreting the points of G as characters of G^*, one gets a morphism $G \longrightarrow U$ which is an immersion and which is closed, since G is finite. Since A is local, U is a product of the multiplicative group $\mathbb{G}_m$ and of a smooth connected unipotent group G'. Since G is unipotent, the morphism $G \longrightarrow U$ yields an embedding of G into G'.

(b) Let us start by collecting some facts on extensions of commutative unipotent algebraic groups by étale groups.

(1) If N is an étale K-group and H is an algebraic K-group, the canonical map

$$\mathrm{Ext}(H, N) \longrightarrow \mathrm{Ext}(H \otimes_K K', N \otimes_K K')$$

is bijective for any radicial field extension K'/K; cf. [SGA I], Exp. IX, 4.10.

(2) Let

$$1 \longrightarrow G_1 \longrightarrow G_2 \longrightarrow G_3 \longrightarrow 1$$

be an extension of smooth commutative unipotent algebraic K-groups. Then the canonical sequence of quasi-algebraic commutative group extensions

$$1 \longrightarrow \operatorname{Ext}(G_3, \mathbb{Q}_p/\mathbb{Z}_p) \longrightarrow \operatorname{Ext}(G_2, \mathbb{Q}_p/\mathbb{Z}_p) \longrightarrow \operatorname{Ext}(G_1, \mathbb{Q}_p/\mathbb{Z}_p) \longrightarrow 1$$

is exact. If G_2 is killed by multiplication with p^n, one can replace $\mathbb{Q}_p/\mathbb{Z}_p$ by $\mathbb{Z}/p^n\mathbb{Z}$. Now, due to (1), we may assume that K is perfect. In this case, the result is provided by Bégueri [1], Prop. 1.21.

(3) If K is not perfect, there exists for each smooth connected commutative unipotent K-group G a commutative extension

$$1 \longrightarrow N \longrightarrow \tilde{G} \longrightarrow G \longrightarrow 1$$

of G by a finite étale group N such that $\tilde{G}$ is K-wound.

Namely, we may assume that G is an extension

$$1 \longrightarrow \mathbb{G}_a \longrightarrow G \longrightarrow G_0 \longrightarrow 1$$

of a smooth connected unipotent K-group G_0 by $\mathbb{G}_a$. Proceeding by induction on the dimension of the group, we may assume, that there exists a commutative extension $\tilde{G}_0$ of G_0 by a finite étale group such that $\tilde{G}_0$ is connected and K-wound. Then, one is easily reduced to the case where G_0 is K-wound. For the group $\mathbb{G}_a$ and each element $x \in K - K^p$, consider the extension

$$1 \longrightarrow \mathbb{Z}/p\mathbb{Z} \longrightarrow \tilde{\mathbb{G}}_a(x) \longrightarrow \mathbb{G}_a \longrightarrow 1$$

where $\tilde{\mathbb{G}}_a(x)$ is defined as a subgroup of $\mathbb{G}_a \times \mathbb{G}_a$ by the p-polynomial

$$T_1^p + xT_2^p - T_1$$

and the map $\tilde{\mathbb{G}}_a(x) \longrightarrow \mathbb{G}_a$ is the second projection. Then, due to (2), there exists an extension $\tilde{G} \longrightarrow G$ by a finite étale group which induces $\tilde{\mathbb{G}}_a(x) \longrightarrow \mathbb{G}_a$ by restriction. Thus, $\tilde{G}$ is K-wound as an extension of K-wound groups.

Using these results, the proof of assertion (b) is easily done. Assume that K is not perfect and let G be connected, unipotent, commutative, and K-wound. Due to (a), there exists an immersion of G into a smooth unipotent commutative connected group G_1. Let H be the quotient of G_1 by G, so we have the exact sequence

$$1 \longrightarrow G \longrightarrow G_1 \longrightarrow H \longrightarrow 1 .$$

Since G_1 is smooth, H is smooth also. Due to (3), there exists a commutative extension

$$1 \longrightarrow N \longrightarrow \tilde{H} \longrightarrow H \longrightarrow 1 .$$

of H by a finite étale group N such that $\tilde{H}$ is K-wound and connected. Pulling back this extension to G_1, one gets a commutative extension

$$1 \longrightarrow N \longrightarrow \tilde{G}_1 \longrightarrow G_1 \longrightarrow 1 .$$

Note that $\tilde{G}_1$ is smooth and unipotent. Denote the identity component of $\tilde{G}_1$ by G'. Hence, one gets an exact sequence

$$1 \longrightarrow G \longrightarrow G' \longrightarrow \tilde{H} \longrightarrow 1 .$$

So the group G' is smooth, unipotent, commutative, and connected, and, as an extension of K-wound unipotent groups, it is K-wound, too. □

Next we want to discuss the compactification of tori. Let T be a torus, denote by M the group of characters of T and by N the group of 1-parameter subgroups of T. Then

$$M = \mathrm{Hom}_{\bar{K}}(T, \mathbb{G}_m) \text{ and } N = \mathrm{Hom}_{\bar{K}}(\mathbb{G}_m, T)$$

are $\mathrm{Gal}(\bar{K}/K)$-modules, where $\bar{K}$ is an algebraic closure of K. There is a perfect pairing

$$M \times N \longrightarrow \mathbb{Z} .$$

Hence, N and M are canonically dual to each other. Recall that T is anisotropic if one of the following equivalent conditions is satisfied:

(i) T does not contain a subgroup of type $\mathbb{G}_m$.
(ii) T does not admit a group of type $\mathbb{G}_m$ as a quotient.
(iii) M does not contain the unit representation.
(iv) N does not contain the unit representation.

Proposition 14. *Let T be an anisotropic torus over K. Then T admits a T-equivariant compactification $\bar{T}$ such that $\bar{T}$ is normal and projective, such that $\bar{T} - T$ does not contain a K-orbit under T, and such that there is an ample line bundle on $\bar{T}$ with a T-linearization on it.*

Proof. Equivariant compactifications of tori are closely related to rational polyhedral cone decompositions of $N_{\mathbb{Q}} = N \otimes_{\mathbb{Z}} \mathbb{Q}$. Over an algebraically closed field, this technique is well documented in the literature; cf. Kempf et al. [1], Chap. I, §§ 1 and 2. So, we will only give advice how to proceed in the case of an arbitrary field.

Consider a finite rational polyhedral cone decomposition $\{\sigma_\alpha\}$ of $N_{\mathbb{Q}}$, which is invariant under $\mathrm{Gal}(\bar{K}/K)$. The vertex of each cone is the origin of $N_{\mathbb{Q}}$. Let $\bar{T}$ be the associated T-equivariant compactification of T. The variety $\bar{T}$ is normal and projective. It has a finite number of orbits under T and these correspond bijectively to the faces of the decomposition $\{\sigma_\alpha\}$; cf. Kempf et al. [1], Chap. I, § 2, Thm. 6. Since $\{\sigma_\alpha\}$ is invariant under $\mathrm{Gal}(\bar{K}/K)$, the Galois group acts on the $\bar{K}$-variety $\bar{T}$ and, hence, by projective descent, $\bar{T}$ is defined over K.

We are going to show that $\bar{T} - T$ does not contain a K-orbit under T. So assume that there is a K-orbit in $\bar{T} - T$. It corresponds to a non-zero face σ of the decomposition $\{\sigma_\alpha\}$ which is stable under $\mathrm{Gal}(\bar{K}/K)$. Consider now the set of the extreme edges of σ which consists of a finite number of half lines $\{L_i, i \in I\}$. This set is invariant under $\mathrm{Gal}(\bar{K}/K)$. Now we can choose non-zero points $x_i \in L_i, i \in I$, such that the set $\{x_i, i \in I\}$ is invariant under $\mathrm{Gal}(\bar{K}/K)$. So the point $x = \sum_{i \in I} x_i$ is a non-zero point of σ which is invariant under $\mathrm{Gal}(\bar{K}/K)$ and, hence, gives rise to a non-zero element of N. Thus, we get a contradiction to T being anisotropic.

It remains to show that there is an ample T-linearized line bundle on $\bar{T}$. Let $\mathscr{L}$ be the ample line bundle on $\bar{T}$. Since the Picard group of $\bar{T}$ is discrete (use Kempf et al. [1]. Chap. I, § 2, Thm. 9), $\mathscr{L}$ is invariant under T. Hence, it is easy to see that a power of $\mathscr{L}$ admits a T-linearization. □

For the dévissage, we need a technique of constructing an equivariant compactification of an extension of groups with given equivariant compactifications. This part works also for not necessarily commutative groups.

So consider an exact sequence

$$1 \longrightarrow G \longrightarrow E \longrightarrow H \longrightarrow 1$$

of algebraic K-groups. In particular, $E \longrightarrow H$ is a torsor over H with respect to the H-group scheme $G_H = G \times_K H$. In order to avoid problems with representability of quotients, we will work with projective equivariant compactifications admitting ample line bundles with linearizations. We have to introduce some more notations:

Let X be a K-scheme with an action of G on X on the left and let L be an ample line bundle on X with a G-linearization. Then G_H acts on $X_H = X \times_K H$ as an H-group scheme and $L_H = L \times_K H$ is an H-ample line bundle on X_H with a G_H-linearization. G_H acts freely on $X \times_K E = X_H \times_H E$ by setting

$$g \circ (x, e) = (g \circ x, ge) .$$

Denote by $(X_H \times_H E)/G_H$ the quotient (in terms of sheaves for the fppf-topology over H) of $(X_H \times_H E)$ with respect to the G_H-action. Introduce similar notations for L instead of X. Due to Lemma 6, $(X_H \times_H E)/G_H$ is an H-scheme and $(L_H \times_H E)/G_H$ is an H-ample line bundle on $(X_H \times_H E)/G_H$.

Furthermore, there is an action of E (on the right)

$$(X \times_K E) \times_K E \longrightarrow (X \times_K E) , \qquad ((x, e), e') \longmapsto (x, ee') .$$

This action is compatible with the left action of G on X. So the E-action on $(X \times_K E)$ induces an E-action on $(X_H \times_H E)/G_H$ in a canonical way. The projection

$$(X_H \times_H E)/G_H \longrightarrow H$$

is E-equivariant where E acts on H by right translation. Similarly, the line bundle $(L_H \times_H E)/G_H$ on $(X_H \times_H E)/G_H$ has a canonical E-linearization with respect to the E-action on $(X_H \times_H E)/G_H$.

Lemma 15. *Consider the exact sequence*

$$1 \longrightarrow G \longrightarrow E \longrightarrow H \longrightarrow 1$$

of algebraic K-groups. Let $\bar{G}$ be an equivariant compactification of G and let L be an ample line bundle on $\bar{G}$ with a G-linearization. Set $Y = (\bar{G}_H \times_H E)/G_H$ and $M = (L_H \times_H E)/G_H$. Then

(a) *Y is a projective H-scheme which contains E as an open subscheme and the canonical action of E on itself by right translation extends to an action on Y and is compatible with the G-action on Y. The projection $p : Y \longrightarrow H$ is E-equivariant where E acts on H by right translation. The line bundle M has an E-linearization and is H-ample. Y is quasi-projective over K.*

If $\bar{G} - G$ does not contain a K-orbit under the action of G, then $Y - E$ does not contain a K-orbit under the action of E.

(b) *Let $\bar{H}$ be an equivariant compactification of H and let N be an ample line bundle on $\bar{H}$ with an H-linearization. Then there is a commutative cartesian diagram*

$$\begin{array}{ccc} Y & \hookrightarrow & \bar{Y} \\ \Big\downarrow p & & \Big\downarrow \bar{p} \\ H & \hookrightarrow & \bar{H} \end{array}$$

such that the following is satisfied: $Y \hookrightarrow \bar{Y}$ is an E-equivariant compactification and $\bar{p}$ is E-equivariant. $\bar{Y}$ is a projective K-scheme and has an ample line bundle with an E-linearization.

If $\bar{G}$-G and $\bar{H}$-H do not contain K-orbits, then $\bar{Y} - Y$ does not contain a K-orbit under the action of E.

Proof. Assertion (a) follows mainly from what has been said before. Y is quasi-projective, since H is quasi-projective. It remains to show that there is no K-orbit contained in $Y - E$. So consider a K-orbit Z of Y under the action of E. Its image $p(Z)$ is a K-orbit of H and, hence, $p(Z) = H$. The E-action on Y induces a right action of G on the fibre over the unit element of H which is canonically isomorphic to $\bar{G}$. This action is related to the left action of G we started with by the relations

$$\bar{g} \cdot f = f^{-1} \cdot \bar{g} , \qquad \bar{g} \in \bar{G} , \qquad f \in G .$$

Thus we see that the intersection of Z with the fibre over the unit element of H is a K-orbit of $\bar{G}$ under the action of G. So it must be G. Then we get $Z = E$.

(b) After replacing L by $L^{\otimes n}$ for a suitable integer n, we may assume that L is very ample and, hence, that M is very H-ample. Since H admits an ample line bundle with H-linearization, it is affine. So, we may assume that M is very ample.

The K-vector space $\Gamma(Y, M)$ has an E-action induced by the E-linearization of M. Now there is a finite-dimensional subspace W of the vectorspace $\Gamma(Y, M)$ which defines an embedding of Y into its associated projective space $P = \mathbb{P}(W)$. Since the smallest subspace which is stable under E and which contains W is also of finite dimension, we may assume that W is stable under E. So E acts on P and there is an E-linearization on $\mathcal{O}_P(1)$. Due to the choice of W, there is an E-equivariant embedding $Y \longrightarrow P$ such that the pull-back of $\mathcal{O}_P(1)$ is isomorphic to M. Now consider the morphism

$$Y \longrightarrow P \times_K \bar{H}$$

induced by $Y \longrightarrow P$ and $Y \longrightarrow H \longrightarrow \bar{H}$. Let $\bar{Y}$ be the schematic image of Y in $P \times_K \bar{H}$. Then $\bar{Y}$ is projective. Since Y is proper over H, the schematic closure $\bar{Y}$ coincides with Y over H. By continuity, the action of E on Y extends to an action on $\bar{Y}$. Let

$$p_1 : \bar{Y} \longrightarrow P , \qquad p_2 : \bar{Y} \longrightarrow \bar{H}$$

be the projections. The restriction $\bar{M}$ of $p_1^*(\mathcal{O}_P(1))$ on $\bar{Y}$ has an E-linearization extending the given E-linearization on M and is $\bar{H}$-ample.

For $n \in \mathbb{N}$, the tensor product $p_2^*(N^{\otimes n}) \otimes \bar{M}$ has a canonical E-linearization with respect to the E-action on $\bar{Y}$ and, for large integers n, it is ample on $\bar{Y}$.

It remains to prove the assertion concerning the orbits. So let Z be a K-orbit of $\bar{Y}$ under the action of E. The projection $p_2(Z)$ is a K-orbit of $\bar{H}$ under the action of H. Due to our assumption, $p_2(Z)$ must be contained in H and, hence, is equal to H. Now we can continue as in part (a) in order to show that Z coincides with E. □

Proof of Theorem 7. We start with the implication (a) $\Longrightarrow$ (d'). Since G is linear, it is an extension of a unipotent group U by a subgroup of multiplicative type M; cf. [SGA 3_{II}], Exp. XVII, Thm. 7.2.1. Due to [SGA 3_{II}], Exp. XVII, Thm. 6.1.1 (A) (ii), the unipotent group U is K-wound. The multiplicative group M is an extension of a finite multiplicative group N by a torus T which is necessarily anisotropic since G does not contain a subgroup of type $\mathbb{G}_m$. Hence, due to Lemma 15 (b), we are reduced to prove the assertion for the groups N, T, and U. It is clear for N. Furthermore, Proposition 14 provides the assertion in the case of T. In the case of U, we may assume that K has characteristic $p > 0$ and, due to Lemma 13, that U is smooth. Using Lemma 15 and Lemma 12, we are reduced to the case where U is killed by the multiplication with p. However, this case has been dealt with in Proposition 11.

Next let us turn to the implication (a) $\Longrightarrow$ (d). It follows from the theorem of Chevalley (cf. 9.2/1) that there exists a connected linear subgroup H of G such that the quotient G/H is an abelian variety. Namely, the kernel F_n of the n-fold Frobenius morphism on G is an affine subgroup of G and, for large integers n, the quotient G/F_n is smooth, cf. [SGA 3_{II}], Exp. XVII, Prop. 4.2.1. Then the assertion follows by Lemma 15 (a) from the implication (a) $\Longrightarrow$ (d'). This concludes the proof, the remaining assertions being trivial. □

The above verification of the implication (a) $\Longrightarrow$ (d') shows that a commutative linear group G which does not contain a subgroup of type $\mathbb{G}_a$ or $\mathbb{G}_m$ admits a G-equivariant compactification $\bar{G}$ together with a G-linearized ample line bundle such that there is no K-orbit contained in $\bar{G} - G$. So, due to Lemma 15 which is valid for not necessarily commutative groups, the construction carries over to the case of solvable groups G; cf. Remark 8. Namely, a K-wound solvable group admits a filtration

$$G = G_0 \supset G_1 \supset \ldots \supset G_n = \{1\}$$

such that G_i is a normal subgroup of G_{i-1} and G_{i-1}/G_i is commutative and K-wound, $i = 1, \ldots, n$; cf. Tits [1], Chap. IV, Prop. 4.1.4.

10.3 The Global Case

Let S be an excellent Dedekind scheme with infinitely many closed points and let K be its ring of rational functions. Let G_K be a smooth commutative algebraic K-group.

The existence of a Néron lft-model (resp. of a Néron model) of G_K over S implies the existence of a Néron lft-model (resp. of a Néron model) over each local ring of S. But, as we have seen in Example 10.1/11, the converse is not true when dealing with Néron models. The example was given in the case where the characteristic of K is positive.

If K has characteristic zero, we claim that the existence of a global Néron lft-model (resp. of a global Néron model) is equivalent to the existence of the local Néron lft-models (resp. of the local Néron models). Namely, due to 10.2/2, the existence of Néron lft-models over each local ring of S is equivalent to the fact that the unipotent radical of G_K is trivial. Then G_K is an extension of an abelian variety by a torus T and, hence, admits a Néron lft-model over S; the latter follows from 10.1/7 by using 10.1/4. Moreover, when the local Néron lft-models are of finite type over each local ring of S, the subtorus T of G_K is trivial. Indeed, T splits over a finite separable field extension K' of K. There exists a closed point of S at which K' is unramified. Since Néron models are compatible with localization and étale extensions, there is a closed point s' of S', where S' is the spectrum of the integral closure of $\mathcal{O}_{S,s}$ in K', such that $G_K \otimes_K K'$ admits a local Néron model at s'. Then, it follows from 10.2/1 that the torus T is trivial. Thus, we see that G_K is an abelian variety and, hence, that G_K has a Néron model over S; cf. 1.4/3.

The existence of Néron lft-models or Néron models over a global base is still an open question when K has positive characteristic. We conjecture that G_K has a Néron lft-model over S if and only if G_K has one over each local ring of S. Using Theorem 10.2/2, we can state this conjecture in the following way.

Conjecture I. *Let S be an excellent Dedekind scheme with ring of rational functions K and let G_K be a smooth commutative algebraic K-group. Then G_K admits a Néron* lft-*model over S if G_K contains no subgroup of type $\mathbb{G}_a$.*

As explained before, the conjecture is true if the characteristic of K is zero, but in the case of positive characteristic it is still an open question.

For the remainder of this section we want to concentrate on the existence of Néron models (of finite type). We can give a criterion for the case where G_K admits a regular compactification. Let us begin with some definitions.

A K-variety X (i.e., a separated K-scheme of finite type which is geometrically reduced and irreducible) is called *rational* (resp. *unirational*) if its field of rational functions is purely transcendental over K (resp. contained in a purely transcendental field extension of K). In geometric terms, the latter means that there is a rational map from $\mathbb{A}_K^n$ to X which is birational (resp. dominant). An algebraic K-group G_K is called rational (resp. unirational) if its underlying scheme is rational (resp. unirational). It is easy to see that unirational groups are smooth and connected. For example, tori are unirational; also the K-group of Example 10.1/11 is unirational. Each unirational subscheme of G_K which contains the origin generates a unirational subgroup of G_K. In particular, G_K contains a *largest unirational subgroup* denoted by $\mathrm{uni}(G_K)$. If G_K is an abelian variety, then $\mathrm{uni}(G_K) = 0$.

Theorem 1. *Let G_K be a smooth algebraic group over a field K, where G_K is connected and commutative. Then the following conditions are equivalent:*

(a) $\mathrm{uni}(G_K) = 0$

(b) *Each K-rational map from the projective line $\mathbb{P}_K^1$ to G_K is constant.*

(c) *For any smooth affine curve C_K over K and for any closed point x of C_K, each morphism of $C_K - \{x\}$ to G_K extends to a morphism from C_K to G_K.*

(d) *For any smooth K-scheme X_K, each K-rational map from X_K to G_K is defined everywhere.*

If, in addition, G_K admits a regular compactification $\bar{G}_K$, these conditions are equivalent to

(e) *The smooth locus of $\bar{G}_K$ coincides with G_K.*

The implications

$$(a) \Longleftrightarrow (b) \Longleftarrow (c) \Longleftarrow (d) \Longrightarrow (e)$$

are quite easy to verify and we leave them to the reader. Also it is not difficult to show the implication (e) $\Longrightarrow$ (c) (if G_K admits a regular compactification) and (c) $\Longrightarrow$ (d). Finally the implication (a) $\Longrightarrow$ (c) requires more efforts.

To start the proof, let us begin with the verification of implication (e) $\Longrightarrow$ (c). Let $\varphi : C_K - \{x\} \longrightarrow G_K$ be a K-morphism. Due to the valuation criterion of properness, φ extends to a K-morphism $\bar{\varphi} : C_K \longrightarrow \bar{G}_K$. Now consider the C_K-scheme $\bar{G}_{C_K} = \bar{G}_K \times_K C_K$ which is regular; cf. 2.3/9. Due to assumption (e), the smooth locus of $\bar{G}_{C_K}$ over C_K coincides with $G_K \times_K C_K$; cf. [EGA IV_4], 17.7.2. By base extension, $\bar{\varphi}$ gives rise to a section $\bar{\varphi}_{C_K}$ of $\bar{G}_{C_K}$. Now it follows from 3.1/2 that $\bar{\varphi}_{C_K}$ factors through the smooth locus of $\bar{G}_{C_K}$ and, hence, $\bar{\varphi}$ maps to G_K.

For the implication (c) $\Longrightarrow$ (d), consider a rational map $\varphi_K : X_K \dashrightarrow G_K$, where X_K is smooth and irreducible of dimension n. Since we consider K-schemes of finite type, φ_K is induced by a T-rational map $\varphi : X \dashrightarrow G$ from a smooth T-scheme X to a smooth and separated T-group scheme G, where T is an irreducible regular scheme of finite type over the ring of integers $\mathbb{Z}$. We may assume that K is the field of rational functions on T. Due to 4.4/1, the complement F of the domain of definition of φ is of pure codimension 1 and, hence, is a relative Cartier divisor. We have to show that F is empty. Proceeding indirectly, let us assume that F is not empty. Then look at the graph Γ_K of φ_K in $X_K \times_K G_K$. It is clear that the image Q_K of Γ_K under the first projection p_1 cannot contain a generic point of F_K as seen by a similar argument as used in the proof of 4.3/4. Since Q_K is constructible, we may assume, after shrinking X_K, that Q_K is disjoint from F_K. Now we will derive a contradiction by constructing a smooth curve C_K contained in X_K, but not in F_K such that C_K meets F_K at a closed point. Namely, due to assumption (c), the curve C_K must be contained in Q_K. Since F is not empty, there exists a closed point x in F. Let t be the image of x in T. The residue field of t is finite and hence perfect. So $k(x)$ is separable over $k(t)$. Then it follows from the Jacobi criterion 2.2/7 that there exist elements $f_2, \ldots, f_n$ in the maximal ideal of the local ring of X at x which, in a neighborhood of x, define an irreducible relative smooth T-curve C. We may assume that F induces a relative Cartier divisor on C. In particular, $C \cap F$ is flat over T. Hence, the generic fibre of

$C \cap F$ is not empty. Now, the induced morphism $C_K \longrightarrow G_K$ yields a contradiction to (c).

The proof of the implication (a) $\Longrightarrow$ (c) is delicate. It will follow from Corollary 3 below which makes use of the theory of Rosenlicht and Serre on rational maps from curves into commutative algebraic groups. In the following we want to sketch the main ideas of this theory.

So let X be a *proper irreducible curve* over K, assumed to be *geometrically reduced*. Denote by U the smooth locus of X, which is open and dense in X. Let G be a smooth commutative algebraic K-group. We want to study rational maps

$$\varphi : X \dashrightarrow G .$$

If V is the domain of definition of φ, then, for any $n \in \mathbb{N}$, there is a canonical morphism of the n-fold symmetric product $V^{(n)}$ to G induced by φ. We will denote it by $\varphi^{(n)} : V^{(n)} \longrightarrow G$. By restriction to $(U \cap V)^{(n)}$ we get a morphism of the set of Cartier divsors of degree n with support in $U \cap V$ to G; cf. Section 9.3. We denote this map by φ, too. A finite subscheme Y of X is called a *conductor for* φ if $\varphi(\mathrm{div}(f)) = 0$ for each rational function f of X which is defined on Y, which induces the constant function with value 1 on Y, and whose associated divisor has support in $U \cap V$.

Now let Y be a finite subscheme of X. If Y is non-empty, it is a rigidificator for $\mathrm{Pic}_{X/K}$. As introduced in Section 8.1, we denote by $(\mathrm{Pic}_{X/K}, Y)$ the rigidified Picard functor. We set $(\mathrm{Pic}_{X/K}, Y) = \mathrm{Pic}_{X/K}$ if Y is empty. Since, for a K-scheme T, any section of $(U - Y) \times_K T$ induces an effective relative Cartier divisor on $U \times_K T$ of degree 1 whose associated invertible sheaf is canonically rigidified along Y by the function 1, there exists a canonical map $(U - Y) \longrightarrow (\mathrm{Pic}_{X/K}, Y)$ and, hence, a rational map

$$\iota_Y : X \dashrightarrow (\mathrm{Pic}_{X/K}, Y) .$$

By construction Y is a conductor for ι_Y. If Y is empty, we will write ι instead of ι_Y.

For the proof of the implication (a) $\Longrightarrow$ (c) we will use the following result.

Theorem 2. *Keeping the notations of above, the following hold:*

(a) *A finite subscheme Y of X is a conductor for φ if and only if there exists a K-morphism of algebraic groups $\Phi : (\mathrm{Pic}_{X/K}, Y) \longrightarrow G$ making the following diagram commutative:*

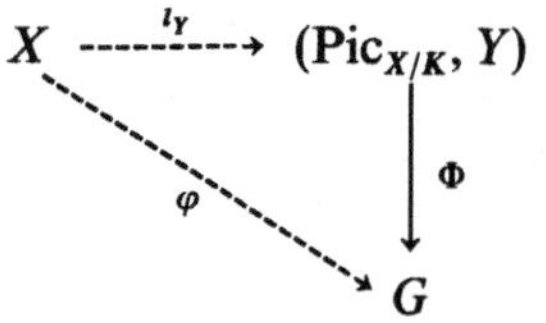

Moreover, the map Φ is uniquely determined.

(b) *There exists a conductor for φ and there even is a smallest one. The latter is called the conductor of φ.*

(c) *Let $\pi : \tilde{X} \longrightarrow X$ be the normalization of X and let x be a closed point of X such that $\pi^{-1}(x)$ is contained in the smooth locus of $\tilde{X}$. If $\varphi \circ \pi$ is defined at $\pi^{-1}(x)$, then x is not contained in the support of the conductor of φ.*

(d) *If X is smooth at x and if x is not contained in the conductor of φ, then φ is defined at x.*

(e) *The conductor of φ commutes with finite separable field extensions.*

Proof. If K is algebraically closed and if X is smooth, the result is classical and is due to Rosenlicht and Serre, cf. Serre [1]; for (a) and (d) see Chap. V, n°9, Thm. 2, for (b) and (c) see Chap. III, n°3, Thm. 1. We want to give some indications on how to proceed in the general case. We may assume that X is geometrically irreducible. Namely, using assertion (e), one can easily reduce to this case.

(a) The if-part is obvious. For the only-if-part, consider first the case where Y is empty. Then the factorization follows from the construction of $\mathrm{Pic}_{X/K}$ via symmetric products à la Weil as explained in Section 9.3. The uniqueness of the factorization is due to the fact that $\mathrm{Pic}_{X/K}$ is generated by the image of ι. Now let Y be a non-empty conductor for φ. There exists a finite birational morphism $X \longrightarrow X'$ which contracts Y to a rational point Y' and which is an isomorphism outside Y and Y'. One easily checks that the canonical map

$$\mathrm{Pic}_{X'/K} = (\mathrm{Pic}_{X'/K}, Y') \longrightarrow (\mathrm{Pic}_{X/K}, Y)$$

is an isomorphism. Thus, the general case is reduced to the case discussed above.

(b) Let Y_1 and Y_2 be finite subschemes of X. Then the diagram

$$\begin{array}{ccc} (\mathrm{Pic}_{X/K}, Y_1 \cup Y_2) & \longrightarrow & (\mathrm{Pic}_{X/K}, Y_2) \\ \downarrow & & \downarrow \\ (\mathrm{Pic}_{X/K}, Y_1) & \longrightarrow & (\mathrm{Pic}_{X/K}, Y_1 \cap Y_2) \end{array}$$

is co-cartesian. Thus, by using the characterization given in (a), we see that the intersection of two conductors is a conductor again. So the existence of a conductor implies the existence of a unique smallest one. Furthermore, one can see by the same argument that the smallest conductor of φ is compatible with finite Galois extensions of the base field; thus assertion (e) is clear. So it remains to show that there is at least one conductor for φ which satisfies assertion (c); hence the smallest one will satisfy (c), too. By what we have said above, we may assume that K is separably closed. Denote by $\pi : \tilde{X} \longrightarrow X$ the normalization of X. Assume for a moment that the base field is algebraically closed. Then, due to Rosenlicht and Serre, there exists a conductor $\tilde{Y}$ for $\varphi \circ \pi$ whose support is disjoint from the domain of definition of $\varphi \circ \pi$. Now let Y be the schematic image of $\tilde{Y}$ in X. Then one shows easily by using the very definition of conductors that Y is a conductor for φ satisfying the assertion (c). When K is not necessarily algebraically closed, we can first work over an algebraic closure $\bar{K}$ of K. So there is a conductor $\bar{Y}$ of $\varphi \otimes_K \bar{K}$. We can replace $\bar{Y}$ by a larger conductor, say Y, without changing its support. Furthermore, we can assume that Y is defined over K, since $\bar{K}$ is radicial over K. So Y fulfills assertion (c).

(d) follows from (a). □

Corollary 3. *Let X be a proper curve over a field K and assume that X is normal and geometrically reduced. Let G be a smooth commutative algebraic K-group. Let*

$\varphi : X \dashrightarrow G$ be a rational map and let Y be the conductor of φ.

(a) *If G does not contain a subgroup of type $\mathbb{G}_a$, then Y is reduced.*

(b) *If* $\mathrm{uni}(G) = 0$, *the conductor of φ is empty and φ decomposes into a composition $\varphi = \Phi \circ \iota$ where $\Phi : \mathrm{Pic}_{X/K} \longrightarrow G$ is a morphism of algebraic groups. In particular, φ is defined on the smooth locus of X.*

Proof. Denote by $\bar{Y}$ the largest reduced subscheme of Y. Then, we get an exact sequence

$$1 \longrightarrow U \longrightarrow V_Y^* \longrightarrow V_{\bar{Y}}^* \longrightarrow 1$$

of algebraic groups where V_Y^* and $V_{\bar{Y}}^*$ are the algebraic groups representing the functor of global units on Y and on $\bar{Y}$; cf. 8.1/10. The kernel U is a unipotent group which is a successive extension of groups of type $\mathbb{G}_a$. Now look at the exact sequence of 8.1/11

$$0 \longrightarrow V_X^* \longrightarrow V_Y^* \longrightarrow (\mathrm{Pic}_{X/K}, Y) \longrightarrow \mathrm{Pic}_{X/K} \longrightarrow 0$$

In the case of assertion (a), the canonical map

$$\Phi : (\mathrm{Pic}_{X/K}, Y) \longrightarrow G$$

induced by φ sends the image of U in $(\mathrm{Pic}_{X/K}, Y)$ to zero. Hence, Φ factors through

$$(\mathrm{Pic}_{X/K}, Y) \longrightarrow (\mathrm{Pic}_{X/K}, \bar{Y}) .$$

Thus, due to Theorem 2, $\bar{Y}$ is also a conductor for φ, hence $Y = \bar{Y}$ is reduced. In the case of assertion (b), the kernel of the map

$$(\mathrm{Pic}_{X/K}, Y) \longrightarrow \mathrm{Pic}_{X/K}$$

is the group of global units on Y modulo K^* which is unirational. Thus, we see that Φ factors through $\mathrm{Pic}_{X/K}$ and that the conductor of φ is empty. Then the assertion follows by Theorem 2. □

Corollary 3 yields the proof of the implication (a) $\Longrightarrow$ (c) of Theorem 1 and thus completes the proof of Theorem 1.

Remark 4. Using the characterization (c) of Theorem 2, one sees immediately that the condition $\mathrm{uni}(G_K) = 0$ is stable under finite separable field extensions.

Conjecture II. *Let S be an excellent Dedekind scheme with ring of rational functions K and let G_K be a smooth commutative algebraic K-group. If* $\mathrm{uni}(G_K) = 0$ *then G_K admits a Néron model over S.*

If one admits Conjecture II, Conjecture I is mainly a problem of unirational groups; use the technique of 7.5/1 (b). Conjecture II is true if K has characteristic zero. Indeed, if $\bar{K}$ is an algebraic closure of K, one has $\mathrm{uni}(G_K \otimes_K \bar{K}) = 0$ due to Remark 4. Then $G_K \otimes_K \bar{K}$ cannot contain a subgroup of type $\mathbb{G}_a$ or $\mathbb{G}_m$ and, hence, G_K is an abelian variety. In the case of positive characteristic, some parts of the conjecture can be proved, provided it is known that G_K admits a regular compactification.

Theorem 5. *Let S be an excellent Dedekind scheme with ring of rational functions K and let G_K be a smooth commutative algebraic K-group.*

(a) *Assume that G_K admits a regular compactification $\bar{G}_K$. If* $\operatorname{uni}(G_K) = 0$, *then G_K admits a Néron model over S.*

(b) *If S is a normal algebraic curve over a field and if G_K admits a Néron model over S, then* $\operatorname{uni}(G_K) = 0$.

Proof. (a) Let us first show that the local Néron models exist. So, we may assume for a moment that S is the affine scheme of a local ring R. Since $\operatorname{uni}(G_K) = 0$, it follows by Remark 4 that $\operatorname{uni}(G_K \otimes_K K^{sh}) = 0$ where K^{sh} is the field of fractions of a strict henselization of R. Then $G_K \otimes_K K^{sh}$ cannot contain a subgroup of type $\mathbb{G}_a$ or of type $\mathbb{G}_m$. Since S is excellent, it follows from 10.2/1 that a Néron model of G_K exists over S. Now let us return to the general situation. It remains to see that there exists a dense open subscheme U of S such that a Néron model of G_K exists over U; cf. 1.4/1. There exists a dense open subscheme U of S such that $\bar{G}_K$ extends to a proper flat U-scheme $\bar{G}_U$. Since S is excellent, the regular locus of $\bar{G}_U$ is open by [EGA IV_2], 7.8.6. So we may assume that $\bar{G}_U$ is regular. Let G_U be the smooth locus of $\bar{G}_U$. Since $\operatorname{uni}(G_K) = 0$, we see by Theorem 1 that the generic fibre of G_U coincides with G_K. After replacing U by a dense open subset, we may assume that G_U is a group scheme over U. Now we claim that G_U is the Néron model of G_K over U. Let $U(s)$ be the spectrum of the strict henselization of the local ring of U at a closed point s of U. Since $\bar{G}_U \times_U U(s)$ is regular, the $U(s)$-valued points of $\bar{G}_U$ factor through the smooth locus G_U by 3.1/2. Then it follows from 7.1/1 that $G_U \times_U \operatorname{Spec} \mathcal{O}_{S,s}$ is the local Néron model of G_K over $\mathcal{O}_{S,s}$ and the assertion follows from 1.2/4.

(b) Let us assume that $\operatorname{uni}(G_K)$ is non-trivial. Due to Theorem 1, there exists an affine smooth curve C_K with a closed point x_K and a morphism

$$\varphi_K : C_K - \{x_K\} \longrightarrow G_K$$

such that φ_K does not extend to C_K. Since we are free to replace S by an étale extension (cf. 1.2/2), we may assume that the residue field $k(x_K)$ is radicial over K. Since C_K is smooth over K, the extension $k(x_K)$ can be generated by one element over K. So, after shrinking S, there exist an element $f \in \Gamma(S, \mathcal{O}_S)$ and a p-power p^n such that $k(x_K)$ is generated by the p^n-th root of f. Now $C_K \longrightarrow \operatorname{Spec}(K)$ is induced by a smooth relative curve $C \longrightarrow S$. Denote by Z the schematic closure of the point x_K in C. We may assume, after shrinking S, that Z is a subscheme of $\mathbb{A}^1_S$ defined by $(T^{p^n} - f)$. It is a general fact that there exist infinitely many closed points s of S such that the polynomial $(T^{p^n} - f)$ has a solution over the residue field $k(s)$; cf. Lemma 6 below. If G_K admits a Néron model G of finite type over S, the morphism φ_K extends to a morphism

$$\varphi : (C - Z) \longrightarrow G .$$

Now look at the graph $\Gamma_\varphi \subset C \times_S G$ of φ viewed as a rational map $C \dashrightarrow G$. So Γ_φ

is closed in $C \times_S G$. Let Q be the image of Γ_φ under the first projection $p_1 : C \times_S G \longrightarrow C$. Since G is of finite type over S, the subset Q is constructible. The point x_K is not contained in Q, because φ_K is not defined at x_K. As x_K is the generic point of Z, we may assume, after shrinking S, that Q is disjoint from Z. Now let z be a point of Z such that the field extension $k(z)/k(s)$ is trivial where s is the image of z in S. So there exist an étale extension $S' \longrightarrow S$ and an S'-valued point x' of C such that z is the image of a point s' of S' under x' and such that x_K does not belong to the image of x'. Due to the Néron mapping property, $x'_K \circ \varphi_K$ extends to an S'-valued point of G. By continuity, x' factors through the graph Γ_φ. Thus, we see that the point z must belong to Q and we get a contradiction. □

In the last proof we have used the following fact.

Lemma 6. *Let k be a field of positive characteristic p and let A be an integral k-algebra of finite type and of dimension $d \geq 1$. Let n be a positive integer and let f be an element of A. Then, for any $n \geq 1$, there exist infinitely many prime ideals $\mathfrak{p}$ of A of codimension 1 such that the equation $T^{p^n} - f = 0$ has a solution modulo $\mathfrak{p}$.*

Proof. It suffices to show that there is at least one such prime ideal. By standard limit arguments, we may assume that k is of finite type over its prime field k_0. Then there exists a smooth and irreducible k_0-scheme R_0 such that k is the field of rational functions of R_0, and there exists an R_0-scheme S_0 of finite type such that the generic fibre of S_0 is isomorphic to S, where S is the affine scheme of A. We may assume that S_0 is affine, irreducible, and reduced. Moreover we may assume that f extends to a global section of $\mathcal{O}_{S_0}$. Now let x be a closed point of S_0. Then $k(x)$ is a finite field and, hence, perfect. So we can write

$$f = g^{p^n} + h$$

where g and h are global sections of $\mathcal{O}_{S_0}$ and where $h(x) = 0$. Since the relative dimension of S over R_0 is $d \geq 1$, we can choose g and h in such a way that the subscheme $V(h)$ defined by h is dominant over R_0. So there is a generic point s of $V(h)$ lying above the generic point of R_0. Let $\mathfrak{p} \subset \Gamma(S_0, \mathcal{O}_{S_0})$ be the prime ideal corresponding to s. Then g is a solution of the equation $T^{p^n} - f = 0$ modulo $\mathfrak{p}$, and $\mathfrak{p}$ gives rise to a prime ideal of A as required. □

If we want to apply Theorem 5(a) to an algebraic K-group G_K, it has to be known that G_K admits a regular compactification $\bar{G}_K$, a question which is related to the resolution of singularities in characteristic >0. Since it is widely accepted that the latter problem should admit a positive answer, we get strong indications for Conjecture II being true. Also note that, for a K-wound unipotent group G_K, Thm. VI.3.1 of Oesterlé [1] implies $\mathrm{uni}(G_K) = 0$ if K is of characteristic p and if $\dim G_K < p - 1$.

Bibliography

Grothendieck, A.

[FGA] *Fondements de la Géométrie Algébrique*
Sém. Bourbaki, exp. n° *149* (1956/57), *182* (1958/59), *190* (1959/60), *195* (1959/60), *212* (1960/61), *221* (1960/61), *232* (1961/62), *236* (1961/62), Benjamin, New York (1966)

Grothendieck, A., and Dieudonné, J.

Eléments de Géométrie Algébrique

[EGA I] Le langage des schémas. Publ. Math. IHES *4* (1960)

[EGA II] Etude globale élémentaire de quelques classes de morphismes. Publ. Math. IHES *8* (1961)

[EGA III] Etude cohomologique des faisceaux cohérents. Publ. Math. IHES *11* (1961), *17* (1963)

[EGA IV] Etude locale des schémas et des morphismes de schémas. Publ. Math. IHES *20* (1964), *24* (1965), *28* (1966), *32* (1967)

Grothendieck, A., et al.

Séminaire de Géométrie Algébrique

[SGA 1] Revêtements Etales et Groupe Fondamental. Lect. Notes Math. *224*, Springer, Berlin-Heidelberg-New York (1971)

[SGA 3] Schémas en Groupes I,II,III. Lect. Notes Math. *151*, *152*, *153*, Springer, Berlin-Heidelberg-New York (1970)

[SGA 4] Théorie des Topos et Cohomologie Etale. Lect. Notes Math. *269*, *270*, *305*, Springer, Berlin-Heidelberg-New York (1972–73)

[SGA 6] Théorie des Intersections et Théorème de Riemann-Roch. Lect. Notes Math. *225*, Springer, Berlin-Heidelberg-New York (1971)

[SGA 7] Groupes de Monodromie en Géométrie Algébrique. Lect. Notes Math. *288*, *340*, Springer, Berlin-Heidelberg-New York (1972–73)

Abhyankar, S. S.

1. Resolution of singularities of arithmetical surfaces. Proc. Conf. on Arithm. Alg. Geometry, Purdue (1963)

Altman, A., and Kleiman, S.

1. Compactifying the Picard Scheme. Adv. Math. *35*, 50–112 (1980)
2. Compactifying the Picard Scheme II. Am. J. Math. *101*, 10–41 (1979)

Artin, M.

1. Some numerical criteria for contractibility of curves on algebraic surfaces. Am. J. Math. *84*, 485–496 (1962)
2. On isolated rational singularities of surfaces. Am. J. Math. *88*, 129–136 (1966)
3. The implicit function theorem in algebraic geometry. Algebraic Geometry, Bombay Colloquium 1968. Oxford University Press, Oxford, 13–34 (1969)
4. Algebraic approximation of structures over complete local rings. Publ. Math. IHES *36*, 23–58 (1969)
5. Algebraization of formal moduli I. Global Analysis, Papers in Honor of K. Kodaira. University of Tokyo Press, Princeton University Press, 21–71 (1969)
6. Algebraization of formal moduli II. Ann. Math. *91*, 88–135 (1970)
7. Versal Deformations and Algebraic Stacks. Invent. Math. *27*, 165–189 (1974)

8. Algebraic structure of power series rings. Algebraist's Homage: Papers in Ring Theory and Related Topics. Proc. Conf. on Algebra in Honor of N. Jacobson, New Haven (1981), Contemp. Math. *13*, AMS, Providence, 223–227 (1982)
9. Néron models. Arithmetic Geometry, edited by G. Cornell and J.H. Silverman. Proc. Storrs 1984, Springer, Berlin-Heidelberg-New York (1986)

Artin, M., and Rotthaus, C.
1. A structure theorem for power series rings. Algebraic Geometry and Commutative Algebra in Honor of Masayoshi Nagata, vol. 1, pp. 35–44. Kinokuniya Tokyo (1987)

Artin, M., and Winters, G.
1. Degenerate fibres and stable reduction of curves. Topology *10*, 373–383 (1971)

Becker, J., Denef, J., Lipshitz, L., van den Dries, L.
1. Ultraproducts and Approximation in Local Rings I. Invent. Math. *51*, 189–203 (1979)

Bégueri, L.
1. Dualité sur un corps local à corps résiduel algébriquement clos. Bull. Soc. Math. Fr., Suppl. Mém. (N.S.) *4* (1980)

Borel, A., and Tits, J.
1. Eléments unipotents et sous-groupes paraboliques de groupes réductifs I. Invent. Math. *12*, 95–104 (1971)

Bosch, S. and Lütkebohmert, W.
1. Stable Reduction and Uniformization of Abelian Varieties I. Math. Ann. *270*, 349–379 (1985)
2. Stable Reduction and Uniformization of Abelian Varieties II. Invent. Math. *78*, 257–297 (1984)
3. Néron models from the rigid analytic viewpoint. J. Reine Angew. Math. *364*, 69–84 (1986)

Bourbaki, N.
1. Algèbre. Chap. 1–7, 10, Hermann, Paris (1947–59), Masson, Paris (1980–81)
2. Algèbre commutative. Chap. 1–9, Hermann, Paris (1961–65), Masson, Paris (1980–85)

Chai, C.-L.
1. Compactification of Siegel moduli schemes. London Mathematical Society Lecture Note Series *107*, Cambridge University Press, Cambridge (1985)

Chevalley, C.
1. La théorie des groupes algébriques. Proc. Int. Cong. Math. 1958, Cambridge University Press, Cambridge, 53–68 (1960)

Deligne, P.
1. Le Lemme de Gabber, Séminaire sur les pinceaux arithmétiques: La conjecture de Mordell, edited by L. Szpiro. Astérisque *127*, 131–150 (1985)

Deligne, P., and Mumford, D.
1. The irreducibility of the space of curves of given genus. Publ. Math. IHES *36*, 75–110 (1969)

Deligne, P., and Rapoport, M.
1. Les schémas de modules de courbes elliptiques. Modular Functions of One Variable II. Lect. Notes Math. *349*, Springer, Berlin-Heidelberg-New York (1973)

Faltings, G.
1. Endlichkeitssätze für abelsche Varietäten über Zahlkörpern. Invent. Math. *73*, 349–366 (1983)

Greenberg, M. J.
1. Schemata over local rings. Ann. Math. *73*, 624–648 (1961)
2. Rational points in henselian discrete valuation rings. Publ. Math. IHES *31*, 59–64 (1966)

Grothendieck, A.
1. Techniques de construction en géométrie analytique. IV: Formalisme général des foncteurs représentables. Sém. Cartan 1960/61, n° *11* (1961)
2. Techniques de construction en géométrie analytique. V: Fibrés vectoriels, fibrés projectifs, fibrés en drapeaux. Sém. Cartan 1960/61, n° *12* (1961)

3. Le groupe de Brauer III. Dix Exposés sur la Cohomologie des Schémas, North-Holland, Amsterdam, 88–188 (1968)

Hironaka, H.

1. An example of a non-Kählerian complex-analytic deformation of Kählerian complex structures. Ann. Math. *75*, 190–208 (1962)
2. Resolution of singularities of an algebraic variety over a field of characteristic zero (I, II). Ann. Math. *79*, 109–203, 205–326 (1964)

Igusa, J.

1. Fibre systems of Jacobian varieties I, II. Am. J. Math. *78*, 171–199 and 745–760 (1956)

Katz, N., and Mazur, B.

1. Arithmetic Moduli of Elliptic Curves. Annals of Mathematics Studies *108*, Princeton University Press, Princeton (1985)

Kempf, G., Knudsen, F., Mumford, D., Saint-Donat, B.

1. Toroidal embeddings I. Lect. Notes Math. *339*, Springer, Berlin-Heidelberg-New York (1973)

Knutson, D.

1. Algebraic spaces. Lect. Notes Math. *203*, Springer, Berlin-Heidelberg-New York (1971)

Kodaira, K.

1. On compact analytic surfaces. Princeton Math. Series *24*, 121–135
2. On compact analytic surfaces II. Ann. Math. *77*, 563–626 (1963)

Koizumi, S.

1. On specialization of the Albanese and Picard varieties. Mem. College Sci. Univ. Kyoto *32*, 371–382 (1960)

Koizumi, S., and Shimura, G.

1. On specialization of abelian varieties. Sci. Papers Coll. Gen. Educ. Univ. Tokyo *9*, 187–211 (1959)

Lang, S.

1. Abelian Varieties. Interscience Publishers, New York (1959)

Lang, S., and Néron, A.

1. Rational points of abelian varieties over function fields. Am. J. Math. *81*, 95–118 (1959)

Lazard, D.

1. Autour de la platitude. Bull. Soc. Math. Fr. *97*, 81–128 (1968)

Lichtenbaum, S.

1. Curves over discrete valuation rings. Am. J. Math. *90*, 380–405 (1968)

Lipman, J.

1. Desingularization of two-dimensional schemes. Ann. Math. *107*, 151–207 (1978)

Lorenzini, D. J.

1. Degenerating curves and their Jacobians. Preprint, Berkeley (1988)

Mazur, B.

1. Modular curves and the Eisenstein ideal. Publ. Math. IHES *47*, 173–186 (1977)

Mazur, B., and Swinnerton-Dyer, P.

1. Arithmetic of Weil curves. Invent. Math. *25*, 1–61 (1974)

McCallum, W. G.

1. The component group of a Néron model. Preprint, Berkeley (1988)

Milne, J. S.

1. Etale Cohomology. Princeton Math. Series 33, Princeton University Press, Princeton (1980)
2. Arithmetic Duality Theorems. Perspectives in Math., Academic Press, London (1986)

Moret-Bailly, L.

1. Groupes de Picard et problèmes de Skolem I. Ann. Sci. Ec. Norm. Super., *22*, 161–179 (1989)

Mumford, D.
1. Geometric Invariant Theory. Ergebnisse *34*, Springer, Berlin-Heidelberg-New York (1965)
2. Lectures on Curves on an Algebraic Surface. Annals of Math. Studies *59*, Princeton University Press, Princeton (1966)
3. Abelian Varieties. Oxford University Press, Oxford (1970)

Murre, J. P.
1. On contravariant functors from the category of preschemes over a field into the category of abelian groups. Publ. Math. IHES *23*, 5–43 (1964)
2. Representation of unramified functors. Applications. (according to unpublished results of A. Grothendieck). Sém. Bourbaki 1964/65, exp. n° *294*, Benjamin, New York (1966)

Nagata, M.
1. Imbedding of an abstract variety in a complete variety. J. Math. Kyoto Univ. *2*, 1–10 (1962)
2. A generalization of the imbedding problem of an abstract variety in a complete variety. J. Math. Kyoto Univ. *3*, 89–102 (1963)

Néron, A.
1. Modèles p-minimaux des variétés abéliennes. Sém. Bourbaki 1961/62, exp. n° *227*, Benjamin, New York (1966)
2. Modèles minimaux des variétés abéliennes. Publ. Math. IHES *21* (1964)
3. Quasi-fonctions et hauteurs sur les variétés abéliennes. Ann. Math. *82*, 249–331 (1965)

Oesterlé, J.
1. Nombres de Tamagawa et groupes unipotents en caractéristique p. Invent. Math. *78*, 13–88 (1984)

Ogg, A. P.
1. Cohomology of abelian varieties over function fields. Ann. Math. *76*, 185–212 (1962)

Oort, F
1. Sur le schéma de Picard. Bull. Soc. Math. Fr. *90*, 1–14 (1962)

Peskine, Ch.
1. Une généralisation du "Main Theorem" de Zariski. Bull. Sci. Math. *90*, 119–127 (1966)

Raynaud, M.
1. Caractéristique d'Euler-Poincaré d'un faisceau et cohomologie des variétés abéliennes. Sém. Bourbaki 1964/65, exp. n° *286*, Benjamin, New York (1966)
2. Modèles de Néron. C. R. Acad. Sci., Paris, Sér. A *262*, 345–347 (1966)
3. Un critère d'effectivité de descente. Sém. Algèbre commutative P. Samuel 1967/68, exp. *V* (1968)
4. Faisceaux amples sur les schémas en groupes et les espaces homogènes. Lect. Notes Math. *119*, Springer, Berlin-Heidelberg-New York (1970)
5. Anneaux locaux henséliens. Lect. Notes Math. *169*, Springer, Berlin-Heidelberg-New York (1970)
6. Spécialisation du foncteur de Picard. Publ. Math. IHES *38*, 27–76 (1970)
7. Schémas en groupes de type (p, ..., p). Bull. Soc. Math. Fr. *102*, 241–280 (1974)

Raynaud, M., and Gruson, L.
1. Critères de platitude et de projectivité. Invent. Math. *13*, 1–89 (1971)

Rego, C. J.
1. The compactified Jacobian. Ann. Sci. Ec. Norm. Super., *13*, 211–223 (1980)

Rosenlicht, M.
1. Generalized Jacobian varieties. Ann. Math. *59*, 505–530 (1954)
2. Some basic theorems on algebraic groups. Am. J. Math. *78*, 401–443 (1956)

Rotthaus, C.
1. On the approximation property of excellent rings. Invent. Math. *88*, 39–63 (1987)

Serre, J -P.
1. Groupes Algébriques et Corps de Classes. Hermann, Paris (1959)
2. Groupes proalgébriques. Publ. Math. IHES *7* (1960)
3. Sur les corps locaux à corps résiduel algébriquement clos. Bull. Soc. Math. Fr. *89*, 105–154 (1961)
4. Corps locaux (2nd ed.). Hermann, Paris (1968)

Serre, J.-P., and Tate, J.
1. Good reduction of abelian varieties. Ann. Math. *88*, 492–517 (1968)

Shafarevich, I. R.
1. Principal homogeneous spaces defined over a function field. Amer. Math. Soc. Transl. *37*, 85–114 (1964)
2. Lectures on minimal models and birational transformations of two-dimensional schemes. Tata Inst. Fund. Research *37*, Bombay (1966)

Shimura, G.
1. Reduction of algebraic varieties with respect to a discrete valuation of the basic field. Am. J. Math. *77*, 134–176 (1955)

Silverman, J. H.
1. The Arithmetic of Elliptic Curves. Grad. Texts in Math. *106*, Springer, Berlin-Heidelberg-New York (1986)

Tate, J.
1. WC-groups over p-adic fields. Sém. Bourbaki 1957/58, exp. n° *156*, Benjamin, New York (1966)
2. Algorithm for determining the Type of a Singular Fiber in an Elliptic Pencil. Modular Functions of One Variable IV. Lect. Notes Math. *476*, Springer, Berlin-Heidelberg-New York (1975)

Tits, J.
1. Lectures on algebraic groups. Yale University, New Haven (1967)

Weil, A.
1. Foundations of Algebraic Geometry. Amer. Math. Soc. Colloquium Publ. *29* (1946) (revised and enlarged edition 1962)
2. Variétés Abéliennes et Courbes Algébriques. Hermann, Paris (1948), republished in Courbes Algébriques et Variétés Abéliennes. Hermann, Paris (1971)

Subject Index

Ergebnisse der Mathematik und ihrer Grenzgebiete, 3. Folge

A Series of Modern Surveys in Mathematics

Volume 1: **A. Fröhlich**

Galois Module Structure of Algebraic Integers

1983. X, 262 pp. ISBN 3-540-11920-5

Volume 2: **W. Fulton**

Intersection Theory

1984. XI, 470 pp. ISBN 3-540-12176-5

Volume 3: **J. C. Jantzen**

Einhüllende Algebren halbeinfacher Lie-Algebren

1983. V, 298 S. ISBN 3-540-12178-1

Volume 4: **W. Barth, C. Peters, A. van de Ven**

Compact Complex Surfaces

1984. X, 304 pp. ISBN 3-540-12172-2

Volume 5: **K. Strebel**

Quadratic Differentials

1984. XII, 184 pp. 74 figs. ISBN 3-540-13035-7

Volume 6: **M. J. Beeson**

Foundations of Constructive Mathematics

Metamathematical Studies

1985. XXIII, 466 pp. ISBN 3-540-12173-0

Volume 7: **A. Pinkus**

n-Widths in Approximation Theory

1985. X, 291 pp. ISBN 3-540-13638-X

Volume 8: **R. Mañé**

Ergodic Theory and Differentiable Dynamics

Translated from the Portuguese by Silvio Levy

1987. XII, 317 pp. 32 figs.
ISBN 3-540-15278-4

Volume 9: **M. Gromov**

Partial Differential Relations

1986. IX, 363 pp. ISBN 3-540-12177-3

Volume 10: **A. L. Besse**

Einstein Manifolds

1986. XII, 510 pp. 22 figs.
ISBN 3-540-15279-2

Volume 11: **M. D. Fried, M. Jarden**

Field Arithmetic

1986. XVII, 458 pp. ISBN 3-540-16640-8

Volume 12: **J. Bochnak, M. Coste, M.-F. Roy**

Géométrie algébrique réelle

1987. X, 373 pp. 44 figs.
ISBN 3-540-16951-2

Volume 13: **E. Freitag, R. Kiehl**

Etale Cohomology and the Weil Conjecture

With an Historical Introduction by J. A. Dieudonné

Translated from the German Manuscript

1987. XVIII, 317 pp. ISBN 3-540-12175-7

Volume 14: **M. R. Goresky, R. D. MacPherson**

Stratified Morse Theory

1988. XIV, 272 pp. 84 figs. ISBN 3-540-17300-5

Springer

GPSR Compliance

The European Union's (EU) General Product Safety Regulation (GPSR) is a set of rules that requires consumer products to be safe and our obligations to ensure this.

If you have any concerns about our products, you can contact us on ProductSafety@springernature.com

In case Publisher is established outside the EU, the EU authorized representative is:

Springer Nature Customer Service Center GmbH
Europaplatz 3
69115 Heidelberg, Germany

Batch number: 10755805

Printed by Printforce, the Netherlands